Josiah Willard Gibbs 1839-1903

Ludwig Boltzmann 1844-1906

Heike Kamerlingh Onnes 1853-1926

Max Planck 1858-1947

Walther Nernst 1864-1941

Constantin Carathéodory 1873-1950

Albert Einstein 1879-1955

Peter Debye 1884-1966

Francis E. Simon 1893-1956

HEAT AND THERMODYNAMICS

An Intermediate Textbook

Sixth Edition

Mark W. Zemansky, Ph.D.

Professor of Physics, Emeritus
The City College of
The City University of New York

Richard H. Dittman, Ph.D.

Associate Professor of Physics
University of Wisconsin-Milwaukee

McGraw-Hill Publishing Company

New York St. Louis San Francisco Auckland Bogotá Caracas
Hamburg Lisbon London Madrid Mexico Milan
Montreal New Delhi Oklahoma City Paris San Juan
São Paulo Singapore Sydney Tokyo Toronto

This book was set in Times Roman.
The editors were Marian D. Provenzano and Scott Amerman;
the production supervisor was Diane Renda.
The drawings were done by ECL Art Associates, Inc.
The cover was designed by Rafael Hernandez.
Fairfield Graphics was printer and binder.

HEAT AND THERMODYNAMICS

13141516 HDHD 99876543210

Library of Congress Cataloging in Publication Data

Zemansky, Mark Waldo, date
Heat and thermodynamics.

Bibliography: p.
Includes index.
1. Heat. 2. Thermodynamics. I. Dittman, Richard, joint author. II. Title.
QC254.2.Z45 1981 536 80-18253
ISBN 0-07-072808-9

To
Adele C. Zemansky
and
Maria M. Dittman

CONTENTS

PREFACE

Early textbooks on "heat" devoted many chapters to discussions of thermometry, calorimetry, and heat engines, included many experimental details, and ended with one or two chapters in which an attempt was made to construct a deductive theory, known as "thermodynamics." In these two divisions only the large-scale characteristics of matter were considered, so that both divisions were strictly "macroscopic." In later years, when molecular theory was better understood, the subject matter of statistical mechanics and of kinetic theory was included in the field. Since these two subjects are microscopic in point of view, a more general term was needed to include the four divisions: heat, thermodynamics, statistical mechanics, and kinetic theory. The term most widely used today for this purpose is "thermal physics." Of course, there is no hard and fast rule concerning the relative weights of each of the divisions of thermal physics. Some authors regard statistical mechanics and kinetic theory of primary importance, with heat and thermodynamics a somewhat necessary evil. Others, feeling guilty about their worship of statistical mechanics, hide behind the misnomer "statistical thermodynamics," which seems to refer to a subject that is both microscopic and macroscopic at the same time.

In the present volume, thermodynamics constitutes about 50 percent, heat about 37 percent, statistical mechanics about 10 percent, and kinetic theory about 3 percent. The main reason for this distribution is the deep conviction on the part of the authors that the fundamental foundation of the subject is thermodynamics, which is well within the abilities of undergraduate students. At the sophomore or junior level students have sufficient physical and mathematical sophistication to prepare them for thermodynamic arguments and proofs, whereas they are not quite ready for the subtleties of ergodic theory or of Gibbsian ensembles. Only the simplest treatment of weakly interacting particles is given and applied to an ideal gas, an electron gas, blackbody radiation, a vibrating lattice, and a paramagnetic-ion subsystem in a crystal. This sixth edition contains about the same amount of engineering, chemistry, and experimental detail as its predecessor, but the subjects of superfluidity and superconductivity have been deleted, on the ground that their treatment is no longer exclusively thermodynamic.

The two main features of this sixth edition are: (1) the almost complete use of SI units in all parts of the subject, with the exception of a few topics in chemical thermodynamics; and (2) the subdivision of the book into two almost-equal sections, the first part being devoted to fundamental concepts, designed to be the core of an introductory course in thermodynamics, and the second part, starting with a chapter on elementary statistical mechanics, designed to enable the teacher to choose those applications which require the use of thermodynamics *and* statistical mechanics, such as specific heats of solids, cryogenics, nuclear paramagnetism, and negative temperatures, to name a few. Part 2 also contains such topics as chemical equilibrium, ideal-gas reactions, phase theory, the third law, fuel cells, thermocouples, blackbody radiation, and negative Kelvin temperatures.

The authors would like to express their appreciation of the expertise, understanding, and kindness displayed by the McGraw-Hill staff in the preparation of this book—particularly Marian Provenzano and Scott Amerman. The endpapers contain eighteen pictures of world-famous pioneers in thermodynamics and statistical mechanics. The authors are very grateful to Professor Peter T. Landsberg of Southampton University and to Joan Warnow of the American Institute of Physics for supplying us with these pictures.

Mark W. Zemansky
Richard H. Dittman

NOTATION

CAPITAL ITALIC		LOWERCASE ITALIC	
A	Area; paramagnetic heat-capacity constant	a	A dimension; $g\mu_B \mu_0 \mathcal{H}/kT$
B	Second virial coefficient; Brillouin function	b	A dimension; a constant
C	Heat capacity; critical point	c	Molar heat capacity; speed of light
D	Debye function; electric displacement	d	Differential sign
E	Electric intensity; energy	e	Naperian logarithm base; electronic charge
F	Helmholtz function	f	Molar Helmholtz function; variance
G	Gibbs function	g	Molar Gibbs function; Landé g factor; degree of degeneracy
H	Enthalpy	h	Molar enthalpy; Planck's constant
I	Current; nuclear quantum number	i	Vapor-pressure constant
J	Electronic quantum number	j	Valence
K	Thermal conductivity; equilibrium constant	k	Boltzmann's constant
L	Length; latent heat; coupling coefficients	l	Latent heat per kilogram or per mole
M	Magnetization; mass	m	Mass of a molecule or electron
N	Number of molecules	n	Number of moles; quantum number
P	Pressure	p	Partial pressure; momentum
Q	Heat	q	Heat per mole

R Universal gas constant; electric resistance; radius
S Entropy
T Kelvin temperature
U Internal energy
V Volume
W Work
X Generalized displacement
Y Generalized force; Young's modulus
Z Electric charge; partition function; compressibility factor

r Radius; number of individual reactions
s Molar entropy
t Celsius temperature; empirical temperature
u Molar energy; radiant-energy density
v Molar volume
w Speed of a wave or a molecule
x Space coordinate; mole fraction
y Space coordinate; fraction
z Space coordinate

SCRIPT CAPITALS

$\mathcal{B}$ Magnetic induction
$\mathcal{E}$ Electromotive force
$\mathcal{F}$ Tension; force
$\mathcal{H}$ Magnetic intensity
$\mathcal{M}$ Molar mass or molecular weight
$\mathcal{R}$ Radiant exitance
$\mathcal{S}$ Surface tension

ROMAN SYMBOLS FOR UNITS

m meter
kg kilogram
s second
atm atmosphere
A ampere
A/m ampere · turn per meter
C coulomb
Hz hertz (cps)
J joule
N newton
Pa Pascal
T tesla
V volt
W watt

SPECIAL SYMBOLS

N_A Avogadro's number
đ Inexact differential
N_F Faraday's constant
T^* Magnetic temperature
C'_C Curie constant

GREEK LETTERS

α Linear expansivity; critical-point exponent
β Volume expansivity; $1/kT$; critical-point exponent
γ Ratio of heat capacities; electronic term in heat capacity; critical-point exponent
Ω Thermodynamic probability; solid angle
δ Energy of a magnetic ion; critical-point exponent
Δ Finite difference
ϵ Degree of reaction; molecular energy; reduced temperature difference; Seebeck coefficient
η Efficiency

θ Ideal-gas temperature; angle
Θ Debye temperature
κ Compressibility
λ Wavelength; Lagrange multiplier; integrating factor
μ Joule-Kelvin coefficient; molecular magnetic moment; chemical potential
μ_0 permeability of vacuum
ν Molecular density; frequency; stoichiometric coefficient
Π Polarization
π Peltier coefficient
ρ Density (mass per unit volume)
σ Thomson coefficient; Stefan-Boltzmann constant; function for isentropic surface
τ Time; period
ϕ Angle; function of temperature
φ Number of phases
ω Coefficient of performance; angular speed
χ Magnetic susceptibility

PART ONE

FUNDAMENTAL CONCEPTS

CHAPTER
ONE
TEMPERATURE

1-1 MACROSCOPIC POINT OF VIEW

The study of any special branch of physics starts with a separation of a restricted region of space or a finite portion of matter from its surroundings. The portion which is set aside (in the imagination) and on which the attention is focused is called the *system*, and everything outside the system which has a direct bearing on its behavior is known as the *surroundings*. When a system has been chosen, the next step is to describe it in terms of quantities related to the behavior of the system or its interactions with the surroundings, or both. There are in general two points of view that may be adopted, the *macroscopic* point of view and the *microscopic* point of view.

Let us take as a system the contents of a cylinder of an automobile engine. A chemical analysis would show a mixture of hydrocarbons and air before being ignited, and after the mixture has been ignited there would be combustion products describable in terms of certain chemical compounds. A statement of the relative amounts of these substances is a description of the *composition* of the system. At any moment, the system whose composition has just been described occupies a certain *volume*, depending on the position of the piston. The volume can be easily measured and, in the laboratory, is recorded automatically by means of an appliance coupled to the piston. Another quantity that is indispensable in the description of our system is the *pressure* of the gases in the cylinder. After igniting the mixture the pressure is large; after exhausting the combustion products the pressure is small. In the laboratory, a pressure gauge may be used to measure the changes of pressure and to make an automatic record as the engine operates. Finally, there is one more quantity without which we should have no adequate idea of the operation of the engine. This quantity is the *temperature;* as we shall see, in many instances, it can be measured just as simply as the other quantities.

We have described the materials in a cylinder of an automobile engine by specifying four quantities: composition, volume, pressure, and temperature. These quantities refer to the gross characteristics, or large-scale properties, of the system and provide a *macroscopic description.* They are therefore called *macroscopic coordinates.* The quantities that must be specified to provide a macroscopic description of other systems are, of course, different; but macroscopic coordinates in general have the following characteristics in common:

1. They involve no special assumptions concerning the structure of matter.
2. They are few in number.
3. They are suggested more or less directly by our sense perceptions.
4. They can in general be directly measured.

In short, a macroscopic description of a system involves the specification of a *few fundamental measurable properties* of a system.

1-2 MICROSCOPIC POINT OF VIEW

From the viewpoint of *statistical mechanics*, a system is considered to consist of an enormous number N of molecules, each of which is capable of existing in a set of states whose energies are $\epsilon_1, \epsilon_2, \ldots$. The molecules are assumed to interact with one another by means of collisions or by forces caused by fields. The system of molecules may be imagined to be isolated or, in some cases, may be considered to be embedded in a set of similar systems, or *ensemble* of systems. Concepts of probability are applied, and the equilibrium state of the system is assumed to be the state of highest probability. The fundamental problem is to find the number of molecules in each of the molecular energy states (known as the *populations* of the states) when equilibrium is reached.

Since statistical mechanics will be treated at some length in Chap. 11, it is not necessary to pursue the matter further at this point. It is evident, however, that a microscopic description of a system involves the following characteristics:

1. Assumptions are made concerning the structure of matter; e.g., the existence of molecules is assumed.
2. Many quantities must be specified.
3. These quantities specified are not suggested by our sense perceptions.
4. They cannot be measured.

1-3 MACROSCOPIC VS. MICROSCOPIC

Although it might seem that the two points of view are hopelessly different and incompatible, there is nevertheless a relation between them; and when both points of view are applied to the same system, they must lead to the same conclusion. The relation between the two points of view lies in the fact that the few directly measurable properties whose specification constitutes the macroscopic description are really averages over a period of time of a large number of microscopic characteristics. For example, the macroscopic quantity, pressure, is the average rate of change of momentum due to all the molecular collisions made on a unit of area. Pressure, however, is a property that is perceived by our senses. We feel the effects of pressure. Pressure was experienced, measured, and used long before physicists had reason to believe in the existence of molecular impacts. If the molecular theory is changed, the concept of pressure will still remain and will still mean the same thing to all normal human beings. Herein lies an important distinction between the macroscopic and microscopic points of view. The few measurable macroscopic properties are as sure as our senses. They will remain unchanged as long as our senses remain the same. The microscopic point of view, however, goes much further than our senses. It postulates the existence of molecules, their motion, their energy states, their interactions, etc. It is constantly being changed, and we can never be sure that the assumptions are justified until we have compared some deduction made on the basis of these assumptions with a similar deduction based on the macroscopic point of view.

1-4 SCOPE OF THERMODYNAMICS

It has been emphasized that a description of the gross characteristics of a system by means of a few of its measurable properties, suggested more or less directly by our sense perceptions, constitutes a macroscopic description. Such descriptions are the starting point of all investigations in all branches of physics. For example, in dealing with the mechanics of a rigid body, we adopt the macroscopic point of view in that only the external aspects of the rigid body are considered. The position of its center of mass is specified with reference to coordinate axes at a particular time. Position and time and a combination of both, such as velocity, constitute some of the macroscopic quantities used in mechanics and are called *mechanical coordinates*. The mechanical coordinates serve to determine the potential and the kinetic energy of the rigid body with reference to the coordinate axes, i.e., the kinetic and the potential energy of the body as a whole. These two types of energy constitute the *external*, or *mechanical*, *energy* of the rigid body. It is the purpose of mechanics to find such relations between the position coordinates and the time as are consistent with Newton's laws of motion.

In thermodynamics, however, the attention is directed to the *interior* of a system. A macroscopic point of view is adopted, and emphasis is placed on those macroscopic quantities which have a bearing on the internal state of a system. It is the function of experiment to determine the quantities that are necessary and sufficient for a description of such an internal state. Macroscopic quantities having a bearing on the internal state of a system are called *thermodynamic coordinates*. Such coordinates serve to determine the *internal energy* of a system. It is the purpose of thermodynamics to find general relations among the thermodynamic coordinates that are consistent with the fundamental laws of thermodynamics.

A system that may be described in terms of thermodynamic coordinates is called a *thermodynamic system*. In engineering, the important thermodynamic systems are a gas, such as air; a vapor, such as steam; a mixture, such as gasoline vapor and air; and a vapor in contact with its liquid, such as liquid and vaporized ammonia. Chemical thermodynamics deals with these systems and, in addition, with solids, surface films, and electric cells. Physical thermodynamics includes, in addition to the above, such systems as stretched wires, electric capacitors, thermocouples, and magnetic substances.

1-5 THERMAL EQUILIBRIUM

We have seen that a microscopic description of a gaseous mixture may be given by specifying such quantities as the composition, the mass, the pressure, and the volume. Experiment shows that, for a given composition and for a constant mass, many different values of pressure and volume are possible. If the pressure is kept constant, the volume may vary over a wide range of values, and vice versa. In other words, the pressure and the volume are independent coordinates. Similarly, experiment shows that, for a wire of constant mass, the tension and the length are independent coordinates, whereas, in the case of a surface film, the surface tension and the area may be varied independently. Some systems that, at first sight, seem quite complicated, such as an electric cell with two different electrodes and an electrolyte, may still be described with the aid of only two independent coordinates. On the other hand, some systems composed of a number of homogeneous parts require the specification of two independent coordinates for each homogeneous part. Details of various thermodynamic systems and their thermodynamic coordinates will be given in Chap. 2. For the present, to simplify our discussion, we shall deal only with systems of constant mass and composition, each requiring *only one pair* of independent coordinates for its description. This involves no essential loss of generality and results in a considerable saving of words. In referring to any nonspecified system, we shall use the symbols Y and X for the pair of independent coordinates.

A state of a system in which Y and X have definite values that remain constant so long as the external conditions are unchanged is called an *equilibrium* state. Experiment shows that the existence of an equilibrium state in one system depends on the proximity of other systems and on the nature of the wall

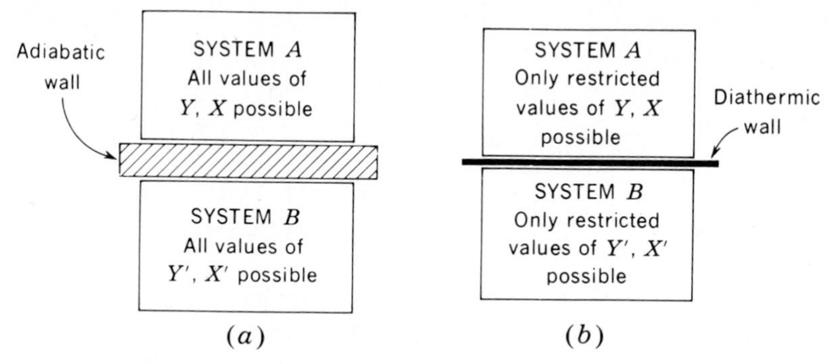

Figure 1-1 Properties of adiabatic and diathermic walls.

separating them. Walls are said to be either adiabatic or diathermic. If a wall is *adiabatic* (see Fig. 1-1*a*), a state Y, X for system A and Y', X' for system B may coexist as equilibrium states for *any* attainable values of the four quantities, provided only that the wall is able to withstand the stress associated with the difference between the two sets of coordinates. Thick layers of wood, concrete, asbestos, felt, styrofoam, etc., are good experimental approximations to adiabatic walls. If the two systems are separated by a *diathermic* wall (see Fig. 1-1*b*), the values of Y, X and Y', X' will change spontaneously until an equilibrium state of the combined system is attained. The two systems are then said to be in *thermal equilibrium* with each other. The most common diathermic wall is a thin metallic sheet. *Thermal equilibrium is the state achieved by two (or more) systems, characterized by restricted values of the coordinates of the systems, after they have been in communication with each other through a diathermic wall.*

Imagine two systems A and B separated from each other by an adiabatic wall but each in contact with a third system C through diathermic walls, the whole assembly being surrounded by an adiabatic wall as shown in Fig. 1-2*a*. Experiment shows that the two systems will come to thermal equilibrium with the third and that no further change will occur if the adiabatic wall separating A and B is then replaced by a diathermic wall (Fig. 1-2*b*). If, instead of allowing both systems A and B to come to equilibrium with C at the same time, we first have equilibrium between A and C and then equilibrium between B and C (the state of system C being the same in both cases), then, when A and B are brought into communication through a diathermic wall, they will be found to be in thermal equilibrium. We shall use the expression "two systems are in thermal equilibrium" to mean that the two systems are in states such that, if the two *were* connected through a diathermic wall, the combined system *would be* in thermal equilibrium.

These experimental facts may then be stated concisely in the following form: *Two systems in thermal equilibrium with a third are in thermal equilibrium with each other*. Following R. H. Fowler, we shall call this postulate the *zeroth law of thermodynamics*.

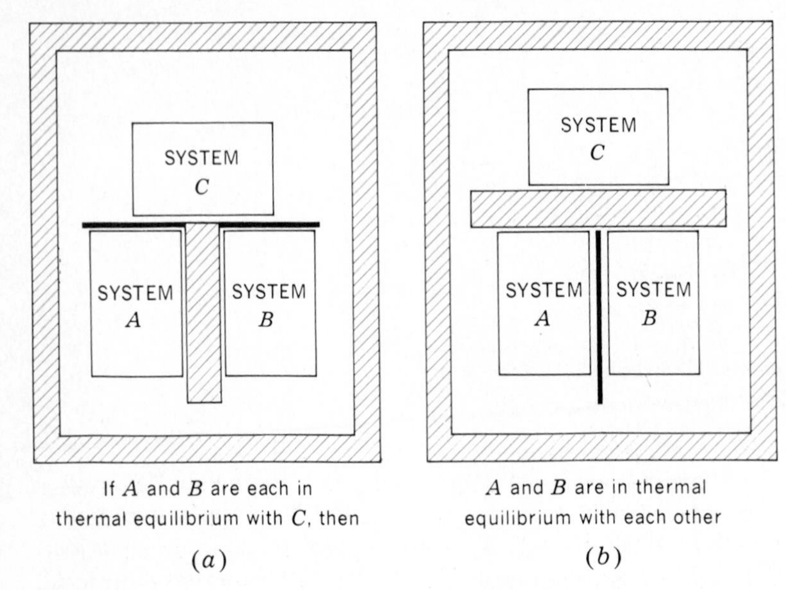

Figure 1-2 The zeroth law of thermodynamics. (Adiabatic walls are designated by cross shading; diathermic walls, by heavy lines.)

1-6 TEMPERATURE CONCEPT

Consider a system A in the state Y_1, X_1 in thermal equilibrium with a system B in the state Y'_1, X'_1. If system A is removed and its state changed, there will be found another state Y_2, X_2 in which it is in thermal equilibrium with the *original* state Y'_1, X'_1 of system B. Experiment shows that there exists a whole set of states Y_1, X_1; Y_2, X_2; Y_3, X_3; etc., every one of which is in thermal equilibrium with this *same* state Y'_1, X'_1 of system B and which, by the zeroth law, are in thermal equilibrium with one another. We shall suppose that *all* such states, when plotted on a YX diagram, lie on a curve such as I in Fig. 1-3, which we shall call an *isotherm. An isotherm is the locus of all points representing states at which a system is in thermal equilibrium with one state of another system.* We make no assumption as to the continuity of the isotherm, although experiments on simple systems indicate usually that at least a portion of an isotherm is a continuous curve.

Similarly, with regard to system B, we find a set of states Y'_1, X'_1; Y'_2, X'_2; etc., all of which are in thermal equilibrium with one state (Y_1, X_1) of system A, and therefore in thermal equilibrium with one another. These states are plotted on the $Y'X'$ diagram of Fig. 1-3 and lie on the isotherm I′. From the zeroth law, it follows that all the states on isotherm I of system A are in thermal equilibrium with all the states on isotherm I′ of system B. We shall call curves I and I′ *corresponding isotherms* of the two systems.

If the experiments outlined above are repeated with different starting condi-

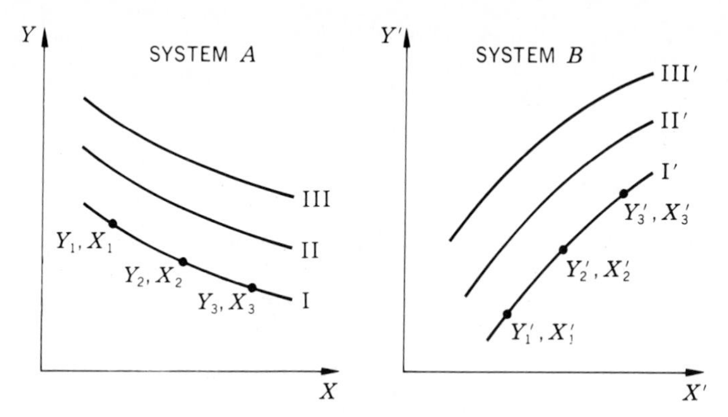

Figure 1-3 Isotherms of two different systems.

tions, another set of states of system A lying on curve II may be found, every one of which is in thermal equilibrium with every state of system B lying on curve II′. In this way, a family of isotherms I, II, III, etc., of system A and a corresponding family I′, II′, III′, etc., of system B may be found. Furthermore, by repeated applications of the zeroth law, corresponding isotherms of still other systems C, D, etc., may be obtained.

All states of corresponding isotherms of all systems have something in common, namely, that they are in thermal equilibrium with one another. The systems themselves, in these states, may be said to possess a property that ensures their being in thermal equilibrium with one another. We call this property *temperature. The temperature of a system is a property that determines whether or not a system is in thermal equilibrium with other systems.*

The concept of temperature may be arrived at in a more concrete manner. When a system A with coordinates Y, X is separated from a system C with coordinates Y'', X'', the approach to thermal equilibrium is indicated by changes in the four coordinates. The final state of thermal equilibrium is denoted by a relation among these coordinates which may be written in the general functional form

$$f_{AC}(Y, X; Y'', X'') = 0. \tag{1-1}$$

For example, if A were a gas with coordinates P (pressure) and V (volume) and obeying Boyle's law, and if C were a similar gas with coordinates P'' and V'', Eq. (1-1) would be

$$PV - P''V'' = 0.$$

Thermal equilibrium between system B, with coordinates Y', X', and system C is similarly denoted by the relation

$$f_{BC}(Y', X'; Y'', X'') = 0, \tag{1-2}$$

where f_{BC} may be quite different from f_{AC} but is also assumed to be a well-behaved function.

Suppose that Eqs. (1-1) and (1-2) are solved for Y''; then

$$Y'' = g_{AC}(Y, X, X''),$$

and

$$Y'' = g_{BC}(Y', X', X''),$$

or

$$g_{AC}(Y, X, X'') = g_{BC}(Y', X', X''). \tag{1-3}$$

Now, according to the zeroth law, thermal equilibrium between A and C and between B and C implies thermal equilibrium between A and B, which is denoted by a relation among *coordinates of systems A and B only;* thus,

$$f_{AB}(Y, X; Y', X') = 0. \tag{1-4}$$

Since Eq. (1-3) also expresses the same two equilibrium situations, it must agree with Eq. (1-4): that is, it must reduce to a relation among $Y, X; Y', X'$ only. The extraneous coordinate X'' in Eq. (1-3) must therefore drop out, and the equation must reduce to

$$h_A(Y, X) = h_B(Y', X').$$

Applying the same argument a second time with systems A and C in equilibrium with B, we get finally, when the three systems are in thermal equilibrium,

$$h_A(Y, X) = h_B(Y', X') = h_C(Y'', X''). \tag{1-5}$$

In other words, *a function of each set of coordinates exists, and these functions are all equal when the systems are in thermal equilibrium with one another.* The common value t of these functions is the *empirical temperature* common to all the systems.

$$\boxed{t = h_A(Y, X) = h_B(Y', X') = h_C(Y'', X'').} \tag{1-6}$$

The relation $t = h_A(Y, X)$ is merely *the equation of an isotherm of system A*, such as curve I of Fig. 1-3. If t is given a different numerical value, a different curve is obtained, such as II in Fig. 1-3.

The temperature of all systems in thermal equilibrium may be represented by a number. The establishment of a temperature scale is merely the adoption of a set of rules for assigning one number to one set of corresponding isotherms and a different number to a different set of corresponding isotherms.

1-7 MEASUREMENT OF TEMPERATURE

To establish an empirical temperature scale, we select some system with coordinates Y and X as a standard, which we call a *thermometer*, and adopt a set of rules for assigning a numerical value to the temperature associated with each of its isotherms. To every other system in thermal equilibrium with the thermometer, we assign the same number for the temperature. The simplest procedure is to

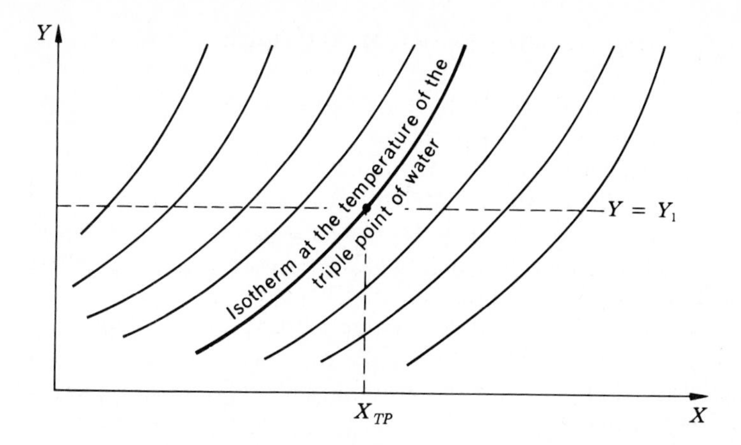

Figure 1-4 Setting up a temperature scale involves assignment of numerical values to the isotherms of an arbitrarily chosen standard system, or thermometer.

choose any convenient path in the YX plane, such as that shown in Fig. 1-4 by the dashed line $Y = Y_1$ which intersects the isotherms at points each of which has the same Y coordinate but a different X coordinate. The temperature associated with each isotherm is then taken to be a convenient function of the X at this intersection point. The coordinate X is called the *thermometric property*, and the form of the *thermometric function* $\theta(X)$ determines the temperature scale. There are six important kinds of thermometer, each with its own thermometric property, as shown in Table 1-1.

Let X stand for any one of the thermometric properties listed in Table 1-1, and let us decide *arbitrarily* to define the temperature scale so that the temperature θ is directly proportional to X. Thus, the temperature common to the thermometer *and to all systems in thermal equilibrium with it* is given by

$$\theta(X) = aX \qquad \text{(const. } Y\text{)}. \tag{1-7}$$

It should be noted that *different* temperature scales usually result when this arbitrary relation is applied to different kinds of thermometers and even when it is applied to different systems of the same kind. One must thus ultimately select, either arbitrarily or in some rational way, one kind of thermometer and one

Table 1-1 Thermometers and thermometric properties

Thermometer	Thermometric property	Symbol
Gas (const. volume)	Pressure	P
Electric resistor (const. tension)	Electric resistance	R'
Thermocouple (const. tension)	Thermal emf	$\mathscr{E}$
Helium vapor (saturated)	Pressure	P
Paramagnetic salt	Magnetic susceptibility	χ
Blackbody radiation	Radiant emittance	$\mathscr{R}_{B,\lambda}$

particular system (or type of system) to serve as the standard thermometric device. But regardless of what standard is chosen, the value of a in Eq. (1-7) must be established; only then does one have a numerical relation between the temperature $\theta(X)$ and the thermometric property X.

Equation (1-7) applies generally to a thermometer placed in contact with a system whose temperature $\theta(X)$ is to be measured. It therefore applies when the thermometer is placed in contact with an arbitrarily chosen standard system in an *easily reproducible state;* such a state of an *arbitrarily chosen standard system* is called *a fixed point*. Since 1954, only one standard fixed point has been in use, the *triple point of water*, the state of pure water existing as an equilibrium mixture of ice, liquid, and vapor. The temperature at which this state exists is arbitrarily assigned the value 273.16 Kelvin, abbreviated 273.16 K. Equation (1-7) solved for a now becomes

$$a = \frac{273.16\ \text{K}}{X_{TP}}, \tag{1-8}$$

where the subscript TP identifies the property value X_{TP} explicitly with the triple-point temperature. In view of Eq. (1-8), the general Eq. (1-7) may be written

$$\theta(X) = 273.16\ \text{K}\,\frac{X}{X_{TP}} \qquad (\text{const. } Y). \tag{1-9}$$

The temperature of the triple point of water is the *standard fixed point* of thermometry. To achieve the triple point, one distills water of the highest purity and of substantially the same isotopic composition of ocean water into a vessel depicted schematically in Fig. 1-5. When all air has been removed, the vessel is

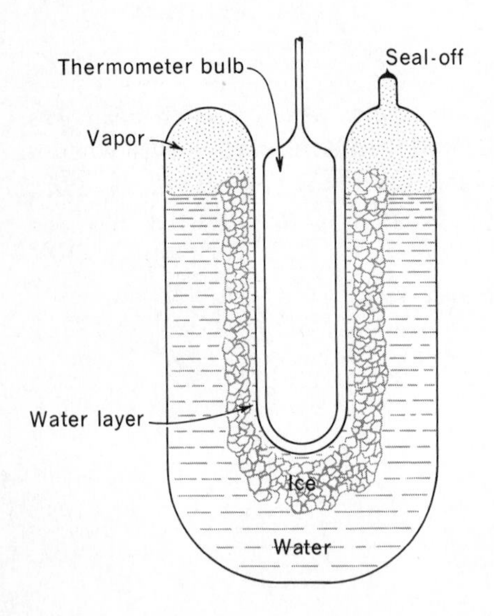

Figure 1-5 Triple-point cell.

sealed off. With the aid of a freezing mixture in the inner well, a layer of ice is formed around the well. When the freezing mixture is replaced by a thermometer bulb, a thin layer of ice is melted nearby. So long as the solid, liquid, and vapor phases coexist in equilibrium, the system is at the triple point.

1-8 COMPARISON OF THERMOMETERS

Applying the principles outlined in the preceding paragraphs to the first three thermometers listed in Table 1-1, we have three different ways of measuring temperature. Thus, for a gas at constant volume,

$$\theta(P) = 273.16 \text{ K} \frac{P}{P_{TP}} \qquad \text{(const. } V\text{);}$$

for an electric resistor,

$$\theta(R') = 273.16 \text{ K} \frac{R'}{R'_{TP}};$$

and for a thermocouple,

$$\theta(\mathcal{E}) = 273.16 \text{ K} \frac{\mathcal{E}}{\mathcal{E}_{TP}}.$$

Now imagine a series of tests in which the temperature of a given system is measured simultaneously with each of the three thermometers. Such a comparison is shown in Table 1-2. The initials NBP stand for the *normal boiling point*,

Table 1-2 Comparison of thermometers

Fixed point	Copper-constantan thermocouple		Platinum resistance thermometer		Constant-volume H_2 thermometer		Constant-volume H_2 thermometer	
	$\mathcal{E}$, mV	$\theta(\mathcal{E})$	R', ohms	$\theta(R')$	P, kPa†	$\theta(P)$	P, kPa†	$\theta(P)$
N_2 (NBP)	0.73	32.0	1.96	54.5	184	73	29	79
O_2 (NBP)	0.95	41.5	2.50	69.5	216	86	33	90
CO_2 (NSP)	3.52	154	6.65	185	486	193	73	196
H_2O (TP)	$\mathcal{E}_{TP} = 6.26$	273	$R'_{TP} = 9.83$	273	$P_{TP} = 689$	273	$P_{TP} = 101$	273
H_2O (NBP)	10.05	440	13.65	380	942	374	139	374
Sn (NMP)	17.50	762	18.56	516	1287	510	187	505

† 1 Pa = 1 N/m^2, 1 atm = 101.3 kPa.

which is the temperature at which a liquid boils at atmospheric pressure; the letters NMP stand for *normal melting point*, NSP for the *normal sublimation point*, and TP for the *triple point*. The numerical values are not meant to be exact, and 273.16 has been written simply 273. If one compares the θ columns, it may be seen that at any fixed point, except the triple point of water, the thermometers disagree. Even the two hydrogen thermometers disagree slightly, but the variation among gas thermometers may be greatly reduced by using low pressures, so that a gas thermometer has been chosen as the standard thermometer in terms of which the empirical temperature scale is defined.

1-9 GAS THERMOMETER

A schematic diagram of a constant-volume gas thermometer is shown in Fig. 1-6. The materials, construction, and dimensions differ in the various bureaus and institutes throughout the world where these instruments are used and depend on the nature of the gas and the temperature range for which the thermometer is intended. The gas is contained in the bulb B (usually made of

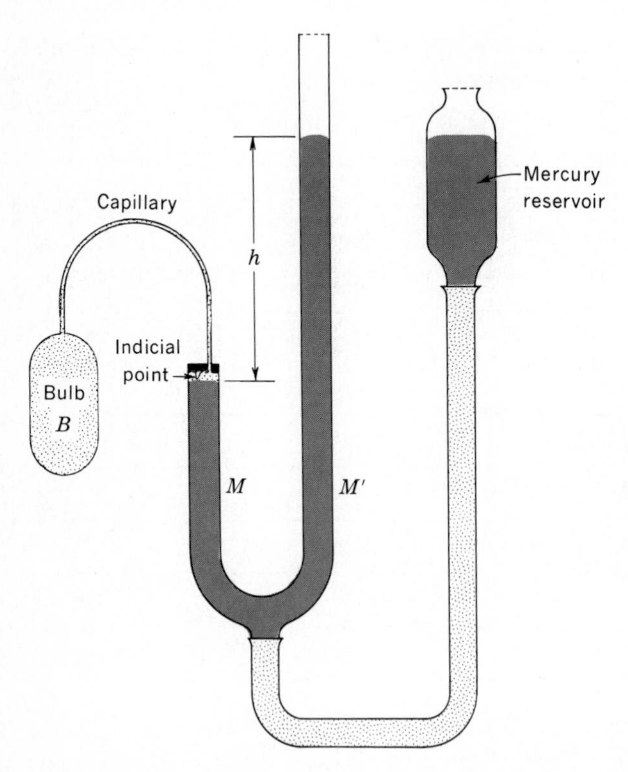

Figure 1-6 Simplified constant-volume gas thermometer. Mercury reservoir is raised or lowered so that meniscus at left always touches indicial point. Bulb pressure equals h plus atmospheric pressure.

platinum or a platinum alloy), which communicates with the mercury column M through a capillary. The volume of the gas is kept constant by adjusting the height of the mercury column M until the mercury level just touches the tip of a small pointer (indicial point) in the space above M, known as the *dead space* or *nuisance volume*. The mercury column M is adjusted by raising or lowering the reservoir. The difference in height h between the two mercury columns M and M' is measured when the bulb is surrounded by the system whose temperature is to be measured, and when it is surrounded by water at the triple point.

The various values of the pressure must be corrected to take account of the following sources of error:

1. The gas present in the dead space (and in any other nuisance volumes) is at a temperature different from that in the bulb.
2. The gas in the capillary connecting the bulb with the manometer has a temperature gradient; i.e., it is not at a uniform temperature.
3. The bulb, capillary, and nuisance volumes undergo changes of volume when the temperature and pressure change.
4. If the diameter of the capillary is comparable with the mean free path of the molecules of the gas, a pressure gradient exists in the capillary (Knudsen effect).
5. Some gas is adsorbed on the walls of the bulb and capillary; the lower the temperature, the greater the adsorption.
6. There are effects due to temperature and compressibility of the mercury in the manometer.

Many great improvements in the design of gas thermometers have been made in recent years. Two of these are depicted schematically in Fig. 1-7. Instead of the thermometric gas in the bulb communicating directly with the mercury in

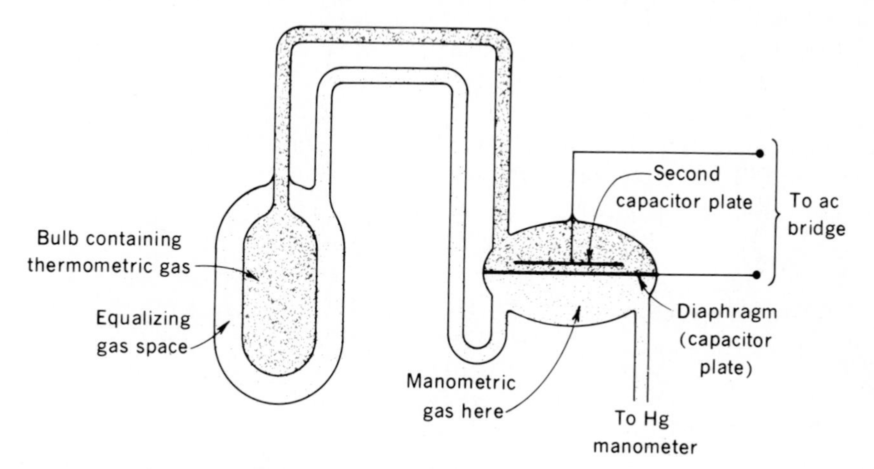

Figure 1-7 Schematic diagram of two improvements in a gas thermometer in use at the U.S. National Bureau of Standards.

the manometer, there are two separate volumes of gas: the thermometric gas, which goes as far as a diaphragm and exerts a pressure on one side of it, and a manometric gas on the other side of the diaphragm leading to the manometer. The diaphragm itself is one plate of a capacitor, with the other plate fixed nearby. A difference of pressure across the diaphragm causes a slight motion of the diaphragm, resulting in a change of capacitance that is observed with the aid of an ac bridge. At standard atmospheric pressure, a pressure differential of 1 part per million is detectable. When the diaphragm shows no deflection, the manometric gas pressure is the same as that of the thermometric gas, and a reading of the manometer gives the gas pressure in the bulb.

Another improvement depicted in Fig. 1-7 is an equalizing gas space surrounding the bulb. The manometric gas is allowed to fill this space. At the moment when a manometer reading is made, there is no net force tending to alter the dimensions of the bulb, and therefore no correction need be made for a variation of bulb volume with pressure.

The greatest improvements have been made in the mercury manometer. The mercury meniscus in each tube is made very flat by widening the tubes, since the dead space does not depend on this width as it did in the older instrument depicted in Fig. 1-6. The position of a mercury meniscus is obtained by using it as one plate of a capacitor, with the other being fixed nearby, and measuring the capacitance with an ac bridge. Gauge blocks are used to measure the difference in height of the two mercury columns. Pressures can be measured exact to a few ten-thousandths of a millimeter of mercury or to a few hundredths of a pascal.†

1-10 IDEAL-GAS TEMPERATURE

Suppose that an amount of gas is introduced into the bulb of a constant-volume gas thermometer so that the pressure P_{TP}, when the bulb is surrounded by water at its triple point, is equal to 120 kPa. Keeping the volume V constant, suppose the following procedures are carried out:

1. Surrounding the bulb with steam condensing at standard atmospheric pressure, determine the gas pressure P_s and calculate

$$\theta(P_s) = 273.16\ \text{K}\,\frac{P_s}{120}.$$

2. Remove some of the gas so that P_{TP} has a smaller value, say, 60 kPa. Determine the new value of P_s and calculate a new value

$$\theta(P_s) = 273.16\ \text{K}\,\frac{P_s}{60}.$$

† 1 Pa = 1 N/m^2 = 10 dynes/cm^2.

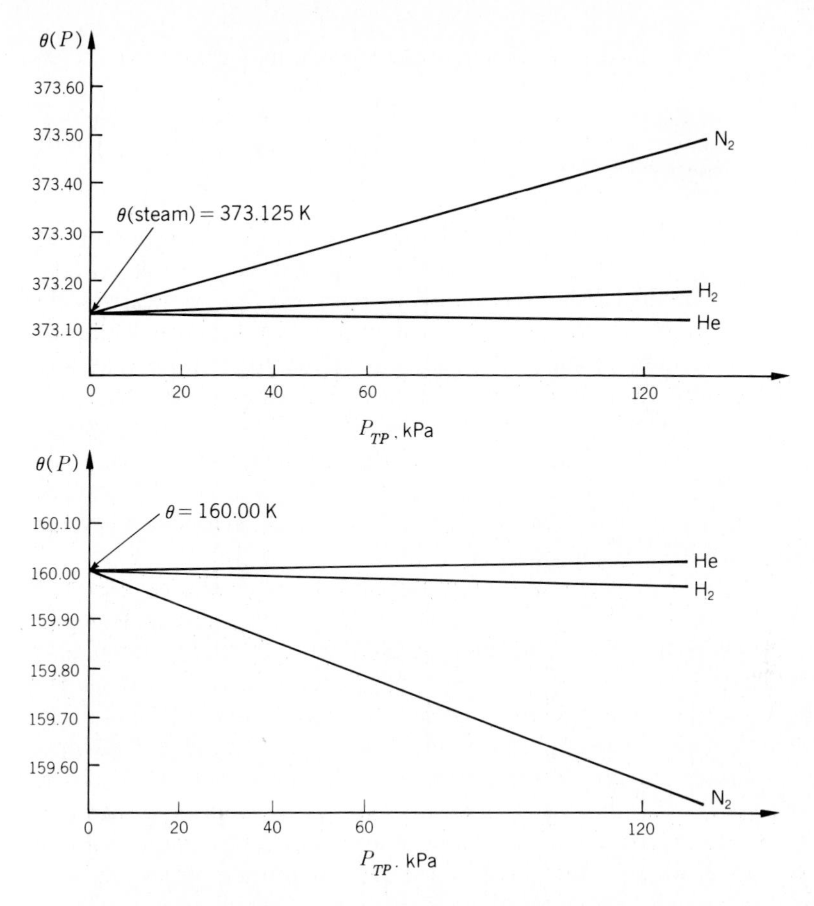

Figure 1-8 Readings of a constant-volume gas thermometer for the temperature of condensing steam and for that of a test object, when different gases are used at various values of P_{TP}.

3. Continue reducing the amount of gas in the bulb so that P_{TP} and P_s have smaller and smaller values, P_{TP} having values of, say, 40 kPa, 20 kPa, etc. At *each* value of P_{TP}, calculate the corresponding $\theta(P_s)$.
4. Plot $\theta(P_s)$ against P_{TP} and extrapolate the resulting curve to the axis where $P_{TP} = 0$. Read from the graph

$$\lim_{P_{TP}\to 0} \theta(P_s).$$

The results of a series of tests of this sort are plotted in Fig. 1-8 for three different gases in order to measure $\theta(P)$ not only of condensing steam but also of another temperature. The graph conveys the information that, although the readings of a constant-volume gas thermometer depend upon the nature of the gas at ordinary values of P_{TP}, *all gases indicate the same temperature as P_{TP} is lowered and made to approach zero.*

We therefore define the *ideal-gas temperature* θ by the equation

$$\theta = 273.16\ \text{K} \lim_{P_{TP} \to 0} \left(\frac{P}{P_{TP}} \right) \quad \text{(const. } V\text{)}. \tag{1-10}$$

Although the ideal-gas temperature scale is independent of the properties of any one particular gas, it still depends on the properties of gases in general. Helium is the most useful gas for thermometric purposes for two reasons. At high temperatures helium does not diffuse through platinum, whereas hydrogen does. Furthermore, helium becomes a liquid at a temperature lower than any other gas, and therefore a helium thermometer may be used to measure temperatures lower than those which can be measured with any other gas thermometer.

The lowest ideal-gas temperature that can be measured with a gas thermometer is about 0.5 K, provided that low-pressure ^{3}He is used. *The temperature* $\theta = 0$ *remains as yet undefined.* In Chap. 7 the Kelvin temperature scale, which is independent of the properties of any particular substance, will be developed. It will be shown that, *in the temperature region in which a gas thermometer may be used, the ideal-gas scale and the Kelvin scale are identical.* In anticipation of this result, we write K after an ideal-gas temperature. It will also be shown in Chap. 7 how the absolute zero of temperature is defined on the Kelvin scale. Until then, the phrase "absolute zero" will have no meaning. It should be remarked that the statement, found in so many textbooks of elementary physics, that, at the temperature $T = 0$, all molecular activity ceases is entirely erroneous. First, such a statement involves an assumption connecting the purely macroscopic concept of temperature and the microscopic concept of molecular motion. If we want our theory to be general, this is precisely the sort of assumption that must be avoided. Second, when it is necessary in statistical mechanics to correlate temperature to molecular activity, it is found that classical statistical mechanics must be modified with the aid of quantum mechanics and that, when this modification is carried out, the molecules of a substance at absolute zero have a *finite* amount of kinetic energy, known as the *zero-point energy*.

1-11 CELSIUS TEMPERATURE SCALE

The Celsius temperature scale employs a degree of the same magnitude as that of the ideal-gas scale, but its zero point is shifted so that *the Celsius temperature of the triple point of water is* 0.01 *degree Celsius*, abbreviated 0.01°C. Thus, if t denotes the Celsius temperature,

$$t(^\circ\text{C}) = \theta(\text{K}) - 273.15. \tag{1-11}$$

Thus, the Celsius temperature t_s at which steam condenses at standard atmospheric pressure is

$$t_s = \theta_s - 273.15,$$

and reading θ_s from Fig. 1-8,

$$t_s = 373.125 - 273.15$$
$$= 99.975°\text{C}.$$

Similar measurements for the ice point (the temperature at which ice and liquid water saturated with air at standard atmospheric pressure are in equilibrium) show this temperature on the Celsius scale to be 0.00°C. It should be noted, however, that these two temperatures are subject to the experimental uncertainty attending the determination of intercepts by extrapolation, as illustrated in Fig. 1-8. The only Celsius temperature which is fixed *by definition* is that of the triple point.

1-12 ELECTRIC RESISTANCE THERMOMETRY

When the resistance thermometer is in the form of a long, fine wire, it is usually wound around a thin frame constructed so as to avoid excessive strains when the wire contracts upon cooling. In special circumstances the wire may be wound on or embedded in the material whose temperature is to be measured. In the very-low-temperature range, resistance thermometers often consist of small carbon-composition radio resistors or a germanium crystal doped with arsenic and sealed in a helium-filled capsule. These may be bonded to the surface of the substance whose temperature is to be measured or placed in a hole drilled for that purpose.

It is customary to measure the resistance by maintaining a known constant current in the thermometer and measuring the potential difference across it with the aid of a very sensitive potentiometer. A typical circuit is shown in Fig. 1-9. The current is held constant by adjusting a rheostat so that the potential difference across a standard resistor in series with the thermometer, as observed with a monitoring potentiometer, remains constant.

The platinum resistance thermometer may be used for very accurate work within the range −253 to 1200°C. The calibration of the instrument involves the measurement of R'_{Pt} at various known temperatures and the representation of the results by an empirical formula. In a restricted range, the following quadratic equation is often used:

$$R'_{\text{Pt}} = R_0(1 + At + Bt^2),$$

where R'_0 is the resistance of the platinum wire when it is surrounded by water at the triple point, A and B are constants, and t is the empirical Celsius temperature.

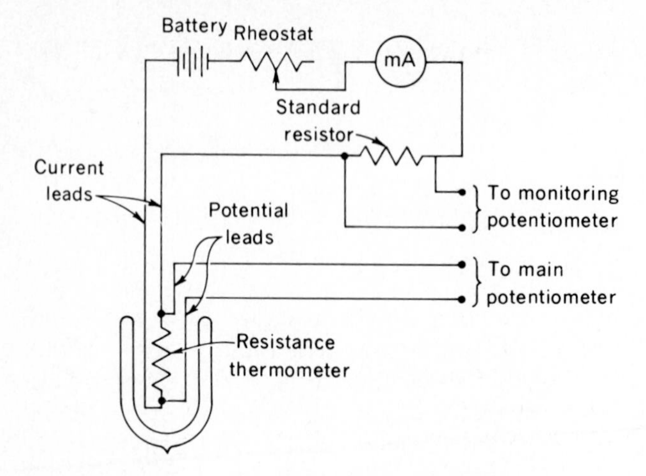

Figure 1-9 Circuit for measuring the resistance of a resistance thermometer through which a constant current is maintained.

1-13 THERMOCOUPLE

The correct use of a thermocouple is shown in Fig. 1-10. The thermal electromotive force (emf) is measured with a potentiometer, which, as a rule, must be placed at some distance from the system whose temperature is to be measured. The reference junction, therefore, is placed near the test junction and consists of two connections with copper wire, maintained at the temperature of melting ice. This arrangement allows the use of copper wires for connection to the potentiometer. The binding posts of the potentiometer are usually made of brass, and

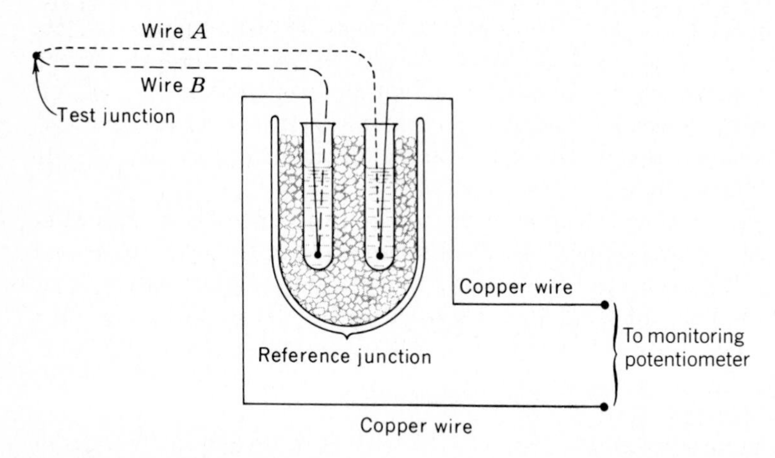

Figure 1-10 Thermocouple of wires *A* and *B* with a reference junction consisting of two junctions with copper, connected to a potentiometer.

therefore at the potentiometer there are two copper-brass thermocouples. If the two binding posts are at the same temperature, these two copper-brass thermocouples introduce no error.

A thermocouple is calibrated by measuring the thermal emf at various known temperatures, the reference junction being kept at 0°C. The results of such measurements on most thermocouples can usually be represented by a cubic equation as follows:

$$\mathcal{E} = a + bt + ct^2 + dt^3,$$

where $\mathcal{E}$ is the thermal emf and the constants a, b, c, and d are different for each thermocouple. Within a restricted range of temperature, a quadratic equation is often sufficient. The range of a thermocouple depends upon the materials of which it is composed. A platinum-10% rhodium/platinum thermocouple has a range of 0 to 1600°C. The advantage of a thermocouple is that it comes to thermal equilibrium quite rapidly with the system whose temperature is to be measured, because its mass is small. It therefore follows temperature changes easily, but it is not so accurate as a platinum resistance thermometer.

1-14 INTERNATIONAL PRACTICAL TEMPERATURE SCALE OF 1968 (IPTS-68)

The use of an ideal-gas thermometer for routine calibrations or for the usual measurement of thermodynamic temperature is impractical. At the Seventh General Conference of Weights and Measures in 1927, an international practical temperature scale was adopted to provide the means for easy and rapid calibration of scientific and industrial instruments. The temperature scale was revised in 1948 and amended in 1960.

The International Practical Temperature Scale of 1968 (IPTS-68), as amended in 1975, consists of a set of fixed points measured with a constant-volume gas thermometer and a set of procedures for interpolation between the fixed points. Although IPTS-68 is not intended to supplant the ideal-gas or Celsius scales, it is constructed so as to provide a very close approximation to them; the differences between the scales are within the limits of accuracy of measurement attained in 1975.

The accurate measurement of temperature with a gas thermometer requires months of painstaking laboratory work and mathematical computation and when completed becomes an international event. Such work is published in a physical journal and eventually is listed in tables of physical constants. The temperatures of the normal boiling points (NBP), normal melting points (NMP), and equilibrium states of a number of materials have been measured, and the results are tabulated in Table 1-3.

The lower temperature limit of IPTS-68 is 13.81 K, the triple point of equilibrium hydrogen. Below this temperature the scale is undefined. Above the freezing point of gold (1337.58 K) an optical method is used in conjunction with the

Table 1-3 Temperatures of fixed points of IPTS-68†

Fixed points‡		T, K	t, °C
Standard	TP of water	273.16	0.01
Defining fixed points	TP of equilibrium hydrogen	13.81	−259.34
	BP of equilibrium hydrogen at 33 330.6 Pa	17.042	−256.108
	NBP of equilibrium hydrogen	20.28	−252.87
	NBP of neon	27.102	−246.048
	TP of oxygen	54.361	−218.789
	TP of argon	83.798	−189.352
	NBP of oxygen	90.188	−182.962
	NBP of water§	373.125	99.975
	NMP of tin§	505.074	231.924
	NMP of zinc§	692.664	419.514
	NMP of silver	1235.08	961.93
	NMP of gold	1337.58	1064.43
Secondary reference points	TP of normal hydrogen	13.965	−259.194
	NBP of normal hydrogen	20.397	−252.753
	TP of neon	24.561	−248.589
	TP of nitrogen	63.146	−210.004
	NBP of nitrogen	77.344	−195.806
	NBP of argon	87.294	−185.856
	NSP of carbon dioxide	194.674	−78.476
	NMP of mercury	234.314	−38.836
	Equilibrium of ice and air-saturated water	273.15	0
	TP of phenoxybenzene	300.02	26.87
	TP of benzoic acid	395.52	122.37
	NMP of indium	429.784	156.634
	NMP of bismuth	544.592	271.442
	NMP of cadmium	594.258	321.108
	NMP of lead	600.652	327.502
	NBP of mercury	629.81	356.66
	NBP of sulfur	717.824	444.674
	NMP of antimony	903.905	630.755
	NMP of aluminum	933.61	660.46
	NMP of copper	1358.03	1084.88
	NMP of nickel	1728	1455
	NMP of cobalt	1768	1495
	NMP of palladium	1827	1554
	NMP of platinum	2042	1769
	NMP of rhodium	2236	1963
	NMP of indium	2720	2447
	NMP of niobium	2750	2477
	NMP of molybdenum	2896	2623
	NMP of tungsten	3695	3422

† *Metrologia*, **12**:7 (1976).

§ *Metrologia*, **13**:177 (1977).

‡ The normal boiling points (NBP) and normal melting points (NMP) occur at a pressure of 101 325 Pa (1 standard atmosphere).

Planck radiation formula. The interval between 13.81 and 1337.58 K is divided into three main parts, as follows:

1. *From 13.81 K to 273.15 K*. A strain-free annealed platinum resistance thermometer is used. The temperature range is divided into four subintervals, and within each interval the resistance is measured at specified fixed points selected from Table 1-3. The differences between the measured resistances and a tabulated reference function are fitted to polynomial equations in T or t; these equations serve as interpolation formulas for conversion of measured values of R to temperature.
2. *From 273.15 K to 903.89 K*. The same platinum resistance thermometer is used as in part 1. A polynomial equation for t as a function of R is employed, with the constants in the equation determined from resistance measurements at the triple point of water, the normal boiling point of water, and the normal freezing point of zinc.
3. *From 903.89 K to 1337.58 K*. A thermocouple, one wire of which is made of platinum of a specified purity and the other of an alloy of 90 percent platinum and 10 percent rhodium, has one junction maintained at 0°C. The electromotive force e is represented by the formula

$$e = a + bt + ct^2$$

where a, b, and c are calculated from measurements of e at 903.89 ± 0.2 K, as determined by a platinum resistance thermometer, and at the normal freezing points of silver and gold.

PROBLEMS

1-1 Systems A, B, and C are gases with coordinates P, V; P', V'; P'', V''. When A and C are in thermal equilibrium, the equation

$$PV - nbP - P''V'' = 0$$

is found to be satisfied. When B and C are in thermal equilibrium, the relation

$$P'V' - P''V'' + \frac{nB'P''V''}{V'} = 0$$

holds. The symbols n, b, and B' are constants.

(*a*) What are the three functions which are equal to one another at thermal equilibrium and each of which is equal to t, where t is the empirical temperature?

(*b*) What is the relation expressing thermal equilibrium between A and B?

1-2 Systems A and B are paramagnetic salts with coordinates $\mathscr{H}$, M and $\mathscr{H}'$, M', respectively. System C is a gas with coordinates P, V. When A and C are in thermal equilibrium, the equation

$$4\pi nRC_C\mathscr{H} - MPV = 0$$

is found to hold. When B and C are in thermal equilibrium, we get

$$nR\Theta M' + 4\pi nRC'_C\mathscr{H}' - M'PV = 0,$$

where n, R, C_C, C'_C, and Θ are constants.

(*a*) What are the three functions that are equal to one another at thermal equilibrium?

(*b*) Set each of these functions equal to the ideal-gas temperature θ, and see whether any of these equations are equations of state as discussed in Chap. 2.

1-3 In the table below, a number in the top row represents the pressure of a gas in the bulb of a constant-volume gas thermometer (corrected for dead space, thermal expansion of bulb, etc.) when the bulb is immersed in a water triple-point cell. The bottom row represents the corresponding readings of pressure when the bulb is surrounded by a material at a constant unknown temperature. Calculate the ideal-gas temperature θ of this material. (Use five significant figures.)

P_{TP}, mm Hg	1000.0	750.00	500.00	250.00
P, mm Hg	1535.3	1151.6	767.82	383.95

1-4 The resistance R' of a particular carbon resistor obeys the equation

$$\sqrt{\frac{\log R'}{\theta}} = a + b \log R',$$

with $a = -1.16$ and $b = 0.675$.

(*a*) In a liquid helium cryostat, the resistance is found to be exactly 1000 Ω. What is the temperature?

(*b*) Make a log-log graph of R' against θ in the resistance range from 1000 to 30,000 Ω.

1-5 The resistance of a doped germanium crystal obeys the equation

$$\log R' = 4.697 - 3.917 \log \theta.$$

(*a*) In a liquid helium cryostat, the resistance is measured to be 218 Ω. What is the temperature?

(*b*) Make a log-log graph of R' against θ in the resistance range from 200 to 30,000 Ω.

CHAPTER

TWO

SIMPLE THERMODYNAMIC SYSTEMS

2-1 THERMODYNAMIC EQUILIBRIUM

Suppose that experiments have been performed on a thermodynamic system and that the coordinates necessary and sufficient for a macroscopic description have been determined. When these coordinates change in any way whatsoever, either spontaneously or by virtue of outside influence, the system is said to undergo a *change of state*.† When a system is not influenced in any way by its surroundings, it is said to be isolated. In practical applications of thermodynamics, isolated systems are of little importance. We usually have to deal with a system that is influenced in some way by its surroundings. In general, the surroundings may exert forces on the system or provide contact between the system and a body at some definite temperature. When the state of a system changes, interactions usually take place between the system and its surroundings.

When there is no unbalanced force in the interior of a system and also none between a system and its surroundings, the system is said to be in a state of *mechanical equilibrium*. When these conditions are not satisfied, either the system alone or both the system and its surroundings will undergo a change of state, which will cease only when mechanical equilibrium is restored.

† This must not be confused with the terminology of elementary physics, where the expression "change of state" is often used to signify a transition from solid to liquid or liquid to gas, etc. Such a change in the language of thermodynamics is called a *change of phase*.

When a system in mechanical equilibrium does not tend to undergo a spontaneous change of internal structure, such as a chemical reaction, or a transfer of matter from one part of the system to another, such as diffusion or solution, however slow, then it is said to be in a state of *chemical equilibrium.* A system not in chemical equilibrium undergoes a change of state that, in some cases, is exceedingly slow. The change ceases when chemical equilibrium is reached.

Thermal equilibrium exists when there is no spontaneous change in the coordinates of a system in mechanical and chemical equilibrium when it is separated from its surroundings by a diathermic wall. In thermal equilibrium, all parts of a system are at the same temperature, and this temperature is the same as that of the surroundings. When these conditions are not satisfied, a change of state will take place until thermal equilibrium is reached.

When the conditions for all three types of equilibrium are satisfied, the system is said to be in a state of *thermodynamic equilibrium;* in this condition, it is apparent that there will be no tendency whatever for any change of state, either of the system or of the surroundings, to occur. *States of thermodynamic equilibrium can be described in terms of macroscopic coordinates that do not involve the time, i.e., in terms of thermodynamic coordinates.* Classical thermodynamics does not attempt to deal with any problem involving the rate at which a process takes place. The investigation of such problems is carried out in other branches of science, as in the kinetic theory of gases, hydrodynamics, and chemical kinetics.

When the conditions for any one of the three types of equilibrium that constitute thermodynamic equilibrium are not satisfied, the system is said to be in a *nonequilibrium state.* Thus, when there is an unbalanced force in the interior of a system or between a system and its surroundings, the following phenomena may take place: acceleration, turbulence, eddies, waves, etc. While such phenomena are in progress, a system passes through nonequilibrium states. If an attempt is made to give a macroscopic description of any one of these nonequilibrium states, it is found that the pressure varies from one part of a system to another. There is no single pressure that refers to the system as a whole. Similarly, in the case of a system at a different temperature from its surroundings a nonuniform temperature distribution is set up and there is no single temperature that refers to the system as a whole. We therefore conclude that, *when the conditions for mechanical and thermal equilibrium are not satisfied, the states traversed by a system cannot be described in terms of thermodynamic coordinates referring to the system as a whole.*

It must not be concluded, however, that we are entirely helpless in dealing with such nonequilibrium states. If we divide the system into a large number of small mass elements, then thermodynamic coordinates may be found in terms of which a macroscopic description of each mass element may be approximated. There are also special methods for dealing with systems in mechanical and thermal equilibrium but not in chemical equilibrium. All these special methods will be considered later. At present we shall deal exclusively with systems in thermodynamic equilibrium.

Imagine, for the sake of simplicity, a constant mass of gas in a vessel so equipped that the pressure, volume, and temperature may be easily measured. If we fix the volume at some arbitrary value and cause the temperature to assume an arbitrarily chosen value, then we shall not be able to vary the pressure at all. Once V and θ are chosen by us, the value of P at equilibrium is determined by nature. Similarly, if P and θ are chosen arbitrarily, then the value of V at equilibrium is fixed. That is, of the three thermodynamic coordinates P, V, and θ, only two are independent variables. This implies that there exists an equation of equilibrium which connects the thermodynamic coordinates and which robs one of them of its independence. Such an equation is called an *equation of state*. Every thermodynamic system has its own equation of state, although in some cases the relation may be so complicated that it cannot be expressed in terms of simple mathematical functions.

An equation of state expresses the individual peculiarities of one system in contradistinction to another and must therefore be determined either by experiment or by molecular theory. A general theory like thermodynamics, based on general laws of nature, is incapable of expressing the behavior of one material as opposed to another. An equation of state therefore is not a theoretical deduction from thermodynamics but is usually an experimental addition to thermodynamics. It expresses the results of experiments in which the thermodynamic coordinates of a system were measured as accurately as possible, within a limited range of values. An equation of state is therefore only as accurate as the experiments that led to its formulation and holds only within the range of values measured. As soon as this range is exceeded, a different form of equation of state may be valid.

No equation of state exists for the states traversed by a system that is not in mechanical and thermal equilibrium, since such states cannot be described in terms of thermodynamic coordinates referring to the system as a whole. For example, if a gas in a cylinder were to expand and to impart to a piston an accelerated motion, the gas might have, at any moment, a definite volume and temperature, but the corresponding pressure could not be calculated from an equation of state. The pressure would not be a thermodynamic coordinate because it would not only depend on the velocity and the acceleration of the piston but would also perhaps vary from point to point.

Any system of constant mass that exerts on the surroundings a uniform hydrostatic pressure, in the absence of surface, gravitational, electric, and magnetic effects, we shall call a *hydrostatic system*. Hydrostatic systems are divided into the following categories:

1. A *pure substance*, which is one chemical constituent in the form of a solid, a liquid, a gas, a mixture of any two, or a mixture of all three.
2. A *homogeneous mixture of different constituents*, such as a mixture of inert gases, a mixture of chemically active gases, a mixture of liquids, or a solution.
3. A *heterogeneous mixture*, such as a mixture of different gases in contact with a mixture of different liquids.

Experiments show that the states of equilibrium† of a hydrostatic system can be described with the aid of three coordinates, namely, the pressure P exerted by the system on the surroundings, the volume V, and the absolute temperature θ. The pressure is measured in newtons per square meter (pascal) and the volume in cubic meters; the most convenient scale of temperature is the ideal-gas scale. Other units of pressure, such as pounds per square inch, atmospheres, and millimeters of mercury are used in various applications of thermodynamics and will be used occasionally in this book. In the absence of any special remarks about units, however, it will be understood that SI units are to be employed.

2-2 *PV* DIAGRAM FOR A PURE SUBSTANCE

If 1 kg of water at about 94°C is introduced into a vessel about 2 cubic meters in volume from which all the air has been exhausted, the water will evaporate completely and the system will be in the condition known as *unsaturated vapor*, the pressure of the vapor being less than standard atmospheric pressure. On the PV diagram shown in Fig. 2-1, this state is represented by the point A. If the vapor is then compressed slowly and isothermally, the pressure will rise until there is *saturated vapor* at the point B. If the compression is continued, condensation takes place, with the pressure remaining constant (*isobaric* process) as long as the temperature remains constant. The straight line BC represents the isothermal isobaric condensation of water vapor, the constant pressure being called the *vapor pressure*. At any point between B and C, water and steam are in equilibrium; at the point C, there is only liquid water, or *saturated liquid*. Since a very large increase of pressure is needed to compress liquid water, the line CD is almost vertical. At any point on the line CD, the water is said to be in the *liquid phase;* at any point on AB, in the *vapor phase;* and at any point on BC, there is equilibrium between the liquid and the vapor phases. $ABCD$ is a typical isotherm of a pure substance on a PV diagram.

At other temperatures the isotherms are of similar character, as shown in Fig. 2-1. It is seen that the lines representing equilibrium between liquid and vapor phases, or *vaporization lines*, get shorter as the temperature rises until a certain temperature is reached—the *critical temperature*—above which there is no longer any distinction between a liquid and a vapor. Above the critical temperature only the *gas phase* exists. The isotherm at the critical temperature is called the *critical isotherm*, and the point that represents the limit of the vaporization lines is called the *critical point*. It is seen that the critical point is a point of inflection on the critical isotherm. The pressure and volume at the critical point are known as the *critical pressure* and the *critical volume*, respectively. All

† In the remainder of this book the word "equilibrium," unmodified by any adjective, will refer to thermodynamic equilibrium.

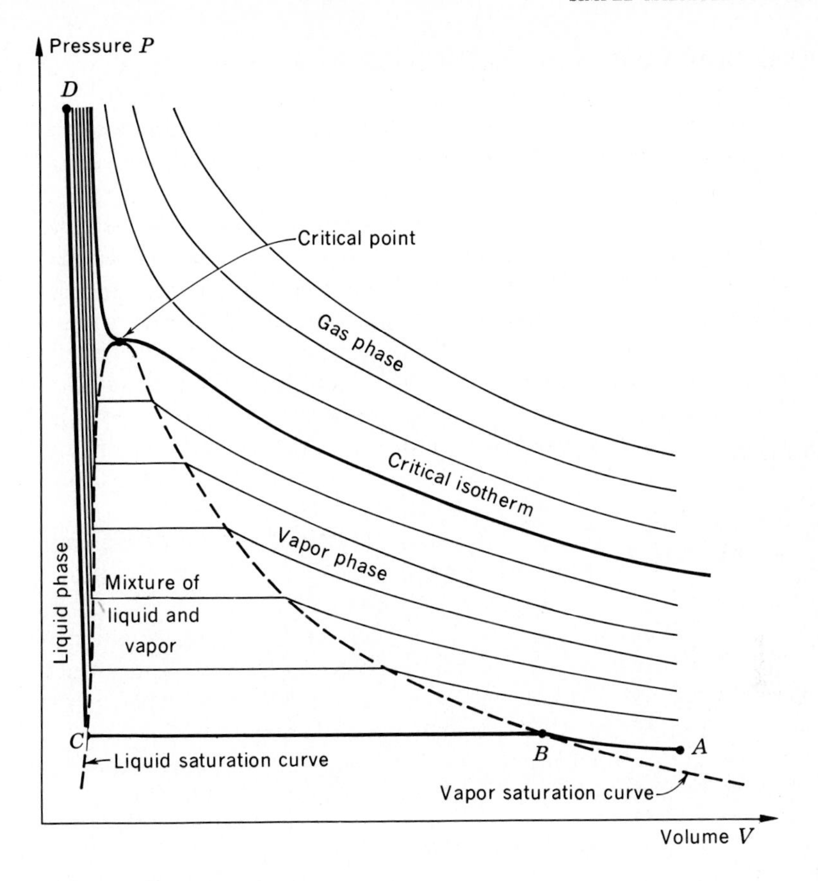

Figure 2-1 Isotherms of a pure substance.

points at which the liquid is saturated lie on the *liquid saturation curve*, and all points representing saturated vapor lie on the *vapor saturation curve*. The two saturation curves denoted by dashed lines meet at the critical point. Above the critical point the isotherms are continuous curves that at large volumes and low pressures approach the isotherms of an ideal gas.

In the PV diagram shown in Fig. 2-1, the low-temperature region representing the solid phase has not been shown. The solid region and the region of equilibrium between solid and vapor are indicated by isotherms of the same general character as those in Fig. 2-1. The horizontal portion of one of these isotherms represents the transition from saturated solid to saturated vapor, or *sublimation*. There is obviously one such line that is the boundary between the liquid-vapor region and the solid-vapor region. This *line* is associated with the *triple point*. In the case of 1 kg ordinary water, the triple point is at a pressure of 611.2 Pa and a temperature of 0.01°C, and the line extends from a volume of 10^{-3} m^3 (saturated liquid) to a volume of 206 m^3 (saturated vapor).

2-3 $P\theta$ DIAGRAM FOR A PURE SUBSTANCE

If the vapor pressure of a solid is measured at various temperatures until the triple point is reached and then that of the liquid is measured until the critical point is reached, the results when plotted on a $P\theta$ diagram appear as in Fig. 2-2. If the substance at the triple point is compressed until there is no vapor left and the pressure on the resulting mixture of liquid and solid is increased, the temperature will have to be changed for equilibrium to exist between the solid and the liquid. Measurements of these pressures and temperatures give rise to a third curve on the $P\theta$ diagram, starting at the triple point and continuing indefinitely. The points representing the coexistence of (1) solid and vapor lie on the *sublimation curve;* (2) liquid and vapor lie on the *vaporization curve;* (3) liquid and solid lie on the *fusion curve.* In the particular case of water, the sublimation curve is called the *frost line*, the vaporization curve is called the *steam line*, and the fusion curve is called the *ice line.*

The slopes of the sublimation and the vaporization curves for all substances are positive. The slope of the fusion curve, however, may be positive or negative. The fusion curve of most substances has a positive slope. Water is one of the important exceptions. When an equation known as the Clapeyron equation is derived as in Chap. 10, it will be seen that any substance, such as water, which contracts upon melting, has a fusion curve with a negative slope, whereas the opposite is true for a substance which expands upon melting.

The triple point is the point of intersection of the sublimation and vaporization curves. It must be understood that only on a $P\theta$ diagram is the triple point

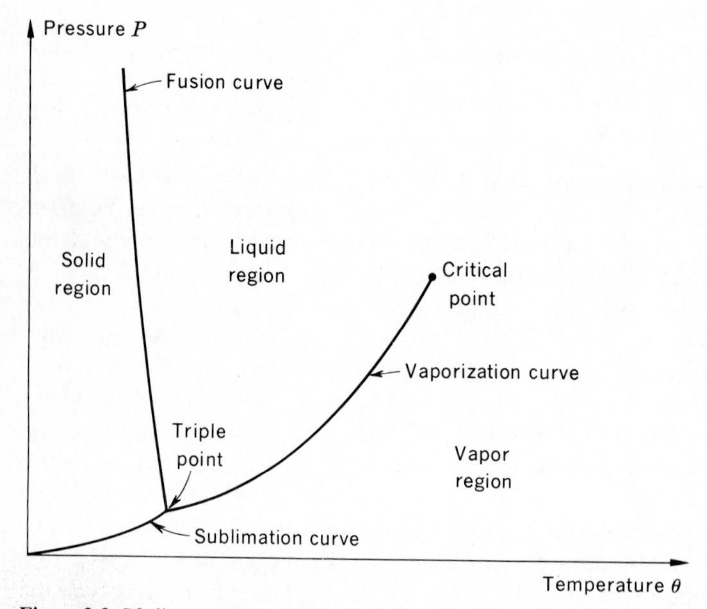

Figure 2-2 $P\theta$ diagram for a substance such as water.

Table 2-1 Triple-point data

Substance	T, K	P, mm Hg	P, Pa
Helium-4 (λ point)	2.177	37.77	5035
Hydrogen (normal)	13.97	52.8	7040
Deuterium (normal)	18.63	128	17,100
Neon	24.56	324	43,200
Oxygen	54.36	1.14	152
Nitrogen	63.15	94	12,500
Ammonia	195.40	45.57	6075
Sulfur dioxide	197.68	1.256	167.5
Carbon dioxide	216.55	3880	517,000
Water	273.16	4.58	611

Table 2-2 Triple points of water

Phases in equilibrium	T, K	P, mm Hg	P, Pa
Ice I, liquid, vapor	273.16	4.584	611.2
Ice I, liquid, ice III	251.15	1.556×10^6	2.075×10^8
Ice, I, ice II, ice III	238.45	1.597×10^6	2.129×10^8
Ice II, ice III, ice V	248.85	2.583×10^6	3.443×10^8
Ice III, liquid, ice V	256.15	2.598×10^6	3.463×10^8
Ice V, liquid, ice VI	273.31	4.694×10^6	6.258×10^8
Ice VI, liquid, ice VII	354.75	1.648×10^7	2.197×10^9

represented by a point. On a PV diagram it is a line. Triple-point data for some interesting substances are given in Table 2-1.

In investigating the ice line of water at very high pressures, Bridgman and Tammann discovered five new modifications of ice, designated as ice II, III, V, VI, and VII—ordinary ice being denoted by ice I. Two other modifications of ice, IV and VIII, were found to be unstable. Equilibrium conditions among these forms of ice and liquid give rise to six other triple points, which, along with the low-pressure triple point, are listed in Table 2-2.

2-4 $PV\theta$ SURFACE

All the information that is represented on both the PV and the $P\theta$ diagrams can be shown on one diagram if the three coordinates P, V, and θ are plotted along rectangular axes. The result is called the *$PV\theta$ surface*. Two such surfaces are shown in Figs. 2-3 and 2-4: the first for a substance like H_2O that contracts upon melting, and the second for a substance like CO_2 that expands upon melting. These diagrams are not drawn to scale, the volume axis being considerably foreshortened. If the student imagines a $PV\theta$ surface projected on the

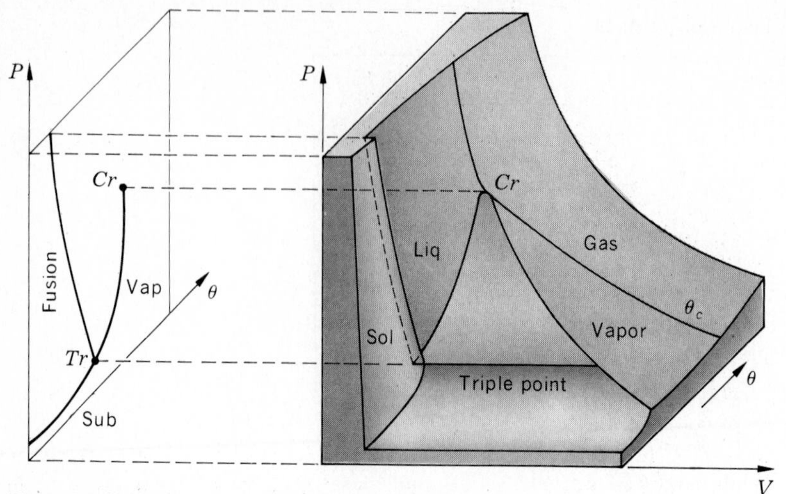

Figure 2-3 Surface for a substance that contracts on melting.

PV plane, the usual PV diagram will be seen. Upon projecting the surface on the $P\theta$ plane, the whole solid-vapor region projects into the sublimation curve, the whole liquid-vapor region projects into the vaporization curve, the whole solid-liquid region projects into the fusion curve, and finally the *triple-point line* projects into the triple point. The critical point is denoted by the letters Cr, and the triple point by Tr. The critical isotherm is marked θ_c. A substance with no free surface and with a volume determined by that of the container is called a *gas* when its temperature is above the critical temperature; otherwise it is called a *vapor*.

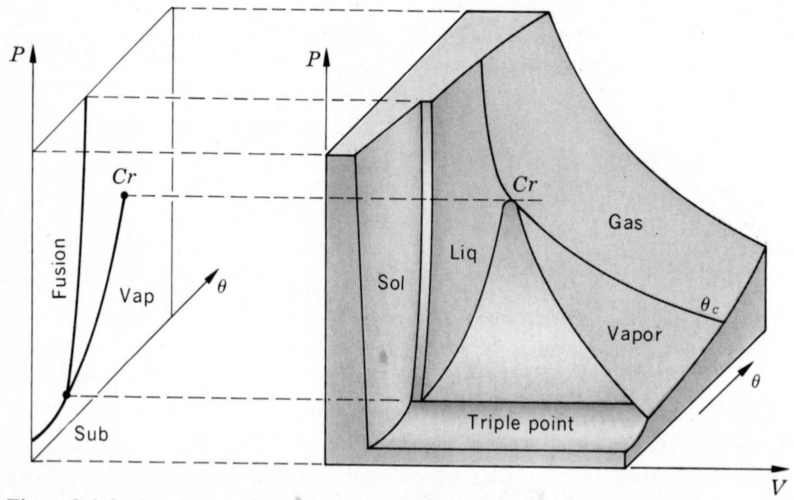

Figure 2-4 Surface for a substance that expands on melting.

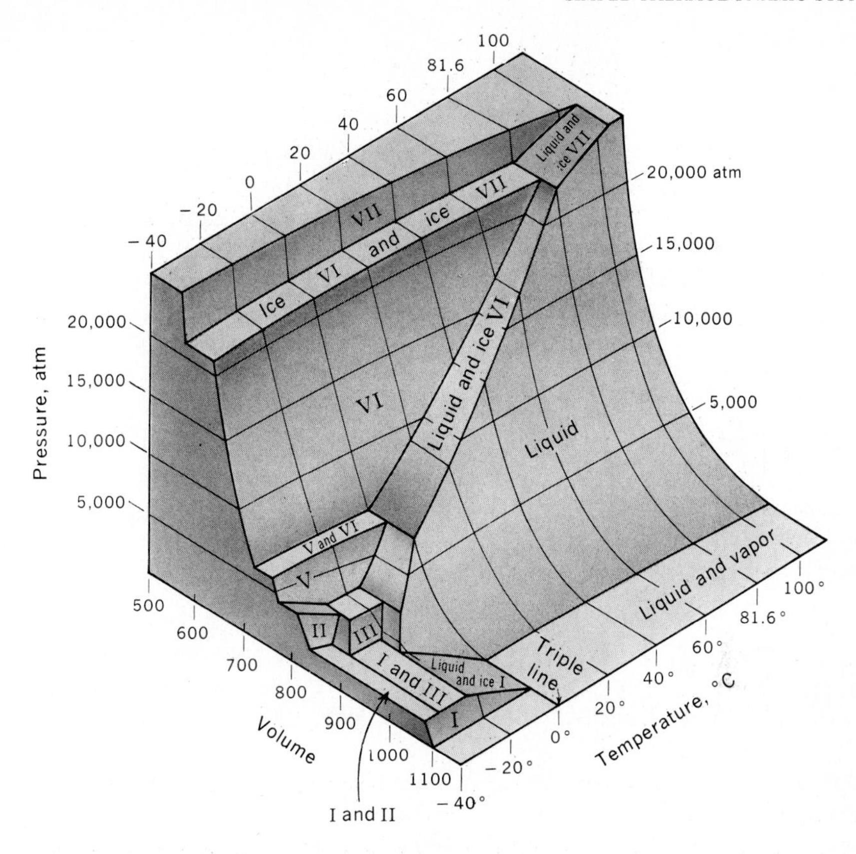

Figure 2-5 Surface for water, showing all the triple points. *(Constructed by Verwiebe on the basis of measurements by Bridgman.)*

All the triple points of water are shown on the PVt surface shown in Fig. 2-5, which was constructed by Verwiebe on the basis of measurements by Bridgman.

2-5 EQUATIONS OF STATE

It is impossible to express the complete behavior of a substance over the whole range of measured values of P, V, and θ by means of one simple equation. There have been over sixty equations of state suggested to represent only the liquid, vapor, and liquid-vapor regions, ranging from the ideal gas equation

$$Pv = R\theta, \tag{2-1}$$

which holds only at low pressures in the vapor and gas regions, to the Beattie-Bridgman equation:

$$P = \frac{R\theta(1-\epsilon)}{v^2}(v+B) - \frac{A}{v^2}, \tag{2-2}$$

where $$A = A_0\left(1 - \frac{a}{v}\right), \qquad B = B_0\left(1 - \frac{b}{v}\right), \qquad \epsilon = \frac{c}{v\theta^3},$$

which, because of its five adjustable constants, represents with some accuracy the whole range above the triple point.

Some of these equations are frankly empirical, designed to represent as closely as possible the measured values of P, V, and θ, while others are theoretical, having been calculated on the basis of the kinetic theory of gases. One of the most famous of the theoretical equations of state, based on assumptions concerning molecular behavior that are still of use today, is the van der Waals equation of state:

$$\left(P + \frac{a}{v^2}\right)(v - b) = R\theta. \tag{2-3}$$

This equation holds fairly well in the liquid region, the vapor region, and near and above the critical point. In all these equations R is a constant called the *universal gas constant*, v is the molar volume (V/n), and n is the number of moles of gas.

2-6 DIFFERENTIAL CHANGES OF STATE

If a system undergoes a small change of state whereby it passes from an initial state of equilibrium to another state of equilibrium very near the initial one, then all three coordinates, in general, undergo slight changes. If the change of, say, V is very small in comparison with V and very large in comparison with the space occupied by a few molecules, then this change of V may be written as a differential dV. If V were a geometrical quantity referring to the volume of *space*, then dV could be used to denote a portion of that space arbitrarily small. Since, however, V is a macroscopic coordinate denoting the volume of *matter*, then, for dV to have a meaning, it must be large enough to include enough molecules to warrant the use of the macroscopic point of view.

Similarly, if the change of P is very small in comparison with P and very large in comparison with molecular fluctuations, then it also may be represented by the differential dP. *Every infinitesimal in thermodynamics must satisfy the requirement that it represents a change in a quantity which is small with respect to the quantity itself and large in comparison with the effect produced by the behavior of a few molecules.* The reason for this is that thermodynamic coordinates such as volume, pressure, and temperature have no meaning when applied to a few molecules. This is another way of saying that thermodynamic coordinates are macroscopic coordinates.

We may imagine the equation of state solved for any coordinate in terms of the other two. Thus,

$$V = \text{function of } (\theta, P).$$

An infinitesimal change from one state of equilibrium to another state of equilibrium involves a dV, a $d\theta$, and a dP, all of which we shall assume satisfy the condition laid down in the previous paragraph. A fundamental theorem in partial differential calculus enables us to write

$$dV = \left(\frac{\partial V}{\partial \theta}\right)_P d\theta + \left(\frac{\partial V}{\partial P}\right)_\theta dP,$$

where each partial derivative is itself a function of θ and P. Both partial derivatives above have an important physical meaning. The student will remember from elementary physics a quantity called the *average coefficient of volume expansion*, or volume expansivity. This was defined as

$$\text{Av. vol. exp.} = \frac{\text{change of vol. per unit vol.}}{\text{change of temp.}}$$

and referred to conditions under which the pressure was constant. If the change of temperature is made infinitesimal, then the change in volume also becomes infinitesimal and we have what is known as the instantaneous volume expansivity, or just *volume expansivity*, which is denoted by β. Thus,

$$\boxed{\beta = \frac{1}{V}\left(\frac{\partial V}{\partial \theta}\right)_P.} \tag{2-4}$$

Strictly speaking, β is a function of θ and P, but experiments to be described later show that there are many substances for which β is quite insensitive to a change in P and varies only slightly with θ. Consequently, within a small temperature range β may, as a rule, be regarded as a constant. β is expressed in units of reciprocal degrees.

The effect of a change of pressure on the volume of a hydrostatic system when the temperature is kept constant is expressed by a quantity called *isothermal compressibility* and is represented by the symbol κ (Greek kappa). Thus,

$$\boxed{\kappa = -\frac{1}{V}\left(\frac{\partial V}{\partial P}\right)_\theta.} \tag{2-5}$$

The dimension of compressibility is reciprocal pressure, which can be measured in units of Pa^{-1} or bar^{-1} (1 bar $= 10^5$ Pa). The value of κ for solids and liquids varies but little with temperature and pressure, so that κ may often be regarded as constant.

If the equation of state is solved for P, then

$$P = \text{function of } (\theta, V),$$

and

$$dP = \left(\frac{\partial P}{\partial \theta}\right)_V d\theta + \left(\frac{\partial P}{\partial V}\right)_\theta dV.$$

Finally, if θ is imagined as a function of P and V,

$$d\theta = \left(\frac{\partial \theta}{\partial P}\right)_V dP + \left(\frac{\partial \theta}{\partial V}\right)_P dV.$$

In all the equations above the system was assumed to undergo an infinitesimal process from an initial state of equilibrium to another. This enabled us to use an equation of equilibrium (equation of state) and to solve it for any coordinate in terms of the other two. The differentials dP, dV, and $d\theta$ therefore are differentials of actual functions and are called *exact differentials*. If dz is an exact differential of a function of, say, x and y, then dz may be written

$$dz = \left(\frac{\partial z}{\partial x}\right)_y dx + \left(\frac{\partial z}{\partial y}\right)_x dy.$$

An infinitesimal that is not the differential of an actual function is called an *inexact differential* and cannot be expressed by an equation of the type shown above. Other distinctions between exact and inexact differentials will be made clear later.

2-7 MATHEMATICAL THEOREMS

There are two simple theorems in partial differential calculus that are used very often in this subject. The proofs are as follows. Suppose that there exists a relation among the three coordinates x, y, and z; thus,

$$f(x, y, z) = 0.$$

Then x can be imagined as a function of y and z, and

$$dx = \left(\frac{\partial x}{\partial y}\right)_z dy + \left(\frac{\partial x}{\partial z}\right)_y dz.$$

Also, y can be imagined as a function of x and z, and

$$dy = \left(\frac{\partial y}{\partial x}\right)_z dx + \left(\frac{\partial y}{\partial z}\right)_x dz.$$

Substituting the second equation into the first, we have

$$dx = \left(\frac{\partial x}{\partial y}\right)_z \left[\left(\frac{\partial y}{\partial x}\right)_z dx + \left(\frac{\partial y}{\partial z}\right)_x dz\right] + \left(\frac{\partial x}{\partial z}\right)_y dz,$$

or

$$dx = \left(\frac{\partial x}{\partial y}\right)_z \left(\frac{\partial y}{\partial x}\right)_z dx + \left[\left(\frac{\partial x}{\partial y}\right)_z \left(\frac{\partial y}{\partial z}\right)_x + \left(\frac{\partial x}{\partial z}\right)_y\right] dz.$$

Now, of the three coordinates, only two are independent. Choosing x and z as the independent coordinates, the equation above must be true for all sets of values of dx and dz. Thus, if $dz = 0$ and $dx \neq 0$, it follows that

$$\left(\frac{\partial x}{\partial y}\right)_z \left(\frac{\partial y}{\partial x}\right)_z = 1,$$

or

$$\boxed{\left(\frac{\partial x}{\partial y}\right)_z = \frac{1}{(\partial y/\partial x)_z}.} \tag{2-6}$$

If $dx = 0$ and $dz \neq 0$, it follows that

$$\left(\frac{\partial x}{\partial y}\right)_z \left(\frac{\partial y}{\partial z}\right)_x + \left(\frac{\partial x}{\partial z}\right)_y = 0,$$

$$\left(\frac{\partial x}{\partial y}\right)_z \left(\frac{\partial y}{\partial z}\right)_x = -\left(\frac{\partial x}{\partial z}\right)_y$$

and

$$\left(\frac{\partial x}{\partial y}\right)_z \left(\frac{\partial y}{\partial z}\right)_x \left(\frac{\partial z}{\partial x}\right)_y = -1. \tag{2-7}$$

In the case of a hydrostatic system, the second theorem yields the result

$$\left(\frac{\partial P}{\partial V}\right)_\theta \left(\frac{\partial V}{\partial \theta}\right)_P = -\left(\frac{\partial P}{\partial \theta}\right)_V.$$

The volume expansivity β and the isothermal compressibility κ were defined as

$$\beta = \frac{1}{V}\left(\frac{\partial V}{\partial \theta}\right)_P,$$

and

$$\kappa = -\frac{1}{V}\left(\frac{\partial V}{\partial P}\right)_\theta.$$

Therefore,

$$\left(\frac{\partial P}{\partial \theta}\right)_V = \frac{\beta}{\kappa}.$$

An infinitesimal change in pressure may now be expressed in terms of these physical quantities. Thus,

$$dP = \left(\frac{\partial P}{\partial \theta}\right)_V d\theta + \left(\frac{\partial P}{\partial V}\right)_\theta dV,$$

or

$$dP = \frac{\beta}{\kappa}\, d\theta - \frac{1}{\kappa V}\, dV. \tag{2-8}$$

At constant volume,

$$dP = \frac{\beta}{\kappa}\, d\theta.$$

If we cause the temperature to change a finite amount from θ_i to θ_f at constant volume, the pressure will change from P_i to P_f, where the subscripts i and f denote the initial and final states, respectively. Upon integrating between these two states, we get

$$P_f - P_i = \int_{\theta_i}^{\theta_f} \frac{\beta}{\kappa}\, d\theta.$$

The right-hand member can be integrated if we know the way in which β and κ vary with θ at constant volume. If the temperature range $\theta_f - \theta_i$ is small, very little error is introduced by assuming that both are constant. With these assumptions we get

$$P_f - P_i = \frac{\beta}{\kappa}(\theta_f - \theta_i),$$

from which the final pressure may be calculated. For example, consider the following problem. A mass of mercury at standard atmospheric pressure and a temperature of 0°C is kept at constant volume. If the temperature is raised to 10°C, what will be the final pressure? From tables of physical constants, β and κ of mercury remain practically constant within the temperature range of 0 to 10°C and have the values

$$\beta = 181 \times 10^{-6}\ \mathrm{K}^{-1},$$

and

$$\kappa = 3.82 \times 10^{-11}\ \mathrm{Pa}^{-1};$$

whence

$$P_f - P_i = \frac{181 \times 10^{-6}\ \mathrm{K}^{-1} \times 10\ \mathrm{K}}{3.82 \times 10^{-11}\ \mathrm{Pa}^{-1}}$$

$$= 473 \times 10^5\ \mathrm{Pa},$$

and

$$P_f = 473 \times 10^5\ \mathrm{Pa} + 1 \times 10^5\ \mathrm{Pa}$$

$$= 474 \times 10^5\ \mathrm{Pa}.$$

2-8 STRETCHED WIRE

Experiments on stretched wires are usually performed under conditions in which the pressure remains constant at standard atmospheric pressure and changes in volume are negligible. For most practical purposes, it is found unnecessary to include the pressure and the volume among the thermodynamic coordinates. A sufficiently complete thermodynamic description of a wire is given in terms of only three coordinates:

1. The tension in the wire $\mathcal{J}$, measured in newtons (N).
2. The length of the wire L, measured in meters (m).
3. The ideal-gas temperature θ.

The states of thermodynamic equilibrium are connected by an equation of state that as a rule cannot be expressed by a simple equation. For a wire at constant temperature within the limit of elasticity, Hooke's law holds; namely,

$$\mathcal{J} = \text{const.}\ (L - L_0),$$

where L_0 is the length at zero tension.

If a wire undergoes an infinitesimal change from one state of equilibrium to another, then the infinitesimal change of length is an exact differential and can be written

$$dL = \left(\frac{\partial L}{\partial \theta}\right)_{\mathcal{J}} d\theta + \left(\frac{\partial L}{\partial \mathcal{J}}\right)_{\theta} d\mathcal{J},$$

where both partial derivatives are functions of θ and $\mathcal{J}$. These derivatives are connected with important physical quantities. We define the *linear expansivity* α as

$$\boxed{\alpha = \frac{1}{L}\left(\frac{\partial L}{\partial \theta}\right)_{\mathcal{J}}.} \tag{2-9}$$

The experimental measurement of α will be considered later. Measurements of α show that it depends only slightly on $\mathcal{J}$ and varies mostly with θ. In a small temperature range, however, it may be regarded as practically constant. α is expressed in reciprocal degrees.

By definition, the *isothermal Young's modulus*, denoted by Y, is

$$\boxed{Y = \frac{L}{A}\left(\frac{\partial \mathcal{J}}{\partial L}\right)_{\theta},} \tag{2-10}$$

where A denotes the area of the wire. The isothermal Young's modulus is found experimentally to depend but little on $\mathcal{J}$ and mostly on θ. For a small temperature range, it may be regarded as practically constant. The unit of Y is 1 kN/m^2.

2-9 SURFACE FILM

The study of surface films is an interesting branch of physical chemistry. There are three important examples of such films:

1. The upper surface of a liquid in equilibrium with its vapor.
2. A soap bubble, or soap film, stretched across a wire framework, consisting of two surface films with a small amount of liquid between.
3. A thin (sometimes monomolecular) oil film on the surface of water.

A surface film is somewhat like a stretched membrane. The surface on one side of any imaginary line pulls perpendicular to this line with a force equal and opposite to that exerted by the surface on the other side of the line. The force acting perpendicularly to a line of unit length is called the *surface tension*. An adequate thermodynamic description of a surface film is given by the specifying three coordinates:

1. The surface tension $\mathcal{S}$, measured in N/m.
2. The area of the film A, measured in m^2.
3. The ideal-gas temperature θ.

In dealing with a surface film, the accompanying liquid must always be considered as part of the system. This may be done, however, without introducing the pressure and volume of the composite system because, as a rule, the pressure remains constant and volume changes are negligible. The surface of a pure liquid in equilibrium with its vapor has a particularly simple equation of state. Experiment shows that the surface tension of such a film does not depend on the area but is a function of the temperature only. For most pure liquids, the equation of state can be written

$$\mathcal{S} = \mathcal{S}_0\left(1 - \frac{\theta}{\theta'}\right)^n,$$

where $\mathcal{S}_0$ is the surface tension at 0°C, θ' is a temperature within a few degrees of the critical temperature, and n is a constant that lies between 1 and 2. It is clear from this equation that the surface tension decreases as θ increases, becoming zero when $\theta = \theta'$.

The equation of state of a monomolecular oil film on water is particularly interesting. If $\mathcal{S}_W$ denotes the surface tension of a clean water surface and $\mathcal{S}$ the surface tension of the water covered by the monolayer, then, within a restricted range of values of A,

$$(\mathcal{S} - \mathcal{S}_W)A = \text{const. } \theta.$$

The difference $\mathcal{S} - \mathcal{S}_W$ is sometimes called the *surface pressure*. Such films can be compressed and expanded and, when deposited on glass, have interesting optical properties.

2-10 REVERSIBLE CELL

A reversible cell consists of two electrodes each immersed in a different electrolyte. The emf depends on the nature of the materials, the concentrations of the electrolytes, and the temperature. In Fig. 2-6 a schematic diagram of a reversible cell, the Daniell cell, is shown. A copper electrode immersed in a saturated $CuSO_4$ solution is separated by a porous wall from a zinc electrode immersed in

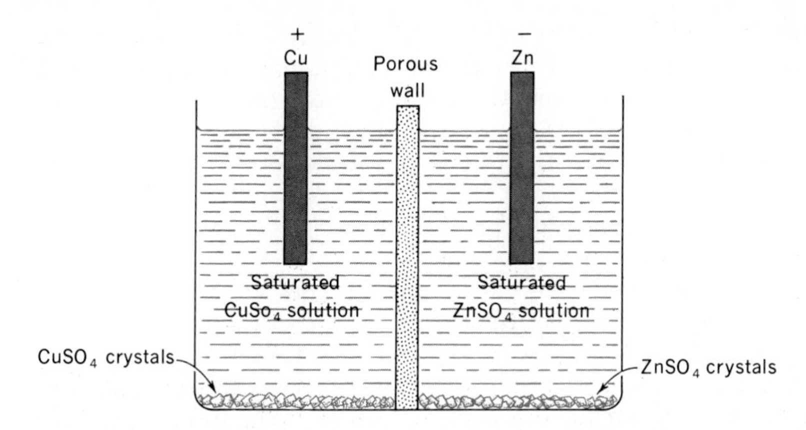

Figure 2-6 The Daniell cell.

a saturated solution of $ZnSO_4$. Experiment shows that the copper electrode is positive with respect to the zinc.

Suppose that the cell is connected to a potentiometer whose potential difference is slightly smaller than the emf of the cell. Under these conditions, the current that exists may be described conventionally as a transfer of positive electricity externally from the copper electrode to the zinc electrode. When this is the case, zinc goes into solution, zinc sulfate is formed, copper is deposited, and copper sulfate is used up. These changes are expressed by the chemical reaction

$$\mathrm{Zn} + \mathrm{CuSO_4} \rightarrow \mathrm{Cu} + \mathrm{ZnSO_4}.$$

When positive electricity is transferred in the opposite direction, i.e., externally from zinc to copper, the reaction proceeds in the reverse direction (hence the name reversible cell); thus,

$$\mathrm{Cu} + \mathrm{ZnSO_4} \rightarrow \mathrm{Zn} + \mathrm{CuSO_4}.$$

The important feature of a reversible cell is that the chemical changes accompanying the transfer of electricity in one direction take place to the same extent in the reverse direction when the same quantity of electricity is transferred in the reverse direction. Furthermore, according to one of Faraday's laws of electrolysis, the simultaneous disappearance of 1 mol of zinc and deposit of 1 mol of copper are accompanied by the transfer of exactly jN_F coulombs of electricity, where j is the valence and N_F is Faraday's constant, or 96,500 C. We may therefore define a quantity Z, called the *charge* of the cell, as a number whose absolute magnitude is of no consequence but whose change is numerically equal to the quantity of electricity that is transferred during the chemical reaction, the change being negative when positive electricity is transferred externally from the positive to the negative electrode. Thus, if Δn moles of zinc disappear and Δn moles of copper are deposited, the charge of the cell changes from Z_i to Z_f, where

$$Z_f - Z_i = -\Delta n j N_F. \tag{2-11}$$

Now, if we limit ourselves to reversible cells in which no gases are liberated and which operate at constant atmospheric pressure, we may ignore the pressure and the volume and describe the cell with the aid of three coordinates only:

1. The emf $\mathcal{E}$, measured in V.
2. The charge Z, measured in C.
3. The ideal-gas temperature θ.

When the cell is on open circuit, there is a tendency for diffusion to take place slowly and the cell is not in equilibrium. If the cell is connected to a potentiometer, however, and the circuit is adjusted until there is no current, then the emf of the cell is balanced and the cell is in mechanical and chemical equilibrium. When thermal equilibrium is also satisfied, the cell is then in thermodynamic equilibrium. The states of thermodynamic equilibrium of a reversible cell are connected by an equation of state among the coordinates $\mathcal{E}$, Z, and θ. If the electrolytes are saturated solutions, a transfer of electricity accompanying the performance of the chemical reaction at constant temperature and pressure will not alter the concentrations of the electrolytes. The emf will therefore remain constant. Experiment shows that the emf of a saturated reversible cell at constant pressure is a function of the temperature only. The equation of state is usually written

$$\mathcal{E} = \mathcal{E}_{20} + \alpha(t - 20°) + \beta(t - 20°)^2 + \gamma(t - 20°)^3,$$

where t is the Celsius temperature, $\mathcal{E}_{20}$ is the emf at 20°C, and α, β, and γ are constants depending on the materials. We shall see later that, once the equation of state of a reversible cell is known, all the quantities of interest to a chemist which refer to the chemical reaction going on in the cell can be determined.

2-11 DIELECTRIC SLAB

Consider a capacitor consisting of two parallel conducting plates of area A whose linear dimensions are large in comparison with their separation l, filled with an isotropic solid or liquid dielectric. If a potential difference is established across the plates an electric field E is caused to exist *in the dielectric* between the plates. If the center of gravity of the + and − charges within each molecule originally coincided, i.e., if the molecules of the dielectric are originally *nonpolar*, the effect of the electric field is to separate the electric charges of each molecule so as to make each molecule *polar* in the direction of the electric field. If the molecules are naturally polar, with the polar axes distributed at random, the effect of the electric field is to produce a partial orientation of the molecular polar axes in the direction of the electric field. The effect is the same in either case, and the degree to which either the natural or induced polar molecules are oriented in the direction of the field is provided by the electric charge induced on either face of the dielectric multiplied by the thickness of the dielectric, a quantity called either the *total electric moment*, or the *total electric polarization*, which we shall designate with the symbol Π (capital pi). If the volume of the dielectric is V, the *electric displacement* of the dielectric D is given by

$$D = \epsilon_0 E = \frac{\Pi}{V}.$$

The polarization Π produced by a field E depends on the nature of the dielectric and upon the temperature. Typically, a dielectric substance experiences insignificantly small volume changes in an experiment performed at constant atmospheric pressure. Therefore, the pressure and volume can be ignored and we may describe a dielectric with the aid of the three thermodynamic coordinates:

1. The electric intensity E, measured in V/m.
2. The electric polarization Π, measured in C · m.
3. The ideal gas temperature θ.

There are many dielectrics whose equation of state at temperatures above about 10 K is given by

$$\frac{\Pi}{V} = \left(a + \frac{b}{\theta}\right)E, \tag{2-12}$$

where a and b are constants.

2-12 PARAMAGNETIC ROD

A paramagnetic substance in the absence of an external magnetic field is not a magnet. Upon being introduced into a magnetic field it becomes slightly magnetized in the direction of the field. Its permeability, however, is still very nearly unity, in contradistinction to a ferromagnetic substance like iron whose permeability may be very large. Certain paramagnetic crystals nevertheless play an interesting and important role in modern physics, particularly at very low temperatures.

Modern experiments on paramagnetic materials are usually performed on samples in the form of cylinders, ellipsoids, or spheres. In these cases, the $\mathcal{H}$ field inside the material is somewhat smaller than the $\mathcal{H}$ field generated by the electric current in the surrounding winding because of the reverse field (demagnetizing field) set up by magnetic poles which form on the surfaces of the samples. In longitudinal magnetic fields, the demagnetizing effect either may be rendered negligible by using cylinders whose length is much larger than the diameter or may be corrected for in a simple way. In transverse magnetic fields, a correction factor must be applied. We shall limit ourselves to long thin cylinders in longitudinal fields where the internal and external $\mathcal{H}$ fields are the same.

When a paramagnetic rod is placed in a solenoid, where the magnetic intensity is $\mathscr{H}$, the needle develops a total magnetic moment M, which is called the *magnetization*, and whose magnitude depends on the chemical composition and the temperature. The *magnetic induction* within the needle $\mathscr{B}$ is given by

$$\mathscr{B} = \mu_0 \left(\mathscr{H} + \frac{M}{V} \right).$$

Most experiments on magnetic rods are performed at constant atmospheric pressure and involve only minute volume changes. Consequently, we may ignore the pressure and the volume and describe a paramagnetic solid with the aid of only three thermodynamic coordinates:

1. The magnetic intensity $\mathscr{H}$, measured in A/m.
2. The magnetization M, measured in $\text{A} \cdot \text{m}^2$.
3. The ideal-gas temperature θ.

The states of thermodynamic equilibrium of a paramagnetic solid can be represented by an equation of state among these coordinates. Experiment shows that the magnetization of many paramagnetic solids is a function of the ratio of the magnetic intensity to the temperature. For small values of this ratio the function reduces to a very simple form, namely,

$$M = C'_C \frac{\mathscr{H}}{\theta}, \tag{2-13}$$

which is known as *Curie's equation*—C'_C being called the *Curie constant*. The unit in which the Curie constant is expressed is, therefore,

$$\text{Unit of } C'_C = \frac{\text{A} \cdot \text{m}^2}{\text{A/m}} \text{K} = \text{m}^3 \cdot \text{K}.$$

Since the Curie constant depends upon the amount of material, its unit may be taken to be any one of the four listed in the accompanying table:

Units of the Curie constant

Total	Per mole	Per kg	Per m³
$\text{m}^3 \cdot \text{K}$	$\frac{\text{m}^3 \cdot \text{K}}{\text{mol}}$	$\frac{\text{m}^3 \cdot \text{K}}{\text{kg}}$	K

Paramagnetic solids are of particular interest in thermodynamics. It will be seen later how these are used to obtain extremely low temperatures.

2-13 INTENSIVE AND EXTENSIVE QUANTITIES

Imagine a system in equilibrium to be divided into two equal parts, each with equal mass. Those properties of each half of the system which remain the same are said to be *intensive;* those which are halved are called *extensive.* The intensive coordinates of a system, such as temperature and pressure, are independent of the mass; the extensive coordinates are proportional to the mass. The thermodynamic coordinates that have been introduced in this chapter are listed in Table 2-3.

Table 2-3 Intensive and extensive quantities

Simple systems	Intensive coordinate		Extensive coordinate	
Hydrostatic system	Pressure	P	Volume	V
Stretched wire	Tension	$\mathcal{J}$	Length	L
Surface film	Surface tension	$\mathcal{S}$	Area	A
Electric cell	Emf	$\mathcal{E}$	Charge	Z
Dielectric slab	Electric intensity	E	Polarization	Π
Paramagnetic rod	Magnetic intensity	$\mathcal{H}$	Magnetization	M

PROBLEMS

2-1 The equation of state of an ideal gas is $Pv = R\theta$. Show that (*a*) $\beta = 1/\theta$ and (*b*) $\kappa = 1/P$.

2-2 An approximate equation of state of a real gas at moderate pressures, devised to take into account the finite size of the molecules, is $P(v - b) = R\theta$, where R and b are constants. Show that:

(*a*) $$\beta = \frac{1/\theta}{1 + bP/R\theta},$$

(*b*) $$\kappa = \frac{1/P}{1 + bP/R\theta}.$$

2-3 An approximate equation of state of a real gas at moderate pressures is given by $Pv = R\theta(1 + B/v)$, where R is a constant and B is a function of θ only. Show that

(*a*) $$\beta = \frac{1}{\theta} \cdot \frac{v + B + \theta(dB/d\theta)}{v + 2B},$$

(*b*) $$\kappa = \frac{1}{P} \cdot \frac{1}{1 + BR\theta/Pv^2}.$$

2-4 A metal whose volume expansivity is $5.0 \times 10^{-5}\ \text{K}^{-1}$ and isothermal compressibility is 1.2×10^{-11} Pa is at a pressure of 1×10^5 Pa and a temperature of 20°C. A thick surrounding cover of invar, of negligible expansivity and compressibility, fits it very snugly.

(*a*) What will be the final pressure if the temperature is raised to 32°C?

(*b*) If the surrounding cover can withstand a maximum pressure of 1.2×10^8 Pa, what is the highest temperature to which the system may be raised?

2-5 A block of the same metal as in Prob. 2-4 at a pressure of 1×10^5 Pa, a volume of 5 liters, and a temperature of 20°C undergoes a temperature rise of 12 deg and an increase in volume of 0.5 cm^3. Calculate the final pressure.

2-6 (*a*) Express the volume expansivity and the isothermal compressibility in terms of the density ρ and its partial derivatives.

(*b*) Derive the equation

$$\frac{dV}{V} = \beta\, d\theta - \kappa\, dP.$$

2-7 The thermal expansivity and the compressibility of liquid oxygen are given in the accompanying table. Draw a graph showing how $(\partial P/\partial \theta)_V$ depends on the temperature.

θ, K	60	65	70	75	80	85	90
β, 10^{-3} K^{-1}	3.48	3.60	3.75	3.90	4.07	4.33	4.60
κ, 10^{-9} Pa^{-1}	0.95	1.06	1.20	1.35	1.54	1.78	2.06

2-8 The thermal expansivity and the compressibility of water are given in the accompanying table. Draw a graph showing how $(\partial P/\partial \theta)_V$ depends on the temperature. If water were kept at constant volume and the temperature were continually raised, would the pressure increase indefinitely?

t, °C	0	50	100	150	200	250	300
β, 10^{-3} K^{-1}	−0.07	0.46	0.75	1.02	1.35	1.80	2.90
κ, 10^{-9} Pa^{-1}	0.51	0.44	0.49	0.62	0.85	1.50	3.05

2-9 At the critical point $(\partial P/\partial V)_T = 0$. Show that, at the critical point, both the volume expansivity and the isothermal compressibility are infinite.

2-10 If a wire undergoes an infinitesimal change from an initial equilibrium state to a final equilibrium state, show that the change of tension is equal to

$$d\mathcal{J} = -\alpha A Y\, d\theta + \frac{AY}{L}\, dL.$$

2-11 A metal wire of cross-sectional area 0.0085 cm^2 under a tension of 20 N and a temperature of 20°C is stretched between two rigid supports 1.2 m apart. If the temperature is reduced to 8°C, what is the final tension? (Assume that α and Y remain constant at the values 1.5×10^{-5} K^{-1} and 2.0×10^9 N/m^2, respectively.)

2-12 The fundamental frequency of vibration of a wire of length L, mass m, and tension $\mathcal{J}$ is given by

$$f_1 = \frac{1}{2L}\sqrt{\frac{\mathcal{J}L}{m}}.$$

With what frequency will the wire in Prob. 2-11 vibrate at 20°C; at 8°C? (The density of the wire is 9.0×10^3 kg/m^3.)

2-13 If, in addition to the conditions mentioned in Prob. 2-11, the supports approach each other by 0.012 cm, what will be the final tension?

2-14 The equation of state of an ideal elastic substance is

$$\mathcal{J} = K\theta\left(\frac{L}{L_0} - \frac{L_0^2}{L^2}\right),$$

where K is a constant and L_0 (the value of L at zero tension) is a function of the temperature only.

(*a*) Show that the isothermal Young's modulus is given by

$$Y = \frac{K\theta}{A}\left(\frac{L}{L_0} + \frac{2L_0^2}{L^2}\right).$$

(*b*) Show that the isothermal Young's modulus at zero tension is given by

$$Y_0 = \frac{3K\theta}{A}.$$

(*c*) Show that the linear expansivity is given by

$$\alpha = \alpha_0 - \frac{\mathcal{J}}{AY\theta} = \alpha_0 - \frac{1}{\theta} \cdot \frac{L^3/L_0^3 - 1}{L^3/L_0^3 = 2},$$

where α_0 is the value of the linear expansivity at zero tension, or

$$\alpha_0 = \frac{1}{L_0}\frac{dL_0}{d\theta}.$$

(*d*) Assume the following values for a certain sample of rubber: $\theta = 300$ K, $K = 1.33 \times 10^{-2}$ N/K, $A = 1 \times 10^{-6}$ m^2, $\alpha_0 = 5 \times 10^{-4}$ K^{-1}. Calculate $\mathcal{J}$, Y, and α for the following values of L/L_0: 0.5, 1.0, 1.5, 2.0. Show graphically how $\mathcal{J}$, Y, and α depend on the ratio L/L_0.

2-15 The equation of state of an ideal paramagnetic material valid for all values of the ratio $\mathcal{H}/\theta$ is given by Brillouin's equation, as follows:

$$M = Ng\mu_B\left[(J + \tfrac{1}{2}) \coth (J + \tfrac{1}{2})\frac{g\mu_B\mathcal{H}}{k\theta} - \tfrac{1}{2} \coth \tfrac{1}{2}\frac{g\mu_B\mathcal{H}}{k\theta}\right],$$

where N, g, μ_B, J, and k are atomic constants.

(*a*) Find out how the hyperbolic cotangent of x behaves as x approaches zero.

(*b*) Show that Brillouin's equation reduces to Curie's equation when $\mathcal{H}/\theta$ approaches zero.

(*c*) Show that the Curie constant is given by

$$C_C' = \frac{Ng^2J(J + 1)\mu_B^2\mu_0}{3k}.$$

CHAPTER

THREE

WORK

3-1 WORK

If a system undergoes a displacement under the action of a force, *work* is said to be done, the amount of work being equal to the product of the force and the component of the displacement parallel to the force. If a system as a *whole* exerts a force on its surroundings and a displacement takes place, the work that is done either by or on the system is called *external work*. Thus, a gas, confined in a cylinder and at uniform pressure, while expanding and imparting motion to a piston does external work on its surroundings. The work done, however, by part of a system on another part is called *internal work*.

Internal work has no place in thermodynamics. Only the work that involves an interaction between a system and its surroundings is significant. When a system does external work, the changes that take place can be described by means of macroscopic quantities referring to the system as a whole, in which case the changes may be imagined to accompany the raising or lowering of a suspended body, the winding or unwinding of a spring, or, in general, the alteration of the position or *configuration* of some external mechanical device. This may be regarded as the ultimate criterion as to whether external work is done or not. It will often be found convenient throughout the remainder of this book to describe the performance of external work in terms of or in conjunction with the operation of a mechanical device such as a system of suspended bodies. *Unless otherwise indicated, the word work, unmodified by any adjective, will mean external work.*

A few examples will be found helpful. If an electric cell is on open circuit, changes that take place in the cell (such as diffusion) are not accompanied by the performance of work. If, however, the cell is connected to an external circuit through which electricity is transferred, the current may be imagined to produce rotation of the armature of a motor, thereby lifting a weight or winding a spring. Therefore, *for an electric cell to do work it must be connected to an external circuit.* As another example, consider a magnet far removed from any external electric conductor. A change of magnetization within the magnet is not accompanied by the performance of work. If, however, the magnet undergoes a change of magnetization while it is surrounded by an electric conductor, eddy currents are set up in the conductor, constituting an external transfer of electricity. Hence, *for a magnetic system to do work it must interact with an electric conductor or with other magnets.*

In mechanics, we are concerned with the behavior of systems acted on by external forces. When the resultant force exerted *on* a mechanical system is in the same direction as the displacement of the system, the work of the force is positive, work is said to be done on the system, and the energy of the system increases.

For thermodynamics to be in conformity with mechanics, we adopt the same sign convention for work that is employed in mechanics. Thus, when the external force acting on a thermodynamic system is in the *same* direction as the displacement of the system, work is done *on* the system; the work is regarded as *positive.* Conversely, when the external force is *opposite* to the displacement, work is done *by* the system; the work is regarded as *negative.*

3-2 QUASI-STATIC PROCESS

A system in thermodynamic equilibrium satisfies the following stringent requirements:

1. *Mechanical equilibrium.* There are no unbalanced forces acting on any part of the system or on the system as a whole.
2. *Thermal equilibrium.* There are no temperature differences between parts of the system or between the system and its surroundings.
3. *Chemical equilibrium.* There are no chemical reactions within the system and no motion of any chemical constituent from one part of a system to another part.

Once a system is in thermodynamic equilibrium and the surroundings are kept unchanged, no motion will take place and no work will be done. If, however, the sum of the external forces is changed so that there is a finite unbalanced force acting on the system, then the condition for mechanical equilibrium is no longer satisfied and the following situations may arise:

1. Unbalanced forces may be created within the system; as a result, turbulence, waves, etc., may be set up. Also, the system as a whole may execute some sort of accelerated motion.

2. As a result of this turbulence, acceleration, etc., a nonuniform temperature distribution may be brought about, as well as a finite difference of temperature between the system and its surroundings.
3. The sudden change in the forces and in the temperature may produce a chemical reaction or the motion of a chemical constituent.

It follows that a finite unbalanced force may cause the system to pass through nonequilibrium states. If it is desired during a process to describe every state of a system by means of thermodynamic coordinates referring to the system as a whole, the process must *not* be brought about by a finite unbalanced force. We are led, therefore, to conceive of an ideal situation in which the external forces acting on a system are varied only slightly so that the unbalanced force is infinitesimal. A process performed in this ideal way is said to be *quasi-static*. *During a quasi-static process, the system is at all times infinitesimally near a state of thermodynamic equilibrium*, and all the states through which the system passes can be described by means of thermodynamic coordinates referring to the system as a whole. An equation of state is valid, therefore, for all these states. A quasi-static process is an idealization that is applicable to all thermodynamic systems, including electric and magnetic systems. The conditions for such a process can never be rigorously satisfied in the laboratory, but they can be approached with almost any degree of accuracy. In the next few articles it will be seen how approximately quasi-static processes may be performed by all the systems treated in Chap. 2.

3-3 WORK OF A HYDROSTATIC SYSTEM

Imagine any hydrostatic system contained in a cylinder equipped with a movable piston on which the system and the surroundings may act. Suppose that the cylinder has a cross-sectional area A, that the pressure exerted *by the system* at the piston face is P, and that the force is PA. The surroundings also exert an opposing force on the piston. The origin of this opposing force is irrelevant; it might be due to friction or a combination of friction and the push of a spring. The system within the cylinder does not have to know how the opposing force originated. The important condition that must be satisfied is that the opposing force must differ only slightly from the force PA. If, under these conditions, the piston moves a distance dx, in a direction opposite to that of the force PA (Fig. 3-1), an infinitesimal amount of work $đW$ (the differential symbol with the line drawn through it will be explained in Art. 3-5), where

$$đW = -PA\,dx.$$

But

$$A\,dx = dV,$$

and hence

$$\boxed{đW = -P\,dV.} \tag{3-1}$$

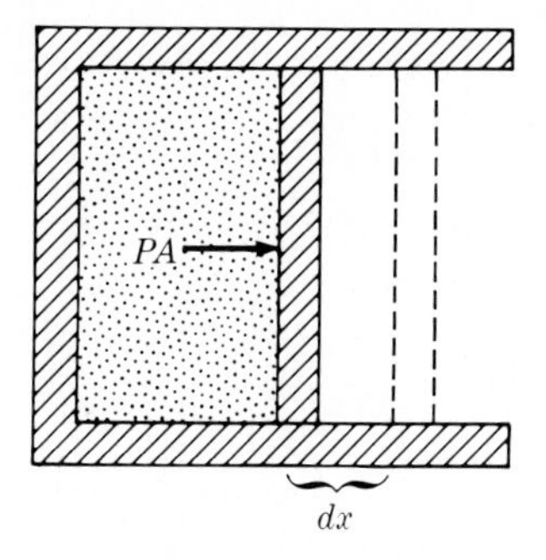

Figure 3-1 Quasi-static contraction of hydrostatic system.

The presence of the minus sign before $P\,dV$ ensures that a positive dV (an expansion) gives rise to negative work and conversely, a negative dV (a compression) is attended by positive work.

During this process, we might have a chemical reaction or a transport of a component from one point to another taking place slowly enough to keep the system near mechanical equilibrium; or we might have some dissipative process such as friction taking place—or even all these processes. The lack of chemical equilibrium (and therefore of complete thermodynamic equilibrium) and the presence of dissipation *do not preclude* writing $đW = -P\,dV$, where P is the system pressure. A lack of mechanical equilibrium, however, such as exists when there are waves or turbulence within the system, definitely precludes expressing $đW = -P\,dV$, because the system pressure is no longer defined. It should be noted that the validity of Eq. (3-1) does not depend upon the piston-and-cylinder device used in its derivation; it can be applied to any expanding or contracting hydrostatic system of arbitrary shape.

In a *finite* quasi-static process in which the volume changes from V_i to V_f, the work is

$$W = -\int_{V_i}^{V_f} P\,dV. \tag{3-2}$$

Since the change in volume is performed quasi-statically, the system pressure P is at all times not only equal to the external pressure, but is also a thermodynamic coordinate. Thus, the pressure can be expressed as a function of θ and V by means of an equation of state. The evaluation of the integral can be accomplished once the behavior of θ is specified, because then P can be expressed as a function of V only. If P is expressed as a function of V, the *path* of integration is defined. Along a particular quasi-static path, the work done on a system in going from a volume V_i to a smaller volume V_f is expressed as

$$W_{if} = -\int_{V_i}^{V_f} P\,dV;$$

whereas in expanding from f to i, along the same path but in the opposite direction, the work done by the system is

$$W_{fi} = -\int_{V_f}^{V_i} P\, dV.$$

When the path is quasi-static,

$$W_{if} = -W_{fi}.$$

Sufficient approximation to a quasi-static process may be achieved in practice by having the external pressure differ from that exerted by the system by only a small finite amount.

The SI unit of P is 1 Pa (1 N/m^2 = 1 Pa) and that of V is 1 m^3. The unit of work is therefore 1 J. It is often convenient to take standard atmospheric pressure (101.325 kPa) as a unit of P and 1 liter as a unit of V. The unit of work is then 1 liter · atm, which is equal to 101 J.

3-4 *PV* DIAGRAM

As the volume of a hydrostatic system changes by virtue of the motion of a piston in a cylinder, the position of the piston at any moment is proportional to the volume. A pen whose motion along the X axis of a diagram follows exactly the motion of the piston will trace out a line every point of which represents an instantaneous value of the volume. If, at the same time, this pen is given a motion along the Y axis such that the Y coordinate is proportional to the pressure, then the pressure and volume changes of the system during expansion or compression are indicated simultaneously on the same diagram. Such a device is called an *indicator*. The diagram in which pressure is plotted along the Y axis and volume along the X axis is called a *PV diagram* (formerly, an *indicator diagram*).

In Fig. 3-2*a*, the pressure and volume changes of a gas during expansion are indicated by curve I. The integral $-\int P\, dV$ for this process is evidently the shaded area under curve I. Similarly, for a compression, the work absorbed by the gas is represented by the shaded area under curve II in Fig. 3-2*b*. In conformity with the sign convention for work, the area under I is regarded as negative, and that under II as positive. In Fig. 3-2*c*, curves I and II are drawn together so that they constitute a series of processes whereby the gas is brought back to its initial state. Such a series of processes, represented by a closed figure, is called a *cycle*. The area within the closed figure is obviously the difference between the areas under curves I and II and therefore represents the *net* work done in the cycle. Notice that the cycle is traversed in a direction such that the net work is negative, and net work is done by the system. If the direction were reversed, the net work would be positive, and net work is done on the system.

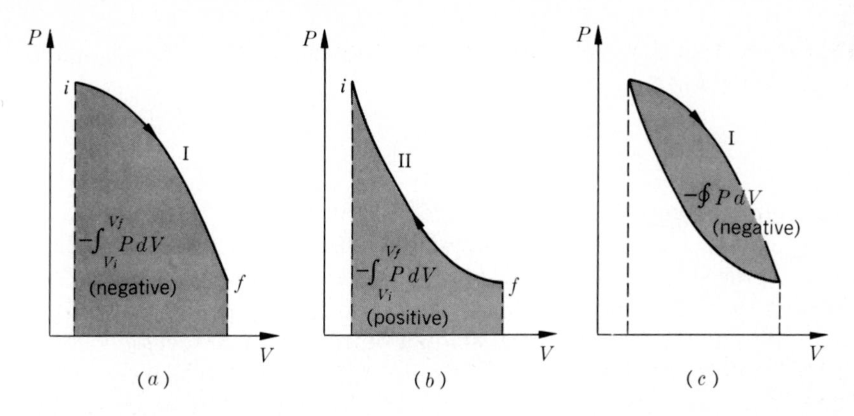

Figure 3-2 *PV* diagram. (*a*) Curve I, expansion; (*b*) curve II, compression; (*c*) curves I and II together constitute a cycle.

3-5 WORK DEPENDS ON THE PATH

On the *PV* diagram depicted in Fig. 3-3, an initial equilibrium state and a final equilibrium state of a hydrostatic system are represented by the two points *i* and *f*, respectively. There are many ways in which the system may be taken from *i* to *f*. For example, the pressure may be kept constant from *i* to *a* (*isobaric process*) and then the volume kept constant from *a* to *f* (*isochoric process*), in which case the work done is equal to the area under the line *ia*, which is equal to $-2P_0 V_0$. Another possibility is the path *ibf*, in which case the work is the area under the line *bf*, or $-P_0 V_0$. The straight line from *i* to *f* represents another path, where the work is $-\frac{3}{2}P_0 V_0$. We can see, therefore, that the *work done by a system depends not only on the initial and final states but also on the intermediate states,*

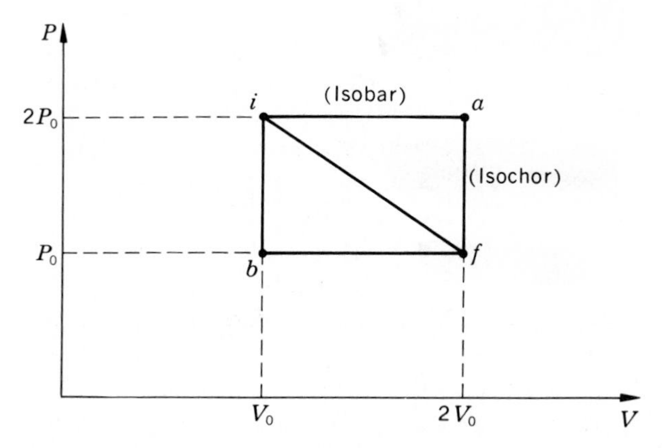

Figure 3-3 Work depends on the path.

i.e., on the path. This is merely another way of saying that, for a quasi-static process, the expression

$$W = -\int_{V_i}^{V_f} P\,dV$$

cannot be integrated until P is specified as a function of V.

The expression $-P\,dV$ is an infinitesimal amount of work and has been represented by the symbol $đW$. There is, however, an important distinction between an infinitesimal amount of work and the other infinitesimals we have considered up to now. An infinitesimal amount of work is an *inexact differential;* i.e., it is *not* the differential of an actual function of the thermodynamic coordinates. There is no function of the thermodynamic coordinates representing the work in a body. (The phrase "work in a body" has no meaning. Work is an external activity or process that leads to a change in a body, namely, the energy in a body.) To indicate that an infinitesimal amount of work is *not* a mathematical differential of a function W and to emphasize at all times that it is an inexact differential, a line is drawn through the differential sign thus: $đW$.

3-6 WORK IN QUASI-STATIC PROCESSES

The preceding ideas will be clarified by the following examples:

Quasi-static isothermal expansion or compression of an ideal gas

$$W = -\int_{V_i}^{V_f} P\,dV,$$

but an ideal gas is one whose equation of state is

$$PV = nR\theta,$$

where n and R are constants. Substituting for P, we get

$$W = -\int_{V_i}^{V_f} \frac{nR\theta}{V}\,dV,$$

and, since θ is also constant,

$$W = -nR\theta \int_{V_i}^{V_f} \frac{dV}{V}$$

$$= -nR\theta \ln \frac{V_f}{V_i},$$

where the symbol ln denotes the natural, or napierian, logarithm. In terms of the common logarithm, denoted by log,

$$W = -2.30nR\theta \log \frac{V_f}{V_i}.$$

If there are 2 kmol of gas kept at a constant temperature of 0°C and if this gas is compressed from a volume of 4 m^3 to 1 m^3, then $n = 2$ kmol, $R = 8.31$ kJ/kmol · K, $\theta = 273$ K (using three significant figures), $V_i = 4$ m^3, $V_f = 1$ m^3, and

$$W = -2.30 \times 2 \text{ kmol} \times 8.31 \frac{\text{kJ}}{\text{kmol} \cdot \text{K}} \times 273 \text{ K} \times \log \frac{1}{4}$$

$$= 6300 \text{ kJ}.$$

The positive value of W indicates that work was done *on* the gas.

Quasi-static isothermal increase of pressure on a solid. Suppose that the pressure on 10^2 kg of solid copper is increased quasi-statically and isothermally at 0°C from 0 to 1000 times standard atmospheric pressure. The work is calculated as follows:

$$W = -\int P\, dV,$$

$$dV = \left(\frac{\partial V}{\partial P}\right)_\theta dP + \left(\frac{\partial V}{\partial \theta}\right)_P d\theta.$$

Since the isothermal compressibility is

$$\kappa = -\frac{1}{V}\left(\frac{\partial V}{\partial P}\right)_\theta,$$

we have, at constant temperature,

$$dV = -\kappa V\, dP.$$

Substituting for dV, we obtain

$$W = \int_{P_i}^{P_f} \kappa V P\, dP.$$

Now the changes in V and κ at constant temperature are so small that they may be neglected. Hence,

$$W \approx \frac{\kappa V}{2}(P_f^2 - P_i^2).$$

Since the volume is equal to the mass m divided by the density ρ,

$$W \approx \frac{m\kappa}{2\rho}(P_f^2 - P_i^2).$$

For copper at 0°C, $\rho = 8930$ kg/m^3, $\kappa = 7.16 \times 10^{-12}$ Pa^{-1}, $m = 100$ kg, $P_i = 0$, and $P_f = 1000$ atm $= 1.013 \times 10^8$ Pa. Hence,

$$W = \frac{100 \text{ kg} \times 7.16 \times 10^{-12} \text{ Pa}^{-1} \times (1.013 \times 10^8 \text{ Pa})^2}{2 \times 8930 \text{ kg/m}^3}$$

$$= 0.411 \times 10^3 \text{ Pa} \cdot \text{m}^3 = 0.411 \times 10^3 \text{ N} \cdot \text{m}$$

$$= 0.411 \text{ kJ}.$$

The positive value of W indicates that work was done *on* the copper. This result, together with that of the first example, indicates that when a gas is compressed we can usually neglect the work done on the material of the container.

3-7 WORK IN CHANGING THE LENGTH OF A WIRE

If the length of a wire on which there is a force $\mathcal{F}$ is changed from L to $(L + dL)$, the infinitesimal amount of work that is done on the wire is equal to

$$\boxed{đW = \mathcal{F}\, dL.} \tag{3-3}$$

A positive value of dL means an extension of the wire, for which work must be done *on* the wire, i.e., positive work. For a finite change of length from L_i to L_f,

$$W = \int_{L_i}^{L_f} \mathcal{F}\, dL,$$

where $\mathcal{F}$ indicates the instantaneous value of the force at any moment during the process. If the wire is undergoing a motion involving large unbalanced forces, the integral cannot be evaluated in terms of thermodynamic coordinates referring to the wire as a whole. If, however, the external force is maintained at all times only slightly different from the tension, the process is sufficiently quasi-static to warrant the use of an equation of state, in which case the integration can be carried out once $\mathcal{F}$ is known as a function of L. When $\mathcal{F}$ is measured in newtons and L in meters, W will be in joules.

3-8 WORK IN CHANGING THE AREA OF A SURFACE FILM

Consider a double surface film with liquid in between, stretched across a wire framework, one side of which is movable, as shown in Fig. 3-4. If the movable wire has a length L and the surface tension is $\mathcal{S}$, the force exerted on both films is $2\mathcal{S}L$. For an infinitesimal displacement dx, the work is

$$dW = 2\mathcal{S}L\, dx;$$

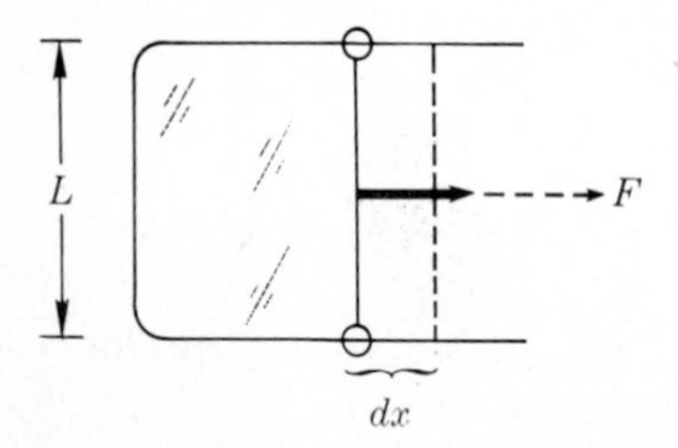

Figure 3-4 Surface film stretched across a wire framework.

but for two films

$$2L\,dx = dA.$$

Hence,

$$\boxed{dW = \mathcal{S}\,dA.} \tag{3-4}$$

For a finite change from A_i to A_f,

$$W = \int_{A_i}^{A_f} \mathcal{S}\,dA.$$

A quasi-static process may be approximated by maintaining the external force at all times only slightly different from that exerted by the film. When $\mathcal{S}$ is expressed in newtons per meter and A in square meters, W is in joules.

3-9 WORK IN CHANGING THE CHARGE OF A REVERSIBLE CELL

The conventional description of an electric current is that it is the motion of positive electricity from a region of higher to a region of lower potential. Although this is opposite to the direction of electron drift, the convention is still used, and it is convenient to adopt it in thermodynamics. Imagine a reversible cell of emf $\mathcal{E}$ to be connected to a potentiometer so that an almost continuous variation of potential difference may be obtained with a sliding contactor. The circuit is shown in Fig. 3-5. The external potential difference may be made equal to, slightly less, or slightly more than $\mathcal{E}$ by sliding the contactor.

If the external potential difference is made infinitesimally smaller than $\mathcal{E}$, then, during the short time this difference exists, there is a transfer of a quantity of charge dZ through the external circuit in a direction from the positive to the negative electrode. In this case, work is done by the cell on the outside. If the external potential difference is made slightly larger than $\mathcal{E}$, electricity is transferred in the opposite direction and work is done on the cell. In either case, the amount of work is

$$\boxed{dW = \mathcal{E}\,dZ.} \tag{3-5}$$

When the cell is discharging through the external circuit, dZ is negative; i.e., there is a quantity Z connected with the *state of charge* of the cell which decreases by an amount dZ, where dZ is the actual quantity of charge transferred. Charging the cell involves an increase in Z or a positive dZ.

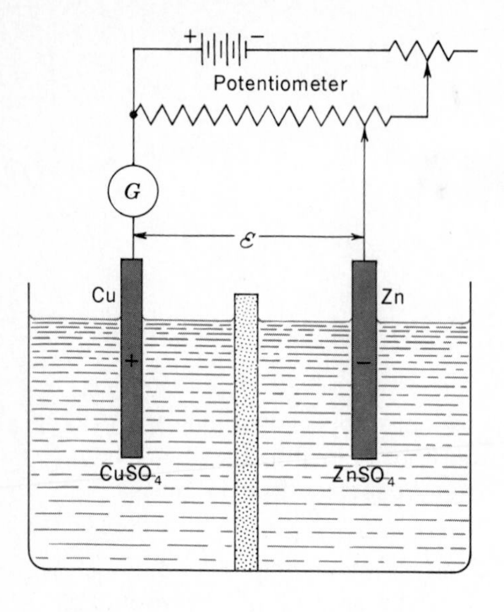

Figure 3-5 Approximately quasi-static transfer of charge in a reversible cell.

If Z changes by a finite amount,

$$W = \int_{Z_i}^{Z_f} \mathcal{E}\, dZ.$$

If the current is i, then in time $d\tau$ the quantity $dZ = i\, d\tau$, and

$$W = \int_i^f \mathcal{E} i\, d\tau.$$

With $\mathcal{E}$ in volts and the charge in coulombs, the work will be expressed in joules.

3-10 WORK IN CHANGING THE POLARIZATION OF A DIELECTRIC SOLID

Consider a slab of isotropic dielectric material placed between the conducting plates of a parallel-plate capacitor, as shown in Fig. 3-6. The surface area A of the capacitor plates has linear dimensions large compared to the separation l. A potential difference $\mathcal{E}$ may be maintained on the plates by a battery.

The effect of a potential difference on the plates is to set up an electric field with electric intensity E between the plates. The electric intensity will be nearly uniform between the plates and given by

$$E = \frac{\mathcal{E}}{l}.$$

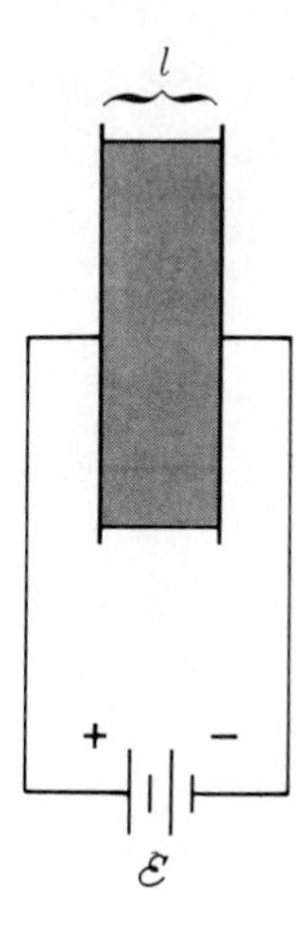

Figure 3-6 Changing the polarization of a dielectric solid.

Moreover, one plate is given a charge $+Z$ and the other a charge $-Z$. When the charge of the capacitor is changed an infinitesimal amount dZ, the work done is

$$\begin{aligned} đW &= \mathcal{E}\, dZ \\ &= El\, dZ. \end{aligned}$$

The charge Z on the plates is equal to

$$Z = DA,$$

where D is the *electric displacement*. Therefore,

$$\begin{aligned} đW &= AlE\, dD \\ &= VE\, dD, \end{aligned} \tag{3-6}$$

where V is the volume of the dielectric material.

If Π is the *total electric moment* of the material (assumed to be isotropic), or the *polarization*, we have the relation

$$D = \epsilon_0 E + \frac{\Pi}{V}, \tag{3-7}$$

where ϵ_0 is the permittivity of vacuum and Π is the total polarization (or its total dipole moment). Therefore,

$$đW = V\epsilon_0 E\, dE + E\, d\Pi.$$

The first term is the work required to increase the electric field strength by dE and would be present even if a vacuum existed between the plates of the capacitor. The second term is the work required to increase the polarization of the dielectric by $d\Pi$; it is zero when no material is present between the capacitor plates. We shall be concerned with changes of the material only, brought about by work done on or by the dielectric material (which is the system) and not in work done in changing the electric field. Consequently, the net work done *on* the dielectric is

$$\boxed{đW = E\,d\Pi} \tag{3-8}$$

Although our derivation has been specific for the case of a dielectric in a parallel-plate capacitor, the result is general for a dielectric in a uniform electric field.

If E is measured in volts per meter and Π in coulomb · meter, then the work is expressed in joules. If the polarization is changed a finite amount from Π_i to Π_f, the work will be

$$W = \int_{\Pi_i}^{\Pi_f} E\,d\Pi.$$

Experiments on dielectric materials are performed on samples of such shapes that the E field is uniform. For solid dielectrics the capacitor plates are plane and parallel, either circular or square. For liquid or gaseous dielectrics the capacitor plates are coaxial right cylinders. Regardless of the electrode configuration there must be guard electrodes extending beyond the measuring electrodes. The guard electrodes minimize the "fringing" of the electric field at the edge of the measuring electrode.

3-11 WORK IN CHANGING THE MAGNETIZATION OF A MAGNETIC SOLID

Consider a sample of magnetic material in the form of a ring of cross-sectional area A and of mean circumference L. Suppose that an insulated wire is wound on top of the sample, forming a toroidal winding of N closely spaced turns, as shown in Fig. 3-7. A current may be maintained in the winding by a battery, and by moving the sliding contactor of a rheostat this current may be changed.

The effect of a current in the winding is to set up a magnetic field with magnetic induction $\mathcal{B}$. If the dimensions are as shown in Fig. 3-7, $\mathcal{B}$ will be nearly uniform over the cross-section of the toroid. Suppose that the current is changed and that in time $d\tau$ the magnetic induction changes by an amount $d\mathcal{B}$. Then, by Faraday's principle of electromagnetic induction, there is induced in the winding a back emf $\mathcal{E}$, where

$$\mathcal{E} = -NA\frac{d\mathcal{B}}{d\tau}.$$

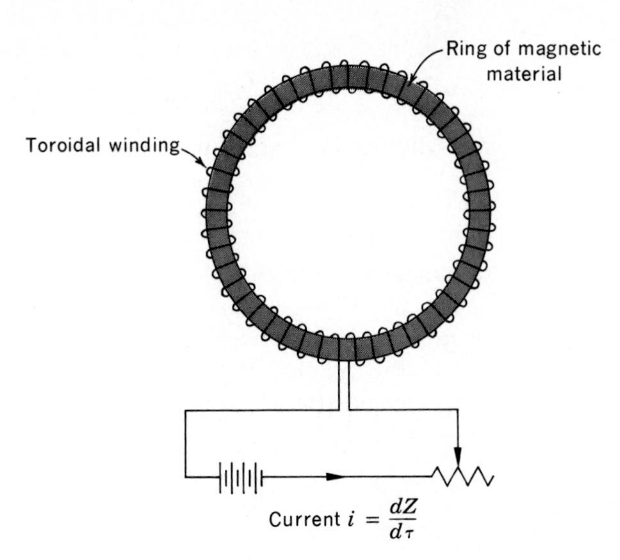

Figure 3-7 Changing the magnetization of a magnetic solid.

During the time interval $d\tau$, a quantity of charge dZ is transferred in the circuit, and the work done *by* the system to maintain the current is calculated as

$$\begin{aligned} đW &= -\mathcal{E}\, dZ \\ &= NA\frac{d\mathcal{B}}{d\tau}\, dZ \\ &= NA\frac{dZ}{d\tau}\, d\mathcal{B} \\ &= NAi\, d\mathcal{B}, \end{aligned}$$

where i, equal to $dZ/d\tau$, is the momentary value of the current.

The magnetic intensity $\mathcal{H}$ due to a current i in a toroidal winding is given by

$$\mathcal{H} = \frac{Ni}{L} = \frac{NAi}{AL} = \frac{NAi}{V},$$

where V is the volume of magnetic material. Therefore,

$$NAi = V\mathcal{H}$$

and

$$đW = V\mathcal{H}\, d\mathcal{B}. \tag{3-9}$$

If M is the *total magnetic moment* of the material (assumed to be isotropic), or *total magnetization*, we have the relation

$$\mathcal{B} = \mu_0 \mathcal{H} + \mu_0 \frac{M}{V}. \tag{3-10}$$

Therefore,

$$dW = V\mu_0 \mathscr{H}\, d\mathscr{H} + \mu_0 \mathscr{H}\, dM.$$

If no material were present within the toroidal winding, M would be zero, $\mathscr{B}$ would equal $\mathscr{H}$, and

$$đW = V\mu_0 \mathscr{H}\, d\mathscr{H} \qquad \text{(vacuum only)}.$$

This is the work necessary to increase the magnetic field in a volume V *of empty space* by an amount $d\mathscr{H}$. The second term, $\mu_0 \mathscr{H}\, dM$, is the work done in increasing the magnetization of the material by an amount dM. We shall be concerned in this book with changes of temperature, energy, etc., of the material only, brought about by work done on or by the material. Consequently, for the purpose of this book,

$$\boxed{đW = \mu_0 \mathscr{H}\, dM.} \tag{3-11}$$

If $\mathscr{H}$ is measured in amperes per meter and M in ampere · square meters, then the work will be expressed in joules. If the magnetization is caused to change a finite amount from M_i to M_f, the work will be

$$W = \mu_0 \int_{M_i}^{M_f} \mathscr{H}\, dM.$$

Experiments on paramagnetic materials are usually performed on samples in the form of cylinders or ellipsoids, not toroids. In these cases, the $\mathscr{H}$ field inside the material is somewhat smaller than the $\mathscr{H}$ field generated by the electric current in the surrounding winding because of the reverse field (demagnetizing field) set up by magnetic poles which form on the surfaces of the samples. In longitudinal magnetic fields, the demagnetizing effect either may be rendered negligible by using cylinders whose length is much larger than the diameter or may be corrected for in a simple way. In transverse magnetic fields, a correction factor must be applied. We shall limit ourselves to toroids or to long thin cylinders in longitudinal fields where the internal and external $\mathscr{H}$ fields are the same.

In any actual case, a change of magnetization is accomplished very nearly quasi-statically, and therefore an equation of state may be used in the integration of the expression denoting the work.

3-12 SUMMARY

The work values of the various simple systems are summarized in Table 3-1. It should be noted that each expression for work is the product of an intensive and an extensive quantity; consequently, *work is an extensive quantity.*

Table 3-1 Work of simple systems

Simple system	Intensive quantity (generalized force)	Extensive quantity (generalized displacement)	Work, J
Hydrostatic system	P, in Pa	V, in m^3	$-P\,dV$
Wire	$\mathcal{F}$, in N	L, in m	$\mathcal{F}\,dL$
Surface film	$\mathcal{S}$, in N/m	A, in m^2	$\mathcal{S}\,dA$
Reversible cell	$\mathcal{E}$, in V	Z, in C	$\mathcal{E}\,dZ$
Dielectric solid	E, in V/m	Π, in C · m	$E\,d\Pi$
Magnetic solid	$\mathcal{H}$, in A/m	M, in A · m^2	$\mu_0\mathcal{H}\,dM$

We have seen that a work diagram is obtained if any one of the intensive coordinates is plotted against its corresponding extensive coordinate. There are therefore as many work diagrams as there are systems. It is desirable at times, for the sake of argument, to formulate a work diagram which does not refer to one system in particular but which represents the behavior of any system. If we designate the intensive quantities P, $\mathcal{F}$, $\mathcal{S}$, $\mathcal{E}$, E, and $\mathcal{H}$ as *generalized forces* and their corresponding extensive quantities V, L, A, Z, Π, and M as *generalized displacements*, we may represent the work done by any simple system on a *generalized work diagram* by plotting the generalized force Y against the generalized displacement X. Conclusions based on such a diagram will hold for any simple system.

3-13 COMPOSITE SYSTEMS

Up to this point we have dealt exclusively with simple systems, whose equilibrium states are described with the aid of three thermodynamic coordinates, one of which is always the temperature. A single equation of state was found to exist in each case, so that only two of the coordinates are independent. The laws of thermodynamics, however, which are to be developed in the next few chapters, must apply to any system no matter how complicated—i.e., to systems having more than three coordinates and more than one equation of state.

Consider the composite system depicted schematically in Fig. 3-8*a* with two different simple hydrostatic systems separated by a diathermic wall, which ensures that both parts have the same temperature. There are five thermodynamic coordinates (P, V, P', V', and θ) and two equations of state, one for each of the

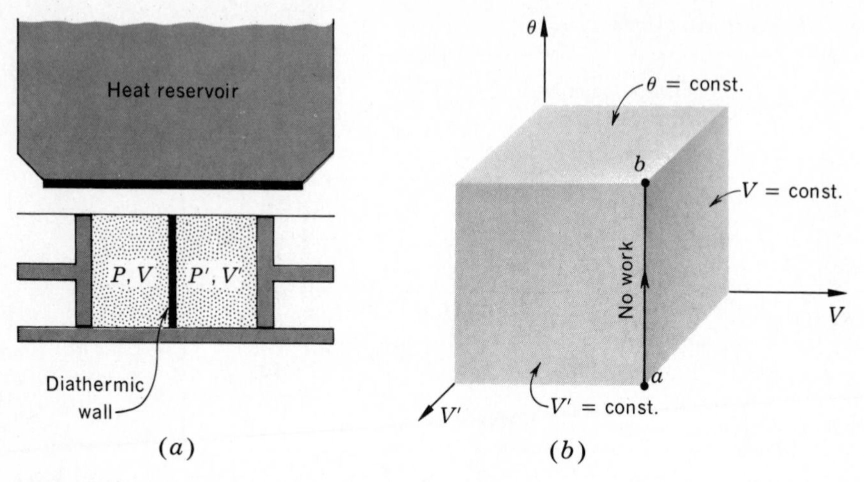

Figure 3-8 (*a*) A composite system whose coordinates are P, V, P', V', and θ. (*b*) A graph of the independent coordinates θ, V, and V'.

simple systems. Consequently, only three of the five coordinates are independent. In any small displacement of each piston, the work is

$$đW = -P\,dV - P'\,dV'.$$

The most convenient diagram to use in demonstrating the features of this system is a three-dimensional diagram with θ, V, and V' plotted along rectangular axes, as shown in Fig. 3-8*b*. A typical isothermal process would be a curve on a plane such as the one marked $\theta =$ const. A curve on a plane such as that marked $V =$ const. would represent a process in which no work is done by the left-hand part. The points a and b lie on a vertical line every point of which refers to a constant V and V'. The straight line ab therefore represents a process in which no work is done by the composite system.

Two simple systems do not have to be separated spatially by a diathermic wall in order to have two equations of state and a common temperature. Consider an ideal paramagnetic gas, such as oxygen at low pressures, as depicted schematically in Fig. 3-9*a*. The oxygen may have its pressure P and volume V varied with the aid of a piston-cylinder combination, and it is immersed in a magnetic field whose intensity $\mathcal{H}$ may be varied by varying the current in the surrounding solenoid. The gas is kept at a uniform temperature θ. The coordinates are P, V, $\mathcal{H}$, M, and θ, only three of which are independent because of the two equations of state: the ideal-gas equation $PV = nR\theta$ and Curie's equation $M/\mathcal{H} = C_C'/\theta$. Since the work done in any infinitesimal process is

$$đW = -P\,dV + \mu_0 \mathcal{H}\,dM,$$

the most convenient independent coordinates are θ, V, and M, which are plotted

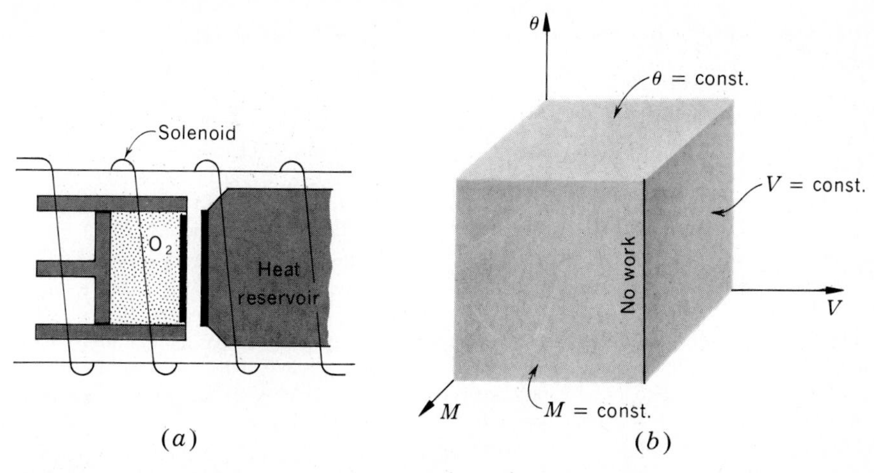

Figure 3-9 (*a*) A composite system whose coordinates are P, V, $\mathcal{H}$, M, and θ. (*b*) A graph of the independent coordinates θ, V, and M.

along rectangular axes in Fig. 3-9*b*. Any vertical line would represent a process in which no work is done.

Later, we shall have occasion to refer to a general five-coordinate system whose coordinates are Y, X, Y', X', and θ and whose work is

$$đW = Y\,dX + Y'\,dX'.$$

The most convenient independent coordinates of this system are θ, X, and X'.

PROBLEMS

3-1 A thin-walled metal bomb of volume V_B contains a gas at high pressure. Connected to the bomb is a capillary tube and stopcock. When the stopcock is opened slightly, the gas leaks slowly into a cylinder equipped with a nonleaking, frictionless piston where the pressure remains constant at the atmospheric value P_0.

(*a*) Show that, after as much gas as possible has leaked out, an amount of work

$$W = -P_0(V_0 - V_B)$$

has been done, where V_0 is the volume of the gas at atmospheric pressure and temperature.

(*b*) How much work would be done if the gas leaked directly into the atmosphere?

3-2 Calculate the work done by 1 mol of gas during a quasi-static, isothermal expansion from an initial volume v_i to a final volume v_f when the equation of state is:

(*a*) $$P(v - b) = R\theta \qquad (R, b = \text{const.}).$$

(*b*) $$Pv = R\theta\left(1 - \frac{B}{v}\right) \qquad [R = \text{const.};\ B = f(\theta)].$$

3-3 During a quasi-static adiabatic expansion of an ideal gas, the pressure at any moment is given by the equation

$$PV^{\gamma} = K,$$

where γ and K are constants. Show that the work done in expanding from a state (P_i, V_i) to a state (P_f, V_f) is

$$W = -\frac{P_i V_i - P_f V_f}{\gamma - 1}.$$

If the initial pressure and volume are 10^6 Pa and 10^{-3} m^3, respectively, and the final values are 2×10^5 Pa and 3.16×10^{-3} m^3, how many joules of work is done by a gas whose $\gamma = 1.4$?

3-4 A vertical cylinder, closed at the bottom, is placed on a spring scale. The cylinder contains a gas whose volume may be changed with the aid of a frictionless, nonleaking piston. The piston is pushed down.

(*a*) How much work is done by an outside agent in compressing the gas an amount dV while the spring scale goes down a distance dy?

(b) If this device is used only to produce effects in the gas—in other words, if the gas is the system—what expression for work is appropriate?

3-5 A stationary vertical cylinder, closed at the top, contains a gas whose volume may be changed with the aid of a heavy, frictionless, nonleaking piston of weight w.

(*a*) How much work is done by an outside agent in compressing the gas an amount dV by raising the piston a distance dy?

(*b*) If this device is used only to produce temperature changes of the gas, what expression for work would be appropriate?

(*c*) Compare this situation with that of Prob. 3-4, and also with that involved in increasing the magnetic induction of a ring of magnetic material.

3-6 The pressure on 0.1 kg of metal is increased quasi-statically and isothermally from 0 to 10^8 Pa. Assuming the density and the isothermal compressibility to remain constant at the values 10^4 kg/m^3 and 6.75×10^{-12} Pa^{-1}, respectively, calculate the work in joules.

3-7 (*a*) The tension in a wire is increased quasi-statically and isothermally from $\mathcal{F}_i$ to $\mathcal{F}_f$. If the length, cross-sectional area, and isothermal Young's modulus of the wire remain practically constant, show that the work done is

$$W = \frac{L}{2AY}(\mathcal{F}_f^2 - \mathcal{F}_i^2).$$

(*b*) The tension in a metal wire 1 m long and 1.0×10^{-7} m^2 in area is increased quasi-statically and isothermally at 0°C from 10 to 100 N. How many joules of work is done? (The isothermal Young's modulus at 0°C is 2.5×10^{11} N/m^2.)

3-8 The equation of state of an ideal elastic substance is

$$\mathcal{F} = K\theta\left(\frac{L}{L_0} - \frac{L_0^2}{L^2}\right),$$

where K is a constant and L_0 (the value of L at zero tension) is a function of temperature only. Calculate the work necessary to compress the substance from $L = L_0$ to $L = L_0/2$ quasi-statically and isothermally.

3-9 Show that the work required to blow a spherical soap bubble of radius R in an isothermal, quasi-static process in the atmosphere is equal to $8\pi\mathcal{S}R^2$.

3-10 A dielectric has an equation of state $\Pi/V = \chi E$, where χ is a function of θ only. Show that the work done in an isothermal, quasi-static change of state is given by

$$W = \frac{1}{2V\chi}(\Pi_f^2 - \Pi_i^2) = \frac{V\chi}{2}(E_f^2 - E_i^2).$$

3-11 Prove that the work done during a quasi-static isothermal change of state of a paramagnetic substance obeying Curie's equation is given by

$$W = \frac{\mu_0 \theta}{8\pi C'_C}(M_f^2 - M_i^2),$$

$$= \frac{2\pi\mu_0 C'_C}{\theta}(\mathcal{H}_f^2 - \mathcal{H}_i^2).$$

3-12 A volume of 2×10^{-4} m^3 of a paramagnetic substance is maintained at constant temperature. A magnetic field is increased quasi-statically and isothermally from 0 to 10^6 A/m. Assuming Curie's equation to hold and the Curie constant per unit volume to be 0.15 deg:

(*a*) How much work would have to be done if no material were present?

(*b*) How much work is done to change the magnetization of the material when the temperature is 300 K and when it is 1 K?

(*c*) How much work is done at both temperatures by the agent supplying the magnetic field?

3-13 A chamber with rigid walls consists of two compartments, one containing a gas and the other evacuated; the partition between the two compartments is destroyed suddenly. Is the work done during any infinitesimal portion of this process (called a *free expansion*) equal to $P\,dV$? Explain.

3-14 A reversible cell consists of an electrolyte, one solid electrode, and one electrode involving gaseous hydrogen.

(*a*) What coordinates are needed to describe the equilibrium states of this system?

(*b*) How many equations of state are there? What are typical examples of these equations?

(*d*) What is the expression for $đW$?

(*d*) Choose convenient independent coordinates.

3-15 A framework such as that shown in Fig. 3-4 is placed in a vessel where the air pressure may be varied at will. Consider the two surface films *and* the liquid between the films as the system.

(*a*) What coordinates are needed?

(*b*) How many equations of state are there?

(*c*) What is the expression for $đW$?

(*d*) Choose convenient independent coordinates.

3-16 Devise a system consisting of an ideal nonmagnetic gas, a paramagnetic solid, and a reversible cell, all separated by diathermic walls. Draw a diagram.

(*a*) What are the coordinates?

(*b*) How many equations of state are there?

(*c*) What is the expression for $đW$?

(*d*) Choose convenient independent coordinates.

CHAPTER

FOUR

HEAT AND THE FIRST LAW OF THERMODYNAMICS

4-1 WORK AND HEAT

It was shown in Chap. 3 how a system could be transferred from an initial to a final state by means of a quasi-static process and how the work done during the process could be calculated. There are, however, other means of changing the state of a system that do not necessarily involve the performance of work. Consider, for example, the three processes depicted schematically in Fig. 4-1. In (*a*) a fluid undergoes an adiabatic expansion in a cylinder-piston combination that is coupled to the surroundings with a suspended body so that, as the expansion takes place, the body is lifted while the fluid remains always close to equilibrium.† In (*b*), a liquid in equilibrium with its vapor is in contact through a diathermic wall with the hot combustion products of a bunsen burner and undergoes vaporization, accompanied by a rise of temperature and of pressure, without the performance of work. In (*c*) a fluid is expanded while in contact with the flame of a bunsen burner.

What happens when two systems at different temperatures are placed together is one of the most familiar experiences of mankind. It is well known that the final temperature reached by both systems is intermediate between the two starting temperatures. Up to the beginning of the nineteenth century, such phenomena, which comprise the subject of *calorimetry*, were explained by postulating the existence of a substance or form of matter termed *caloric*, or heat, in

† The clever coupling device in Fig. 4-1 is due to E. Schmidt, *Thermodynamics: Principles and Applications to Engineering*, Dover, New York, 1966.

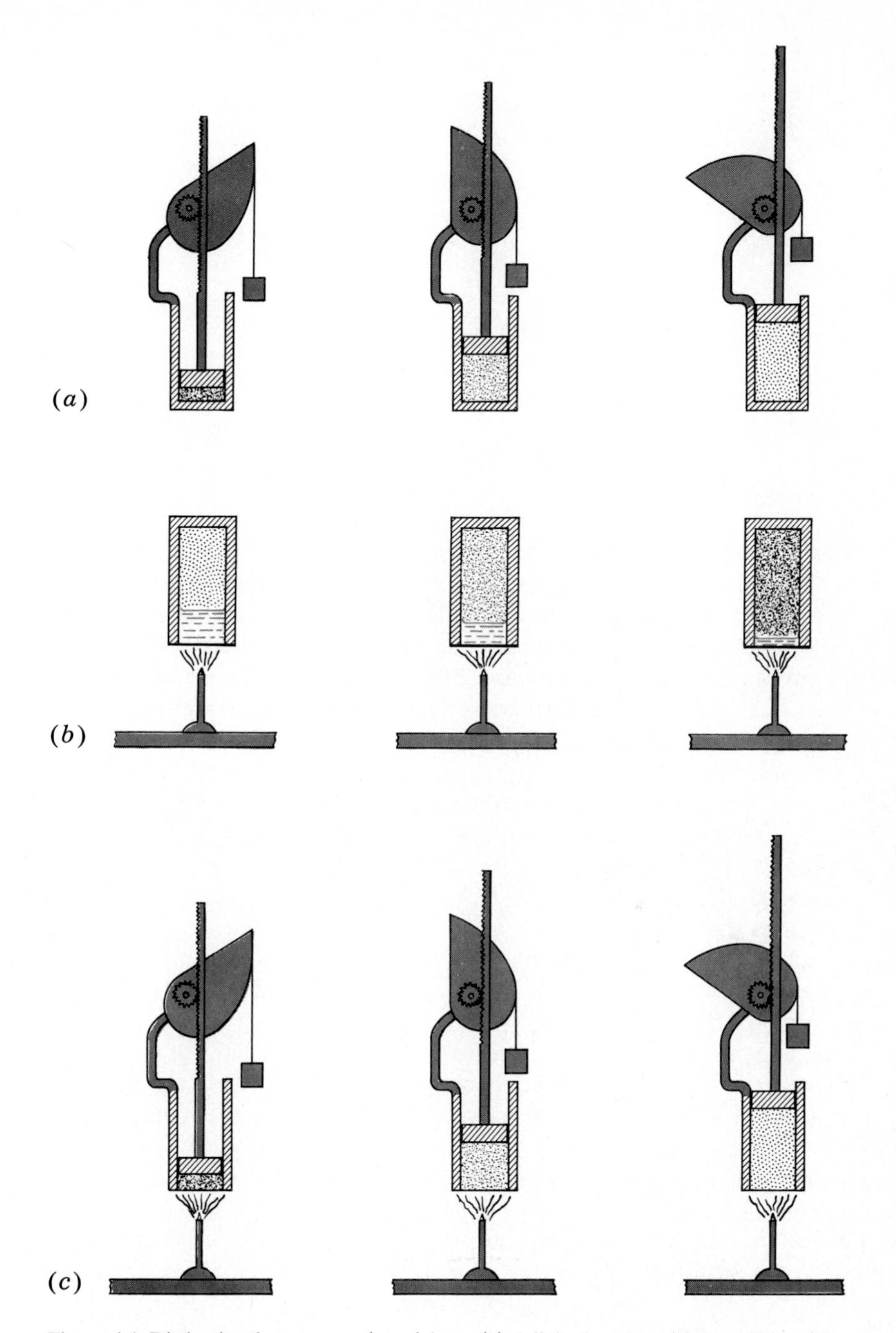

Figure 4-1 Distinction between work and heat. (*a*) Adiabatic work; (*b*) heat flow without work; (*c*) work and heat.

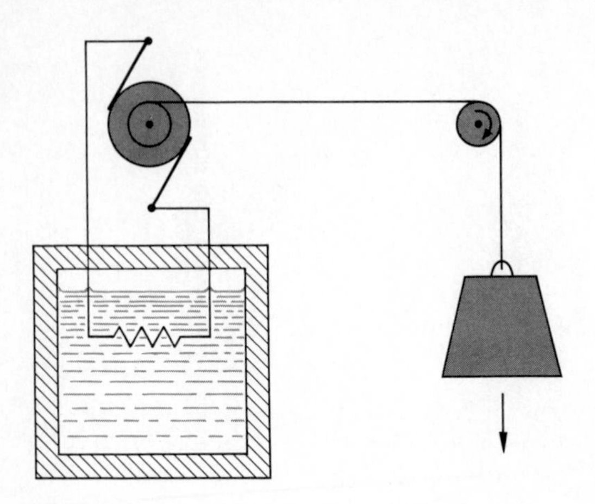

Figure 4-2 Whether a process is designated as a work or a heat interaction depends upon the choice of system.

every body. It was believed that a body at a high temperature contained much caloric and that one at a low temperature had only a little. When the two bodies were put together, the body rich in caloric lost some to the other, and thus the final temperature was intermediate. Although we now know that heat is not a substance whose total amount remains constant, nevertheless we ascribe the changes that take place in Fig. 4-1*b* and *c* to the transfer of "something" from the body at the higher temperature to the one at the lower, and this something we call *heat*. We therefore adopt as a *calorimetric* definition of heat *that which is transferred between a system and its surroundings by virtue of a temperature difference only*. It is obvious that an adiabatic wall is one which is impervious to heat, or a *heat insulator*, and that a diathermic wall is a *heat conductor*.

It is important to observe that the decision as to whether a particular change of state involves the performance of work or the transfer of heat requires first an unequivocal answer to these questions: What is the system, and what are the surroundings? For example, in Fig. 4-2, a resistor immersed in water carries a current provided by an electric generator that is rotated with the aid of a descending body. If we assume the absence of friction in the shafts of the pulleys and the absence of electrical resistance in the generator and the connecting wires, we have a device whereby the thermodynamic state of a system composed of *water and the resistor* is changed by purely mechanical means, i.e., by the performance of work. If, however, the resistor is regarded as the system and the water as the surroundings, then there is a *transfer of heat from the resistor* by virtue of the temperature difference between the resistor and the water. Also, if a small part of the water is regarded as the system, with the rest of the water being considered the surroundings, then again there is a transfer of heat. Regarding the composite system comprising both the water and the resistor, however, the surroundings do not contain any object whose temperature differs from that of the system, and hence no heat is transferred between *this composite system* and its surroundings.

4-2 ADIABATIC WORK

Figures 4-1*a* and 4-2 show that a system completely surrounded by an adiabatic envelope may still be coupled to the surroundings so that work may be done. Three other simple examples of adiabatic work are shown in Fig. 4-3. It is an important fact of experience that the state of a system may be caused to change from a given initial state to a final state by the performance of adiabatic work *only*.

Consider the composite system shown in Fig. 4-4*a*, consisting of a hydrostatic fluid *and* an immersed resistor on each side of a diathermic wall. This system can undergo an adiabatic work interaction with its surroundings in two ways. It may be done by moving one or both of the pistons in or out, either slowly (a quasi-static process) so that $W = -\int P\,dV$, with P being equal to the equilibrium value, or very rapidly (a non-quasi-static process) so that the pressure at the piston face is less than the equilibrium value. If the piston is pulled out at a faster rate than the velocity of the molecules of the fluid, the fluid will do no work on the piston at all. Such a process is called a *free expansion*, and it will be discussed in some detail in the next chapter. To repeat, one way of doing work is by slow or rapid motion, in or out, of either or both of the pistons. Another way in which work may be done on the system is by dissipation of electrical energy in the resistors, in which currents are maintained by generators actuated by the descending bodies. (Exactly the same effects could be produced by the dissipation of mechanical energy in the fluids by irregular churning of the fluids with paddle wheels actuated by the descending bodies.)

As in Chap. 3, the most convenient independent coordinates of this system are θ, the common temperature, and the two volumes V and V'. States i and f of the system, shown on a $\theta VV'$ diagram in Fig. 4-4*b*, are arbitrarily chosen, and it is purely a coincidence that f corresponds to a higher temperature than i. In the path iaf, the dashed curve ia represents a frictionless, quasi-static adiabatic compression accomplished with one of the pistons. It is drawn on a surface cutting the two isothermal planes. The existence of such a reversible adiabatic surface will be proved in Art. 8-7, but it should be noted at this point that, since ia is accomplished only through a frictionless slow motion of the piston, it may be

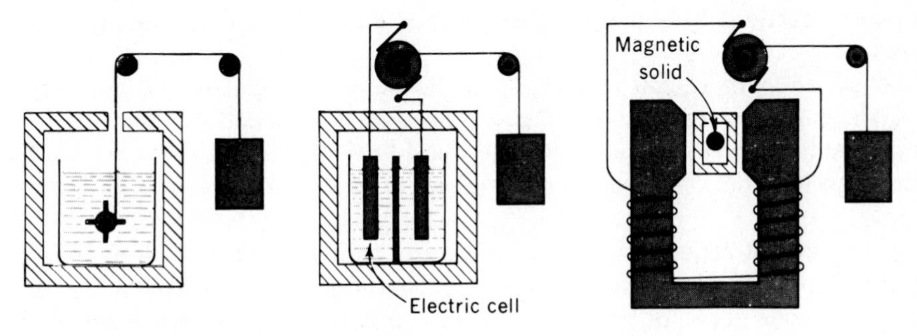

Figure 4-3 Adiabatic work.

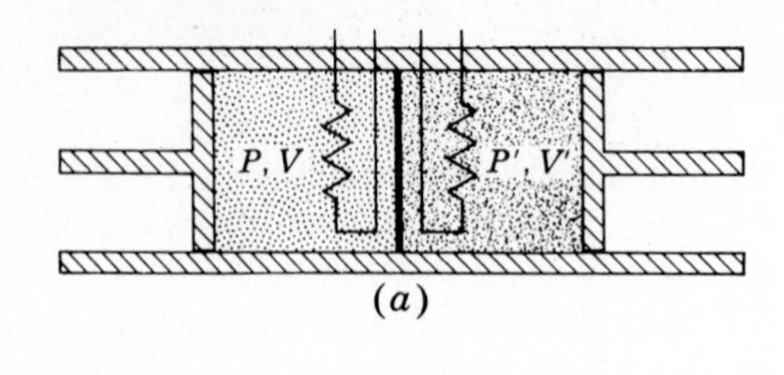

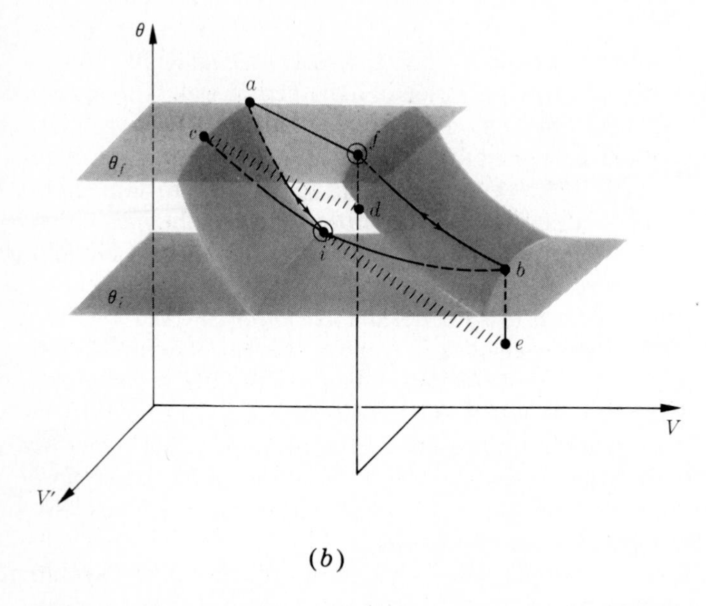

Figure 4-4 (*a*) A composite system on which adiabatic work may be done in two ways. (*b*) Joining two states *i* and *f* by several different adiabatic paths.

performed in *either* the direction *ia* or *ai*. The curve *af* represents the adiabatic dissipation of electrical energy in conjunction with piston movements that keep the system at constant temperature. In other words, the line *af* represents a process that is *both* adiabatic and isothermal! There is, however, an important distinction between this process and the previous one: the process *af* can go in *only one* direction. You can add energy with a current in a resistor, but you cannot extract it.

The path *ibf* represents another adiabatic way of changing the system from *i* to *f*. The curve *ib* represents the dissipation process accomplished with the resistors, and the curve *bf* stands for the quasi-static process achieved with frictionless pistons only. As before, *bf* could go in either direction, but *ib* in only one.

There are, of course, many other adiabatic paths joining *if*, such as *icdf*, in which the process *cd* is a non-quasi-static expansion accomplished by a rapid outward motion of one or both of the pistons, and the process *df* is performed by keeping both pistons motionless and by dissipating electrical energy in either or both of the resistors. Another possible adiabatic path consists of rapid outward motion of the pistons, producing a non-quasi-static expansion *ie*, followed by isochoric dissipation of electric energy *eb*, followed by a quasi-static compression *bf*. Although accurate measurements of adiabatic work along different paths between the same two states have never been made, indirect experiments indicate that the adiabatic work is the same along all such paths. The generalization of this result is known as the *first law of thermodynamics:*

If a system is caused to change from an initial state to a final state by adiabatic means only, the work done is the same for all adiabatic paths connecting the two states.

Whenever a quantity is found to depend only on the initial and final states, and not on the path connecting them, an important conclusion can be drawn. The student will recall from mechanics that, in moving an object from a point in a gravitational field to another point, in the absence of friction the work done depends only on the positions of the two points and not on the path through which the body was moved. It was concluded from this that there exists a function of the space coordinates of the body whose final value minus its initial value is equal to the work done. This function was called the *potential-energy function.* Similarly, the work done in moving an electric charge from one point in an electric field to another is also independent of the path and therefore is also expressible as the value of a function (the electric potential function) at the final state minus its value at the initial state. It therefore follows from the first law of thermodynamics that there exists a function of the coordinates of a thermodynamic system whose value at the final state minus its value at the initial state is equal to the adiabatic work in going from one state to the other. This function is known as the *internal-energy function.*

Denoting the internal-energy function by U, we have

$$\boxed{W_{i \to f}(\text{adiabatic}) = U_f - U_i,} \tag{4.1}$$

where the signs are such that, if positive work is done on the system, its energy increases.

4-3 INTERNAL-ENERGY FUNCTION

The physical interpretation of the difference $U_f - U_i$ is the change in energy of the system. The equality, therefore, of the change of energy and the adiabatic work expresses the principle of the conservation of energy. It should be emphasized, however, that the equation expresses more than the principle of the conservation of energy. It states that *there exists an energy function*, the difference between two values of which is the energy change of the system.

The internal energy is a function of as many thermodynamic coordinates as are necessary to specify the state of a system. The equilibrium states of a hydrostatic system, for example, describable by means of three thermodynamic coordinates P, V, and θ are completely determined by only two, since the third is fixed by the equation of state. Therefore, the internal energy may be thought of as a function of only two (any two) of the thermodynamic coordinates. This is true for each of the simple systems described in Chap. 2. It is not always possible to write this function in simple mathematical form. Very often the exact form of the function is unknown. It must be understood, however, that it is not necessary to know actually what the internal-energy function is, so long as we can be sure that such a function exists.

If the coordinates characterizing the two states differ from each other only infinitesimally, the change of internal energy is dU, where dU is an exact differential, since it is the differential of an actual function. In the case of a hydrostatic system, if U is regarded as a function of θ and V, then

$$dU = \left(\frac{\partial U}{\partial \theta}\right)_V d\theta + \left(\frac{\partial U}{\partial V}\right)_\theta dV,$$

or, regarding U as a function of θ and P,

$$dU = \left(\frac{\partial U}{\partial \theta}\right)_P d\theta + \left(\frac{\partial U}{\partial P}\right)_\theta dP.$$

The student should realize that the two partial derivatives $(\partial U/\partial \theta)_V$ and $(\partial U/\partial \theta)_P$ are not equal. The first is a function of θ and V, and the second a function of θ and P. They are different mathematically and also have a different physical meaning.

4-4 MATHEMATICAL FORMULATION OF THE FIRST LAW OF THERMODYNAMICS

We have been considering, up to now, processes wherein a system undergoes a change of state through the performance of adiabatic work only. Such experiments must be performed in order to measure the change in the energy function of a system, but they are not the usual processes that are carried out in the laboratory. In Fig. 4-5 there are depicted two examples of processes involving changes of state that take place nonadiabatically. In (*a*) a gas is in contact with a bunsen flame whose temperature is higher than that of the gas and at the same time is allowed to expand. In (*b*) the magnetization of a paramagnetic solid is increased while it is in contact with liquid helium, the temperature of which is lower than that of the solid. As a matter of fact, some of the helium boils away during the magnetization.

Let us now imagine two different experiments performed on the same system. In one we measure the adiabatic work necessary to change the state of the system from i to f. This is $U_f - U_i$. In the other we cause the system to undergo the *same* change of state, but nonadiabatically, and measure the work

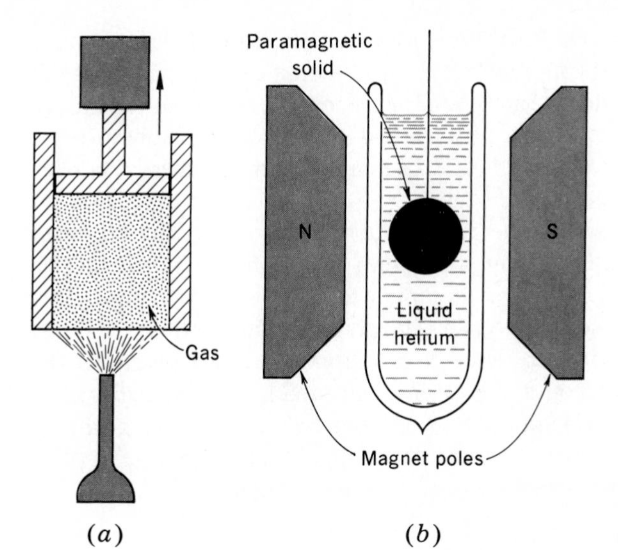

Figure 4-5 Nonadiabatic processes.

done. The result of all such experiments is that *the nonadiabatic work is* not *equal to* $U_f - U_i$. In order that this result shall be consistent with the principle of the conservation of energy, we are forced to conclude that energy has been transferred by means *other than* the performance of work. This energy, whose transfer between the system and its surroundings is required by the principle of the conservation of energy and which has taken place only by virtue of the temperature difference between the system and its surroundings, is what we have previously called heat. We therefore give the following as our *thermodynamic* definition of heat: *When a system whose surroundings are at a different temperature and on which work may be done undergoes a process, the energy transferred by nonmechanical means, equal to the difference between the internal-energy change and the work done, is called heat.* Denoting this difference by Q, we have

$$Q = U_f - U_i - (W),$$

or

$$\boxed{U_f - U_i = Q + W} \tag{4-2}$$

where the convention has been adopted that Q is positive when it enters a system and negative when it leaves a system. The preceding equation is known as the *mathematical formulation of the first law of thermodynamics.*

It should be emphasized that the mathematical formulation of the first law contains three related ideas: (1) the existence of an internal-energy function; (2) the principle of the conservation of energy; (3) the definition of heat as *energy* in transit by virtue of a temperature difference.

It was many years before it was understood that heat is energy. The first really conclusive evidence that heat could not be a substance was given by Benjamin Thompson, an American from Woburn, Massachusetts, who later became Count Rumford of Bavaria. In 1798, Rumford observed the temperature rise in brass chips produced during the boring of cannon and concluded that the work of boring was responsible for the flow of heat. One year later, Sir Humphry Davy tried to show that two pieces of ice could be melted by rubbing them together. His idea was to show that heat is a manifestation of energy, but his experiment was highly inconclusive.

The idea that heat is a form of energy was put forward in 1839 by Séguin, a French engineer, and in 1842 by Mayer, a German physician, but no conclusive experiments were performed by either. It remained for Joule, an independent investigator with a private laboratory, in the period from 1840 to 1849, to convince the world by performing a series of admirable experiments on the relation between heat and work and to establish once and for all the equivalence of these two quantities. Von Helmholtz recognized the epoch-making importance of Joule's work and wrote a brilliant paper in 1847, in which he applied Joule's ideas to the sciences of physical chemistry and physiology.

4-5 CONCEPT OF HEAT

Heat is internal energy in transit. It flows from one part of a system to another, or from one system to another, by virtue of only a temperature difference. During the flow you do not know the whole process, namely, the final state. The heat is not known during the process. The known quantity during the process is the rate of heat flow $\dot{Q}$, which is a function of time. Hence, the heat is

$$Q = \int_{\tau_1}^{\tau_2} \dot{Q}\, d\tau$$

and can be determined only when the time $\tau_2 - \tau_1$ has elapsed. Only after the flow is over can one refer to the heat—internal energy that has been transferred from a system at a higher temperature to a system at a lower temperature.

It would be incorrect to refer to the "heat in a body" as it would be incorrect to speak of the "work in a body." The performance of work and the flow of heat are methods whereby the internal energy of a system is changed. It is impossible to separate or divide the internal energy into a mechanical and a thermal part.

We have seen that, in general, the work done on or by a system is not a function of the coordinates of the system but depends on the path by which the system was brought from the initial to the final state. Exactly the same is true of the heat transferred to or from a system. Q is not a function of the thermodynamic coordinates but depends on the path. An infinitesimal amount of heat, therefore, is an inexact differential and is represented by the symbol $đQ$.

Imagine a system A in thermal contact with a system B, the two systems being surrounded by adiabatic walls. For system A alone,

$$U_f - U_i = Q + W;$$

and for system B alone,

$$U'_f - U'_i = Q' + W'.$$

Adding, we get

$$(U_f + U'_f) - (U_i + U'_i) = Q + Q' + W + W'.$$

Since $(U_f + U'_f) - (U_i + U'_i)$ is the change in energy of the composite system and $W + W'$ is the work done by the composite system, it follows that $Q + Q'$ is the heat transferred by the composite system. Since the composite system is surrounded by adiabatic walls,

$$Q + Q' = 0,$$

and

$$Q = -Q'. \tag{4-3}$$

In other words, *under adiabatic conditions, the heat lost (or gained) by system A is equal to the heat gained (or lost) by system B.*

4-6 DIFFERENTIAL FORM OF THE FIRST LAW OF THERMODYNAMICS

A process involving only infinitesimal changes in the thermodynamic coordinates of a system is known as an *infinitesimal process*. For such a process the first law becomes

$$\boxed{dU = đQ + đW.} \tag{4-4}$$

If the infinitesimal process is quasi-static, then dU and $đW$ can be expressed in terms of thermodynamic coordinates only. An infinitesimal quasi-static process is one in which the system passes from an initial equilibrium state to a neighboring equilibrium state.

For an infinitesimal quasi-static process of a hydrostatic system, the first law becomes

$$dU = đQ - P\,dV, \tag{4-5}$$

where U is a function of any two of the three thermodynamic coordinates and P is, of course, a function of V and θ. A similar equation may be written for each of the other simple systems as shown in Table 4-1.

To deal with more complicated systems, it is merely necessary to replace $đW$ in the first law by two or more expressions. For example, in the case of a composite system consisting of two hydrostatic parts separated by a diathermic wall, we may express dQ as follows:

$$đQ = dU + P\,dV + P'\,dV', \tag{4-6}$$

Table 4-1 The First Law of Thermodynamics for simple systems

System	First law	U is a function of *any two* of
Hydrostatic system	$dU = đQ - P\,dV$	P, V, θ
Wire	$dU = đQ + \mathcal{F}\,dL$	$\mathcal{F}, L, \theta$
Surface film	$dU = đQ + \mathcal{S}\,dA$	$\mathcal{S}, A, \theta$
Electric cell	$dU = đQ + \mathcal{E}\,dZ$	$\mathcal{E}, Z, \theta$
Dielectric slab	$dU = đQ + E\,d\Pi$	E, Π, θ
Paramagnetic rod	$dU = đQ + \mu_0 \mathcal{H}\,dM$	$\mathcal{H}, M, \theta$

whereas for a paramagnetic gas

$$đQ = dU + P\,dV - \mu_0 \mathcal{H}\,dM. \tag{4-7}$$

The right-hand members of Eqs. (4-5), (4-6), and (4-7) are known as *Pfaffian differential forms*, and the question of their integrability is an interesting and important one that will be studied later. At this point, however, it is worthwhile to state a fundamental difference between Eq. (4-5) and Eq. (4-6), in order to justify the seemingly undue emphasis on systems of more than two independent coordinates. Since the left-hand expression in Eqs. (4-5) and (4-6) represents an infinitesimal amount of heat $đQ$, and since the heat transferred depends on the path, $đQ$ is an inexact differential and the Pfaffian differential forms are inexact differentials. An inexact differential, however, may often be made exact by multiplying it by a function, known as an *integrating factor*. The Pfaffian differential form representing $đQ$ of a simple system with two independent coordinates has the mathematical property that *an integrating factor can always be found*. This is not the result of a law of nature; it is a purely mathematical result of the fact that there are only two independent coordinates.

When there are three or more independent coordinates, however, the situation is entirely different. In general, a Pfaffian differential form containing three differentials does *not* admit of an integrating factor! But, because of the existence of a new law of nature (the second law of thermodynamics), the Pfaffian differential form representing $đQ$ *does* have an integrating factor. It is a most remarkable circumstance that the integrating factor for $đQ$ which is found for systems with *any* number of independent variables is *an arbitrary function of the empirical temperature only, which is the same function for all systems*. This enables us to define an absolute thermodynamic (or Kelvin) temperature, as shown in Chap. 7.

4-7 HEAT CAPACITY AND ITS MEASUREMENT

When heat is absorbed by a system, a change of temperature may or may not take place, depending on the process. If a system undergoes a change of temperature from θ_i to θ_f during the transfer of Q units of heat, the average *heat capacity* of the system is defined as the ratio

$$\text{Average heat capacity} = \frac{Q}{\theta_f - \theta_i}.$$

As both Q and $(\theta_f - \theta_i)$ get smaller, this ratio approaches the instantaneous value of the *heat capacity* C thus:

$$C = \lim_{\theta_f \to \theta_i} \frac{Q}{\theta_f - \theta_i},$$

$$\boxed{C = \frac{đQ}{d\theta}.} \tag{4-8}$$

In dealing with extensive quantities (see Art. 2-13), such as volume or internal energy, it is often convenient to divide by the mass of the sample and specify the volume per unit mass or the internal energy per unit mass. These quantities are called *specific* quantities, the adjective "specific" meaning "per unit mass." Heat capacity is an extensive quantity, and the "specific heat capacity," abbreviated "specific heat," is measured in J/kg · K or kJ/kg · K. When the specific heat capacities of different substances are compared, no interesting regularities appear. When, however, *a unit of substance* (a different mass for each difference substance) called *a mole* is used, wonderful regularities (to be explained in Chap. 9) occur.

A mole (abbreviated "mol") is defined as the amount of substance that contains as many elementary entities (molecules, atoms, ions, etc.) as there are atoms in 0.012 kg of carbon-12. This number of atoms of carbon-12 is called Avogadro's number N_A and is equal to 6.023×10^{23} particles/mol. If the mass of an atom is m, then the mass of a mole of atoms is mN_A. This quantity, the *molar mass*, is what has been called in the past the "molecular weight." Designating the molar mass by the script capital $\mathcal{M}$, we have

$$\mathcal{M} = mN_A,$$

and the number of moles n is given by

$$n = \frac{\text{total mass}}{\mathcal{M}}.$$

If C is the heat capacity of n moles, then the *molar heat capacity* c is given by

$$c = \frac{C}{n} = \frac{1}{n}\frac{đQ}{d\theta}$$

and is measured in J/mol · K or kJ/kmol · K.

The heat capacity may be negative, zero, positive, or infinite, depending on the process the system undergoes during the heat transfer. It has a definite value

Table 4-2 Heat capacities of simple systems

System	Heat capacities	Symbol
Hydrostatic	At constant pressure At constant volume	C_P C_V
Linear	At constant tension At constant length	$C_{\mathcal{J}}$ C_L
Surface	At constant surface tension At constant area	$C_{\mathcal{S}}$ C_A
Electric	At constant emf At constant charge	$C_{\mathcal{E}}$ C_Z
Dielectric	At constant electric field At constant polarization	G_E G_Π
Magnetic	At constant magnetic field At constant magnetization	$C_{\mathcal{H}}$ C_M

only for a definite process. In the case of a hydrostatic system, the ratio $đQ/d\theta$ has a unique value when the pressure is kept constant. Under these conditions, C is called the *heat capacity at constant pressure* and is denoted by the symbol C_P, where

$$C_P = \left(\frac{đQ}{d\theta}\right)_P. \tag{4-9}$$

In general, C_P is a function of P and θ. Similarly, the heat capacity *at constant volume* is

$$C_V = \left(\frac{đQ}{d\theta}\right)_V \tag{4-10}$$

and depends on both V and θ. In general, C_P and C_V are different. Both will be thoroughly discussed throughout the book. Each simple system has its own heat capacities as shown in Table 4-2.

Each heat capacity is a function of two variables. Within a small range of variation of these coordinates, however, the heat capacity may be regarded as practically constant. Very often, one heat capacity can be set equal to another without much error. Thus, the $C_{\mathcal{H}}$ of a paramagnetic solid is at times very nearly equal to C_P.

The measurement of the heat capacity of solids is one of the most important experimental projects of modern physics, because numerical values of heat capacity provide one of the most direct means of verifying the calculations of theoretical physicists and of deciding on the validity of the assumptions constituting some of the modern theories. An electrical method of measurement is used almost invariably. If a resistance wire is wound around a cylindrical sample of material and if both the wire and the sample are regarded as the system, then the electrical energy dissipated in the wire is interpreted as work. When the wire is not included as part of the system, however, the energy which is dissipated within the wire and which flows into the sample by virtue of the temperature difference between the wire and the sample (however small) is designated as heat. The wire is often called a *heating coil.* If the current in the wire is I and the potential difference across it is $\mathcal{E}$, then the heat $đQ$ that leaves the heating coil over a time $d\tau$ is calculated as

$$đQ = \mathcal{E} I \, d\tau.$$

If $\mathcal{E}$ is measured in volts, I in amperes, and τ in seconds, the heat will be expressed in joules. The shape, size, and construction of the calorimeter, heating coils, thermometers, etc., depend on the nature of the material to be studied and the temperature range desired. It is impossible to describe one calorimeter that suffices for all purposes. In general, the measurement of any heat capacity is a research problem requiring all the ability of a trained physicist or physical chemist, the facilities of a good workshop, and the skill of an expert glass blower.

In modern calorimetry, particularly in the case of solids at low temperatures, the sample is suspended in a highly evacuated space by means of fine threads of nylon or some other poorly conducting material. A heating coil is wound around the sample, and a thermocouple or a resistance thermometer (platinum, carbon, or germanium, depending on the temperature range) is placed in a small hole drilled for that purpose. The connecting wires for the heater, for the current in the thermometer, and for the potential difference across the thermometer are made very thin so as not to allow much heat to be transferred between the sample and its surroundings. The temperature of the sample is measured as a function of the time; when plotted as in Fig. 4-6, this gives the line AB, marked "foreperiod." At the time corresponding to point B, a switch is closed and a current is established in the heater at the same moment that an electric stop clock is started. After a short interval of time $\Delta\tau$, the switch is opened and the stop clock is stopped. Then the temperature is again measured as a function of time and is plotted as the line DE, marked "afterperiod" in Fig. 4-6.

As a rule, no reading of temperature or time is attempted while the stop clock is on, that is, from B to D. A vertical line is drawn through the center C of the line BD, and both the foreperiod and the afterperiod lines are extrapolated to this vertical line, giving the points F and G, as shown. The molar heat capacity c at the temperature corresponding to point C is then given by

$$c = \frac{\mathcal{E} I \, \Delta\tau}{n \, \Delta\theta}.$$

Sometimes $\Delta\theta$ is made as small as 0.01 deg. Strictly speaking, the graph shown in Fig. 4-6 is not a graph of θ versus τ but of the resistance R' versus τ, and it is possible to have the entire R'_τ curve drawn automatically with the aid of a

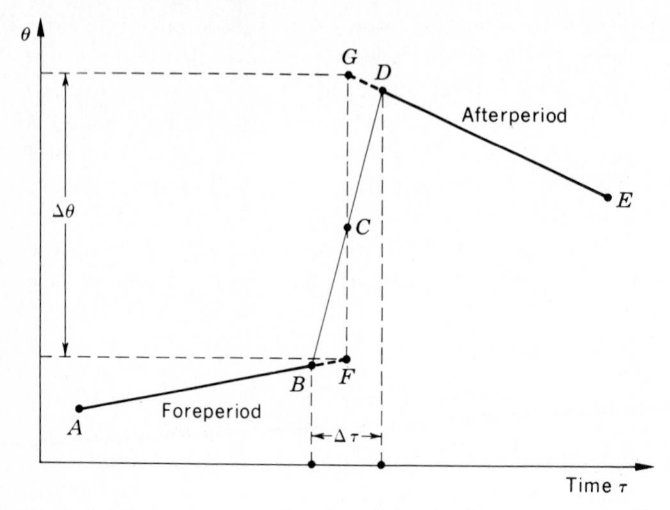

Figure 4-6 Temperature as a function of time in the measurement of heat capacity.

recording potentiometer. When many values of R' and $\Delta R'$ are read from recorder graphs, the corresponding θ's, $\Delta\theta$'s, and c's are obtained with the aid of a computer.

4-8 HEAT CAPACITY OF WATER; THE CALORIE

When the subject of calorimetry was first presented in the middle of the eighteenth century, measurements were confined to the temperature range between the freezing and boiling points of water. The unit of heat found most convenient was called the *calorie* and was defined as the amount of heat required to raise the temperature of 1 g of water 1 Celsius degree. To measure the amount of heat transferred between a system and some water, it was necessary merely to make two measurements: of the mass of water and of its temperature change. Later, as measurements became more precise and more elaborate corrections were made, it was discovered that the heat necessary to change 1 g of water from 0 to 1°C was different from the heat necessary to go from, say, 30 to 31°C. The calorie was then defined to be the heat necessary to go from 14.5 to 15.5°C (the 15-deg calorie).

The amount of work that had to be dissipated in water—either by maintaining a current in a resistor immersed in water or by churning the water in an irregular manner—per unit mass of water in going from 14.5 to 15.5°C was called the *mechanical equivalent of heat*, which was measured to be 4.1860 J/cal. In the 1920s it was recognized that the measurement of this mechanical equivalent of heat was really a measurement of the specific heat of water, with the joule as the unit of heat. Since heat is a form of energy and the joule is a universal unit of energy, the calorie seemed superfluous. Therefore, in a large and important

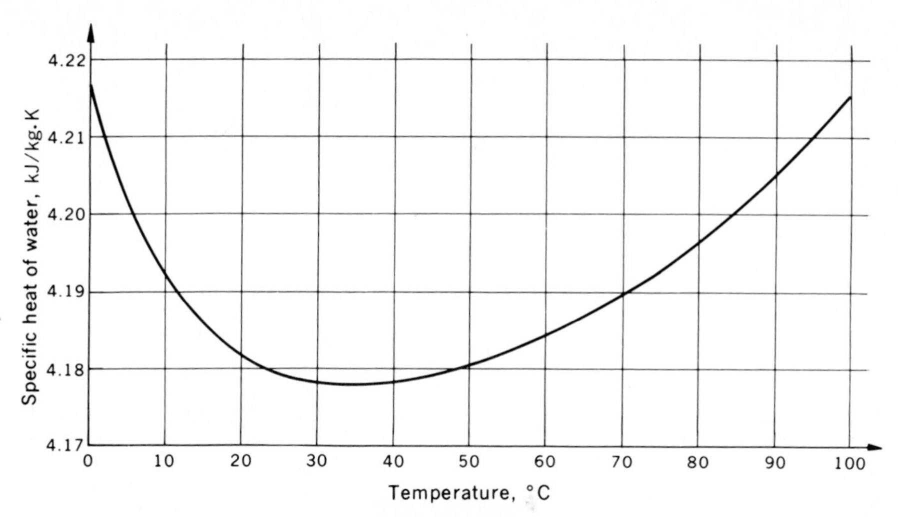

Figure 4-7 Specific heat of water.

collection of tables of physical constants published at that time, which were called the International Critical Tables, all thermal quantities such as specific and molar heat capacities were expressed in terms of joules. This move proved to be somewhat premature. Physicists and chemists preferred to think in terms of calories, and even when the electrical method of calorimetry was used and thermal quantities were actually measured in joules, the measurements were then converted to calories.

Among physicists and chemists today, the calorie is being dropped, and all thermal quantities—at least at very low and very high temperatures, where water is not used but electrical methods are employed exclusively—are expressed in joules. There is no mechanical equivalent of heat, but instead there is the specific heat of water, expressed in kJ/kg · K, whose temperature variation in the range of 0 to 100°C is shown in Fig. 4-7.

4-9 EQUATIONS FOR A HYDROSTATIC SYSTEM

The mathematical formulation of the first law for a hydrostatic system is

$$đQ = dU + P\,dV,$$

where U is a function of any two of P, V, and θ. Choosing θ and V, we have

$$dU = \left(\frac{\partial U}{\partial \theta}\right)_V d\theta + \left(\frac{\partial U}{\partial V}\right)_\theta dV.$$

Therefore, the first law becomes

$$đQ = \left(\frac{\partial U}{\partial \theta}\right)_V d\theta + \left[\left(\frac{\partial U}{\partial V}\right)_\theta + P\right] dV. \tag{4-11}$$

Dividing by $d\theta$, we get

$$\frac{đQ}{d\theta} = \left(\frac{\partial U}{\partial \theta}\right)_V + \left[\left(\frac{\partial U}{\partial V}\right)_\theta + P\right]_\theta \frac{dV}{d\theta}. \tag{4-12}$$

This equation is true for *any* process involving any temperature change $d\theta$ and any volume change dV.

(*a*) If V is constant, $dV = 0$, and

$$\left(\frac{đQ}{d\theta}\right)_V = \left(\frac{\partial U}{\partial \theta}\right)_V.$$

But the ratio on the left, by definition, is the heat capacity at constant volume C_V; therefore,

$$\boxed{C_V = \left(\frac{\partial U}{\partial \theta}\right)_V.} \tag{4-13}$$

If U is calculated mathematically by making special assumptions about the atoms of a particular material, one of the first methods of checking these assumptions is to differentiate U with respect to θ at constant V and to compare the resulting quantity with the experimentally measured value of C_V.

(*b*) If P is constant, Eq. (4-12) becomes

$$\left(\frac{đQ}{d\theta}\right)_P = \left(\frac{\partial U}{\partial \theta}\right)_V + \left[\left(\frac{\partial U}{\partial V}\right)_\theta + P\right]\left(\frac{\partial V}{\partial \theta}\right)_P.$$

But, by definition, $(đQ/d\theta)_P = C_P$ and also $(\partial V/\partial \theta)_P = V\beta$. Hence,

$$C_P = C_V + \left[\left(\frac{\partial U}{\partial V}\right)_\theta + P\right] V\beta,$$

or

$$\left(\frac{\partial U}{\partial V}\right)_\theta = \frac{C_P - C_V}{V\beta} - P. \tag{4-14}$$

Although this equation is not important in its present form, it is a good example of an equation that relates a quantity $(\partial U/\partial V)_\theta$, which is ordinarily not measured, with quantities such as C_P, C_V, and β, which may be measured.

4-10 QUASI-STATIC FLOW OF HEAT; HEAT RESERVOIR

It was shown in Chap. 3 that a process caused by a finite unbalanced force is attended by phenomena such as turbulence and acceleration which cannot be handled by means of thermodynamic coordinates that refer to the system as a whole. A similar situation exists when there is a finite difference between the temperature of a system and that of its surroundings. A nonuniform temperature distribution is set up in the system, and the calculation of this distribution and its variation with time is in most cases an elaborate mathematical problem. During a quasi-static process, however, the difference between the temperature of a system and that of its surroundings is infinitesimal. As a result, the temperature of the system is at any moment uniform throughout, and its changes are infinitely slow. The flow of heat is also infinitely slow and may be calculated in a simple manner in terms of thermodynamic coordinates referring to the system as a whole.

Suppose that a system is in good thermal contact with a body of extremely large mass and that a quasi-static process is performed. A finite amount of heat flow during this process will not bring about an appreciable change in the temperature of the surrounding body if the mass is large enough. For example, a cake of ice of ordinary size, if thrown into the ocean, will not produce a drop in temperature of the ocean. No ordinary flow of heat into the outside air will produce a rise of temperature of the air. The ocean and the outside air are approximate examples of an ideal body called a *heat reservoir*. *A heat reservoir is a body of such a large mass that it may absorb or reject an unlimited quantity of heat without suffering an appreciable change in temperature or in any other thermodynamic coordinate.* It is not to be understood that there is no change in the thermodynamic coordinates of a heat reservoir when a finite amount of heat flows in or out. There is a change, but an extremely small one, too small to be measured. In other words: in any small unit of mass, a change in a physical property is infinitesimal, but there is an infinite number of such units of mass in the heat reservoir.

Any quasi-static process of a system in contact with a heat reservoir is bound to be isothermal. To describe a quasi-static flow of heat involving a change of temperature, one could conceive of a system placed in contact successively with a series of reservoirs. Thus, if we imagine a series of reservoirs ranging in temperature from θ_i to θ_f placed successively in contact with a system at constant pressure of heat capacity C_P, in such a way that the difference in temperature between the system and the reservoir with which it is in contact is infinitesimal, the flow of heat will be quasi-static and can be calculated as follows: by definition,

$$C_P = \left(\frac{đQ}{d\theta}\right)_P,$$

and therefore

$$Q_P = \int_{\theta_i}^{\theta_f} C_P \, d\theta.$$

For example, the heat absorbed by water from a series of reservoirs varying in temperature from θ_i to θ_f during a quasi-static isobaric process is

$$Q_P = \int_{\theta_i}^{\theta_f} C_P \, d\theta.$$

If the heat capacity C_P is assumed to remain practically constant,

$$Q_P = C_P(\theta_f - \theta_i).$$

For a quasi-static isochoric process

$$Q_V = \int_{\theta_i}^{\theta_f} C_V \, d\theta.$$

Similar considerations hold for other systems and other quasi-static processes.

4-11 HEAT CONDUCTION

When two parts of a material substance are maintained at different temperatures and the temperature of each small volume element of the intervening substance is measured, experiment shows a continuous distribution of temperature. The transport of energy between neighboring volume elements by virtue of the temperature difference between them is known as *heat conduction*. The fundamental law of heat conduction is a generalization of the results of experiments on the linear flow of heat through a slab perpendicular to the faces. A piece of material is made in the form of a slab of thickness Δx and of area A. One face is maintained at the temperature θ and the other at $\theta + \Delta\theta$. The heat Q that flows perpendicular to the faces for a time τ is measured. The experiment is repeated with other slabs of the same material but with different values of Δx and A. The results of such experiments show that, for a given value of $\Delta\theta$, Q is proportional to the time and to the area. Also, for a given time and area, Q is proportional to the ratio $\Delta\theta/\Delta x$, provided that both $\Delta\theta$ and Δx are small. These results may be written

$$\frac{Q}{\tau} \propto A \frac{\Delta\theta}{\Delta x},$$

which is only approximately true when $\Delta\theta$ and Δx are finite but which is rigorously true in the limit as $\Delta\theta$ and Δx approach zero. If we generalize this result for an infinitesimal slab of thickness dx, across which there is a temperature difference $d\theta$, and introduce a constant of proportionality K, the fundamental law of heat conduction becomes

$$\boxed{\dot{Q} = \frac{đQ}{d\tau} = -KA\frac{d\theta}{dx}.} \tag{4-15}$$

The derivative $d\theta/dx$ is called the *temperature gradient*. The minus sign is introduced in order that the positive direction of the flow of heat should coincide with the positive direction of x. For heat to flow in the positive direction of x, this must be the direction in which θ decreases. K is called the *thermal conductivity*. A substance with a large thermal conductivity is known as a *thermal conductor*, and one with a small value of K as a *thermal insulator*. It will be shown in the next article that the numerical value of K depends upon a number of factors, one of which is the temperature. Volume elements of a conducting material may therefore differ in thermal conductivity. If the temperature difference between parts of a substance is small, however, K can be considered practically constant throughout the substance. This simplification is usually made in practical problems.

4-12 THERMAL CONDUCTIVITY

When the substance to be investigated is a metal, it is made into the form of a bar, and one end is heated electrically while the other end is cooled with a stream of water. The surface of the bar is thermally insulated, and the heat lost through the insulation is calculated by subtracting the rate at which heat enters the water from the rate at which electrical energy is supplied. In the case of most metals, the heat lost from the surface is very small in comparison with that which flows through the bar. The temperature is measured with suitable thermocouples at two places a distance L apart, and the equation

$$K = \frac{L}{A(\theta_1 - \theta_2)} \dot{Q}$$

is used to determine the average thermal conductivity within the given temperature range. If $\theta_1 - \theta_2$ is small, K is practically equal to the thermal conductivity at the mean temperature. K has the units of W/m · K.

When the substance to be investigated is a nonmetal, it is made into the form of a thin disk or plate, and the same general method is used. The substance is contained between two copper blocks, one of which is heated electrically and the other cooled by running water. The thermal contact between the copper blocks and the substance is improved by smearing them with glycerin. In most cases, the rate at which heat is supplied is almost equal to the rate at which heat enters the water, showing that there is little loss of heat through the edges.

Experiments show that the thermal conductivity of a metal is extraordinarily sensitive to impurities. The slightest trace of arsenic in copper reduces the thermal conductivity by a factor of 3. A change in internal structure brought about by continued heating or a large increase in pressure also affects the value of K. No appreciable change in the K of solids and liquids takes place, however, under moderate changes of pressure. Liquefaction always produces a decrease in the thermal conductivity, and the thermal conductivity of a liquid usually increases as the temperature is raised. Nonmetallic solids behave in a manner similar to that of liquids. At ordinary temperatures these are poor conductors of heat; in general, the thermal conductivity increases as the temperature is raised. In the low-temperature range, however, the behavior is quite different, as shown in Fig. 4-8, where it may be seen that the thermal conductivity of sapphire rises to a

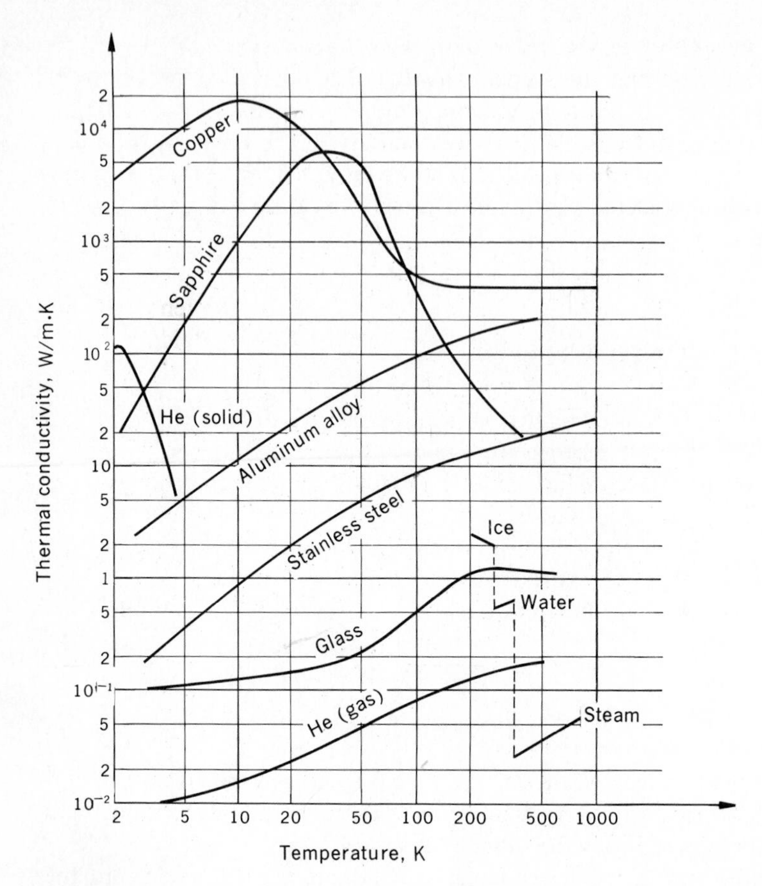

Figure 4-8 Typical curves showing temperature dependence of thermal conductivity. *(R. L. Powell, AIP Handbook, McGraw-Hill, 1972.)*

maximum of over 5000 W/m · K at 30 K (four times the conductivity of silver at room temperature). The thermal conductivity of some metals remains quite constant over a wide temperature range. Thus, silver, copper, and gold have thermal conductivities that remain practically constant in the temperature range from 100 to 1000 K. As a general rule, the thermal conductivity of metals increases as the temperature is lowered, until a maximum is reached. Further reduction of temperature causes a decrease toward zero, as shown in the case of copper in Fig. 4-8.

Gases are by far the poorest heat conductors. At pressures above a certain value, depending on the nature of the gas and the dimensions of the containing vessel, the thermal conductivity is independent of the pressure. Under the usual laboratory conditions, this limiting pressure is considerably below atmospheric pressure. The thermal conductivity of a gas always increases as the temperature is raised, as is evident in Fig. 4-8.

4-13 HEAT CONVECTION

A current of liquid or gas that absorbs heat at one place and then moves to another place, where it mixes with a cooler portion of the fluid and rejects heat, is called a *convection current*. If the motion of the fluid is caused by a difference in density that accompanies a temperature difference, the phenomenon is called *natural convection*. If the fluid is made to move by the action of a pump or a fan, it is called *forced convection*.

Consider a fluid in contact with a flat or curved wall whose temperature is higher than that of the main body of the fluid. Although the fluid may be in motion, there is a relatively thin film of stagnant fluid next to the wall, the thickness of the film depending upon the character of the motion of the main body of fluid. The more turbulent the motion, the thinner the film. Heat is transferred from the wall to the fluid by a combination of conduction through the film and convection in the fluid. Neglecting the transfer of heat by radiation (which must be taken into account separately), we may define a convection coefficient h that includes the combined effect of conduction through the film and convection in the fluid. Thus,

$$\boxed{\dot{Q} = hA\,\Delta\theta,} \tag{4-16}$$

where $\dot{Q}$ is the rate of heat transfer by convection, A is the area of the wall, and $\Delta\theta$ or Δt is the temperature difference between the surface of the wall and the main body of the fluid. The fundamental problem of heat convection is to find the value of h that is appropriate to a particular piece of equipment.

Experiment shows that the convection coefficient depends on the following factors:

1. Whether the wall is flat or curved.
2. Whether the wall is horizontal or vertical.
3. Whether the fluid in contact with the wall is a gas or a liquid.
4. The density, viscosity, specific heat, and thermal conductivity of the fluid.
5. Whether the velocity of the fluid is small enough to give rise to laminar flow or large enough to cause turbulent flow.
6. Whether evaporation, condensation, or formation of scale takes place.

Since the physical properties of the fluid depend upon temperature and pressure, it is clear that the rigorous calculation of a convection coefficient appropriate to a given wall and fluid is an enormously complicated problem. It is only in recent years that solutions of the problem good enough for practical purposes have been achieved with the aid of dimensional analysis. Such analysis yields an expression for h containing the physical properties and velocity of the fluid and unknown constants and exponents. The constants and exponents are then evaluated by experiment.

4-14 THERMAL RADIATION; BLACKBODY

A substance may be stimulated to emit electromagnetic radiation in a number of ways:

1. An electric conductor carrying a high-frequency alternating current emits radio waves.
2. A hot solid or liquid emits thermal radiation.
3. A gas carrying an electric discharge may emit visible or ultraviolet radiation.
4. A metal plate bombarded by high-speed electrons emits X rays.
5. A substance whose atoms are radioactive may emit γ rays.
6. A substance exposed to radiation from an external source may emit fluorescent radiation.

All these radiations are electromagnetic waves, differing only in wavelength. We shall be concerned in this article only with thermal radiation, i.e., the radiation emitted by a solid, liquid, or gas by virtue of its temperature. When thermal radiation is dispersed by a suitable prism, a continuous spectrum is obtained. The distribution of energy among the various wavelengths is such that, at temperatures below about 500°C, most of the energy is associated with infrared waves; at higher temperatures some visible radiation is emitted. In general, the higher the temperature of a body, the greater the total energy emitted.

The loss of energy due to the emission of thermal radiation may be compensated in a variety of ways. The emitting body may be a source of energy itself, such as the sun; or there may be a constant supply of electrical energy from the outside, as in the case of the filament of an electric lamp. Energy may be supplied also by heat conduction or by the performance of work on the emitting body. In the absence of these sources of supply, the only other way in which a body may receive energy is by the absorption of radiation from surrounding bodies. In the case of a body that is surrounded by other bodies, the internal energy of the body will remain constant when the rate at which radiant energy is emitted is equal to that at which it is absorbed.

Experiment shows that the rate at which a body emits thermal radiation depends on the temperature and on the nature of the surface. The total radiant power emitted per unit area is called the *radiant exitance* of the body. For example, the radiant exitance of tungsten at 2177°C is 500 kW/m^2. When thermal radiation is incident upon a body equally from all directions, the radiation is said to be *isotropic*. Some of the radiation may be absorbed, some reflected, and some transmitted. In general, the fraction of the incident isotropic radiation of all wavelengths that is absorbed depends on the temperature and the nature of the surface of the absorbing body. This fraction is called the *absorptivity*. At 2477°C the absorptivity of tungsten is approximately 0.25. To summarize:

Radiant exitance = $\mathcal{R}$ = total radiant power emitted per unit area.

Absorptivity = α = fraction of the total energy of isotropic radiation that is absorbed.

There are some substances, such as lampblack, whose absorptivity is very nearly unity. For theoretical purposes it is useful to conceive of an ideal substance capable of absorbing all the thermal radiation falling on it. Such a substance is called a *blackbody*. If a blackbody is indicated by the subscript B, we have

$$\alpha_B = 1.$$

A very good experimental approximation to a blackbody is provided by a cavity for which the interior walls are maintained at a uniform temperature and which communicates with the outside by means of a hole having a diameter small in comparison with the dimensions of the cavity. Any radiation entering the hole is partly absorbed and partly diffusely reflected a large number of times at the interior walls, with only a negligible fraction eventually finding its way out of the hole. *This is true regardless of the materials of which the interior walls are composed.*

The radiation emitted by the interior walls is similarly absorbed and diffusely reflected a large number of times, so that the cavity is filled with isotropic radiation. Let us define as the *irradiance* within the cavity the radiant energy falling in unit time upon unit area of any surface within the cavity. Suppose a blackbody whose temperature is the same as that of the walls is introduced into the cavity. Then, denoting the irradiance† by H,

$$\text{Radiant power absorbed per unit area} = \alpha_B H = H,$$

and $$\text{Radiant power emitted per unit area} = \mathcal{R}_B.$$

Since the temperature of the blackbody remains constant, the rate at which the energy is absorbed must equal the rate at which it is emitted; whence

$$\boxed{H = \mathcal{R}_B,} \qquad (4\text{-}17)$$

or *the irradiance within a cavity whose walls are at the temperature θ is equal to the radiant exitance of a blackbody at the same temperature*. For this reason, the radiation within a cavity is called *blackbody radiation*. Such radiation is studied by allowing a small amount to escape from a small hole leading to the cavity. Since H is independent of the materials of which the interior walls are composed, it follows that *the radiant exitance of a blackbody is a function of the temperature only*.

† In books on electromagnetic radiation the radiant flux leaving a surface is the radiant exitance (symbol: M) and the radiant flux incident on a surface is the irradiance (symbol: E). We adopt the symbols $\mathcal{R}$ and H to avoid confusion with the magnetization and electric field intensity.

4-15 KIRCHHOFF'S LAW; RADIATED HEAT

The radiant emittance of a non-blackbody depends as much on the nature of the surface as on the temperature, according to a simple law that we may derive as follows: Suppose that a non-blackbody at the temperature θ, with radiant emittance $\mathcal{R}$ and absorptivity α, is introduced into a cavity whose interior walls are at the same temperature and where the irradiance is H. Then,

$$\text{Radiant power absorbed per unit area} = \alpha H,$$

and

$$\text{Radiant power emitted per unit area} = \mathcal{R}.$$

Since the non-blackbody is in equilibrium,

$$\mathcal{R} = \alpha H.$$

But, from Eq. (4-17), $H = \mathcal{R}_B$; hence,

$$\boxed{\mathcal{R} = \alpha \mathcal{R}_B,} \tag{4-18}$$

or *the radiant exitance of any body at any temperature is equal to a fraction of the radiant exitance of a blackbody at that temperature, this fraction being the absorptivity at that temperature.*

This equation, known as *Kirchhoff's law*, shows that the absorptivity of a body may be determined experimentally by measuring the radiant emittance of the body and dividing it by that of a blackbody at the same temperature. Values of the absorptivity of various surfaces, measured in this way, are given in Table 4-3. It should be emphasized that the tabulated values of absorptivity refer to the thermal radiation appropriate to the temperature listed in column 1. Thus, the absorptivity of ice is 0.97 not for visible radiation but for the long infrared waves associated with matter at 0°C.

It should be noticed that the word "heat" has not appeared as yet. If there is a temperature difference between a body and its surroundings, then in a given interval of time the body loses an amount of internal energy equal to the energy radiated minus the energy absorbed, whereas the surroundings gain an amount of internal energy equal to the energy absorbed minus the energy radiated. The gain of one equals the loss of the other. *The gain or loss of internal energy, equal to the difference between the energy of the thermal radiation which is absorbed and that which is radiated, is called heat.* This statement is in agreement with the original definition of heat, since a gain or loss of energy by radiation and absorption will take place *only if there is a difference in temperature* between a body and its surroundings. If the two temperatures are the same, there is no net gain or loss of internal energy of either the body or its surroundings, and there is therefore no transfer of heat.

Imagine a cavity whose interior walls are maintained at a constant temperature θ_W. Suppose that a non-blackbody at a temperature θ different from that of the walls is placed in the cavity. If the body is small compared with the size of

Table 4-3 Approximate absorptivities of various surfaces, as compiled by Hottel

(Values at intermediate temperatures may be obtained by linear interpolation)

Material	Temperature range, °C	Absorptivity
Polished metals:		
Aluminum	250– 600	0.039–0.057
Brass	250– 400	0.033–0.037
Chromium	50– 550	0.08 –0.26
Copper	100	0.018
Iron	150–1000	0.05 –0.37
Nickel	20– 350	0.045–0.087
Zinc	250– 350	0.045–0.053
Filaments:		
Molybdenum	750–2600	0.096–0.29
Platinum	30–1200	0.036–0.19
Tantalum	1300–3000	0.19 –0.31
Tungsten	30–3300	0.032–0.35
Other materials:		
Asbestos	40– 350	0.93 –0.95
Ice (wet)	0	0.97
Lampblack	20– 350	0.95
Rubber (gray)	25	0.86

the cavity, then the character of the radiation in the cavity will not be appreciably affected by its introduction. Let H, as before, denote the irradiance within the cavity, and $\mathcal{R}$ and α the radiant exitance and absorptivity, respectively, of the body. Then, as before,

$$\text{Radiant power absorbed per unit area} = \alpha H,$$

and

$$\text{Radiant power emitted per unit area} = \mathcal{R};$$

but now *these two rates are not equal.* The difference between them is the heat transferred by radiation per second per unit area. If $đQ$ is the heat transferred in time $d\tau$ to the whole body whose area is A, then

$$\dot{Q} = \frac{đQ}{d\tau} = A(\alpha H - \mathcal{R}), \tag{4-19}$$

where, it must be remembered, α and $\mathcal{R}$ refer to the temperature θ and H to the temperature θ_W. Now,

$$H = \mathcal{R}_B(\theta_W),$$

and

$$\mathcal{R} = \alpha \mathcal{R}_B(\theta).$$

Hence,

$$\dot{Q} = A\alpha[\mathcal{R}_B(\theta_W) - \mathcal{R}_B(\theta)], \tag{4-20}$$

or the rate at which heat is transferred by radiation is proportional to the difference between the radiant exitances of a blackbody at the two temperatures in question.

4-16 STEFAN-BOLTZMANN LAW

The first measurements of the heat transferred by radiation between a body and its surroundings were made by Tyndall. On the basis of these experiments, it was concluded by Stefan in 1879 that the heat radiated was proportional to the difference of the fourth powers of the absolute temperatures. This purely experimental result was later derived thermodynamically by Boltzmann, who showed that the radiant emittance of a blackbody at any temperature θ is equal to

$$\boxed{\mathscr{R}_B(\theta) = \sigma\theta^4.} \tag{4-21}$$

This law is now known as the *Stefan-Boltzmann law*, and σ is called the Stefan-Boltzmann constant.

Referring to Eq. (4-20), we have for the heat transferred by radiation between a body at the temperature θ and walls at θ_W,

$$\boxed{\dot{Q} = A\alpha\sigma(\theta_W^4 - \theta^4),} \tag{4-22}$$

where α refers to the temperature θ.

Two simple methods may be employed for the determination of the Stefan-Boltzmann constant:

1. *Nonequilibrium method.* A blackened silver disk is placed in the center of a large blackened copper hemisphere. The silver disk is covered and shielded from radiation until the copper hemisphere achieves the temperature of condensing steam; this temperature is measured with a thermocouple. Then the disk is uncovered, and its temperature is measured as a function of the time. From the resulting heating curve, the slope $d\theta/d\tau$ is obtained. Assuming the silver disk to be a blackbody and putting $đQ = C_P\, d\theta$, where C_P is the heat capacity at constant pressure, we have

$$\frac{C_P\, d\theta}{d\tau} = A\sigma(\theta_W^4 - \theta^4);$$

whence

$$\sigma = \frac{C_P}{A(\theta_W^4 - \theta^4)} \frac{d\theta}{d\tau}.$$

2. *Equilibrium method.* A hollow blackened copper sphere is provided with an electric heater and a thermocouple and is suspended inside a vessel whose walls are maintained at a constant temperature θ_W. Electrical energy is

supplied at a constant rate $\mathscr{E}i$ until the sphere achieves an equilibrium temperature θ at which the rate of supply of energy is equal to the rate of emission of radiation. Assuming the sphere to be a blackbody, we have at equilibrium

$$\mathscr{E}i = A\sigma(\theta^4 - \theta_W^4);$$

whence

$$\sigma = \frac{\mathscr{E}i}{4\pi r^2(\theta^4 - \theta_W^4)},$$

where r is the radius of the sphere. The best measurements, to date, have yielded the value

$$\boxed{\sigma = 56.703 \frac{\text{nW}}{\text{m}^2 \cdot \text{K}^4}.} \tag{4-23}$$

PROBLEMS

4-1 A gas contained in a cylinder surrounded by a thick layer of felt is quickly compressed, the temperature rising several hundred degrees. Has there been a transfer of heat? Has the "heat of the gas" been increased?

4-2 A combustion experiment is performed by burning a mixture of fuel and oxygen in a constant-volume "bomb" surrounded by a water bath. During the experiment the temperature of the water is observed to rise. If we regard the mixture of fuel and oxygen as the system:

(*a*) Has heat been transferred?
(*b*) Has work been done?
(*c*) What is the sign of ΔU?

4-3 A liquid is irregularly stirred in a well-insulated container and thereby undergoes a rise in temperature. If we regard the liquid as the system:

(*a*) Has heat been transferred?
(*b*) Has work been done?
(*c*) What is the sign of ΔU?

4-4 The amount of water in a lake may be increased by action of underground springs, by inflow from a river, and by rain. It may be decreased by various outflows and by evaporation.

(*a*) Is it correct to ask: How much rain is there in the lake?
(*b*) Would it be preferable or sensible to ask: How much water in the lake is due to rain?
(*c*) What concept is analogous to "rain in the lake"?

4-5 A vessel with rigid walls and covered with asbestos is divided into two parts by a partition. One part contains a gas, and the other is evacuated. If the partition is suddenly broken, show that the initial and final internal energies of the gas are equal.

4-6 A gas is enclosed within a cylinder-piston combination. Embedded in the gas is a junction of two dissimilar metals (a *thermojunction*) whose connecting wires pass through the walls of the cylinder and lead to a reversing switch and an electric generator rotated by means of a descending weight. As the weight descends, the current generated may be caused to exist in the thermojunction in either direction. Owing to the *Peltier effect*, the thermojunction undergoes a rise of temperature when the current is in one direction and a drop when in the opposite direction. The entire system is adiabatically shielded so that *all work interactions are adiabatic*. Assume that θ and U refer to the entire system, composed of gas plus thermojunction.

(*a*) Draw a schematic diagram of the apparatus.

(*b*) What is the result of allowing the piston to go out, with no current in the thermojunction? What is the sign of ΔU?

(*c*) Keeping the piston stationary, how could one produce a rise of temperature? What is the sign of ΔU?

(*d*) If the piston is kept stationary, is it possible to produce a drop in temperature? If so, what is the sign of ΔU?

(*e*) How could one produce an adiabatic, isothermal process?

4-7 When an electric current is maintained in an electrolytic cell of acidulated water and 1 mol of water is electrolyzed into hydrogen and oxygen, 2 faradays of electricity is transferred through a seat of emf $\mathcal{E}$ (1 faraday = 96,500 *C*). The energy change of the system is +286,500 J, and 50,000 J of heat is absorbed. What is $\mathcal{E}$?

4-8 A cylindrical tube with rigid walls and covered with asbestos is divided into two parts by a rigid insulating wall with a small hole in it. A frictionless insulating piston is held against the perforated partition, thus preventing the gas that is on the other side from seeping through the hole. The gas is maintained at a pressure P_i by another frictionless insulating piston. Imagine both pistons to move simultaneously in such a way that, as the gas streams through the hole, the pressure remains at the constant value P_i on one side of the dividing wall and at a constant lower value P_f on the other side, until all the gas is forced through the hole. Prove that

$$U_i + P_i V_i = U_f + P_f V_f.$$

4-9 An evacuated chamber with nonconducting walls is connected through a valve to the atmosphere, where the pressure is P_0. The valve is opened, and air flows into the chamber until the pressure within the chamber is P_0. Prove that $u_0 + P_0 v_0 = u_f$, where u_0 and v_0 are the molar energy and molar volume of the air at the temperature and pressure of the atmosphere and u_f is the molar energy of the air in the chamber. (*Hint:* Connect to the chamber a cylinder equipped with a frictionless nonleaking piston. Suppose the cylinder to contain exactly the amount of atmospheric air that will enter the chamber when the valve is opened. As soon as the first small quantity of air enters the chamber, the pressure in the cylinder is reduced a small amount below atmospheric pressure, and the outside air forces the piston in.)

4-10 A bomb of volume V_B contains n moles of gas at high pressure. Connected to the bomb is a capillary tube through which the gas may slowly leak out in the atmosphere, where the pressure is P_0. Surrounding the bomb and capillary is a water bath, in which is immersed an electrical resistor. The gas is allowed to leak slowly through the capillary into the atmosphere while electrical energy is dissipated in the resistor at such a rate that the temperature of the gas, the bomb, the capillary, and the water is kept equal to that of the outside air. Show that, after as much gas as possible has leaked out during time τ, the change of internal energy is

$$\Delta U = \mathcal{E} i \tau + P_0(n v_0 - V_B),$$

where v_0 is the molar volume of the gas at atmospheric pressure, $\mathcal{E}$ is the pd across the resistor, and i is the current in the resistor.

4-11 A thick-walled insulated metal chamber contains n_i mol of helium at high pressure P_i. It is connected through a valve with a large, almost empty gasholder in which the pressure is maintained at a constant value P', very nearly atmospheric. The valve is opened slightly, and the helium flows slowly and adiabatically into the gas holder until the pressure on the two sides of the valve is equalized. Prove that

$$\frac{n_f}{n_i} = \frac{h' - u_i}{h' - u_f},$$

where n_f = number of moles of helium left in the chamber,
u_i = initial molar energy of helium in the chamber,
u_f = final molar energy of helium in the chamber, and
$h' = u' + P'v'$ (where u' = molar energy of helium in the gasholder; v' = molar volume of helium in the gasholder).

4-12 The molar heat capacity at constant pressure of a gas varies with the temperature according to the equation

$$c_P = a + b\theta - \frac{c}{\theta^2},$$

where a, b, and c are constants. How much heat is transferred during an isobaric process in which n mol of gas undergoes a temperature rise from θ_i to θ_f?

4-13 The molar heat capacity of a metal at low temperature varies with the temperature according to the equation

$$c = \frac{a}{\Theta^3}\theta^3 + b\theta,$$

where a, Θ, and b are constants. How much heat per mole is transferred during a process in which the temperature changes from 0.01Θ to 0.02Θ?

4-14 Regarding the internal energy of a hydrostatic system to be a function of θ and P, derive the equations:

(a) $$đQ = \left[\left(\frac{\partial U}{\partial \theta}\right)_P + P\left(\frac{\partial V}{\partial \theta}\right)_P\right] d\theta + \left[\left(\frac{\partial U}{\partial P}\right)_\theta + P\left(\frac{\partial V}{\partial P}\right)_\theta\right] dP.$$

(b) $$\left(\frac{\partial U}{\partial \theta}\right)_P = C_P - PV\beta.$$

(c) $$\left(\frac{\partial U}{\partial P}\right)_\theta = PV\kappa - (C_P - C_V)\frac{\kappa}{\beta}.$$

4-15 Taking U to be a function of P and V, derive the following equations:

(a) $$đQ = \left(\frac{\partial U}{\partial P}\right)_V dP + \left[\left(\frac{\partial U}{\partial V}\right)_P + P\right] dV.$$

(b) $$\left(\frac{\partial U}{\partial P}\right)_V = \frac{C_V \kappa}{\beta}.$$

(c) $$\left(\frac{\partial U}{\partial V}\right)_P = \frac{C_P}{V\beta} - P.$$

4-16 Derive the equations listed in the accompanying table.

System	Heat capacity at constant extensive variable	Heat capacity at constant intensive variable
Stretched wire	$C_L = \left(\frac{\partial U}{\partial \theta}\right)_L$	$C_{\mathcal{J}} = \left(\frac{\partial U}{\partial \theta}\right)_{\mathcal{J}} - \mathcal{J}L\alpha$
Paramagnetic solid obeying Curie's equation	$C_M = \left(\frac{\partial U}{\partial \theta}\right)_M$	$C_{\mathcal{H}} = \left(\frac{\partial U}{\partial \theta}\right)_{\mathcal{H}} + \frac{M^2}{C_C}$

4-17 One mole of a gas obeys the equation of state

$$\left(P + \frac{a}{v^2}\right)(v - b) = R\theta,$$

where v is the molar volume, and its molar internal energy is given by

$$u = c\theta - \frac{a}{v},$$

where a, b, c, and R are constants. Calculate molar heat capacities c_V and c_P.

4-18 The equation of state of a monatomic solid is

$$Pv + f(V) = \Gamma(u - u_0),$$

where v is the molar volume, and Γ and u_0 are constants. Prove that

$$\Gamma = \frac{\beta v}{c_V \kappa},$$

where κ is the isothermal compressibility. This relation, first derived by Grüneisen, plays a role in the theory of the solid state.

4-19 In the case of a paramagnetic gas:

(*a*) Derive the equation

$$đQ = \left(\frac{\partial U}{\partial \theta}\right)_{V,M} d\theta + \left[\left(\frac{\partial U}{\partial V}\right)_{M,\theta} + P\right] dV + \left[\left(\frac{\partial U}{\partial M}\right)_{\theta,V} - \mu_0 \mathscr{H}\right] dM.$$

(*b*) Derive expressions for $C_{V,M}$, $C_{V,\mathscr{H}}$, $C_{P,M}$, and $C_{P,\mathscr{H}}$.

4-20 Suppose that heat conduction occurs at a constant rate of $\dot{Q}$ through the wall of a hollow cylinder with an inner radius r_1 at temperature θ_1 and an outer radius r_2 at temperature θ_2. Show that for a cylinder of length L and constant thermal conductivity K, the temperature difference between the two surfaces of the wall is given by

$$\theta_1 - \theta_2 = \frac{\dot{Q}}{2\pi LK} \ln \frac{r_2}{r_1}.$$

4-21 Heat flows radially outward through a cylindrical insulator of outside radius r_2 surrounding a steam pipe of outside radius r_1. The temperature of the inner surface of the insulator is θ_1, and that of the outer surface is θ_2. At what radial distance from the center of the pipe is the temperature exactly halfway between θ_1 and θ_2?

4-22 Suppose that heat conduction occurs at a constant rate of $\dot{Q}$ in a hollow sphere with an inner radius r_1 at temperature θ_1 and an outer radius r_2 at temperature θ_2. Show that for constant thermal conductivity K, the temperature difference between the two surfaces is given by

$$\theta_1 - \theta_2 = \frac{\dot{Q}}{4\pi K}\left(\frac{1}{r_1} - \frac{1}{r_2}\right).$$

4-23 Two thin concentric spherical shells of radius 0.05 and 0.15 m, respectively, have their annular cavity filled with charcoal. When energy is supplied at the steady rate of 10.8 W to a heater at the center, a temperature difference of 50.0°C is set up between the spheres. Find the thermal conductivity of charcoal.

4-24 A wall, maintained at a constant temperature t_W, is coated with a layer of insulating material of thickness x and of thermal conductivity K. The outside of the insulation is in contact with the air at temperature t_A. Heat is transferred by conduction through the insulation and by natural convection through the air.

(*a*) Show that, in the steady state,

$$\frac{\dot{Q}}{A} = \bar{U}(t_W - t_A),$$

where $\bar{U}$, the *overall coefficient of heat-transfer*, is given by

$$\frac{1}{\bar{U}} = \frac{x}{K} + \frac{1}{h}.$$

(*b*) How do you determine t, the temperature of the outer surface of the insulation?

4-25 The air above the surface of a freshwater lake is at a temperature θ_A, while the water is at its freezing point θ_i (with $\theta_A < \theta_i$). After a time τ has elapsed, ice of thickness y has formed. Assuming that the heat which is liberated when the water freezes flows up through the ice by conduction and thence into the air by natural convection, prove that

$$\frac{y}{h} + \frac{y^2}{2K} = \frac{\theta_i - \theta_A}{\rho l}\tau,$$

where h is the convection coefficient per unit area and is assumed constant while the ice forms, K is the thermal conductivity of ice, l is the heat of fusion of ice, and ρ is the density of ice. (*Hint:* The temperature θ of the upper surface of the ice is variable. Assume the ice to have a thickness y, and imagine an infinitesimal thickness dy to form in time $d\tau$.)

4-26 A solid cylindrical copper rod 0.10 m long has one end maintained at a temperature of 20.00 K. The other end is blackened and exposed to thermal radiation from the body at 300 K, with no energy lost or gained elsewhere. When equilibrium is reached, what is the temperature difference between the two ends? (*Note:* Refer to Fig. 4-8.)

4-27 A cylindrical metal can blackened on the outside, 0.10 m high and 0.05 m in diameter, contains liquid helium at its normal boiling point of 4.2 K, at which its heat of vaporization is 21 kJ/kg. Completely surrounding the helium can are walls maintained at the temperature of liquid nitrogen (78 K), and the intervening space is evacuated. How much helium is lost per hour?

4-28 The operating temperature of a tungsten filament in an incandescent lamp is 2460 K, and its absorptivity is 0.35. Find the surface area of the filament of a 100-W lamp.

4-29 A copper wire of length 1.302 m and diameter 3.26×10^{-4} m is blackened and placed along the axis of an evacuated glass tube. The wire is connected to a battery, a rheostat, an ammeter, and a voltmeter, and the current is increased until, at the moment the wire is about to melt, the ammeter reads 12.8 A and the voltmeter 20.2 V. Assuming that all the energy supplied was radiated and that the radiation of the glass tube is negligible, calculate the melting temperature of copper.

4-30 The *solar constant* is the energy falling per unit of time on a unit area of a surface placed at right angles to a sunbeam just outside the earth's atmosphere. Measurements by Abbot have yielded the value 1.35 kW/m^2. The area of a sphere with a radius of 93,000,000 mi is 2.806×10^{23} m^2, and the surface area of the sun is 6.07×10^{18} m^2. Assuming the sun to be a blackbody, calculate its surface temperature.

4-31 (*a*) A small body with temperature θ and absorptivity α is placed in a large evacuated cavity whose interior walls are at a temperature θ_W. When $\theta_W - \theta$ is small, show that the rate of heat transfer by radiation is

$$\dot{Q} = 4\theta_W^3 A\alpha\sigma(\theta_W - \theta).$$

(*b*) If the body remains at constant pressure, show that the time for the temperature of the body to change from θ_1 to θ_2 is given by

$$\tau = \frac{C_P}{4\theta_W^3 A\alpha\sigma} \ln \frac{\theta_W - \theta_1}{\theta_W - \theta_2}.$$

(*c*) Two small blackened spheres of identical size, one of copper, the other of aluminum, are suspended by silk threads within a large hole in a block of melting ice. It is found that it takes 10 min for the temperature of the aluminum to drop from 3 to 1°C, and 14.2 min for the copper to undergo the same temperature change. What is the ratio of specific heats of aluminum and copper? (The densities of Al and Cu are 2.7×10^3 and 8.9×10^3 kg/m^3, respectively.)

4-32 A blackened solid copper sphere with a radius of 0.02 m is placed in an evacuated enclosure whose walls are kept at 100°C. In what time does its temperature change from 103 to 102°C? ($c_P = 3.81$ kJ/kg · K; $\rho = 8.93 \times 10^3$ kg/m^3.)

CHAPTER

FIVE

IDEAL GASES

5-1 EQUATION OF STATE OF A GAS

It was emphasized in Chap. 1 that a gas is the best-behaved thermometric substance because of the fact that the ratio of the pressure P of a gas at any temperature to the pressure P_{TP} of the same gas at the triple point, as both P and P_{TP} approach zero, approaches a value independent of the nature of the gas. The limiting value of this ratio, multiplied by 273.16 K, was defined to be the ideal-gas temperature θ of the system at whose temperature the gas exerts the pressure P. The reason for this regular behavior may be found by investigating the way in which the product PV of a gas depends on the density or, if the mass is constant, on the reciprocal of the volume.

Suppose that the pressure P and the volume V of n moles of gas held at any constant temperature are measured over a wide range of values of the pressure, and the product Pv, where $v = V/n$, is plotted as a function of $1/v$. Nowadays, such measurements are made at many bureaus of standards and universities. The relation between Pv and $1/v$ may be expressed by means of a power series (or *virial expansion*) of the form

$$Pv = A\left(1 + \frac{B}{v} + \frac{C}{v^2} + \frac{D}{v^3} + \cdots\right), \tag{5-1}$$

where A, B, C, etc., are called *virial coefficients* (A being the first virial coefficient, B the second, etc.) and depend on the temperature and on the nature of the gas. In the pressure range from 0 to about 40 standard atmospheres, the relation between Pv and $1/v$ is practically linear, so that only the first two terms in the

Table 5-1 Virial coefficients for nitrogen

T, K	B, 10^{-3} m³/kmol	C, 10^{-4} m⁶/kmol²	D, 10^{-5} m⁹/kmol³	E, 10^{-6} m¹²/kmol⁴	F, 10^{-7} m¹⁵/kmol⁵
80	−250.80	210	−2000		
90	−200.50	135	−1000		
100	−162.10	85	−600		
110	−131.80	65	−200		
120	−114.62	48	−27		
150	−71.16	22	13	−12.3	4.1
200	−34.33	12	14	−11.8	3.6
273	−9.50	8.2	16	−7.5	−1.6

expansion are significant. In general, the greater the pressure range, the larger the number of terms in the virial expansion.

The virial coefficients play an important role not only in practical thermodynamics but also in theoretical physics, where they are related to molecular properties. Except at very low temperatures, the virial coefficients are quite small, as shown in Table 5-1, where the virial coefficients are given for nitrogen in the temperature range 80 to 273 K and in the pressure range from 0 to 200 standard atmospheres. These data, obtained by Friedman, White, and Johnston of the Ohio State Cryogenic Laboratory, are listed in the *American Institute of Physics Handbook* (McGraw-Hill Book Company, New York, 1972)—hereafter referred to simply as *AIP Handbook*.

The remarkable property of gases that makes them so valuable in thermometry is displayed in Fig. 5-1, where the product Pv is plotted against P for four different gases, all at the temperature of boiling water in the top graph, all at the triple point of water in the next lower graph, and all at the temperature of solid CO_2 in the lowest. In each case, it is seen that as the pressure approaches zero the product Pv approaches the same value for all gases at the same temperature. As the pressure of a constant mass of gas approaches zero, the volume approaches infinity, and according to the virial expansion [Eq. (5-1)], the product Pv approaches the first virial coefficient A. Thus,

$$\lim_{P \to 0} (Pv) = A = \begin{Bmatrix} \text{function of temperature only,} \\ \text{independent of gas.} \end{Bmatrix} \tag{5-2}$$

The ideal-gas temperature θ is defined as

$$\theta = 273.16 \text{ K} \lim \frac{P}{P_{TP}} \qquad (\text{const. } V \text{ and } n),$$

so that

$$\theta = 273.16 \text{ K} \lim \frac{PV/n}{P_{TP}V/n} = 273.16 \text{ K} \frac{\lim (Pv)}{\lim (Pv)_{TP}},$$

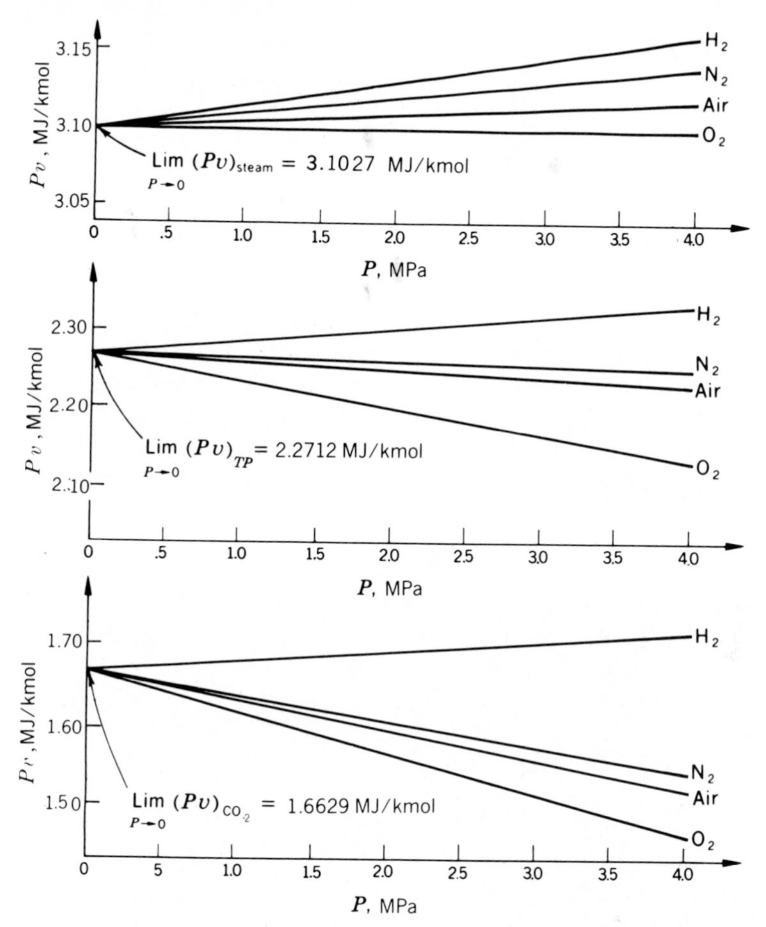

Figure 5-1 Fundamental property of gases is that $\lim_{P \to 0} (Pv)_\theta$ is independent of the nature of the gas and depends only on θ.

and

$$\lim (Pv) = \left[\frac{\lim (Pv)_{TP}}{273.16 \text{ K}}\right] \theta.$$

The bracketed term is called the molar *universal gas constant* and is denoted by R. Thus,

$$R = \frac{\lim (Pv)_{TP}}{273.16 \text{ K}}. \tag{5-3}$$

In 1972 Batuecas determined $\lim (Pv)_{0^\circ\text{C}}$ for oxygen to be 22.4132 liter · atm/mol (2.27102 kJ/mol). Hence, the gas constant $R = 8.31441$ J/mol · K by the method of limiting density. Another independent method of determining the gas constant will be explained in Art. 5-8.

Finally, substituting for v its value V/n, we may write the equation of state of a gas in the limit of low pressures in the form

$$\boxed{\lim (PV) = nR\theta.} \tag{5-4}$$

Since $\lim (PV) = A = R\theta$, the virial expansion may therefore be written

$$\frac{Pv}{R\theta} = 1 + \frac{B}{v} + \frac{C}{v^2} + \cdots.$$

The ratio $Pv/R\theta$ is called the *compressibility factor* and is denoted by Z. It is tabulated for many values of temperature and pressure in several reports and circulars published by the National Bureau of Standards. A condensed set of tables of Z values for the important gases is given in the *AIP Handbook*.

The general behavior of the compressibility factor may be displayed graph-

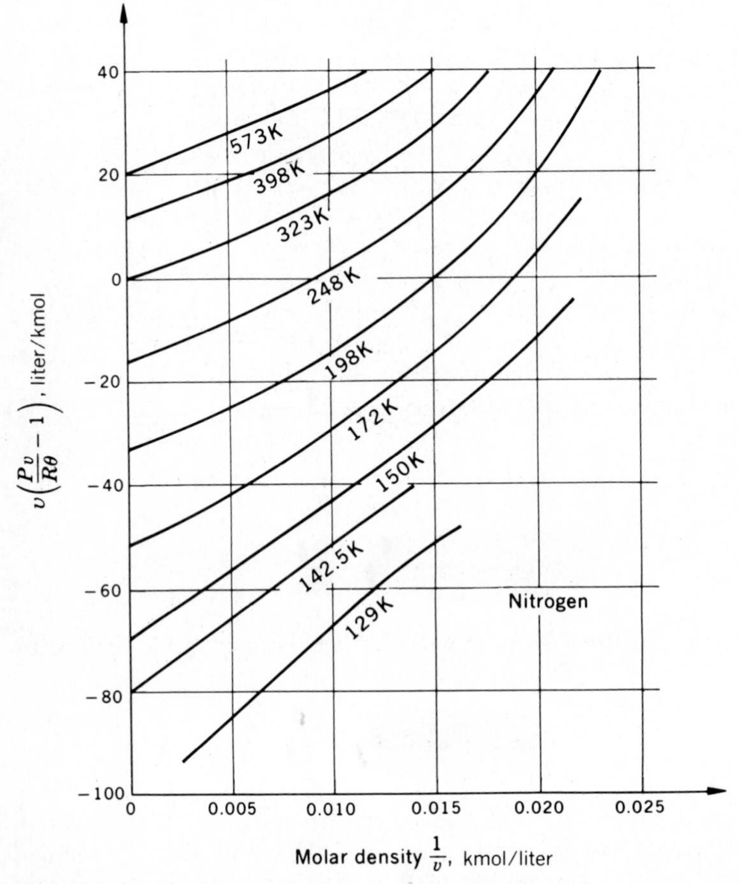

Figure 5-2 Graphical representation of the virial equation for nitrogen. *(Adapted from U.S. National Bureau of Standards, Circular 564.)*

ically by plotting $v(Pv/R\theta - 1)$ against $1/v$, since

$$v\left(\frac{Pv}{R\theta} - 1\right) = B + \frac{C}{v} + \cdots.$$

In the region of small values of $1/v$, the left-hand member should be linear in $1/v$, with B equal to the y intercept and C equal to the slope. A few typical curves for nitrogen are shown in Fig. 5-2, where it may be seen that the graphs are linear in the low-density region and cut the y axis at values in agreement with the B values in Table 5-1.

5-2 INTERNAL ENERGY OF A GAS

Imagine a thermally insulated vessel with rigid walls, divided into two compartments by a partition. Suppose that there is a gas in one compartment and that the other is empty. If the partition is removed, the gas will undergo what is known as a *free expansion* in which no work is done and no heat is transferred. From the first law, since both Q and W are zero, it follows that *the internal energy remains unchanged during a free expansion.* The question of whether or not the temperature of a gas changes during a free expansion and, if it does, of the magnitude of the temperature change has engaged the attention of physicists for over a hundred years. Starting with Joule in 1843, many attempts have been made to measure either the quantity $(\partial\theta/\partial V)_U$, which may be called the *Joule coefficient*, or related quantities that are all a measure, in one way or another, of the effect of a free expansion—or, as it is often called, the *Joule effect*.

In general, the energy of a gas is a function of any two of the coordinates P, V, and θ. Considering U as a function of θ and V, we have

$$dU = \left(\frac{\partial U}{\partial \theta}\right)_V d\theta + \left(\frac{\partial U}{\partial V}\right)_\theta dV.$$

If no temperature change ($d\theta = 0$) takes place in a free expansion ($dU = 0$), then it follows that

$$\left(\frac{\partial U}{\partial V}\right)_\theta = 0;$$

or, in other words, U does not depend on V. Considering U to be a function of θ and P, we have

$$dU = \left(\frac{\partial U}{\partial \theta}\right)_P d\theta + \left(\frac{\partial U}{\partial P}\right)_\theta dP.$$

If no temperature change ($d\theta = 0$) takes place in a free expansion ($dU = 0$), then it follows that

$$\left(\frac{\partial U}{\partial P}\right)_\theta = 0;$$

or, in other words, U does not depend on P. It is apparent then that, if no temperature change takes place in a free expansion, U is independent of V and of P, and therefore *U is a function of θ only.*

In order to study the free expansion of a gas Joule connected two vessels by a short tube and stopcock, which were immersed in a water bath. One vessel contained air at high pressure, and the other was evacuated. The temperature of the water was measured before and after the expansion, the idea being to infer the drop in temperature of the gas from the decrease in temperature of the water. Since the heat capacity of the vessels and the water was approximately one thousand times as large as the heat capacity of the air, Joule was unable to detect any temperature change of the water, although, in the light of our present knowledge, the air must have undergone a temperature decrease of several degrees.

A direct measurement of the temperature change associated with a free expansion is so difficult that it seems necessary to give up the idea of a precise measurement of the Joule coefficient. Modern methods of attacking the problem of the internal energy of a gas involve the measurement of the quantity $(\partial u/\partial P)_\theta$ by having the gas undergo an isothermal expansion in which heat is transferred and work is done. The most extensive series of measurements of this kind was performed by Rossini and Frandsen in 1932 at the National Bureau of Standards. The apparatus is shown in Fig. 5-3. A bomb B contains n moles of gas at a pressure P and communicates with the atmosphere through a long coil wrapped around the bomb. The whole apparatus is immersed in a water bath whose temperature can be maintained constant at exactly the same value as that of the surrounding atmosphere.

The experiment is performed as follows: When the stopcock is opened slightly, the gas flows slowly through the long coil and out into the air. At the

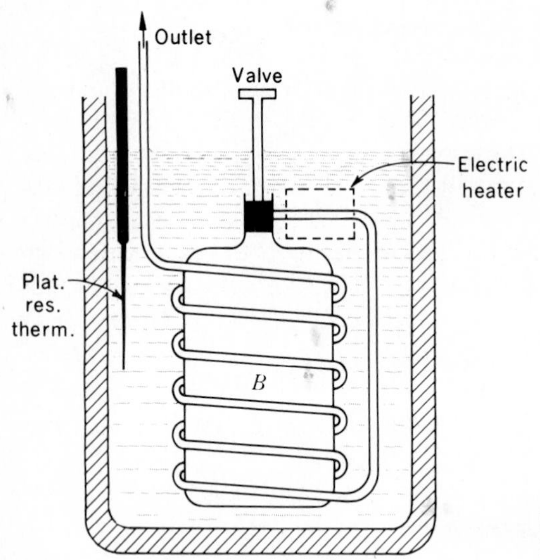

Figure 5-3 Apparatus of Rossini and Frandsen for measuring $(\partial u/\partial P)_\theta$ of a gas.

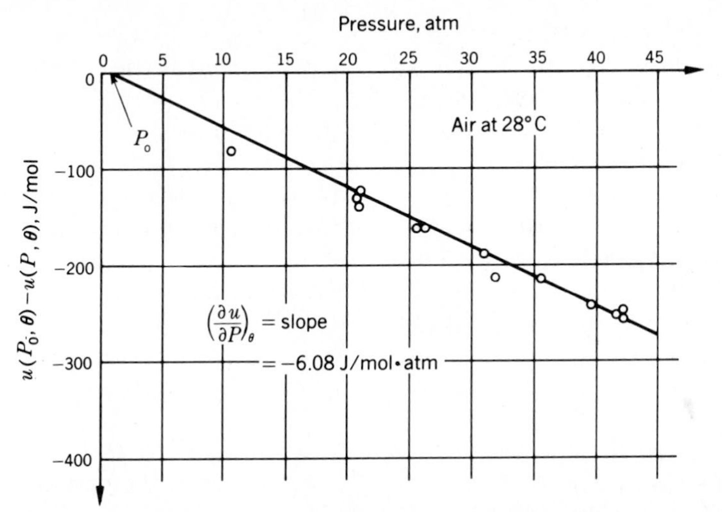

Figure 5-4 Dependence of internal energy of a gas on pressure. One atmosphere equals 101.3 kPa.

same time, the temperature of the gas, the bomb, the coils, and the water is maintained constant by an electric heating coil immersed in the water. The electrical energy supplied to the water is therefore the heat Q absorbed by the gas during the expansion. The work done by the gas is evidently

$$W = -P_0(nv_0 - V_B),$$

where P_0 is atmospheric pressure, v_0 is the molar volume at atmospheric temperature and pressure, V_B is the volume of the bomb, and nv_0 is larger than V_B.

If $u(P, \theta)$ is the molar energy at pressure P and temperature θ and if $u(P_0, \theta)$ is the molar energy at atmospheric pressure and the same temperature, then, from the first law,

$$u(P_0, \theta) - u(P, \theta) = \frac{Q + W}{n},$$

provided that corrections have been made to take account of the energy changes due to the contraction of the walls of the bomb. In this way, the molar internal energy change was measured for various values of the initial pressure and was plotted against the pressure, as shown in Fig. 5-4. Since $u(P_0, \theta)$ is constant, it follows that the slope of the resulting straight line at any value of P is equal to $(\partial u/\partial P)_\theta$. Within the pressure range of 1 to 40 standard atmospheres, it is seen that $(\partial u/\partial P)_\theta$ *is independent of the pressure*, depending only on the temperature. Thus,

$$\left(\frac{\partial u}{\partial P}\right)_\theta = f(\theta),$$

and

$$u = f(\theta)P + F(\theta),$$

where $F(\theta)$ is another function of the temperature only.

Rossini and Frandsen's experiments with air, oxygen, and mixtures of oxygen and carbon dioxide led to the conclusion that the internal energy of a gas is a function of both temperature and pressure. They found no pressure or temperature range in which the quantity $(\partial u/\partial P)_\theta$ was equal to zero.

Their experiment has somewhat the same disadvantage as Joule's original experiment, in that the heat capacity of the gas is much smaller than that of the calorimeter and water bath. To keep the temperature of the gas constant within reasonable limits, the temperature of the water must be kept constant to within less than a thousandth of a degree. In Rossini and Frandsen's measurements, the final precision was estimated to be $2\frac{1}{2}$ percent.

5-3 IDEAL GAS

We have seen that, in the case of a real gas, only in the limit as the pressure approaches zero does the equation of state assume the simple form $PV = nR\theta$. Furthermore, the internal energy of a real gas is a function of pressure as well as of temperature. It is convenient at this point to define an *ideal gas* whose properties, while not corresponding to those of any existing gas, are approximately those of a real gas at low pressures. By definition, an ideal gas satisfies the equations

$$\boxed{\begin{aligned} PV &= nR\theta \\ \left(\frac{\partial U}{\partial P}\right)_\theta &= 0 \end{aligned}} \qquad \text{(ideal gas).} \tag{5-5}$$

The requirement that $(\partial U/\partial P)_\theta = 0$ may be written in other ways. Thus,

$$\left(\frac{\partial U}{\partial V}\right)_\theta = \left(\frac{\partial U}{\partial P}\right)_\theta \left(\frac{\partial P}{\partial V}\right)_\theta,$$

and since $(\partial P/\partial V)_\theta = -nR\theta/V^2 = -P/V$, and therefore is not zero, whereas $(\partial U/\partial P)_\theta$ is zero, it follows that for an ideal gas

$$\boxed{\left(\frac{\partial U}{\partial V}\right)_\theta = 0} \qquad \text{(ideal gas).} \tag{5-6}$$

Finally, since both $(\partial U/\partial P)_\theta$ and $(\partial U/\partial V)_\theta$ are zero,

$$\boxed{U = f(\theta) \text{ only}} \qquad \text{(ideal gas).} \tag{5-7}$$

Whether an actual gas may be treated as an ideal gas depends upon the error that may be tolerated in a given calculation. An actual gas at pressures below about twice standard atmospheric pressure may be treated as an ideal gas without introducing an error greater than a few percent. Even in the case of a saturated vapor in equilibrium with its liquid, the ideal-gas equation of state may be used with only a small error if the vapor pressure is low.

For an infinitesimal quasi-static process of a hydrostatic system, the first law is

$$đQ = dU + P\,dV,$$

and the heat capacity at constant volume is given by

$$C_V = \left(\frac{\partial U}{\partial \theta}\right)_V.$$

In the special case of an ideal gas, U is a function of θ only; therefore, the partial derivative with respect to θ is the same as the total derivative. Consequently,

$$C_V = \frac{dU}{d\theta},$$

and

$$\boxed{đQ = C_V\,d\theta + P\,dV.} \tag{5-8}$$

Now, all equilibrium states are represented by the ideal-gas equation

$$PV = nR\theta,$$

and, for an infinitesimal quasi-static process,

$$P\,dV + V\,dP = nR\,d\theta.$$

Substituting the above in Eq. (5-8), we get

$$đQ = (C_V + nR)\,d\theta - V\,dP,$$

and dividing by $d\theta$ yields

$$\frac{đQ}{d\theta} = C_V + nR - V\frac{dP}{d\theta}.$$

At constant pressure, the left-hand member becomes C_P; whence

$$\boxed{C_P = C_V + nR} \quad \text{(ideal gas)}. \tag{5-9}$$

We have the result, therefore, that the heat capacity at constant pressure of an ideal gas is always larger than that at constant volume, the difference remaining constant and equal to nR.

Since U is a function of θ only, it follows that

$$C_V = \frac{dU}{d\theta} = \text{a function of } \theta \text{ only},$$

and

$$C_P = C_V + nR = \text{a function of } \theta \text{ only}.$$

One more useful equation can be obtained. Since

$$đQ = (C_V + nR)\,d\theta - V\,dP,$$

we get

$$\boxed{đQ = C_P\,d\theta - V\,dP.} \tag{5-10}$$

5-4 EXPERIMENTAL DETERMINATION OF HEAT CAPACITIES

The heat capacities of gases are measured by the electrical method. To measure C_V, the gas is contained in a thin-walled steel flask with a heating wire wound around it. By maintaining an electric current in the wire, an equivalent amount of heat is supplied to the gas, and the specific heat at constant volume is obtained by measuring the temperature rise of the gas. The same method is used to measure C_P except that, instead of confining the gas to a constant volume, the gas is allowed to flow at constant pressure through a calorimeter, where it receives electrically a known equivalent heat per unit of time. From the initial (inlet) and final (outlet) temperatures, the rate of supply of heat, and the rate of flow of gas, the value of C_P is calculated. The results of such measurements on gases at *low pressures* (approximately ideal gases) can be stated in a simple manner in terms of molar heat capacities.

1. *All gases:*
   (*a*) c_V is a function of θ only.
   (*b*) c_P is a function of θ only, and $> c_V$.
   (*c*) $c_P - c_V = \text{const.} = R$.
   (*d*) $\gamma = c_P/c_V =$ a function of θ only, and > 1.
2. *Monatomic gases*, such as He, Ne, and A, and most metallic vapors, such as the vapors of Na, Cd, and Hg:
   (*a*) c_V is constant over a wide temperature range and is very nearly equal to $\frac{3}{2}R$.
   (*b*) c_P is constant over a wide temperature range and is very nearly equal to $\frac{5}{2}R$.
   (*c*) γ is constant over a wide temperature range and is very nearly equal to $\frac{5}{3}$.
3. *So-called permanent diatomic gases*, namely, air, H_2, D_2, O_2, N_2, NO, and CO:
   (*a*) c_V is constant at ordinary temperatures, being equal to about $\frac{5}{2}R$, and increases as the temperature is raised.
   (*b*) c_P is constant at ordinary temperatures, being equal to about $\frac{7}{2}R$, and increases as the temperature is raised.
   (*c*) γ is constant at ordinary temperatures, being equal to about $\frac{7}{5}$, and decreases as the temperature is raised.

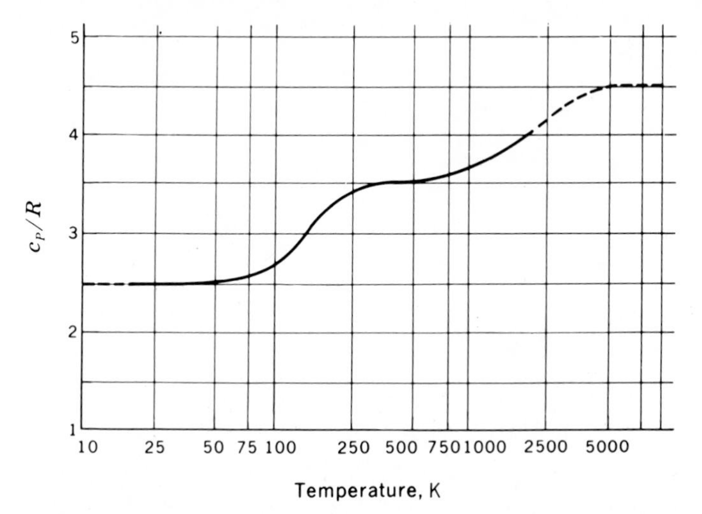

Figure 5-5 Experimental values of c_P/R for hydrogen as a function of temperature, plotted on a logarithmic scale.

4. *Polyatomic gases and gases that are chemically active*, such as CO_2, NH_3, CH_4, Cl_2, and Br_2:
   c_P, c_V, and c_P/c_V vary with the temperature, the variation being different for each gas.

These experimental results indicate that the universal gas constant R (8.31 kJ/kmol · K) is a natural unit with which to express the molar heat capacity of a gas. It is a very interesting consequence of theory that the universal gas constant is also the natural unit for solids. In the remainder of this book we shall specify not the specific heat capacities themselves but the ratios c_V/R and c_P/R.

The behavior of hydrogen is quite exceptional, as shown in Fig. 5-5. At very low temperatures, c_P/R drops to a value of $\frac{5}{2}$, appropriate to a *monatomic* gas. For all other diatomic gases, c_P/R may always be written

$$\frac{c_P}{R} = \frac{7}{2} + f(\theta),$$

where $f(\theta)$ is often one or more functions of the type

$$\left(\frac{b}{\theta}\right)^2 \frac{e^{b/\theta}}{(e^{b/\theta} - 1)^2}.$$

Exact equations of the above type are difficult to handle and are not suitable for the practical calculations of the laboratory scientist; consequently, approximate empirical equations are used. Empirical equations for c_P/R of some of the most important gases, compiled by H. M. Spencer, are given in Table 5-2, within the temperature range 300 to 1500 K.

Table 5-2 c_P/R of important gases

$c_P/R = a + b\theta + c\theta^2$ (from 300 to 1500 K)

Gas	a	b, (kK)$^{-1}$	c, (kK)$^{-2}$
H_2	3.495	−0.101	0.243
O_2	3.068	1.638	−0.512
Cl_2	3.813	1.220	−0.486
Br_2	4.240	0.490	−0.179
N_2	3.247	0.712	−0.041
CO	3.192	0.924	−0.141
HCl	3.389	0.218	0.186
HBr	3.311	0.481	0.079
CO_2	3.206	5.082	−1.714
H_2O	3.634	1.195	0.135
NH_3	3.116	3.970	−0.366
H_2S	3.214	2.871	−0.608
CH_4	1.702	9.083	−2.164

5-5 QUASI-STATIC ADIABATIC PROCESS

When an ideal gas undergoes a quasi-static adiabatic process, the pressure, volume and temperature change in a manner that is described by a relation between P and V, θ and V, or P and θ. In order to derive the relation between P and V, we start with Eqs. (5-8) and (5-10) of Art. 5-4. Thus,

$$đQ = C_V\,d\theta + P\,dV,$$

and

$$đQ = C_P\,d\theta - V\,dP.$$

Since, in an adiabatic process, $đQ = 0$,

$$V\,dP = C_P\,d\theta,$$

and

$$P\,dV = -C_V\,d\theta.$$

Dividing the first by the second, we obtain

$$\frac{dP}{P} = -\frac{C_P}{C_V}\frac{dV}{V},$$

and denoting the ratio of the heat capacities by the symbol γ, we have

$$\frac{dP}{P} = -\gamma\frac{dV}{V}.$$

This equation cannot be integrated until we know something about the behavior of γ. We have seen that for monatomic gases γ is constant, whereas for diatomic and polyatomic gases it may vary with the temperature. It requires, however, a very large change of temperature to produce an appreciable change in γ. For example, in the case of carbon monoxide, a temperature rise from 0 to 2000°C produces a decrease in γ from 1.4 to 1.3. Most adiabatic processes that we deal with do not involve such a large temperature change. We are therefore entitled, in an adiabatic process that involves only a moderate temperature change, to neglect the small accompanying change in γ. Regarding γ, therefore, as constant and integrating, we obtain

$$\ln P = -\gamma \ln V + \ln \text{const.},$$

or

$$\boxed{PV^{\gamma} = \text{const.}} \tag{5-11}$$

This equation holds at all equilibrium states through which an ideal gas passes during a quasi-static adiabatic process. It is important to understand that a free expansion is an adiabatic process but is not quasi-static. It is therefore entirely fallacious to attempt to apply Eq. (5-11) to the states traversed by an ideal gas during a free expansion.

A family of curves representing quasi-static adiabatic processes may be plotted on a PV diagram by assigning different values to the constant in Eq. (5-11). The slope of any adiabatic curve is

$$\left(\frac{\partial P}{\partial V}\right)_S = -\gamma \text{ const. } V^{-\gamma-1}$$

$$= -\gamma \frac{P}{V},$$

where the subscript S is used to denote an adiabatic process.

Quasi-static isothermal processes are represented by a family of equilateral hyperbolas obtained by assigning different values to θ in the equation $PV = nR\theta$.

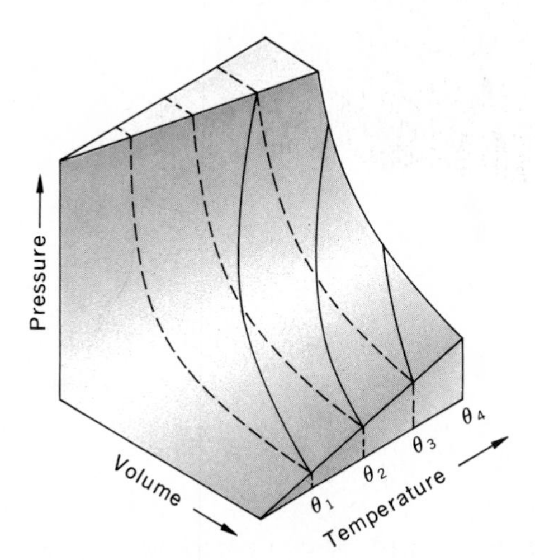

Figure 5.6 Surface for an ideal gas. (Isotherms are represented by dashed curves, and adiabatics by full curves.)

Since

$$\left(\frac{\partial P}{\partial V}\right)_\theta = -\frac{P}{V},$$

it follows that an adiabatic curve has a steeper negative slope than does an isothermal curve at the same point.

The isothermal curves and adiabatic curves of an ideal gas may be shown in a revealing way on a $PV\theta$ surface. If P, V, and θ are plotted along rectangular axes, the resulting surface is shown in Fig. 5-6, where it may be seen that the adiabatic curves cut across the isotherms.

5-6 RÜCHHARDT'S METHOD OF MEASURING γ

An ingenious method of measuring γ developed by Rüchhardt in 1929 makes use of elementary mechanics. The gas is contained in a large jar of volume V. Fitted to the jar (see Fig. 5-7) is a glass tube with an accurate bore of cross-sectional area A, into which a metal ball of mass m fits snugly like a piston. Since the gas is slightly compressed by the steel ball in its equilibrium position, its pressure P is slightly larger than atmospheric pressure P_0. Thus, neglecting friction,

$$P = P_0 + \frac{mg}{A}.$$

If the ball is given a slight downward displacement and then let go, it will oscillate with a period τ. Friction will cause the ball to come to rest eventually.

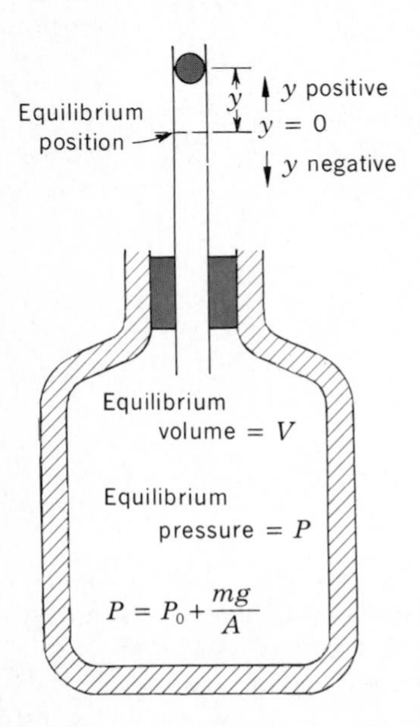

Figure 5-7 Rüchhardt's apparatus for measurement of γ.

Let the displacement of the ball from its equilibrium position at any moment be denoted by y, where y is positive when the ball is above the equilibrium position and negative below. A small positive displacement causes an increase in volume which is very small compared with the equilibrium volume V and which therefore can be denoted by dV, where

$$dV = yA.$$

Similarly, a small positive displacement causes a decrease in pressure which is very small compared with the equilibrium pressure P and which therefore can be denoted by dP, where dP is a negative quantity. The resultant force $\mathcal{F}$ acting on the ball is equal to $A\ dP$ is we neglect friction, or

$$dP = \frac{\mathcal{F}}{A}.$$

Notice that, when y is positive, dP is negative and therefore $\mathcal{F}$ is negative; that is, $\mathcal{F}$ is a restoring force.

Now, as the ball oscillates fairly rapidly, the variations of P and V are adiabatic. Since the variations are also quite small, the states through which the gas passes can be considered to be approximately states of equilibrium. We may therefore assume that the changes of P and V represent an approximately quasi-static adiabatic process, and we may write

$$PV^{\gamma} = \text{const.},$$

and

$$P\gamma V^{\gamma-1}\ dV + V^{\gamma}\ dP = 0.$$

Substituting for dV and dP, we get

$$\mathcal{F} = -\frac{\gamma P A^2}{V} y.$$

This equation expresses the fact that the restoring force is directly proportional to the displacement and is in the opposite direction, which is Hooke's law. This is precisely the condition for *simple harmonic motion*, for which the period τ is

$$\tau = 2\pi \sqrt{\frac{m}{-\mathcal{F}/y}}.$$

Consequently,

$$\tau = 2\pi \sqrt{\frac{mV}{\gamma P A^2}},$$

and

$$\gamma = \frac{4\pi^2 mV}{A^2 P \tau^2}.$$

The mass of the ball, the volume, the cross-sectional area of the tube, and the pressure are all known beforehand, and only the period has to be measured to obtain γ. The values obtained by Rüchhardt for air and for CO_2 were in good agreement with those obtained from calorimetric measurements.

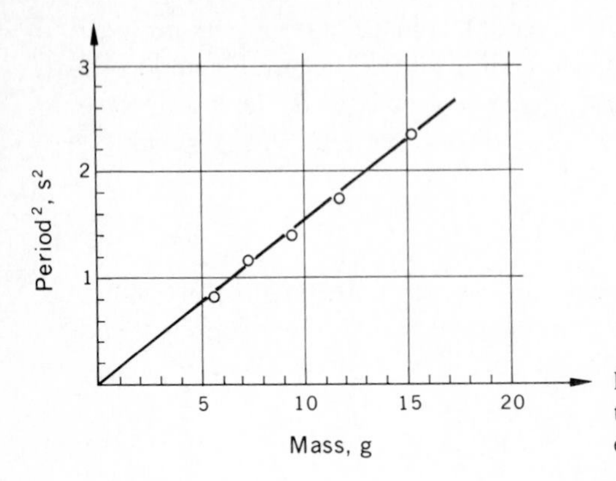

Figure 5-8 The square of the period against the mass for the oscillating ball in Rüchhardt's experiment. *(H. C. Jensen, 1963.)*

Rüchhardt's method involves errors due to three simplifying assumptions: (1) that the gas is ideal, (2) that there is no friction, and (3) that volume changes are strictly adiabatic. It is estimated that the second assumption is responsible for the largest error, amounting to about 3 percent.

The large frictional damping in the usual Rüchhardt method may be avoided by using a glass tube with a slight taper, with the wider diameter at the top. A slow flow of gas maintains a steady oscillation of a ball about an equilibrium position. The flow is controlled by a throttle valve between the experimental apparatus and a high-pressure gas tank. With this method, Flammersfeld found γ to be 1.659 for argon, 1.404 for air, and 1.300 for carbon dioxide. If a small rod is attached to the underside of the oscillating ball, the total mass of the oscillating system may be varied by attaching metal washers to the rod, and the variation of period with mass may be measured. The results of H. C. Jensen are shown in Fig. 5-8, where it is seen that the line passes almost through the origin, indicating that one has here the equivalent of a nearly massless spring.

A modification of Rüchhardt's experiment in which accurate account is taken of the real equation of state of the gas, the friction present, and the departure from strict adiabatic conditions was achieved by Clark and Katz in 1940. The method was adapted for an undergraduate teaching laboratory by D. G. Smith in 1979. A steel piston at the center of a cylindrical tube divides the gas into two equal parts, as shown in Fig. 5-9. It is set in vibration at any desired frequency by external coils in which an alternating current of suitable frequency is maintained. The cylinder is kept in a horizontal position, and friction between the piston and the cylinder is reduced by balancing the weight of the piston by the attraction of an electromagnet.

The amplitude of vibration of the piston is measured with a microscope equipped with a micrometer eyepiece, at a number of values of the frequency of the impressed alternating current, and the resonance curve is plotted. From the resonance frequency and very elaborate calculations not involving the assumptions made by Rüchhardt, γ is calculated. Since friction was reduced to a great extent by

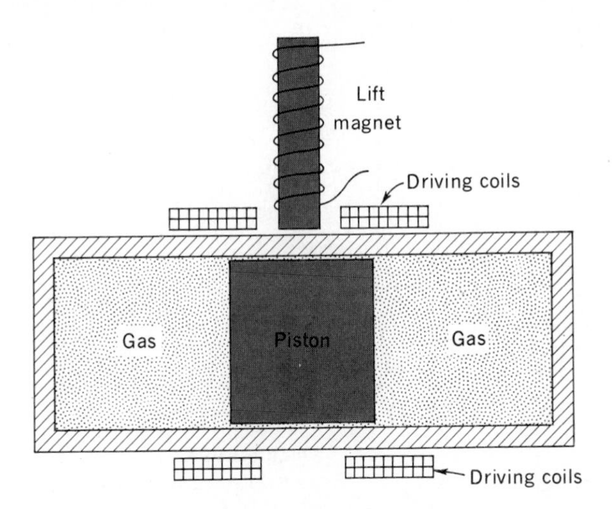

Figure 5-9 Apparatus of Clark and Katz for measuring γ of a real gas as a function of pressure.

Table 5-3 Pressure variation of $\gamma (\gamma = a + bP + cP^2)$

Gas	Temp., K	a	b, $(\text{GPa})^{-1}$	c, $(\text{MPa})^{-2}$	γ extrapolated to zero pressure (ideal gas)
He	296.25	1.6669	−1.97	0	1.667
Ar	297.35	1.6667	34.8	0	1.667
H_2	296.55	1.4045	2.47	0	1.405
N_2	296.15	1.4006	21.8	0	1.401
CO_2	303.05	1.2857	62.1	0	1.286
N_2O	298.45	1.2744	22.2	0.0948	1.274
CH_4	298.25	1.3029	−10.4	0.0472	1.303

the lift magnet, the corrections amounted to only about 1 percent. The authors measured γ at various pressures from 1 to 25 standard atmospheric pressures and expressed the results in the form of empirical equations, as shown in Table 5-3.

5-7 SPEED OF A LONGITUDINAL WAVE

If a compression is produced at one place in a substance, it will travel with a constant speed w, depending on certain properties of the substance that we shall now proceed to determine. Directing our attention to a column of material of cross section A, let us suppose that a piston, on the left in Fig. 5-10, actuated by an agent exerting a force $A(P + \Delta P)$ moves to the right with a constant velocity w_0. This sets up a compression traveling with a constant velocity w, so that in time τ the compression has traveled a distance $w\tau$ while the piston has traveled a distance $w_0 \tau$.

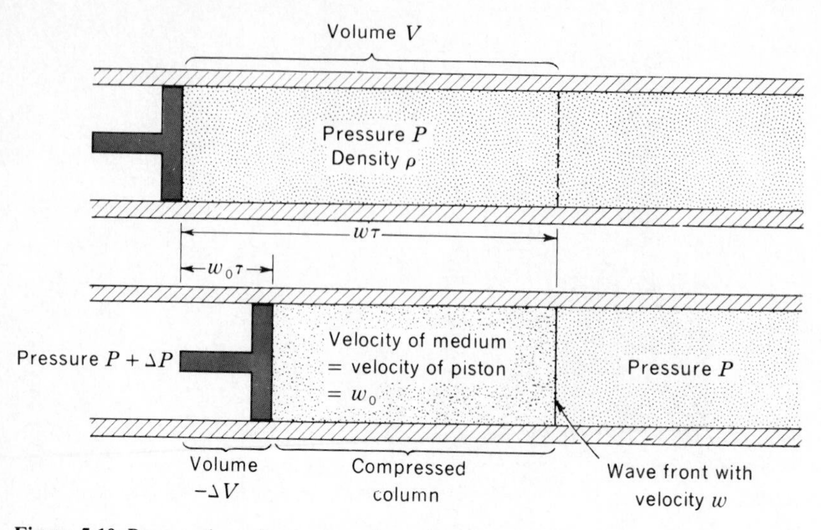

Figure 5-10 Propagation of a compression with constant velocity w by motion of a piston with constant velocity w_0. Upper diagram at the start; lower diagram after time τ.

Consider as a "free body" the compressed column whose length, if uncompressed, would be $w\tau$ and whose uncompressed volume $V = Aw\tau$. If ρ is the density of the normal or uncompressed material, the mass of the free body is $\rho Aw\tau$. At time τ, the

$$\left.\begin{array}{r}\text{Rate of increase of mass of}\\ \text{the compressed column}\end{array}\right\} = \frac{\rho Aw\tau}{\tau} = \rho Aw.$$

The entire compressed column has a velocity w_0 equal to that of the piston. Therefore,

$$\left.\begin{array}{r}\text{Rate of increase of momentum}\\ \text{of the compressed column}\end{array}\right\} = \rho Aww_0.$$

The free body is acted on by a force $A(P + \Delta P)$ to the right and a force AP to the left. Therefore,

$$\left.\begin{array}{r}\text{Unbalanced force on the}\\ \text{compressed column}\end{array}\right\} = A\,\Delta P.$$

From Newton's second law the unbalanced force is equal to the rate of change of momentum,

$$A\,\Delta P = \rho Aww_0,$$

or

$$\Delta P = \rho w^2 \frac{w_0}{w}.$$

The "uncompressed free body" of volume $V = Aw\tau$ has undergone a compression $(-\Delta V) = Aw_0\tau$. That is,

$$\frac{-\Delta V}{V} = \frac{Aw_0\tau}{Aw\tau} = \frac{w_0}{w}.$$

Therefore,

$$\Delta P = \rho w^2\left(\frac{-\Delta V}{V}\right),$$

which may be written

$$w = \sqrt{\frac{-1}{\rho\ \Delta V/V\ \Delta P}}.$$

This formula was first obtained by Newton, who regarded the quantity $\Delta V/V\ \Delta P$ as the isothermal compressibility. It was shown later by Laplace that the expression is really the adiabatic compressibility. To see why this is so, let us consider a column of material of cross section A, bounded by two planes, one at the center of a compression and the other at the center of a rarefaction, a distance $\lambda/2$ apart, where λ is the wavelength. Let us suppose that the temperature at the center of the compression exceeds the temperature at the center of the rarefaction by an amount $\Delta\theta$. Then the heat conducted a distance $\lambda/2$ in the time $\lambda/2w$ (time for the wave to travel the distance $\lambda/2$) is given by

$$\left.\begin{array}{r}\text{Heat conducted in the time for the}\\ \text{wave to travel a distance } \lambda/2\end{array}\right\} = KA\frac{\Delta\theta}{\lambda/2}\frac{\lambda}{2w} = KA\frac{\Delta\theta}{w},$$

where K is the thermal conductivity of the medium. The mass of material between the compression and rarefaction is $\rho A\lambda/2$, and the heat necessary to raise the temperature of this mass by the amount $\Delta\theta$ is

$$\left.\begin{array}{r}\text{Heat necessary to raise temperature}\\ \text{of mass } \rho A\lambda/2 \text{ by } \Delta\theta\end{array}\right\} = \rho A\frac{\lambda}{2}c_V\ \Delta\theta,$$

where c_V is the specific heat at constant volume.

The propagation of the wave would be adiabatic if the conducted heat were much too small to raise the temperature of the mass $\rho A\lambda/2$ by the amount $\Delta\theta$, or

$$\frac{KA\ \Delta\theta}{w} \ll \rho A\frac{\lambda}{2}c_V\ \Delta\theta \qquad \text{(adiabatic condition).}$$

This may be written

$$\frac{2K}{w\rho c_V} \ll \lambda \qquad \text{(adiabatic condition).}$$

The usual range of wavelengths of compressional waves is from a few centimeters to a few hundred centimeters. Let us compare these values with

$2K/w\rho c_V$. Taking a gas like air at 0°C as a typical case, we have, roughly,

$$K = 0.0237 \text{ W/m} \cdot \text{K},$$
$$w = 331 \text{ m/s},$$
$$\rho = 1.293 \text{ kg/m}^3,$$
$$c_V = 0.716 \text{ kJ/kg} \cdot \text{K},$$

and

$$\frac{2K}{w\rho c_V} = \frac{2 \times 0.237 \text{ W/m} \cdot \text{K}}{(331 \text{ m/s})(1.293 \text{ kg/m}^3)(0.176 \text{ kJ/kg} \cdot \text{K})}$$
$$= 155 \text{ nm}.$$

In the case of a metal, K would be much larger, but this would be compensated by the much larger values of w and ρ, and the quantity $2K/w\rho c_V$ would be still smaller than 155 nm. This quantity is therefore seen to be so much smaller than the usual value of a wavelength of a compressional wave (155 nm is the wavelength of ultraviolet light) that the adiabatic condition is well fulfilled. We therefore conclude that, in view of the properties of ordinary matter, *the volume changes which take place under the influence of a longitudinal wave at ordinary frequencies are adiabatic*, not isothermal.

Returning now to the expression for the velocity of a longitudinal wave and identifying $(-\Delta V/V\,\Delta P)$ as the *adiabatic* compressibility κ_S, we have finally

$$w = \sqrt{\frac{1}{\rho \kappa_S}}.$$

From Art. 5-5 we have, for an ideal gas,

$$\kappa_S = -\frac{1}{V}\left(\frac{\partial V}{\partial P}\right)_S = \frac{1}{\gamma P},$$

and

$$\rho = \frac{\mathcal{M}}{v},$$

where $\mathcal{M}$ is the molar mass and v is the molar volume. Hence,

$$w = \sqrt{\frac{\gamma P v}{\mathcal{M}}},$$

or

$$\boxed{w = \sqrt{\frac{\gamma R \theta}{\mathcal{M}}}.} \tag{5-12}$$

Equation (5-12) enables us to calculate γ from experimental measurements of w and θ. For example, the speed of sound in air at 0°C is about 331 m/s. Therefore, using the values

$$w = 331 \text{ m/s},$$

$$\theta = 273 \text{ K},$$

$$R = 8.31 \text{ kJ/kmol} \cdot \text{K},$$

$$\mathcal{M} = 28.96 \text{ kg/kmol},$$

we get

$$\gamma = \frac{\mathcal{M}w^2}{R\theta},$$

$$= \frac{(28.96 \text{ kg/kmol})(331 \text{ m/s})^2}{(8.31 \text{ kJ/kmol} \cdot \text{K})(273 \text{ K})},$$

$$= 1.40.$$

The speed of a sound wave in a gas can be measured roughly by means of Kundt's tube. The gas is admitted to a cylinder tube closed at one end and supplied at the other end with a movable piston capable of being set in vibration parallel to the axis of the tube. In the tube is a small amount of light powder. For a given frequency, a position of the piston can be found at which standing waves are set up. Under these conditions small heaps of powder pile up at the nodes. The distance between any two adjacent nodes is one-half a wavelength, and the speed of the waves is the product of the frequency and the wavelength.

5-8 ACOUSTICAL THERMOMETRY

Much greater accuracy is achieved with an acoustic interferometer, at one end of which is a source of waves such as a piezoelectric crystal and at the other a receiver. When the distance between the two is kept constant and the frequency varied, the various resonances corresponding to different numbers of antinodes are noted. The frequency of the compressional waves can be varied from audible to ultrasonic frequency, but corrections for errors due to viscosity, heat conduction, and boundary layer absorption must be applied. Equation (5-12) can be used to determine the ideal gas temperature θ by plotting the square of the speed of sound as a function of pressure. Then

$$\theta = \frac{w_0^2 \mathcal{M}}{\gamma R}, \tag{5-13}$$

where w_0^2 is the extrapolation of the square of the speed of sound to zero pressure, which assures ideal gas conditions.

Due to inconsistencies in primary and secondary temperature measures below 20 K, Plumb and Cataland at the National Bureau of Standards developed the acoustical thermometer, which is capable of a reproducibility of

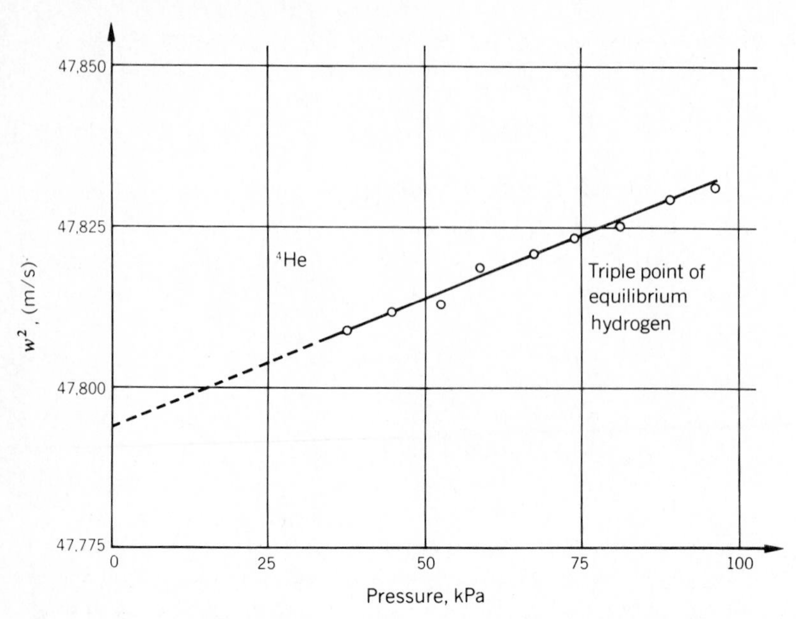

Figure 5-11 Square of speed of an ultrasonic wave in helium as a function of pressure. *(Cataland and Plumb, 1973.)*

0.001 K at 2 K and 0.005 K at 20 K. Their results for the triple point of equilibrium hydrogen using ^{4}He as the measuring gas are shown in Fig. 5-11.

The use of Eq. (5-13) requires independent knowledge of the gas constant R in order to determine an ideal gas temperature. A. R. Colclough and colleagues at the National Physical Laboratory in Great Britain used an acoustic interferometer operating at the only defined temperature, namely, the triple point of water at 273.16 K, to determine the universal gas constant R. Using argon as the measuring gas, they determined a value of R in 1979 of 8.31448 J/mol · K, which agrees within the limits of random uncertainties with Batuecas' value from limiting density experiments discussed earlier in this chapter.

5-9 THE MICROSCOPIC POINT OF VIEW

We have emphasized that the point of view of classical thermodynamics is entirely macroscopic. Systems are described with the aid of their gross, or large-scale, properties. The first law of thermodynamics is a relation among the fundamental physical quantities of work, internal energy, and heat. When the first law is applied to a class of systems, a general relation is obtained which holds for *any* member of the class but which contains no quantities or properties of a particular system that would distinguish it from another. Equation (4-13), for example,

$$C_V = \left(\frac{\partial U}{\partial \theta}\right)_V,$$

is true for all hydrostatic systems whether solid, liquid, or gas. It enables one to calculate C_V of a hydrostatic system, *provided that one knows the internal energy* as a function of θ and V. The heat transferred during an isochoric process (Art. 4-10), which is

$$Q_V = \int_{\theta_i}^{\theta_f} C_V \, d\theta,$$

may be calculated once the C_V of the particular system under consideration is known as a function of θ. But there is nothing in classical thermodynamics that provides detailed information concerning U or C_V.

Another example of the limitation of classical thermodynamics is its inability to provide the equation of state of any desired system. To make use of any thermodynamic equation involving P, V, θ, and the derivatives $(\partial P/\partial V)_\theta$, $(\partial V/\partial \theta)_P$, and $(\partial \theta/\partial P)_V$, one must have an equation of state. Experimental values are very often useful, but there are occasions when it is not feasible to perform the necessary experiments. If an experiment is performed on, let us say, oxygen, the numerical constants in the equation of state of oxygen only are obtained, and no clue is at hand concerning the values of the constants for any other gas.

To obtain detailed information concerning the thermodynamic coordinates and thermal properties of systems without having to resort to experimental measurements, we require calculations based on the properties and behavior of the molecules of the system. There are two such microscopic theories; one is called *kinetic theory*, and the other is *statistical mechanics*. Both theories deal with molecules, their internal and external motion, their collisions with one another and with any existing walls, and their forces of interaction. Making use of the laws of mechanics and the theory of probability, kinetic theory concerns itself with the details of molecular motion and impact and is capable of dealing with the following *nonequilibrium* situations:

1. Molecules escaping from a hole in a container, a process known as *effusion*.
2. Molecules moving through a pipe under the action of a pressure difference, a motion called *laminar flow*.
3. Molecules with momentum moving across a plane and mixing with molecules of lesser momentum—a molecular process responsible for *viscosity*.
4. Molecules with kinetic energy moving across a plane and mixing with molecules of lesser energy—a process responsible for *heat conduction*.
5. Molecules of one sort moving across a plane and mixing with molecules of another sort, a process known as *diffusion*.
6. Chemical combination between two or more kinds of molecules, which takes place at a finite rate and is known as *chemical kinetics*.

7. Inequality of molecular impacts made on various sides of a very small object suspended in a fluid, a difference that give rise to a haphazard zigzag motion of the suspended particle that is known as *Brownian motion*.

Statistical mechanics avoids the mechanical details concerning molecular motions and deals only with the energy aspects of the molecules. It relies heavily on the theory of probability but is mathematically simpler than kinetic theory, although more subtle conceptually. Only equilibrium states can be handled—but in a uniform, straightforward manner, so that once the energy levels of the molecules or of systems of molecules are understood a program of calculations may be carried out through which the equation of state, the energy, and other thermodynamic functions may be obtained.

In this chapter we shall limit ourselves to a small part of the kinetic theory of an ideal gas. A simplified treatment of statistical mechanics will be reserved for Chap. 11.

5-10 EQUATION OF STATE OF AN IDEAL GAS

The fundamental hypotheses of the kinetic theory of an ideal gas are as follows:

1. Any small sample of gas consists of an enormous number of molecules N. For any one chemical species, all molecules are identical. If m is the mass of each molecule, then the total mass is mN. If $\mathcal{M}$ denotes the molar mass in kilograms per kilomole (formerly called the *molecular weight*), then the number of moles n is given by

$$n = \frac{mN}{\mathcal{M}}.$$

The number of molecules per mole of gas is called *Avogadro's number* N_A, where

$$N_A = \frac{N}{n} = \frac{\mathcal{M}}{m} = 6.0225 \times 10^{23} \frac{\text{molecules}}{\text{mol}}.$$

Since a mole of ideal gas at 273 K and at standard atmospheric pressure occupies a volume of 2.24×10^4 cm^3, there are approximately 3×10^{19} molecules in a volume of only 1 cm^3 and 3×10^{16} molecules per mm^3, and even a volume as small as a cubic micrometer contains as many as 3×10^7 molecules.

2. The molecules of an ideal gas are supposed to resemble small hard spheres that are in perpetual random motion. Within the temperature and pressure range of an ideal gas, the average distance between neighboring molecules is large compared with the size of a molecule. The diameter of a molecule is of the order of 2 or 3×10^{-10} m. Under standard conditions, the average distance between molecules is about 50 times their diameter.

3. The molecules of an ideal gas are assumed to exert no forces of attraction or repulsion on other molecules except when they collide with one another and with a wall. Between collisions, they therefore move with uniform rectilinear motion.
4. The portion of a wall with which a molecule collides is considered to be smooth, and the collision is assumed to be perfectly elastic. If w is the speed of a molecule approaching a wall, only the perpendicular component of velocity $w_\perp$ is changed upon collision with the wall, from $w_\perp$ to $-w_\perp$, or a total change of $-2w_\perp$.
5. When there is no external field of force, the molecules are distributed uniformly throughout a container. The *molecular density* N/V is assumed constant, so that in any small element of volume dV there are dN molecules, where

$$dN = \frac{N}{V}\,dV.$$

   The infinitesimal dV must satisfy the same conditions in kinetic theory as in thermodynamics, namely, that it is small compared with V but large enough to make dN a large number. If, for example, a volume of 1 cm^3 contains 10^{19} molecules, then one-millionth of a cubic centimeter would still contain 10^{13} molecules and would qualify as a differential volume element.
6. There is no preferred direction for the velocity of any molecule, so that at any moment there are as many molecules moving in one direction as in another.
7. Not all molecules have the same speed. A few molecules at any moment move slowly and a few move very rapidly, so that speeds may be considered to cover the range from zero to the speed of light. Since most molecular speeds are so far below the speed of light, no error is introduced in integrating the speed from 0 to ∞. If dN_w represents the number of molecules with speeds between w and $w + dw$, it is assumed that dN_w remains constant at equilibrium, even though the molecules are perpetually colliding and changing their speeds.

Since the velocity vectors of the molecules of gas have no preferred direction, consider an arbitrary velocity vector $\mathbf{w}$ directed from the point O in Fig. 5-12 to the elementary area dA'. It is important to know how many molecules have velocity vectors in the neighborhood of $\mathbf{w}$. The calculation of this quantity involves the concept of a *solid angle*. Taking O as the origin of polar coordinates r, θ, and ϕ, we construct a sphere of radius r. The area dA' on the surface of this sphere, formed by two circles of latitude differing by $d\theta$ and two circles of longitude differing by $d\phi$, has the magnitude

$$dA' = (r\,d\theta)(r \sin\theta\,d\phi).$$

The solid angle $d\Omega$, formed by lines radiating from O and touching the edge of dA', is by definition

$$d\Omega = \frac{dA'}{r^2} = \frac{(r\,d\theta)(r\sin\theta\,d\phi)}{r^2},$$

or

$$d\Omega = \sin\theta\,d\theta\,d\phi. \tag{5-14}$$

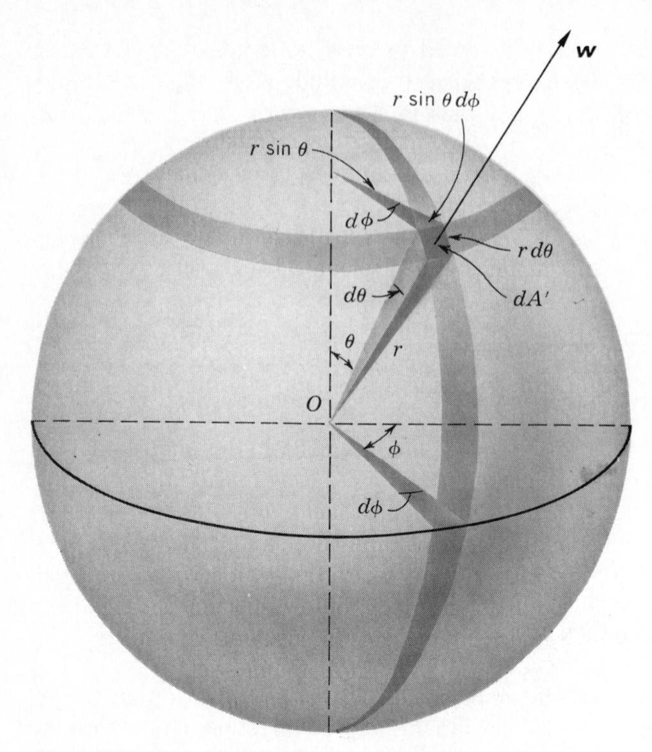

Figure 5-12 The solid angle $d\Omega = \sin\theta\, d\theta\, d\phi$.

Since the largest area on the surface of the sphere is that of the entire sphere $4\pi r^2$, the maximum solid angle is 4π sr (*steradians*).

The fraction of molecules with velocity vectors in the neighborhood of $\mathbf{w}$ will have speeds between w and $w + dw$ and directions within the solid angle $d\Omega$ about $\mathbf{w}$. If dN_w is the number of molecules with speeds between w and $w + dw$, then the fraction of *these* molecules whose directions lie within the solid angle $d\Omega$ is $d\Omega/4\pi$, so that the number of molecules within the speed range dw, in the θ range of $d\theta$ and the ϕ range of $d\phi$, is given by

$$d^3N_{w,\,\theta,\,\phi} = dN_w \frac{d\Omega}{4\pi}, \tag{5-15}$$

an equation expressing the fact that *molecular velocities have no preferred direction.*

Now consider this group of molecules approaching a small area dA of the wall of the containing vessel. Many of these molecules will undergo collisions along the way, but if we consider only those members of the group that lie within the cylinder (Fig. 5-13) whose side is of length $w\, d\tau$, where $d\tau$ is such a short time interval that no collisions are made, then *all* the $d^3N_{w,\,\theta,\,\phi}$ molecules *within this cylinder* will collide with dA. The volume of the cylinder dV is

$$dV = w\, d\tau \cos\theta\, dA, \tag{5-16}$$

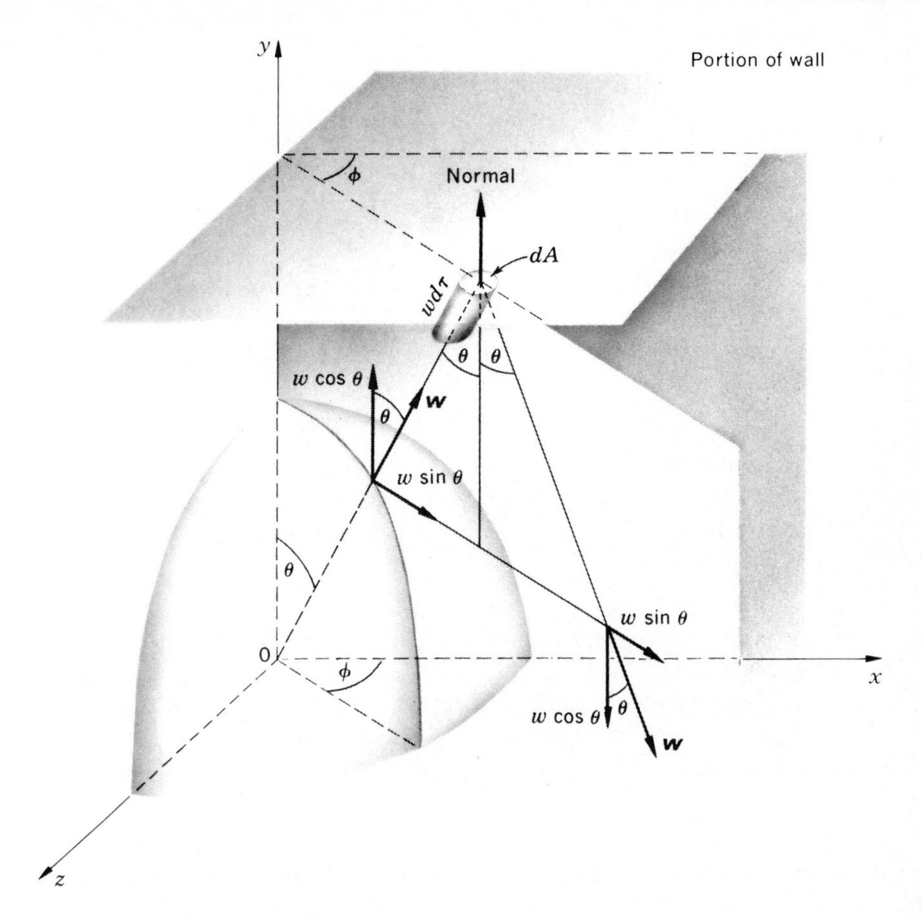

Figure 5-13 All the molecules in the cylinder of length $w\,d\tau$ strike the area dA at the angle θ to the normal. The perpendicular component of velocity $w \cos \theta$ is reversed, but the parallel component $w \sin \theta$ is unchanged.

and if V is the total volume of the container, only the fraction dV/V of the molecules will be contained within the cylinder. Therefore the number of molecules (speed range, dw; θ range, $d\theta$; ϕ range, $d\phi$) striking dA in time $d\tau$ is expressed as

$$\text{No. of } w, \theta, \phi \text{ molecules striking } dA \text{ in time } d\tau = d^3N_{w,\theta,\phi}\frac{dV}{V}, \tag{5-17}$$

which expresses the fact that *molecules have no preferred location.*

According to our fundamental assumptions, a molecular collision is perfectly elastic. It follows, therefoıe, that a molecule moving with speed w in a direction making an angle θ with the normal to a wall will undergo a change only in its perpendicular component of velocity, as shown in Fig. 5-13. Furthermore, it follows that the total change in momentum per collision is

$$\text{Change of momentum per collision} = -2mv \cos \theta. \tag{5-18}$$

Now,

$$\text{Total change of momentum} = \begin{bmatrix}\text{No. of molecules}\\ \text{of speed } w \text{ in}\\ \text{solid angle } d\Omega\end{bmatrix} \begin{bmatrix}\text{Fraction of}\\ \text{these molecules}\\ \text{striking } dA\\ \text{in time } d\tau\end{bmatrix} \begin{bmatrix}\text{Change in}\\ \text{momentum}\\ \text{per collision}\end{bmatrix}$$

$$= \left(dN_w \frac{d\Omega}{4\pi}\right) \qquad \left(\frac{dV}{V}\right) \qquad (-2mw \cos\theta)$$

$$= \left(\frac{dN_w}{4\pi} \sin\theta \, d\theta \, d\phi\right) \left(\frac{1}{V} w \, d\tau \cos\theta \, dA\right) (-2mw \cos\theta).$$

[Eq. (5-14)] [Eq. (5-16)] [Eq. (5-18)]

The change in momentum per unit time and per unit area due to collisions from all directions is the pressure dP_w exerted by the wall on the dN_w gas molecules. Reversing the sign of the momentum change, we get the pressure dP_w exerted by the dN_w molecules *on the wall:*

$$dP_w = mw^2 \frac{dN_w}{V} \left(\frac{1}{2\pi} \int_0^{2\pi} d\phi \int_0^{\pi/2} \cos^2\theta \sin\theta \, d\theta\right). \tag{5-19}$$

The quantity in parentheses may be integrated at sight and is found to be $\frac{1}{3}$, so that the total pressure due to molecules of all speeds is given by

$$PV = \tfrac{1}{3}m \int_0^\infty w^2 \, dN_w.$$

The average of the square of the molecular speeds $\langle w^2 \rangle$ is defined to be

$$\langle w^2 \rangle = \frac{1}{N} \int_0^\infty w^2 \, dN_w, \tag{5-20}$$

so that we have

$$PV = \tfrac{1}{3}Nm\langle w^2 \rangle,$$

or

$$PV = \tfrac{2}{3}N(\tfrac{1}{2}m\langle w^2 \rangle). \tag{5-21}$$

The quantity $\frac{1}{2}m\langle w^2 \rangle$ is the average kinetic energy per molecule. Comparing our theoretical equation of state with the experimental one, we see that an association must be made between the average kinetic energy per molecule and the ideal gas temperature. Thus,

$$\tfrac{2}{3}N(\tfrac{1}{2}m\langle w^2 \rangle) = nR\theta.$$

But N/n is the number of molecules per mole, or Avogadro's N_A, so that

$$\frac{1}{2}m\langle w^2 \rangle = \frac{3}{2}\frac{R}{N_A}\theta = \frac{3}{2}k\theta, \tag{5-22}$$

where k is *Boltzmann's constant*, given by

$$k = \frac{R}{N_A} = \frac{8.3143 \text{ kJ/kmol} \cdot \text{K}}{6.0225 \times 10^{26} \text{ molecules/kmol}} = 1.3805 \times 10^{-23} \text{ J/K}.$$

In this derivation, the average energy per molecule, $\frac{1}{2}m\langle w^2 \rangle$, is wholly kinetic energy of translation. This is the only kind of energy that a hard spherical molecule, uninfluenced by its neighbors, can possess. We have therefore limited ourselves to monatomic molecules only. Diatomic and polyatomic molecules can also rotate and vibrate and may therefore be expected to possess kinetic energy of rotation and vibration as well as potential energy of vibration, even though there are no forces among neighbors.

It is worth while at this point to compare the symbolism used in our treatment of kinetic theory with that used in thermodynamics. This is shown in Table 5-4. The molecular form of the ideal-gas equation has another simple form. Since $PV = \frac{2}{3}N(\frac{1}{2}m\langle w^2 \rangle)$ and $\frac{1}{2}m\langle w^2 \rangle = \frac{3}{2}k\theta$, we get

$$PV = (\tfrac{2}{3}N)(\tfrac{3}{2}k\theta) = Nk\theta$$

and

$$P = \frac{N}{V}k\theta. \tag{5-23}$$

Table 5-4 Comparison of symbols

Thermodynamics	Kinetic theory
M' = Mass of system	N = Number of molecules
n = Number of moles	m = Mass of a molecule
$\mathcal{M}$ = Mass per mole = M'/n (molar mass; molecular weight)	N_A = Molecules per mole = N/n (Avogadro's number)
R = Universal gas constant	k = Boltzmann's constant = R/N_A
V = Volume	V = Volume
ρ = Mass density = M'/V	Molecular density = N/V

PROBLEMS

5-1 A stream of air moves past a stationary obstacle with a speed w. Assume that a mass m of air is stopped adiabatically by the obstacle.

(*a*) Prove that the rise of temperature of this mass of air is given by

$$\Delta T = \frac{w^2}{5R/\mathcal{M}},$$

where is the molecular weight of air.

(*b*) Calculate ΔT when $w = 600$ mi/hr.

(*c*) Apply the equation in (*a*) to a meteorite moving through a stationary atmosphere at a speed of 20 mi/sec. What would happen?

5-2 A vertical cylindrical tank of length greater than 0.76 m has its top end closed by a tightly fitting frictionless piston of negligible weight. The air inside the cylinder is at an absolute pressure of 1 atm. (1 atm = 1.013×10^5 Pa.) The piston is depressed by pouring mercury on it slowly, so that the temperature of the air is maintained constant. What is the length of the air column when mercury starts to spill over the top of the cylinder?

5-3 Mercury is poured into the open end of a J-shaped glass tube which is closed at the short end, trapping the air in that end. Assuming air to act like an ideal gas, how much mercury can be poured in before it overflows? The long and short ends are 1 m and 0.5 m long, respectively, and effects due to the curvature of the bottom may be neglected. Take atmospheric pressure to be 75 cm Hg.

5-4 A cylindrical highball glass 15 cm high and 35 cm^2 in cross section contains water up to the 10-cm mark. A card is placed over the top and held there while the glass is inverted. When the support is removed, what mass of water must leave the glass in order that the rest of the water should remain in the glass, if one neglects the weight of the card? (*Caution:* Try this over a sink.)

5-5 Two bulbs containing air, one of which has a volume three times the other, are connected by a capillary of negligible volume and are initially at the same temperature. To what temperature must the air in the larger bulb be raised in order that the pressure be doubled? Neglect heat conduction through the air in the capillary.

5-6 (*a*) Prove that the volume expansivity β of a real gas and the compressibility factor Z are connected by the following relation:

$$\beta = \frac{1}{\theta} + \frac{1}{Z}\left(\frac{\partial Z}{\partial \theta}\right)_P.$$

(*b*) Prove that the isothermal compressibility κ of a real gas is given by the relation

$$\kappa = \frac{1}{P} - \frac{1}{Z}\left(\frac{\partial Z}{\partial P}\right)_\theta.$$

5-7 Expand the following equations in the form

$$Pv = R\theta\left(1 + \frac{B}{v} + \frac{C}{v^2} + \frac{D}{v^3} + \cdots\right),$$

and determine the second virial coefficient in each case:

(*a*) $$\left(P + \frac{a}{v^2}\right)(v - b) = R\theta \qquad \text{(van der Waals)}.$$

(*b*) $$P = \frac{R\theta(1 - c/v\theta^3)}{v^2}\left[v + B_0\left(1 - \frac{b}{v}\right)\right] - \frac{A_0(1 - a/v)}{v^2} \qquad \text{(Beattie-Bridgeman)}.$$

(*c*) $$Pv = R\theta + B'P + C'P^2 + D'P^3 + \cdots \qquad \text{(another type of virial expansion)}.$$

(*d*) $$P = \frac{R\theta}{v - b}e^{-a/R\theta v} \qquad \text{(Dieterici)}.$$

5-8 An ideal gas is contained in a cylinder equipped with a frictionless, nonleaking piston of area A. When the pressure is atmospheric P_0, the piston face is at a distance l from the closed end. The gas is compressed by moving the piston a distance x. Calculate the spring constant, or force constant, $\mathcal{F}/x$, under (*a*) isothermal conditions and (*b*) under adiabatic conditions. (*c*) In what respect is a gas cushion superior to a steel spring? (*d*) Using Eq. (4-14), show that $C_P - C_V = nR$ for an ideal gas.

5-9 The temperature of an ideal gas in a capillary of constant cross-sectional area varies linearly from one end ($x = 0$) to the other ($x = L$), according to the equation

$$\theta = \theta_0 + \frac{\theta_L - \theta_0}{L}x.$$

If the volume of the capillary is V and the pressure P is uniform throughout, show that the number of moles of gas n is given by

$$PV = nR\frac{\theta_L - \theta_0}{\ln(\theta_L/\theta_0)}.$$

Show that, when $\theta_L = \theta_0 = \theta$, the equation above reduces to the obvious one, $PV = nR\theta$.

5-10 Prove that the work done by an ideal gas with constant heat capacities during a quasi-static adiabatic expansion is equal to

(*a*) $$W = C_V(\theta_i - \theta_f).$$

(*b*) $$W = \frac{P_f V_f - P_i V_i}{\gamma - 1}.$$

(*c*) $$W = \frac{P_f V_f}{\gamma - 1}\left[1 - \left(\frac{P_i}{P_f}\right)^{(\gamma-1)/\gamma}\right].$$

5-11 (*a*) Show that the heat transferred during an infinitesimal quasi-static process of an ideal gas can be written

$$đQ = \frac{C_V}{nR} V\, dP + \frac{C_P}{nR} P\, dV.$$

Applying this equation to an adiabatic process, show that $PV^\gamma = \text{const}$.

(*b*) An ideal gas of volume 0.05 ft^3 and pressure 120 lb/in.^{2} undergoes a quasi-static adiabatic expansion until the pressure drops to 15 lb/in^2. Assuming γ to remain constant at the value 1.4, what is the final volume? How much work is done?

5-12 (*a*) Derive the following formula for a quasi-static adiabatic process of an ideal gas, assuming γ to be constant:

$$\theta V^{\gamma-1} = \text{const}.$$

(*b*) At about 100 ms after detonation of a uranium fission bomb, the "ball of fire" consists of a sphere of gas with a radius of about 50 ft and a temperature of 300,000 K. Making very rough assumptions, estimate at what radius its temperature would be 3000 K.

5-13 (*a*) Derive the following formula for a quasi-static adiabatic process of an ideal gas, assuming γ to be constant:

$$\frac{\theta}{P^{(\gamma-1)/\gamma}} = \text{const}.$$

(*b*) Helium ($\gamma = \frac{5}{3}$) at 300 K and 1 atm pressure is compressed quasi-statically and adiabatically to a pressure of 5 atm. Assuming that the helium behaves like an ideal gas, what is the final temperature?

5-14 A horizontal insulated cylinder contains a frictionless nonconducting piston. On each side of the piston is 54 liters of an inert monatomic ideal gas at 1 atm and 273 K. Heat is slowly supplied to the gas on the left side until the piston has compressed the gas on the right side to 7.59 atm.

(*a*) How much work is done on the gas on the right side?
(*b*) What is the final temperature of the gas on the right side?
(*c*) What is the final temperature of the gas on the left side?
(*d*) How much heat is added to the gas on the left side?

5-15 An evacuated bottle with nonconducting walls is connected through a valve to a gasholder, where the pressure is P_0 and the temperature θ_0. The valve is opened slightly, and helium flows into the bottle until the pressure within the bottle is P_0. Assuming the helium to behave like an ideal gas with constant heat capacities, show that the final temperature of the helium in the bottle is $\gamma\theta_0$.

5-16 A thick-walled insulated chamber contains n_i mol of helium at high pressure P_i. It is connected through a valve with a large, almost empty gasholder in which the pressure is maintained at a constant value P_0, very nearly atmospheric. The valve is opened slightly, and the helium flows slowly and adiabatically into the gasholder until the pressures on the two sides of the valve are equalized. Assuming the helium to behave like an ideal gas with constant heat capacities, show that:

(*a*) The final temperature of the gas in the chamber is

$$\theta_f = \theta_i \left(\frac{P_f}{P_i} \right)^{(\gamma - 1)/\gamma}.$$

(*b*) The number of moles of gas left in the chamber is

$$n_f = n_i \left(\frac{P_f}{P_i} \right)^{1/\gamma}.$$

(*c*) The final temperature of the gas in the gasholder is

$$\theta' = \frac{\theta_i}{\gamma} \frac{1 - P_f/P_i}{1 - (P_f/P_i)^{1/\gamma}}.$$

(*Hint:* See Prob. 4-11.)

5-17 (*a*) If y is the height above sea level, show that the decrease of atmosphere pressure due to a rise dy is given by

$$\frac{dP}{P} = -\frac{\mathcal{M} g}{R\theta} \, dy,$$

where $\mathcal{M}$ is the molecular weight of the air, g the acceleration of gravity, and θ the absolute temperature at the height y.

(*b*) If the decrease of pressure in (*a*) is due to an adiabatic expansion, show that

$$\frac{dP}{P} = \frac{\gamma}{\gamma - 1} \frac{d\theta}{\theta}.$$

(*c*) From (*a*) and (*b*), using some of the numerical data of Art. 5-7, calculate $d\theta/dy$ in degrees per kilometer.

5-18 A steel ball of mass 10 g is placed in a tube of cross-sectional area 1 cm^2. The tube is connected to an air tank of 5 liters capacity, the pressure of the air being 76 cm Hg.

(*a*) With what period will the ball vibrate?

(*b*) If the ball is held originally at a position where the gas pressure is exactly atmospheric and is then allowed to drop, how far will it go before it starts to come up?

5-19 A steel ball of mass 8 g is placed in a tube of cross-sectional area 1.2 cm^2. The tube is connected to an air tank of 6 liters capacity, the pressure of the air being 76 cm Hg.

(*a*) With what period will the ball vibrate?

(*b*) If the ball is held originally at a position where the gas pressure is exactly atmospheric and is then allowed to drop, how far will it go before it starts to come up?

5-20 Carbon dioxide is contained in a vessel whose volume is 5270 cm^3. A ball of mass 16.65 g, placed in a tube of cross-sectional area 2.01 cm^2, vibrates with a period of 0.834 s. What is γ when the barometer reads 72.3 cm?

5-21 Mercury is poured into a U tube open at both ends until the total length of the mercury is h.

(*a*) If the level of mercury on one side of the tube is depressed and then the mercury is allowed to oscillate with small amplitude, show that, neglecting friction, the period τ_1 is given by

$$\tau_1 = 2\pi \sqrt{\frac{h}{2g}}.$$

(*b*) One end of the U tube is now closed so that the length of the entrapped air column is L, and again the mercury is caused to oscillate. Assuming friction to be negligible, the air to be ideal, and the changes of volume to be adiabatic, show that the period τ_2 is now

$$\tau_2 = 2\pi \sqrt{\frac{h}{2g + \gamma h_0 g/L}},$$

where h_0 is the height of the barometric column.

(*c*) Show that

$$\gamma = \frac{2L}{h_0}\left(\frac{\tau_1^2}{\tau_2^2} - 1\right).$$

5-22 Prove that the expression for the speed of a longitudinal wave in an ideal gas may be transformed to

$$w = \sqrt{\left(\frac{\partial P}{\partial \rho}\right)_S}.$$

5-23 What is the speed of a longitudinal wave in argon at 20°C?

5-24 A standing wave of frequency 1100 Hz in a column of methane at 20°C produces nodes that are 20 cm apart. What is γ?

5-25 The speed of a longitudinal wave in a mixture of helium and neon at 300 K was found to be 758 m/s. What is the composition of the mixture?

5-26 The atomic weight of iodine is 127. A standing wave in iodine vapor at 400 K produces nodes that are 6.77 cm apart when the frequency is 1000 Hz. Is iodine vapor monatomic or diatomic?

5-27 Determine the temperature of the helium gas in the acoustical thermometer from which the data of Fig. 5-11 were obtained.

5-28 An open glass tube of uniform bore is bent into the shape of an L. One arm is immersed in a liquid of density ρ', and the other arm, of length L, remains in the air in a horizontal position. The tube is rotated with constant angular speed ω about the axis of the vertical arm. Prove that the height y to which the liquid rises in the vertical arm is equal to

$$y = \frac{P_0(1 - e^{-\omega^2 L^2 \mathcal{M}/2R\theta})}{g\rho'},$$

where P_0 is atmospheric pressure, $\mathcal{M}$ is the molecular weight of air, and g is the acceleration of gravity.

5-29 One mole of an ideal paramagnetic gas obeys Curie's law, with a Curie constant C_C'. Assume that the internal energy U is a function of θ only, so that $dU = C_{V,M}\, d\theta$, where $C_{V,M}$ is a constant.

(*a*) Show that the equation of the family of adiabatic surfaces is

$$\frac{C_{V,M}}{R} \ln \theta + \ln V = \frac{\mu_0 M^2}{8\pi n R C_C'} + \ln A,$$

where A is a constant for one surface.

(*b*) Sketch one of these surfaces on a θVM diagram.

5-30 The definition of the average speed of a molecule of an ideal gas is

$$\langle w \rangle = \frac{1}{N}\int_0^\infty w \, dN_w .$$

Prove that the number of molecules striking unit area of the container wall per unit time is equal to

$$\frac{N\langle w\rangle}{4V}.$$

5-31 The root-mean-square speed w_{rms} is defined as $\sqrt{\langle w^2\rangle}$. Show that:

(*a*) $$w_{\text{rms}} = \sqrt{\frac{3k\theta}{m}}.$$

(*b*) $w_{\text{rms}} = \sqrt{3/\gamma}$ times the speed of sound.

CHAPTER

SIX

ENGINES, REFRIGERATORS, AND THE SECOND LAW OF THERMODYNAMICS

6-1 CONVERSION OF WORK INTO HEAT, AND VICE VERSA

When two stones are rubbed together under water, the work done against friction is transformed into internal energy tending to produce a rise of temperature of the stones. As soon as the temperature of the stones rises above that of the surrounding water, however, there is a flow of heat into the water. If the mass of water is large enough or if the water is continually flowing, there will be no appreciable rise of temperature, and the water can be regarded as a heat reservoir. Since the state of the stones is the same at the end of the process as at the beginning, the net result of the process is merely the conversion of mechanical work into heat. Similarly, when an electric current is maintained in a resistor immersed either in running water or in a very large mass of water, there is also a conversion of electrical work into heat, without any change in the thermodynamic coordinates of the wire. In general, work of any kind W may be done upon a system in contact with a reservoir, giving rise to a flow of heat Q without altering the state of the system. The system acts merely as an intermediary. It is apparent from the first law that the work W is equal to the heat Q; or in other words, the transformation of work into heat is accomplished with 100 percent efficiency. Moreover, this transformation can be continued indefinitely.

To study the converse process, namely, the conversion of heat into work, we must also have at hand a process or series of processes by means of which such a conversion may continue indefinitely without involving any resultant changes in the state of any system. At first thought, it might appear that the isothermal expansion of an ideal gas might be a suitable process to consider in discussing the conversion of heat into work. In this case there is no internal-energy change since the temperature remains constant, and therefore $-W = Q$, or heat has been converted completely into work. This process, however, involves a change of state of the gas. The volume increases and the pressure decreases until atmospheric pressure is reached, at which point the process stops. It therefore cannot be used indefinitely.

What is needed is a series of processes in which a system is brought back to its initial state, i.e., a *cycle*. Each of the processes that constitute a cycle may involve a flow of heat to or from the system and the performance of work by or on the system. For one complete cycle, let

the total number of units of heat absorbed by the system be designated by the symbol $|Q_H|$,
the total number of units of heat rejected by the system be designated by the symbol $|Q_C|$, and
the total number of units of work done by the system be designated by the symbol $|W|$.

All three quantities $|Q_H|$, $|Q_C|$, and $|W|$ must be expressed *in the same units*. When this is done they are *absolute values* represented by *positive numbers* only.

In all chapters of this book *except the present* chapter, the symbols Q and W are algebraic quantities which may take on negative as well as positive values. When, therefore, one calculates the heat transferred as a result of an isothermal magnetization, for example, and obtains the result that $Q = -100$ J, one sees not only that $|Q|$ is 100 J but also that this heat was *rejected* by the system to the reservoir in contact with the system. In this chapter we shall deal with engines and refrigerators where we shall know at all times the direction of flow of Q and W, and are interested only in the absolute value of Q and W.

If $|Q_H|$ is larger than Q_C and if W is done by the system, the mechanical device by whose agency the system is caused to undergo the cycle is called a *heat engine*. The purpose of a heat engine is to deliver work continuously to the outside by performing the same cycle over and over again. The net work in the cycle is the output, and the heat absorbed by the working substance is the input. The *thermal efficiency* of the engine η is defined as

$$\text{Thermal efficiency} = \frac{\text{work output, in any energy units}}{\text{heat input, in the same energy units}},$$

or

$$\eta = \frac{|W|}{|Q_H|}. \tag{6-1}$$

Applying the first law to one complete cycle and remembering that there is no net change of internal energy, we get

$$|Q_H| - |Q_C| = |W|,$$

and therefore

$$\eta = \frac{|Q_H| - |Q_C|}{|Q_H|},$$

or

$$\boxed{\eta = 1 - \frac{|Q_C|}{|Q_H|}.} \tag{6-2}$$

It is seen from this equation that η will be unity (efficiency of 100 percent) when Q_C is zero. In other words, if an engine can be built to operate in a cycle in which there is no outflow of heat from the system, there will be 100 percent conversion of the absorbed heat into work.

The transformation of heat into work is usually accomplished in practice by two general types of engine: the *external-combustion engine,* such as the Stirling engine and the steam engine, and the *internal-combustion engine,* such as the gasoline engine and the diesel engine. In both types, a gas or a mixture of gases contained in a cylinder undergoes a cycle, thereby causing a piston to impart to a shaft a motion of rotation against an opposing force. It is necessary in both engines that, at some time in the cycle, the gas in the cylinder be raised to a high temperature and pressure. In the Stirling and steam engines this is accomplished by an outside furnace. The high temperature and pressure achieved in the internal-combustion engine, however, are produced by a chemical reaction between a fuel and air that takes place in the cylinder itself. In the gasoline engine, the combustion of the gasoline and air takes place explosively through the agency of an electric spark. The diesel engine, however, uses oil as a fuel, the combustion of which is accompanied more slowly by spraying the oil into the cylinder at a convenient rate.

6-2 THE STIRLING ENGINE

In 1816, before the science of thermodynamics had even been begun, a minister of the Church of Scotland named Robert Stirling designed and patented a hot-air engine that could convert some of the energy liberated by a burning fuel into work. The Stirling engine remained useful and popular for many years but, with the development of steam engines and internal-combustion engines, finally became obsolete. The Stirling engine is being developed by the engineers of the Philips Corporation in The Netherlands and the Ford Motor Company for use in automobiles. The Stirling engine has low exhaust emissions and high engine efficiencies for varying loads, but high manufacturing costs.

The steps in the operation of a somewhat idealized Stirling engine are shown schematically in Fig. 6-1*a*. Two pistons, an expansion piston on the left and a

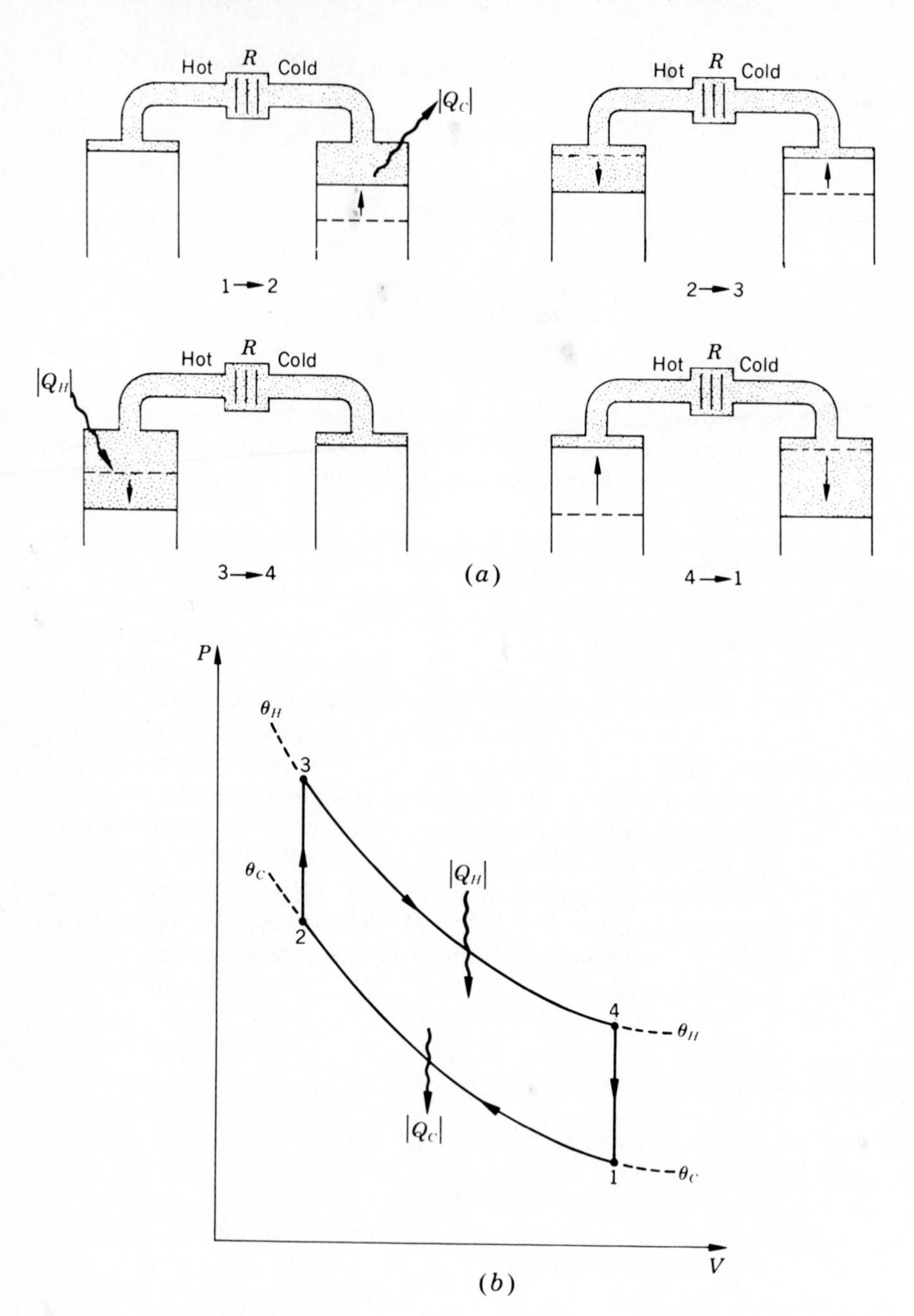

Figure 6-1 (*a*) Schematic diagram of steps in the operation of an idealized Stirling engine. (The numbers under each diagram refer to the processes shown on the *PV* diagram in Fig. 6-1*b*.) (*b*) Idealized Stirling engine cycle on a *PV* diagram.

compression piston on the right, are connected to the same shaft. As the shaft rotates, these pistons move in different phase, with the aid of suitable connecting linkages. The space between the two pistons is filled with gas, and the left-hand portion of the space is kept in contact with a hot reservoir (burning fuel), while the right-hand portion is in contact with a cold reservoir. Between the two portions of gas is a device R, called a *regenerator*, consisting of a packing of steel wool or a series of metal baffles, whose thermal conductivity is low enough to support the temperature difference between the hot and cold ends without appreciable heat conduction. The Stirling cycle consists of four processes depicted schematically in Fig. 6-1*a* and involving pressure and volume changes plotted (as though ideal conditions existed) on the PV diagram of Fig. 6-1*b*.

$1 \rightarrow 2$ While the left piston remains at the top, the right piston moves halfway up, compressing cold gas while in contact with the cold reservoir and therefore causing heat $|Q_C|$ to leave. This is an approximately isothermal compression and is depicted as a rigorously isothermal process at the temperature θ_C in Fig. 6-1*b*.

$2 \rightarrow 3$ The left piston moves down and the right piston up, so that there is no change in volume, but gas is forced through the regenerator from the cold side to the hot side and enters the left-hand side at the higher temperature θ_H. To accomplish this, the regenerator supplied heat $|Q_R|$ to the gas. Note that the process $2 \rightarrow 3$ in Fig. 6-1*b* is at constant volume.

$3 \rightarrow 4$ The right piston now remains stationary as the left piston continues moving down while in contact with the hot reservoir, causing the gas to undergo an approximately isothermal expansion, during which heat $|Q_H|$ is absorbed at the temperature θ_H, as shown in Fig. 6-1*b*.

$4 \rightarrow 1$ Both pistons move in opposite directions, thereby forcing gas through the regenerator from the hot to the cold side and giving up approximately the same amount of heat $|Q_R|$ to the regenerator that is absorbed in the process $2 \rightarrow 3$. This process takes place at practically constant volume.

The net result of the cycle is the absorption of heat $|Q_H|$ at the high temperature θ_H, the rejection of heat $|Q_C|$ at the low temperature θ_C, and the

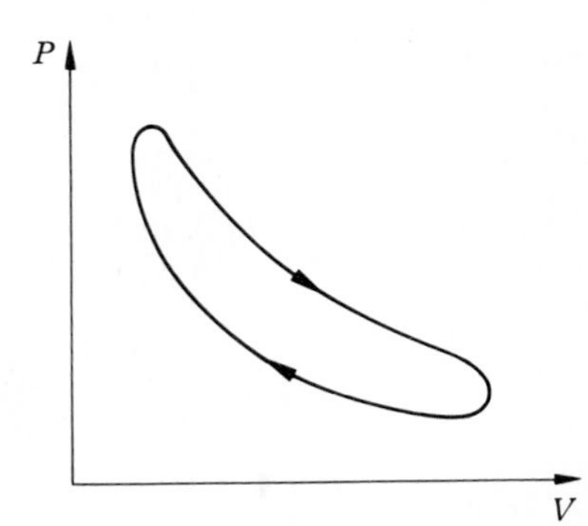

Figure 6-2 PV diagram of an actual Stirling engine.

delivery of work $|W| = |Q_H| - |Q_C|$ to the surroundings, with no net heat transfer resulting from the two constant-volume processes. It must be emphasized that Fig. 6-1*b* is based on the assumptions that (1) the gas is ideal, (2) no leakage of gas takes place, (3) no heat is lost or gained through cylinder walls, (4) no heat is conducted through the regenerator, and (5) there is no friction. An indicator diagram of an actual Stirling cycle would look much more like the diagram in Fig. 6-2. Even if these idealizations could be realized in practice, there would still be some heat $|Q_C|$ rejected at the lower temperature, and therefore all the input $|Q_H|$ could not be converted into work.

6-3 THE STEAM ENGINE

A schematic diagram of an elementary steam power plant is shown in Fig. 6-3*a*. The operation of such a plant can be understood by following the pressure and volume changes of a small constant mass of water as it is conveyed from the condenser, through the boiler, into the expansion chamber, and back to the condenser. The water in the condenser is at a pressure less than atmospheric and at a temperature less than the normal boiling point. By means of a pump it is introduced into the boiler, which is at a much higher pressure and temperature.

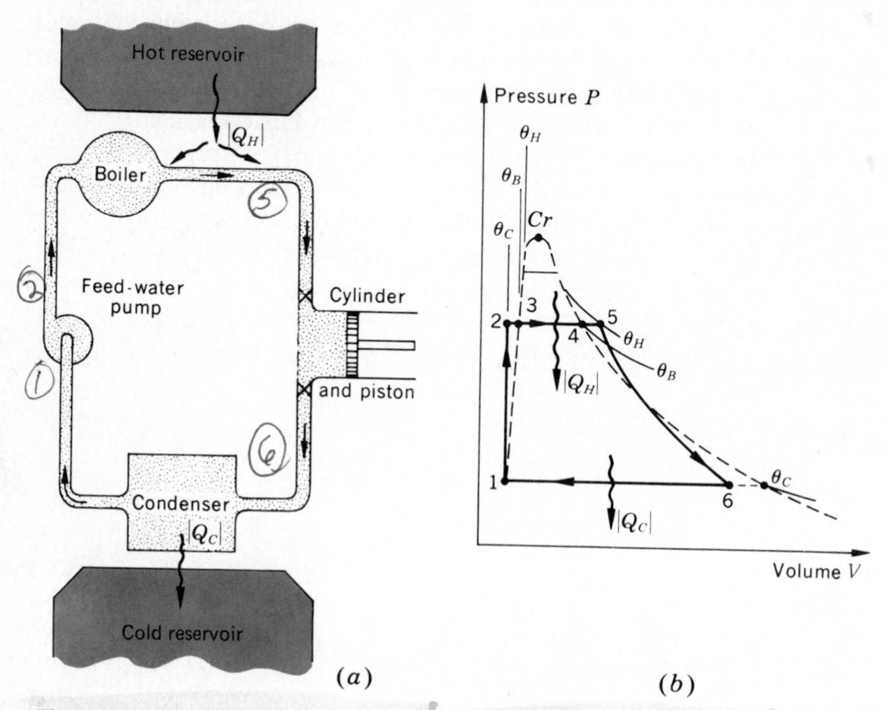

Figure 6-3 (*a*) Elementary steam power plant. (*b*) PV diagram of Rankine cycle.

In the boiler the water is first heated to its boiling point and then vaporized, both processes taking place approximately at constant pressure. The steam is then superheated at the same pressure. It is then allowed to flow into a cylinder, where it expands approximately adiabatically against a piston or a set of turbine blades, until its pressure and temperature drop to that of the condenser. In the condenser, finally, the steam condenses into the water at the same temperature and pressure as at the beginning, and the cycle is complete.

In the actual operation of the steam plant there are several processes that render an exact analysis difficult. These are: (1) acceleration and turbulence caused by the pressure difference required to cause the flow of the steam from one part of the apparatus to another, (2) friction, (3) conduction of heat through the walls during expansion of the steam, (4) heat transfers due to a finite temperature difference between the furnace and the boiler.

A first approximation to the solution of the problem of the steam plant may be made by introducing some simplifying assumptions which, although in no way realizable in practice, provide at least an upper limit to the efficiency of such a plant and which define a cycle called the *Rankine cycle*, in terms of which the actual behavior of a steam plant may be discussed.

In Fig. 6-3*b* three isotherms of water are shown on a PV diagram: one at θ_C corresponding to the temperature of the condenser, another at θ_B for the temperature of the boiler, and a third at a still higher temperature θ_H. The dashed curves are the liquid and the vapor saturation curves, respectively. In the Rankine cycle all processes are assumed to be well behaved; complications that arise from acceleration, turbulence, friction, and heat losses are thus eliminated. Starting at the point 1, representing the state of 1 lb of saturated liquid water at the temperature and pressure of the condenser, the Rankine cycle comprises the following six processes:

$1 \rightarrow 2$ Adiabatic compression of water to the pressure of the boiler (only a very small change of temperature takes place during this process).

$2 \rightarrow 3$ Isobaric heating of water to the boiling point.

$3 \rightarrow 4$ Isobaric, isothermal vaporization of water into saturated steam.

$4 \rightarrow 5$ Isobaric superheating of steam into superheated steam at temperature θ_H.

$5 \rightarrow 6$ Adiabatic expansion of steam into wet steam.

$6 \rightarrow 1$ Isobaric, isothermal condensation of steam into saturated water at the temperature θ_C.

During the processes $2 \rightarrow 3$, $3 \rightarrow 4$, and $4 \rightarrow 5$, heat $|Q_H|$ enters the system from a hot reservoir; whereas during the condensation process $6 \rightarrow 1$, heat $|Q_C|$ is rejected by the system to a reservoir at θ_C. This condensation process *must* exist in order to bring the system back to its initial state 1. Since heat is always rejected during the condensation of water, $|Q_C|$ cannot be made equal to zero, and therefore the input $|Q_H|$ cannot be converted completely into work.

6-4 INTERNAL-COMBUSTION ENGINES

In the gasoline engine, the cycle involves the performance of six processes, four of which require motion of the piston and are called *strokes:*

1. *Intake stroke.* A mixture of gasoline vapor and air is drawn into the cylinder by the suction of the piston. The pressure of the outside is greater than that of the mixture by an amount sufficient to cause acceleration and to overcome friction.
2. *Compression stroke.* The mixture of gasoline vapor and air is compressed until its pressure and temperature rise considerably. This is accomplished by the compression stroke of the piston, in which friction, acceleration, and heat loss by conduction are present.
3. *Ignition.* Combustion of the hot mixture is caused to take place very rapidly by an electric spark. The resulting combustion products attain a very high pressure and temperature, but the volume remains unchanged. The piston does not move during this process.
4. *Power stroke.* The hot combustion products expand and push the piston out, thus suffering a drop in pressure and temperature. This is the power stroke of the piston and is also accompanied by friction, acceleration, and heat conduction.
5. *Valve exhaust.* The combustion products at the end of the power stroke are still at a higher pressure and temperature than the outside. An exhaust valve allows some gas to escape until the pressure drops to that of the atmosphere. The piston does not move during this process.
6. *Exhaust stroke.* The piston pushes almost all the remaining combustion products out of the cylinder by exerting a pressure sufficiently larger than that of the outside to cause acceleration and overcome friction. This is the exhaust stroke.

In the processes described above there are several phenomena that render an exact mathematical analysis almost impossible. Among these are friction, acceleration, loss of heat by conduction, and the chemical reaction between gasoline vapor and air. A drastic but useful simplification is provided by eliminating these troublesome effects. When this is done, we have a sort of idealized gasoline engine that performs a cycle known as an *Otto cycle.*

The behavior of a gasoline engine can be approximated by assuming a set of ideal conditions as follows: (1) The working substance is at all times air, which behaves like an ideal gas with constant heat capacities. (2) All processes are quasi-static. (3) There is no friction. On the basis of these assumptions the air-standard Otto cycle is composed of six simple processes of an ideal gas, which are plotted on a PV diagram in Fig. 6-4*a*.

$5 \rightarrow 1$ represents a quasi-static isobaric intake at atmospheric pressure. There is no friction and no acceleration. The volume varies from zero to V_1 as the

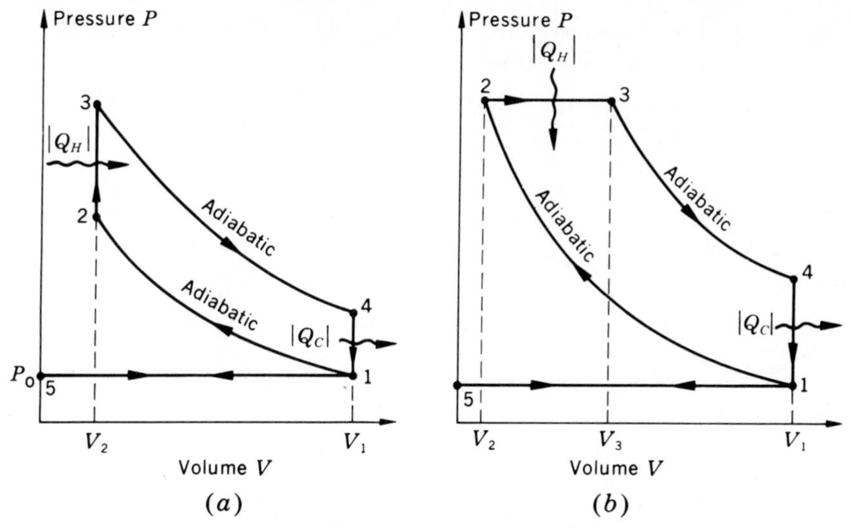

Figure 6-4 (*a*) Air-standard Otto cycle. (*b*) Air-standard Diesel cycle.

number of moles varies from zero to n_1 according to the equation

$$P_0 V = bR\theta_1,$$

where P_0 is atmospheric pressure and θ_1 is the temperature of the outside air.

$1 \to 2$ represents a quasi-static, adiabatic compression of n_1 moles of air. There is no friction and no loss of heat through the cylinder wall. The temperature rises from θ_1 to θ_2 according to the equation

$$\theta_1 V_1^{\gamma-1} = \theta_2 V_2^{\gamma-1}.$$

$2 \to 3$ represents a quasi-static isochoric increase of temperature and pressure of n_1 moles of air, brought about by an absorption of heat $|Q_H|$ from a series of external reservoirs whose temperatures range from θ_2 to θ_3. If there were only one reservoir at temperature θ_3, the flow of heat would not be quasi-static. This process is meant to approximate the effect of the explosion in a gasoline engine.

$3 \to 4$ represents a quasi-static adiabatic expansion of n_1 moles of air, involving a drop in temperature from θ_3 to θ_4 according to the equation

$$\theta_3 V_2^{\gamma-1} = \theta_4 V_1^{\gamma-1}.$$

$4 \to 1$ represents a quasi-static isochoric drop in temperature and pressure of n_1 moles of air, brought about by a rejection of heat $|Q_C|$ to a series of external reservoirs ranging in temperature from θ_4 to θ_1. This process is meant to approximate the drop to atmospheric pressure upon opening the exhaust valve.

$1 \to 5$ represents a quasi-static isobaric exhaust at atmospheric pressure. The volume varies from V_1 to zero as the number of moles varies from n_1 to zero, with the temperature remaining constant at the value θ_1.

The two isobaric processes $5 \to 1$ and $1 \to 5$ obviously cancel each other and need not be considered further. Of the four remaining processes, only two involve a flow of heat. There occur an absorption of $|Q_H|$ units of heat at high temperatures from $2 \to 3$ and a rejection of $|Q_C|$ units of heat at lower temperatures from $4 \to 1$, as indicated in Fig. 6-4*a*.

Assuming C_V to be constant along the line $2 \to 3$, we get

$$|Q_H| = \int_{\theta_2}^{\theta_3} C_V \, d\theta = C_V(\theta_3 - \theta_2).$$

Similarly, for process $4 \to 1$

$$|Q_C| = \int_{\theta_4}^{\theta_1} C_V \, d\theta = C_V(\theta_4 - \theta_1).$$

The thermal efficiency is therefore

$$\eta = 1 - \frac{|Q_C|}{|Q_H|} = 1 - \frac{\theta_4 - \theta_1}{\theta_3 - \theta_2}.$$

The two adiabatic processes provide the equations

$$\theta_4 V_1^{\gamma - 1} = \theta_3 V_2^{\gamma - 1}$$

and

$$\theta_1 V_1^{\gamma - 1} = \theta_2 V_2^{\gamma - 1},$$

which, after subtraction, yield

$$(\theta_4 - \theta_1) V_1^{\gamma - 1} = (\theta_3 - \theta_2) V_2^{\gamma - 1},$$

or

$$\frac{\theta_4 - \theta_1}{\theta_3 - \theta_2} = \left(\frac{V_2}{V_1} \right)^{\gamma - 1}.$$

Denoting the ratio V_1/V_2 by r, where r is called the *compression ratio* or the *expansion ratio*, we have finally

$$\eta = 1 - \frac{1}{(V_1/V_2)^{\gamma - 1}} = 1 - \frac{1}{r^{\gamma - 1}}. \qquad (6\text{-}3)$$

In an actual gasoline engine, r cannot be made greater than about 10 because, if r is larger, the rise of temperature upon compression of the mixture of gasoline and air is great enough to cause combustion before the advent of the spark. This is called *preignition*. Taking r equal to 9 and γ equal to 1.5 (for air, γ is more nearly 1.4),

$$\eta = 1 - \frac{1}{\sqrt{9}}$$
$$= 0.67 = 67 \text{ percent}.$$

All the troublesome effects present in an actual gasoline engine, such as acceleration, turbulence, and heat conduction by virtue of a finite temperature difference, are such as to make the efficiency much lower than that of the air-standard Otto cycle.

In the Diesel engine, only air is admitted on the intake. The air is compressed adiabatically until the temperature is high enough to ignite oil that is sprayed into the cylinder after the compression. The rate of supply of oil is adjusted so that combustion takes place approximately isobarically, the piston moving out during combustion. The rest of the cycle—namely, power stroke, valve exhaust, and exhaust stroke—is exactly the same as in the gasoline engine. The usual troublesome effects, such as chemical combination, friction, acceleration, and heat losses, take place in the Diesel engine as in the gasoline engine. Eliminating these effects by making the same assumptions as before, we are left with a sort of idealized Diesel engine that performs a cycle known as the *air-standard Diesel cycle*. If the line $2 \rightarrow 3$ in Fig. 6-4*a* is imagined horizontal instead of vertical, the resulting cycle will be the air-standard Diesel cycle. This is shown in Fig. 6-4*b*.

It is a simple matter to show that the efficiency of an engine operating in an idealized Diesel cycle is given by

$$\eta = 1 - \frac{1}{\gamma} \frac{(1/r_E)^\gamma - (1/r_C)^\gamma}{(1/r_E) - (1/r_C)}, \tag{6-4}$$

where $$r_E = \frac{V_1}{V_3} = \text{expansion ratio,}$$

and $$r_C = \frac{V_1}{V_2} = \text{compression ratio.}$$

In practice, the compression ratio of a Diesel engine can be made much larger than that of a gasoline engine, because there is no fear of preignition since only air is compressed. Taking, for example, $r_C = 15$, $r_E = 5$, and $\gamma = 1.5$,

$$\begin{aligned} \eta &= 1 - \frac{2}{3} \frac{1/5^{3/2} - 1/15^{3/2}}{\frac{1}{5} - \frac{1}{15}} \\ &= 1 - 5(0.0895 - 0.0172) \\ &= 64 \text{ percent.} \end{aligned}$$

The efficiencies of actual Diesel engines are still lower, of course, for the reasons mentioned in connection with the gasoline engine.

In the Diesel engine just considered, four strokes of the piston are needed for the execution of a cycle, and only one of the four is a power stroke. Since only air is compressed in the Diesel engine, it is possible to do away with the exhaust and intake strokes and thus complete the cycle in two strokes. In the two-stroke-cycle Diesel engine, every other stroke is a power stroke, and thus the power is doubled. The principle is very simple: At the conclusion of the power stroke, when the cylinder is full of combustion products, the valve opens, exhaust takes place until the combustion products are at atmospheric pressure, and then, instead of using the piston itself to exhaust the remaining gases, fresh air is blown into the cylinder, replacing the combustion products. A blower, operated by the engine itself, is used for this purpose, and thus it accomplishes in one simple operation what formerly required two separate piston strokes.

6-5 KELVIN-PLANCK STATEMENT OF THE SECOND LAW OF THERMODYNAMICS

In the preceding pages, four different heat engines have been briefly and somewhat superficially described. There are, of course, more types of enginers and a tremendous number of structural details, methods of increasing thermal efficiency, mathematical analyses, etc., which constitute the subject matter of engineering thermodynamics. Thermodynamics owes its origin to the attempt to convert heat into work and to develop the theory of operation of devices for this purpose. It is therefore fitting that one of the fundamental laws of thermodynamics is based upon the operation of heat engines. Reduced to their simplest terms, the important characteristics of heat-engine cycles may be summed up as follows:

1. There is some process or series of processes during which there is an absorption of heat from an external reservoir at a high temperature (called simply the *hot reservoir*).
2. There is some process or series of processes during which heat is rejected to an external reservoir at a lower temperature (called simply the *cold reservoir*).
3. There is some process or series of processes during which work is delivered to the surroundings.

This is represented schematically in Fig 6-5. No engine has ever been developed that converts the heat extracted from one reservoir into work without

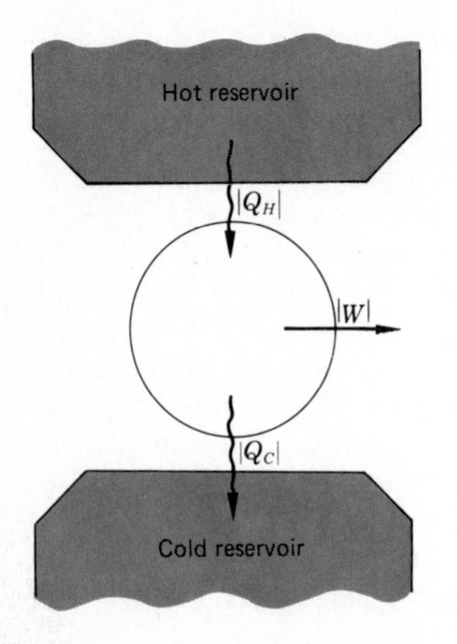

Figure 6-5 Symbolic representation of heat engine.

rejecting some heat to a reservoir at a lower temperature. This negative statement, which is the result of engineering experience, constitutes the *second law of thermodynamics* and has been formulated in several ways. The original statement of Kelvin is: "It is impossible by means of inanimate material agency to derive mechanical effect from any portion of matter by cooling it below the temperature of the coldest of the surrounding objects." In the words of Planck, "It is impossible to construct an engine which, working in a complete cycle, will produce no effect other than the raising of a weight and the cooling of a heat reservoir." We may combine these statements into one equivalent statement, to which we shall refer hereafter as the *Kelvin-Planck statement of the second law*, thus:

No process is possible whose sole *result is the absorption of heat from a reservoir and the conversion of this heat into work.*

If the second law were not true, it would be possible to drive a steamship across the ocean by extracting heat from the ocean or to run a power plant by extracting heat from the surrounding air. The student should notice that neither of these "impossibilities" violates the first law of thermodynamics. After all, both the ocean and the surrounding air contain an enormous store of internal energy, which, in principle, may be extracted in the form of a flow of heat. There is nothing in the first law to preclude the possibility of converting this heat completely into work. The second law, therefore, is not a deduction from the first but stands by itself as a separate law of nature, referring to an aspect of nature different from that contemplated by the first law. The first law denies the possibility of creating or destroying energy; the second denies the possibility of utilizing energy in a particular way. A machine that creates its own energy and thus violates the first law is called *perpetual motion machine of the first kind*. A machine that utilizes the internal energy of only one heat reservoir, thus violating the second law, is called *perpetual motion machine of the second kind*.

6-6 THE REFRIGERATOR

We have seen that a heat engine is a device by which a system is taken through a cycle in such a direction that some heat is absorbed while the temperature is high, a smaller amount is rejected at a lower temperature, and a net amount of work is done on the outside. If we imagine a cycle performed in a direction opposite to that of an engine, the net result would be the absorption of some heat at a low temperature, the rejection of a *larger* amount at a higher temperature, and a net amount of work done *on* the system. A device that performs a cycle in this direction is called a *refrigerator*, and the system undergoing the cycle is called a *refrigerant*.

The Stirling cycle is capable of being reversed and, when reversed, it gives rise to one of the most useful types of refrigerator. The operation of an ideal Stirling refrigerator may best be understood with the aid of the schematic diagrams shown in Fig. 6-6*a*, and the accompanying *PV* diagram of Fig. 6-6*b*.

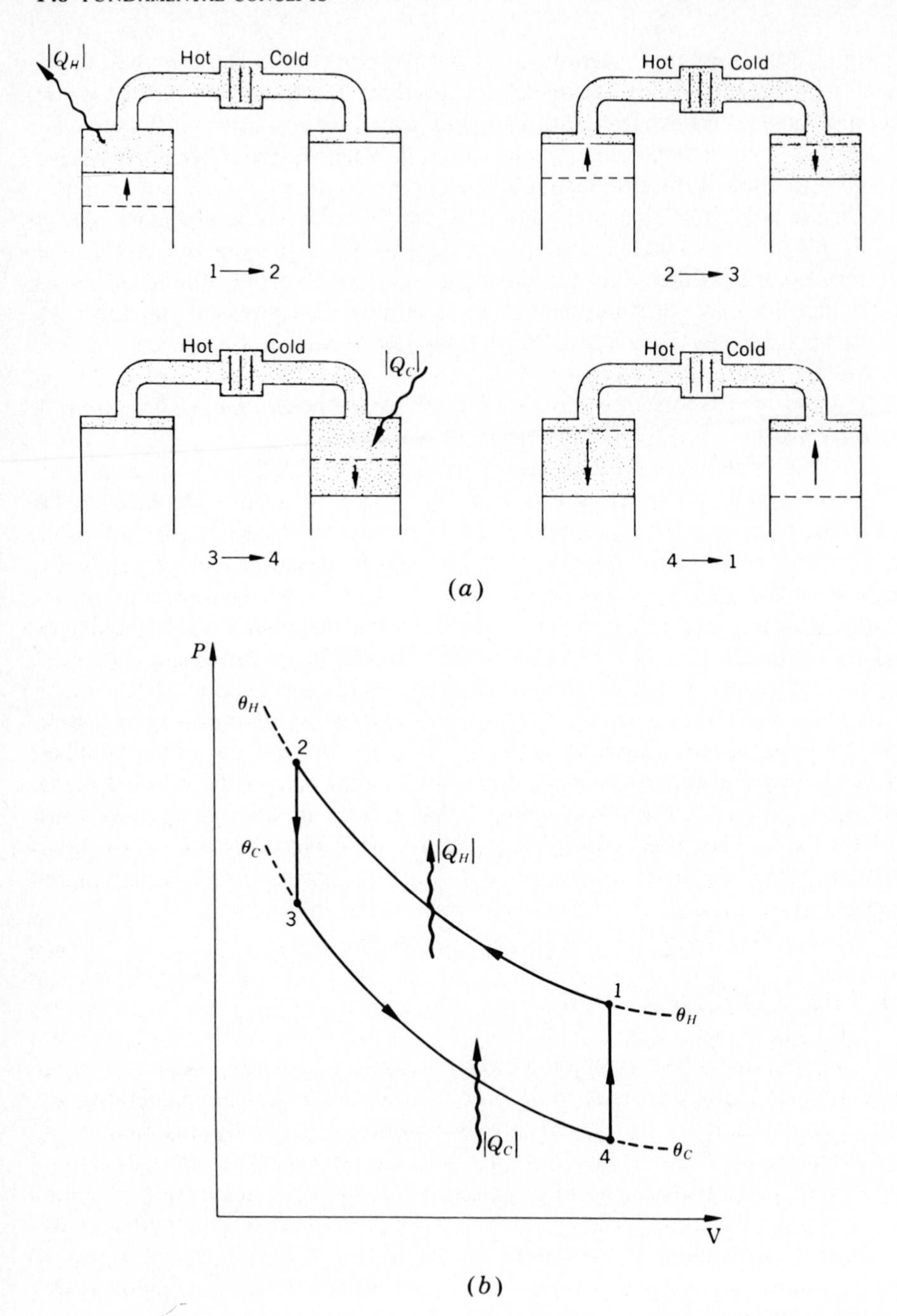

Figure 6-6 (*a*) Schematic diagram of steps in the operation of idealized Stirling refrigerator. (*b*) Idealized Stirling refrigeration cycle on a PV diagram.

$1 \to 2$ While the right piston remains stationary, the left piston moves up, compressing the gas isothermally at the temperature θ_H and rejecting heat $|Q_H|$ to the hot reservoir.

$2 \to 3$ Both pistons move the same amount simultaneously, forcing gas through the regenerator, giving up some heat $|Q_R|$ to the regenerator, and emerging cold in the right-hand space. This takes place at constant volume.

$3 \to 3$ While the left piston remains stationary, the right piston moves down and causes an isothermal expansion at the low temperature θ_C, during which heat $|Q_C|$ is absorbed by the gas from the cold reservoir.

$4 \to 1$ Both pistons move and force gas at constant volume from the cold to the hot end through the regenerator, thereby taking up approximately the same heat $|Q_R|$ that was supplied to the regenerator in process $2 \to 3$.

The Stirling refrigeration cycle has been utilized in recent years by a number of engineering firms in the construction of practical refrigerators for the production of low temperatures, from 90 K down to 12 K. As in the case of the Stirling engine, the Philips Company in Holland has been in the forefront in the design and construction of large industrial installations. (Miniature units suitable for academic laboratories are available from North American Philips, A. D. Little and Hughes Aircraft.)

In order to gain a little more insight into the working of a refrigerator, let us consider some of the details of a commercial refrigeration plant that are reflected in most electric home refrigerators. The schematic diagram in Fig. 6-7*a* shows the path of a constant mass of refrigerant as it is conveyed from the liquid storage, where it is at the temperature and pressure of the condenser, through the throttling valve, through the evaporator, into the compressor, and finally back to the condenser.

In the condenser the refrigerant is at a high pressure and at as low a temperature as can be obtained with air or water cooling. The refrigerant is always of such a nature that, at this pressure and temperature, it is a saturated liquid. When a fluid passes through a narrow opening (a needle valve) from a region of constant high pressure to a region of constant lower pressure adiabatically, it is said to undergo a *throttling process* (see Prob. 4-8), or a *Joule-Thomson* or *Joule-Kelvin expansion.* This process will be considered in some detail in Chap. 9. It is a property of saturated liquids (not of gases) that a throttling process always produces cooling and partial vaporization. In the evaporator the fluid is completely vaporized, with the heat of vaporization being supplied by the materials to be cooled. The vapor is then compressed adiabatically, thereby increasing in temperature. In the condenser, the vapor is cooled until it condenses and becomes completely liquefied.

The ideal refrigeration cycle depicted on a *PV* diagram in Fig. 6-7*b* is the result of ignoring the usual difficulties due to turbulence, friction, heat losses, etc. In Fig. 6-7*b*, two isotherms of a fluid such as ammonia or freon are shown on a *PV* diagram—one at θ_H, the temperature of the condenser, and the other at θ_C,

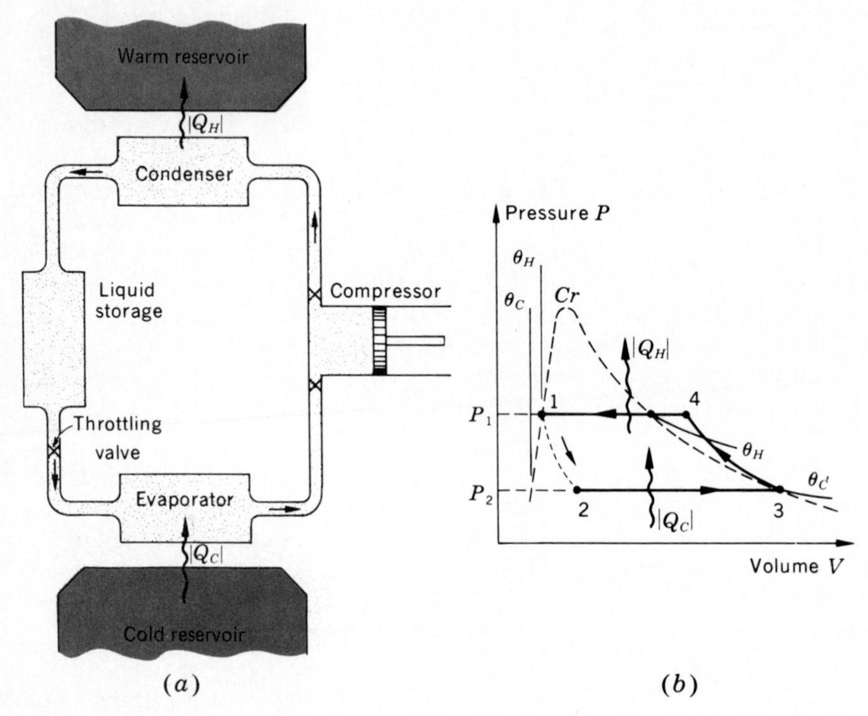

Figure 6-7 (*a*) Elementary refrigeration plant. (*b*) PV diagram of commercial refrigeration cycle.

the temperature of the evaporator. Starting at the point 1, representing the state of the saturated liquid at the temperature and pressure of the condenser, the commercial refrigeration cycle comprises the following processes:

$1 \to 2$ Throttling process involving a drop of pressure and temperature. The states between the initial and final states of a fluid during a throttling process cannot be described with the aid of thermodynamic coordinates referring to the system as a whole and, therefore, cannot be represented by points on a PV diagram. Hence the series of dashes between 1 and 2.

$2 \to 3$ Isothermal, isobaric vaporization in which heat Q_C is absorbed by the refrigerant at the low temperature θ_C, thereby cooling the materials of the cold reservoir.

$3 \to 4$ Adiabatic compression of the vapor to a temperature higher than that of the condenser θ_H.

$4 \to 1$ Isobaric cooling and condensation at θ_H.

The purpose of any refrigerator is to extract as much heat as possible from a cold reservoir with the expenditure of as little work as possible. The "output," so to speak, is the heat extracted from the cold reservoir, and the "input" is work. A convenient measure, therefore, of the performance of a refrigerator is expressed by the *coefficient of performance* ω (also called the *cooling energy ratio*), where

$$\omega = \frac{\text{heat extracted from cold reservoir}}{\text{work done on refrigerant}}$$

If, in one cycle, heat Q_C is absorbed by the refrigerant from the cold reservoir and work $|W|$ is done by the electric motor that operates the refrigerator, then

$$\omega = \frac{|Q_C|}{|W|} = \frac{|Q_C|}{|Q_H| - |Q_C|}$$

The coefficient of performance of a refrigerator may be considerably larger than unity. If, for the sake of argument, one assumes the value 5,

$$\omega = \frac{|Q_C|}{|W|} = 5.$$

But

$$|Q_C| = |Q_H| - |W|;$$

hence,

$$\frac{|Q_H| - |W|}{|W|} = 5,$$

$$\frac{|Q_H|}{|W|} = 6,$$

and therefore the *heat liberated at the higher temperature is equal to six times the work done*. If the work is supplied by an electric motor, for every joule of electrical energy supplied, 6 J of heat will be liberated; whereas if 1 J of electrical energy were dissipated in a resistor, one could obtain at most 1 J of heat. Consequently, it would seem to be highly advantageous to heat a house by refrigerating the outdoors.

This was first pointed out in 1852 by Lord Kelvin, who designed a machine for the purpose. The device was never built, however, and it remained for Haldane, about 75 years later, to utilize the principle and heat his house in Scotland by refrigerating the outdoor air, supplemented by city water. Many devices known as "heat pumps" have appeared on the market for warming the house in winter by refrigerating either the ground, the outside air, or the water supplied in the mains. By turning a valve and reversing the flow of refrigerant, the heat pump may also be used to cool the house in summer, as shown in Fig. 6-8. Various commercial units have coefficients of performance ranging from about 2 to 7. The design, installation, and operation of such units are now an important branch of engineering.

The operation of a refrigerator may be symbolized by the schematic diagram shown in Fig. 6-9, which should be compared with the corresponding engine diagram of Fig. 6-5. Work is always necessary to transfer heat from a cold to a hot reservoir. In household refrigerators, this work is usually done by an electric motor, whose cost of operation appears regularly on the monthly bill. It would

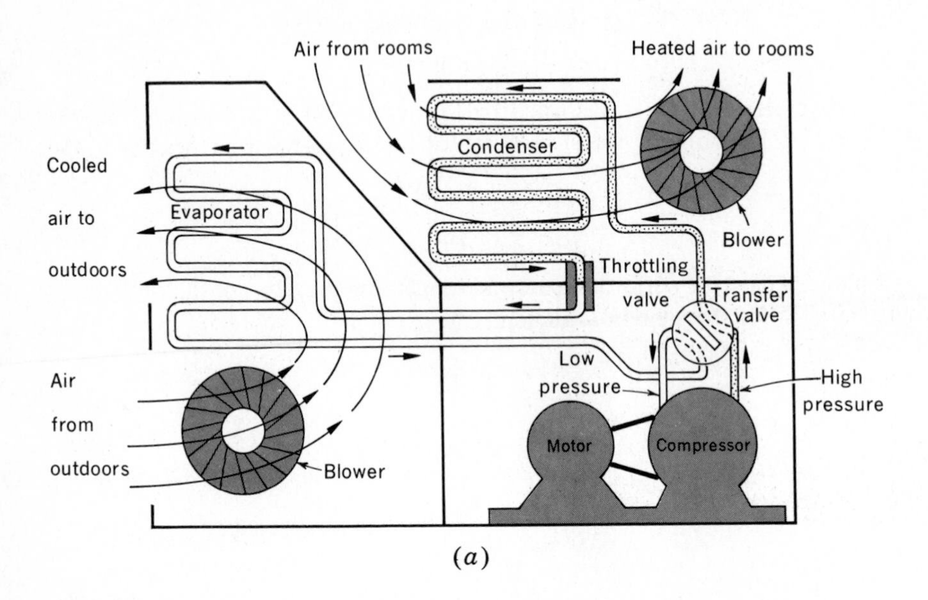

(*a*)

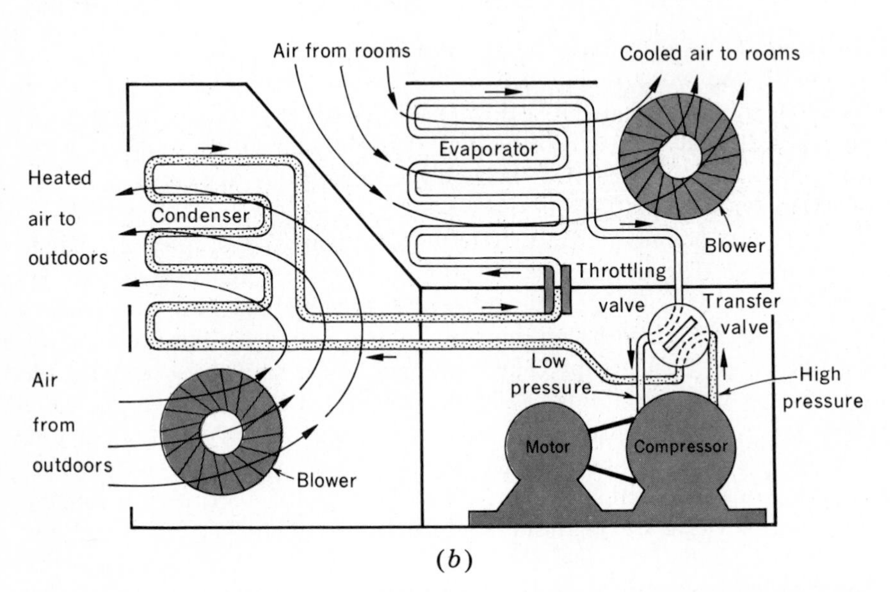

(*b*)

Figure 6-8 (*a*) Heating the house by refrigerating the outside air. (*b*) Cooling the house by heating the outside air.

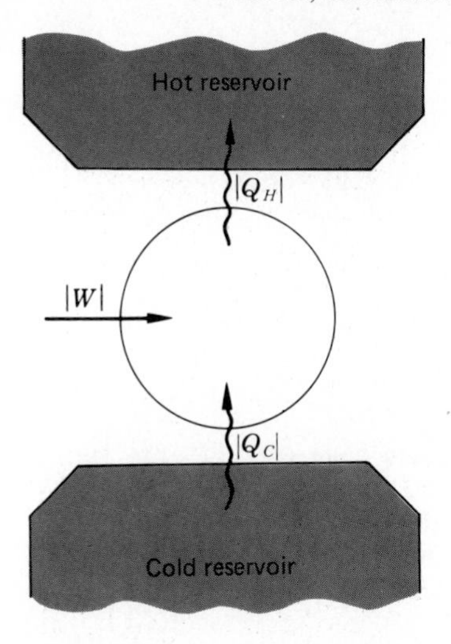

Figure 6-9 Symbolic representation of refrigerator.

be a boon to mankind if no external supply of energy were needed, but it must certainly be admitted that experience indicates the contrary. This negative statement leads us to the *Clausius*† *statement of the second law:*

No process is possible whose sole *result is the transfer of heat from a cooler to a hotter body.*

At first sight, the Kelvin-Planck and the Clausius statements appear to be quite unconnected, but we shall see immediately that they are in all respects equivalent.

6-7 EQUIVALENCE OF KELVIN-PLANCK AND CLAUSIUS STATEMENTS

Let us adopt the following notation:

K = truth of the Kelvin-Planck statement,
$-K$ = falsity of the Kelvin-Planck statement,
C = truth of the Clausius statement,
$-C$ = falsity of the Clausius statement.

Two propositions or statements are said to be equivalent when the truth of one implies the truth of the second and the truth of the second implies the truth of the first. Using the symbol $\supset$ to mean "implies" and the symbol $\equiv$ to denote equivalence, we have, by definition,

$$K \equiv C$$

when $$K \supset C \qquad \text{and} \qquad C \supset K.$$

† R. J. Clausius (pronounced *Klow'-zee-oos*).

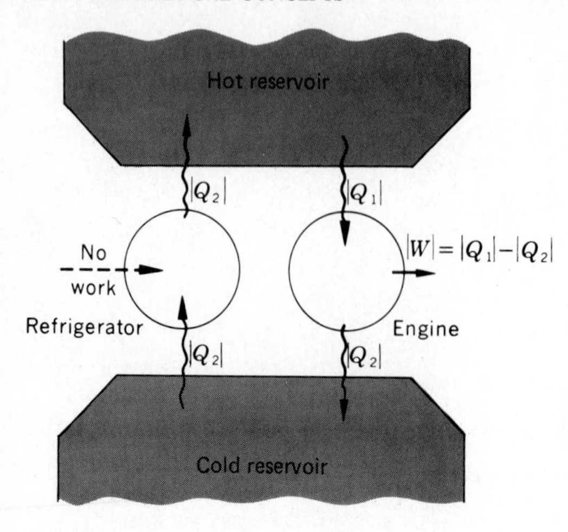

Figure 6-10 Proof that $-C \supset -K$. The refrigerator on the left is a violation of C; the refrigerator and engine acting together constitute a violation of K.

Now, it may easily be shown that

$$K \equiv C$$

also when $\quad -K \supset -C \quad$ and $\quad -C \supset -K.$

Thus, in order to demonstrate the equivalence of K and C, we have to show that a violation of one statement implies a violation of the second, and vice versa.

1. To prove that $-C \supset -K$, consider a refrigerator, shown in the left-hand side of Fig. 6-10, which requires *no work* to transfer $|Q_2|$ units of heat from a cold reservoir to a hot reservoir and which therefore violates the Clausius statement. Suppose that a heat engine (on the right) also operates between the same two reservoirs in such a way that heat Q_2 is delivered to the cold reservoir. The engine, of course, does not violate any law, but the refrigerator and engine *together* constitute a self-acting device whose sole effect is to take heat $Q_1 - Q_2$ from the hot reservoir and to convert *all* this heat into work.

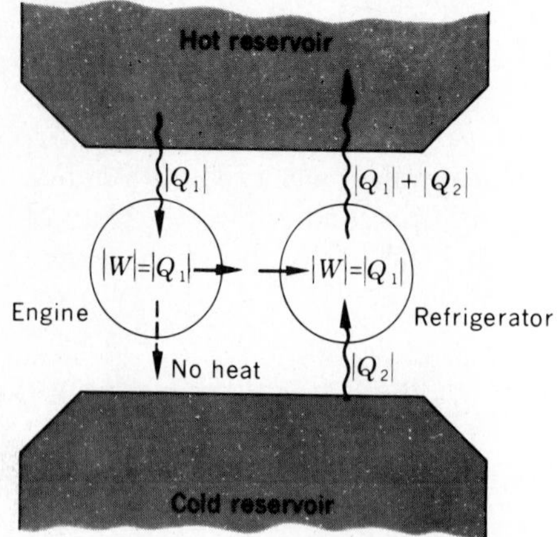

Figure 6-11 Proof that $-K \supset -C$. The engine on the left is a violation of K; the engine and refrigerator acting together constitute a violation of C.

Therefore the refrigerator and engine together constitute a violation of the Kelvin-Planck statement.

2. To prove that $-K \supset -C$, consider an engine, shown on the left-hand side of Fig. 6-11, which rejects no heat to the cold reservoir and which therefore violates the Kelvin-Planck statement. Suppose that a refrigerator (on the right) also operates between the same two reservoirs and uses up all the work liberated by the engine. The refrigerator violates no law, but the engine and refrigerator *together* constitute a self-acting device whose sole effect is to transfer heat $|Q_2|$ from the cold to the hot reservoir. Therefore, the engine and refrigerator together constitute a violation of the Clausius statement.

We therefore arrive at the conclusion that both statements of the second law are equivalent. It is a matter of indifference which one is used in a particular argument.

PROBLEMS

6-1 Figure P6-1 represents a simplified PV diagram of the Joule ideal-gas cycle. All processes are quasi-static, and C_P is constant. Prove that the thermal efficiency of an engine performing this cycle is

$$1 - \left(\frac{P_1}{P_2}\right)^{(\gamma-1)/\gamma}.$$

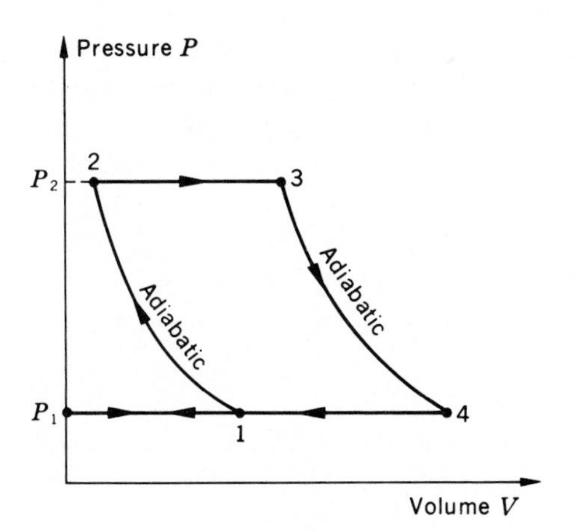

Figure P6-1 Joule ideal-gas cycle.

6-2 Figure P6-2 represents a simplified PV diagram of the Sargent ideal-gas cycle. All processes are quasi-static, and the heat capacities are constant. Prove that the thermal efficiency of an engine performing this cycle is

$$1 - \gamma\frac{\theta_4 - \theta_1}{\theta_3 - \theta_2}.$$

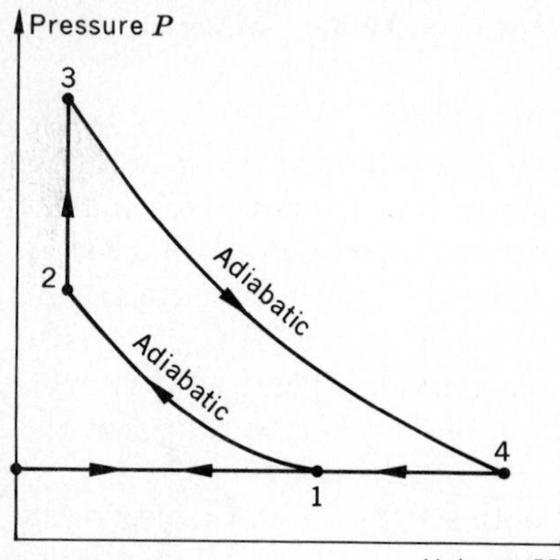

Figure P6-2 Sargent ideal-gas cycle.

6-3 Figure P6-3 represents an imaginary ideal-gas engine cycle. Assuming constant heat capacities, show that the thermal efficiency is

$$\eta = 1 - \gamma \frac{(V_1/V_2) - 1}{(P_3/P_2) - 1}.$$

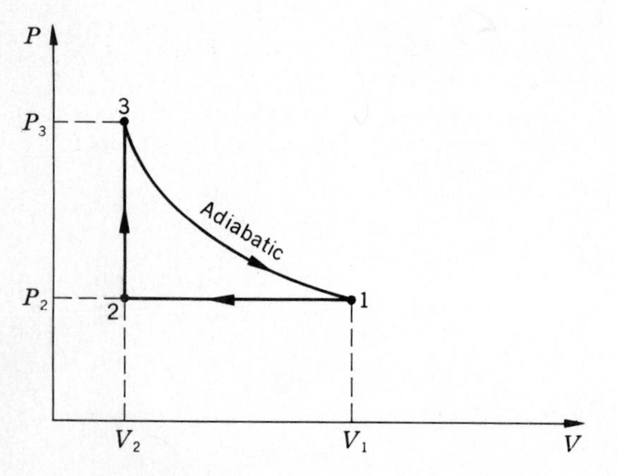

Figure P6-3

6-4 An ideal-gas engine operates in a cycle which, when represented on a PV diagram, is a rectangle. Call P_1, P_2 the lower and higher pressures, respectively; call V_1, V_2 the lower and higher volumes, respectively.

(*a*) Calculate the work done in one cycle.

(*b*) Indicate which parts of the cycle involve heat flow into the gas, and calculate the amount of heat flowing into the gas in one cycle. (Assume constant heat capacities.)

(*c*) Show that the efficiency of this engine is

$$\eta = \frac{\gamma - 1}{\dfrac{\gamma P_2}{P_2 - P_1} + \dfrac{V_1}{V_2 - V_1}}.$$

6-5 A vessel contains 10^{-3} m^3 of helium gas at 3 K and 10^3 Pa. Take the zero of internal energy of helium to be at this point.

(*a*) The temperature is raised at constant volume to 300 K. Assuming helium to behave like an ideal monatomic gas, how much heat is absorbed, and what is the internal energy of the helium? Can this energy be regarded as stored heat or stored work?

(*b*) The helium is now expanded adiabatically to 3 K. How much work is done, and what is the new internal energy? Has heat been converted into work without compensation, thus violating the second law?

(*c*) The helium is now compressed isothermally to its original volume. What are the quantities of heat and work in this process? What is the efficiency of the cycle? Plot on a PV diagram.

6-6 Derive the expression for the efficiency of an ideal engine operating in an air-standard Diesel cycle, Eq. (6-4).

6-7 (*a*) Assume an ideal gas to undergo the idealized Stirling engine cycle shown in Fig. 6-1*b* with perfect regeneration. Show that

$$\eta = 1 - \frac{\theta_C}{\theta_H}.$$

(*b*) If this cycle is reversed, calculate the coefficient of performance.

6-8 In the tropics the water near the surface of the ocean is warmer than the deep water. Would an engine operating between these two levels violate the second law?

6-9 A storage battery is connected to a motor, which is used to lift a weight. The battery remains at constant temperature by receiving heat from the outside air. Is this a violation of the second law? Why?

6-10 There are many paramagnetic solids whose internal energy is a function of temperature only, like an ideal gas. In an isothermal decrease of the magnetic field, heat is absorbed from one reservoir and converted completely into work. Is this a violation of the second law? Why?

6-11 The initial state of 0.1 mol of an ideal monatomic gas is $P_0 = 32$ Pa and $V_0 = 8$ m^3. The final state is $P_1 = 1$ Pa and $V_1 = 64$ m^3. Suppose that the gas undergoes a process along the *straight line* joining these two points whose equation is $P = aV + b$, where $a = -31/56$ and $b = 255/7$. Plot this line to scale on a PV diagram. Calculate:

(*a*) θ as a function of V along this line.
(*b*) The value of V at which θ is a maximum.
(*c*) The values of θ_0, θ_{max}, and θ_1.

Calculate:

(*d*) The heat transferred Q from the volume V_0 to any other volume V along the line.
(*e*) The value of P and V at which Q is a maximum.
(*f*) The heat transferred along the line from V_0 to $V(Q = Q_{max})$.
(*g*) The heat transferred from $V(Q = \max)$ to V_1.

6-12 Show that the two end points specified in Prob. 6-11 lie on an adiabat. A cycle described by Willis and Kirwan of the University of Rhode Island and called the "Sadly Cannot" cycle is obtained by proceeding from 0 to 1 along the line of Prob. 6-11 and from 1 back to 0 along the adiabat. Draw a PV diagram of this cycle to scale. Calculate:

(*a*) The work done on the gas during the adiabat.
(*b*) The net work done in the cycle.
(*c*) The net heat transferred *to* the gas.
(*d*) The efficiency of the cycle.
(*e*) The efficiency of a Carnot cycle operating between a reservoir at the maximum temperature in the cycle and a reservoir at the minimum temperature in the cycle.

CHAPTER

SEVEN

REVERSIBILITY AND THE KELVIN TEMPERATURE SCALE

7-1 REVERSIBILITY AND IRREVERSIBILITY

In thermodynamics, work is a macroscopic concept. The performance of work may always be described in terms of the raising or lowering of an object or the winding or unwinding of a spring, i.e., by the operation of a device that serves to increase or decrease the potential energy of a mechanical system. Imagine, for the sake of simplicity, a suspended object coupled by means of suitable pulleys to a system so that any work done by or on the system can be described in terms of the raising or lowering of the object. Imagine, further, a series of reservoirs which may be put in contact with the system and in terms of which any flow of heat to or from the system may be described. We shall refer to the suspended object and the series of reservoirs as the *local surroundings* of the system. The local surroundings are, therefore, those parts of the surroundings which interact *directly* with the system. Other mechanical devices and reservoirs which are accessible and which *might* interact with the system constitute the *auxiliary surroundings* of the system—or, for want of a better expression, the *rest of the universe.* The word "universe" is used here in a very restricted technical sense, with no cosmic or celestial implications. The universe merely means a finite portion of the world consisting of the system and those surroundings which may interact with the system.

Now suppose that a process occurs in which (1) the system proceeds from an initial state i to a final state f; (2) the suspended object is lowered to an extent that W units of work are performed; and (3) a transfer of heat Q takes place from the system to the series of reservoirs. If, at the conclusion of this process, the system may be restored to its initial state i, the object lifted to its former level, and the reservoirs caused to part with the same amount of heat Q, without producing any changes in any other mechanical device or reservoir in the universe, the original process is said to be *reversible*. In other words, *a reversible process is one that is performed in such a way that, at the conclusion of the process, both the system and the local surroundings may be restored to their initial states, without producing any changes in the rest of the universe*. A process that does not fulfill these stringent requirements is said to be *irreversible*.

The question immediately arises as to whether natural processes, i.e., the familiar processes of nature, are reversible or not. We shall show that it is a consequence of the second law of thermodynamics that all natural processes are irreversible. By considering representative types of natural processes and examining the features of these processes which are responsible for their irreversibility, we shall then be able to state the conditions under which a process may take place reversibly.

7-2 EXTERNAL MECHANICAL IRREVERSIBILITY

There is a large class of processes involving the isothermal transformation of work through a system (which remains unchanged) into internal energy of a reservoir. This type of process is depicted schematically in Fig. 7-1 and is illustrated by the following five examples:

1. Irregular stirring of a viscous liquid in contact with a reservoir.
2. Coming to rest of a rotating or vibrating liquid in contact with a reservoir.
3. Inelastic deformation of a solid in contact with a reservoir.
4. Transfer of electricity through a resistor in contact with a reservoir.
5. Magnetic hysteresis of a material in contact with a reservoir.

In order to restore the system and its local surroundings to their initial states without producing changes elsewhere, Q units of heat would have to be extracted from the reservoir and converted completely into work. Since this would involve a violation of the second law (Kelvin statement), all processes of the above type are irreversible.

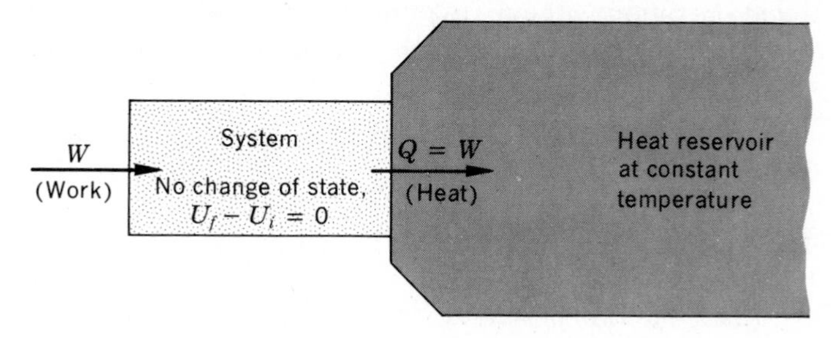

Figure 7-1 Isothermal transformation of work through a system (which remains unchanged) into internal energy of a reservoir.

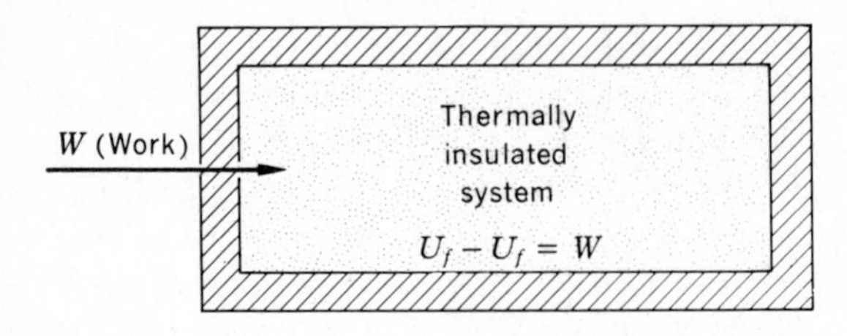

Figure 7-2 Adiabatic transformation of work into internal energy of a system.

Another set of processes involves the adiabatic transformation of work into internal energy of a system. This is depicted schematically in Fig. 7-2 and is illustrated by the following examples, similar to the preceding list:

1. Irregular stirring of a viscous thermally insulated liquid.
2. Coming to rest of a rotating or vibrating thermally insulated liquid.
3. Inelastic deformation of a thermally insulated solid.
4. Transfer of electricity through a thermally insulated resistor.
5. Magnetic hysteresis of a thermally insulated material.

A process of this type is accompanied by a rise of temperature of the system from, say, θ_i to θ_f. In order to restore the system and its local surroundings to their initial states without producing changes elsewhere, the internal energy of the system would have to be decreased by extracting $U_f - U_i$ units of heat, thus lowering the temperature from θ_f to θ_i, and this heat would have to be completely converted into work. Since this violates the second law, all processes of the above type are irreversible.

The transformation of work into internal energy either of a system or of a reservoir is seen to take place through the agency of such phenomena as viscosity, friction, inelasticity, electric resistance, and magnetic hysteresis. These effects are known as *dissipative effects*, and the work is said to be dissipated. Processes involving the dissipation of work into internal energy are said to exhibit *external mechanical irreversibility*. It is a matter of everyday experience that dissipative effects, particularly friction, are always present in moving devices. Friction, of course, may be reduced considerably by suitable lubrication, but experience has shown that it can never be completely eliminated. If it could, a movable device could be kept in continual operation without violating either of the two laws of thermodynamics. Such a device is known as a *perpetual motion machine of the third kind*.

7-3 INTERNAL MECHANICAL IRREVERSIBILITY

The following very important natural processes involve the transformation of internal energy of a system into mechanical energy and then back into internal energy again:

1. Ideal gas rushing into a vacuum (free expansion).
2. Gas seeping through a porous plug (throttling process).
3. Snapping of a stretched wire after it is cut.
4. Collapse of a soap film after it is pricked.

We shall prove the irreversibility of only the first.

During a free expansion, no interactions take place, and hence there are no local surroundings. The only effect produced is a change of state of an ideal gas from a volume V_i and temperature θ to a larger volume V_f and the same temperature θ. To restore the gas to its initial state, it would have to be compressed isothermally to the volume V_i. If the compression were performed quasi-statically and there were no friction between the piston and cylinder, an amount of work W would have to be done by some outside mechanical device, and an equal amount of heat would have to flow out of the gas into a reservoir at the temperature θ. If the mechanical device and the reservoir are to be left unchanged, the heat would have to be extracted from the reservoir and converted completely into work. Since this last step is impossible, the process is irreversible.

In a free expansion, immediately after the stopcock is opened, there is a transformation of some of the internal energy into kinetic energy of "mass motion" or "streaming," and then this kinetic energy is dissipated through viscosity into internal energy again. Similarly, when a stretched wire is cut, there is first a transformation of internal energy into kinetic energy of irregular motion and of vibration and then the dissipation of this energy through inelasticity into internal energy again. In all the processes, the first energy transformation takes place as a result of mechanical instability, and the second by virtue of some dissipative effect. A process of this sort is said to exhibit *internal mechanical irreversibility*.

7-4 EXTERNAL AND INTERNAL THERMAL IRREVERSIBILITY

Consider the following processes involving a transfer of heat between a system and a reservoir by virtue of a *finite* temperature difference:

1. Conduction or radiation of heat from a system to a cooler reservoir.
2. Conduction or radiation of heat through a system (which remains unchanged) from a hot reservoir to a cooler one.

To restore, at the conclusion of a process of this type, both the system and its local surroundings to their initial states without producing changes elsewhere, heat would have to be transferred by means of a self-acting device from a cooler to a hotter body. Since this violates the second law (Clausius statement), all processes of this type are irreversible. Such processes are said to exhibit *external thermal irreversibility*.

A process involving a transfer of heat between parts of the same system because of nonuniformity of temperature is also obviously irreversible by virtue of the Clausius statement of the second law. Such a process is said to exhibit *internal thermal irreversibility*.

7-5 CHEMICAL IRREVERSIBILITY

Some of the most interesting processes that go on in nature involve a spontaneous change of internal structure, chemical composition, density, crystal form, etc. Some important examples follow.

Formation of new chemical constituents:
1. All chemical reactions.

Mixing of two different substances:
2. Diffusion of two dissimilar inert ideal gases.
3. Mixing of alcohol and water.

Sudden change of phase:
4. Freezing of supercooled liquid.
5. Condensation of supersaturated vapor.

Transport of matter between phases in contact:
6. Solution of solid in water.
7. Osmosis

Such processes are by far the most difficult to handle and must, as a rule, be treated by special methods. Such methods constitute what is known as chemical thermodynamics and are discussed in Chapters 14, 15, and 16. It can be shown that the diffusion of two dissimilar inert ideal gases is equivalent to two independent free expansions. Since a free expansion is irreversible, it follows that diffusion is irreversible. The student will have to accept at present the statement that the processes described above are irreversible. Processes that involve a spontaneous change of chemical structure, density, phase, etc., are said to exhibit *chemical irreversibility*.

7-6 CONDITIONS FOR REVERSIBILITY

Most processes that occur in nature are included among the general types of process listed in the preceding articles. Living processes, such as cell division, tissue growth, etc., are no exception. If one takes into account all the interactions that accompany living processes, such processes are irreversible. It is a direct consequence of the second law of thermodynamics that *all natural spontaneous processes are irreversible.*

A careful inspection of the various types of natural process shows that all involve one or both of the following features:

1. The conditions for mechanical, thermal, or chemical equilibrium, i.e., thermodynamic equilibrium, are not satisfied.
2. Dissipative effects, such as viscosity, friction, inelasticity, electric resistance, and magnetic hysteresis, are present.

For a process to be reversible, it must *not* possess these features. If a process is performed quasi-statically, the system passes through states of thermodynamic equilibrium, which may be traversed just as well in one direction as in the opposite direction. If there are no dissipative effects, all the work done by the system during the performance of a process in one direction can be returned to the system during the reverse process. We are led, therefore, to the conclusion that a process will be reversible when (1) it is performed quasi-statically and (2) it is not accompanied by any dissipative effects.

Since it is impossible to satisfy these two conditions perfectly, it is obvious that a reversible process is purely an ideal abstraction, extremely useful for theoretical calculations (as we shall see) but quite devoid of reality. In this sense, the assumption of a reversible process in thermodynamics resembles the assumptions made so often in mechanics, such as those which refer to weightless strings, frictionless pulleys, and point masses.

A heat reservoir was defined as a body of very large mass capable of absorbing or rejecting an unlimited supply of heat without suffering appreciable changes in its thermodynamic coordinates. The changes that do take place are so very slow and so minute that dissipative actions never develop. *Therefore, when heat enters or leaves a reservoir, the changes that take place* in the reservoir *are the same as those which would take place if the same quantity of heat were transferred reversibly.*

It is possible in the laboratory to approximate the conditions necessary for the performance of reversible processes. For example, if a gas is confined in a cylinder equipped with a well-lubricated piston and is allowed to expand very slowly against an opposing force provided either by an object suspended from a frictionless pulley or by an elastic spring, the gas undergoes an approximately reversible process. Similar considerations apply to a wire and to a surface film.

A reversible transfer of electricity through an electric cell may be imagined as follows. Suppose that a motor whose coils have a negligible resistance is caused to rotate until its back emf is only slightly different from the emf of the cell. Suppose further that the motor is coupled either to an object suspended from a frictionless pulley or to an elastic spring. If neither the cell itself nor the connecting wires to the motor have appreciable resistance, a reversible transfer of electricity takes place.

In order to arrive at conclusions concerning the equilibrium states of thermodynamic systems, it is often necessary to invoke some sort of process in which the system passes through these states. To assume the process to be quasi-static only, often is not sufficient, for if dissipative processes are present there may be heat flows or internal energy changes of neighboring systems (envelopes, containers, surroundings) that may limit the validity of the argument. In order to ensure that equilibrium states of the system *only* are considered—without having to take account of the effect of dissipated work in the system itself or in some other neighboring body—it is useful to invoke the concept of a *reversible process*, even though this assumption may at times seem a bit drastic.

7-7 EXISTENCE OF REVERSIBLE ADIABATIC SURFACES

Up to this point, the only consequence of the second law of thermodynamics that has been drawn is the irreversibility of natural, spontaneous processes. To develop further consequences, it has been customary to proceed along either of two lines: the engineering method, due to Carnot, Kelvin, and Clausius; and the axiomatic method, due to the Greek mathematician C. Caratheodory.† The engineering method is based upon the Kelvin-Planck formulation of the second law or its equivalent, the Clausius statement. One starts first by defining a particularly simple reversible cycle called the *Carnot cycle* and then proving that an engine operating in this cycle between reservoirs at two different temperatures is more efficient than any other engine operating between the same two reservoirs. After proving that all Carnot engines operating between the same two reservoirs have the same efficiency, regardless of the substance undergoing the cycle, the Kelvin temperature scale is defined so as to be independent of the properties of any particular kind of thermometer. A theorem called the *Clausius theorem* is then derived, and from it the existence of the entropy function. The engineering method of developing the consequences of the Kelvin-Planck or Clausius statements of the second law is rigorous and general. If one is interested in the design and manufacture of engines and refrigerators, it is essential to employ principles that hold regardless of the nature of the materials involved. If, however, one is interested in the behavior of systems, their coordinates, their equations of state, their properties, their processes, etc., *apart* from their use in the cylinders of engines and refrigerators, then it is desirable to adopt a method more closely associated with the coordinates and equations of actual systems.

In the first decade of the twentieth century Caratheodory, to replace the Kelvin-Planck and Clausius statements of the second law, presented this axiom: *In the neighborhood (however close) of any equilibrium state of a system of any number of thermodynamic coordinates, there exist states that cannot be reached (are inaccessible) by reversible adiabatic processes.* He showed how to derive the

† C. Caratheodory, *Math. Ann.*, **67**:355 (1909) (in German).

Kelvin temperature scale from this axiom and how to derive every other consequence of the engineering method. Physicists (Born, Ehrenfest, Landé) recognized the importance of Caratheodory's work, but since the mathematics needed to deal with Caratheodory's axiom presented much more difficulty than the simple manipulations involving outputs and inputs of engines and refrigerators, other physicists were slow in adopting his methods. In recent years, mainly because of the activities of Pippard, Turner, Landsberg, and Sears,† the mathematical machinery of the Caratheodory method has been considerably simplified, and now it appears that *the axiom itself can be dispensed with entirely.* All the consequences of the Caratheodory axiom follow directly from the Kelvin-Planck statement of the second law.

Let us first consider a simple system. It has been emphasized that $đW$ and $đQ$ are inexact differentials; there are *no* functions W and Q representing, respectively, the work and heat of a body. When a system can be described with the aid of three thermodynamic coordinates, say, a temperature t (on any scale) and a generalized displacement X, then if Y is the generalized force, the first law can be written

$$đQ = dU - Y\,dX.$$

Only two of the thermodynamic coordinates are independent if the equation of state is also known.

Regarding U as a function of t and X, we get

$$đQ = \left(\frac{\partial U}{\partial t}\right)_X dt + \left[-Y + \left(\frac{\partial U}{\partial X}\right)_t\right] dX,$$

where $(\partial U/\partial t)_X$, Y, and $(\partial U/\partial X)_t$ are known functions of t and X. A reversible adiabatic process for this system is represented by the equation

$$\left(\frac{\partial U}{\partial t}\right)_x dt + \left[-Y + \left(\frac{\partial U}{\partial X}\right)_t\right] dX = 0 \tag{7-1}$$

Solving for dt/dX, we get

$$\left(\frac{dt}{dX}\right)_{\text{ad}} = \frac{Y - (\partial U/\partial X)_t}{(\partial U/\partial t)_X}.$$

The right-hand member is known as a function of t and X, and therefore the derivative dt/dX, representing the slope of an adiabat on a tX diagram, is known at all points. Equation (7-1) has therefore a solution consisting of a family of curves, and the curve through any one point may be written

$$\sigma(t, X) = \text{const.}$$

A set of curves is obtained when different values are assigned to the constant. *The existence of the family of curves $\sigma(t, X) = \text{const.}$, representing reversible*

† A. B. Pippard, *Elements of Classical Thermodynamics*, Cambridge University Press, New York, 1957, p. 38; L. A. Turner, *Am. J. Phys.*, **28**:781 (1960); P. T. Landsberg, *Nature*, **201**:485 (1964); F. W. Sears, *Am. J. Phys.*, **31**:747 (1963).

adiabatic processes, follows from the fact that there are only two independent coordinates used to describe a system obeying the first law of thermodynamics.

When three or more independent coordinates are needed to describe a system, the situation is quite different. The second law of thermodynamics is *needed* to enable us to conclude that: *Through any arbitrary initial-state point, all reversible adiabatic processes lie on a surface, and reversible adiabatics through other initial states determine a family of nonintersecting surfaces.*

Consider a system described with the aid of five thermodynamic coordinates: the empirical temperature t, measured on *any* scale whatsoever; two generalized forces Y and Y'; and two corresponding generalized displacements X and X'. For such a system, the first law for a reversible process is

$$đQ = dU - Y\,dX - Y'\,dX',$$

and because of the existence of two equations of state, only three of the coordinates are independent. At first, let us choose these coordinates to be U, X, and X'. A system of three independent variables is chosen for two reasons: (1) it enables us to use simple three-dimensional graphs, and (2) all conclusions concerning the mathematical properties of the differential $đQ$ will hold equally well for all systems with *more or fewer* than three independent variables.

In Fig. 7-3, the three independent variables U, X, and X' are plotted along three rectangular axes, and an arbitrarily chosen equilibrium state i is indicated. Let f_1 be an equilibrium state that the system can reach by means of a reversible

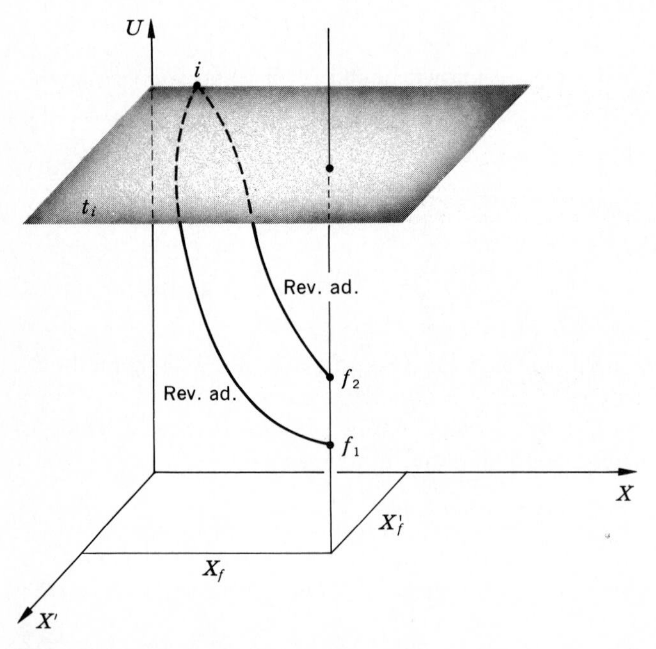

Figure 7-3 Both f_1 and f_2, lying on a line of constant X and X', *cannot* be reached by reversible adiabatic processes from i.

adiabatic process. Through f_1 draw a vertical line for which the values of X and X' are constant at every point. Let f_2 be any other equilibrium state on this vertical line. We now proceed to prove that *both states f_1 and f_2 cannot be reached by reversible adiabatic processes* from i. Assume that it *is* possible for the system to proceed along either of the two reversible adiabatic paths $i \to f_1$ or $i \to f_2$. Let the system start at i, proceed to f_1, then to f_2, and then back to i along $f_2 \to i$, which, being a reversible path, can be traversed in either direction. Since f_2 lies above f_1, the system undergoes an *increase* of energy at constant X and X', during which process no work is done. It follows from the first law that *heat Q must be absorbed* in the process $f_1 \to f_2$. In the reversible adiabatic processes, however, no heat is transferred but *work W is done*. In the entire cycle $if_1 f_2 i$, there is no energy change, and therefore $Q = W$. The system has thus performed a cycle in which the sole effect is the absorption of heat and the conversion of this heat completely into work. Since this violates the Kelvin-Planck statement of the second law, it follows that *both* f_1 and f_2 cannot be reached by reversible adiabatic processes. *Only one point on the line of constant X and X' can be reached by a reversible adiabatic process from i.*

For a different line (different X_f and X'_f), there would be another single point accessible from i by a reversible adiabatic process, and so on. A few such points, f_1, f_2, etc., are shown in Fig. 7-4. *The locus of all points accessible from i by reversible adiabatic processes is a space of dimensionality one less than three; in*

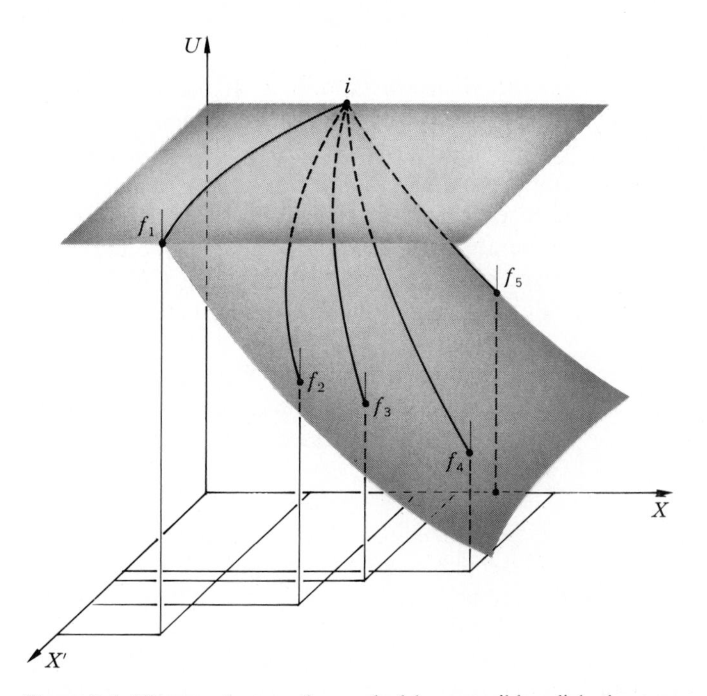

Figure 7-4 All states that can be reached by reversible adiabatic processes starting at i lie on a surface.

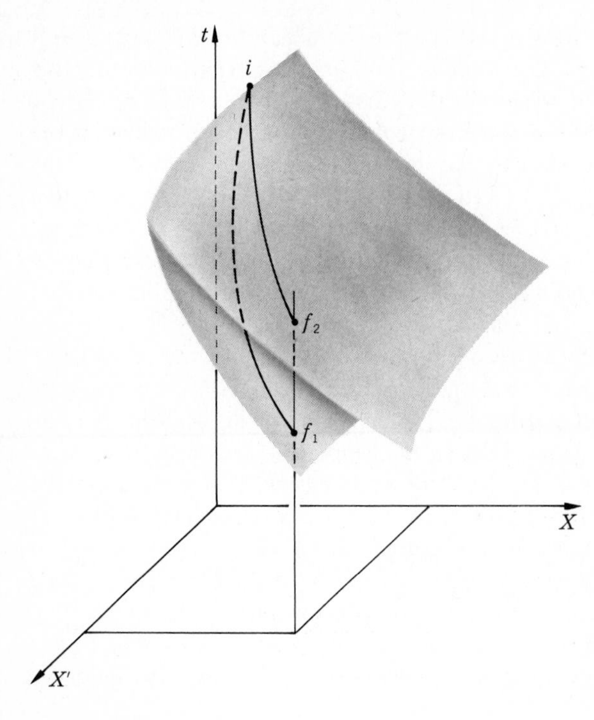

Figure 7-5 If two reversible adiabatic surfaces could intersect, it would be possible to violate the second law by performing the cycle $if_1 f_2 i$.

other words, these points lie on a two-dimensional surface. If the system were described with the aid of four independent coordinates, the states accessible from any given equilibrium state i by reversible adiabatic processes would lie on a three-dimensional hypersurface, and so on.

In what is to follow, it is more convenient to choose as one of the independent coordinates the empirical temperature t instead of the energy U. Since, for a given i, a reversible adiabatic surface has been shown to exist in a UXX' space, such a surface must also exist in a tXX' space, although its shape might be quite different.

With a system of three independent coordinates t, X, and X', the two-dimensional surface comprising all the equilibrium states that are accessible from i by reversible adiabatic processes may be expressed by the equation

$$\sigma(t, X, X') = \text{const.,} \tag{7-2}$$

where σ represents some as yet undetermined function. Surfaces corresponding to other initial states would be represented by different values of the constant.

Reversible adiabatic surfaces cannot intersect, for if they did it would be possible, as shown in Fig. 7-5, to proceed from an initial equilibrium state i on the curve of intersection of two different final states f_1 and f_2, having the same X_f and X'_f, along reversible adiabatic paths. We have just shown that this is impossible.

8-8 INTEGRABILITY of $đQ$

Consider a system whose coordinates are the empirical temperature t, two generalized forces Y and Y', and two generalized displacements X and X'. The first law for a reversible process is expressed by the equation

$$đQ = dU - Y\,dX - Y'\,dX', \tag{7-3}$$

where U, Y, and Y' are functions of t, X, and X'. Since the t, X, X' space is subdivided into a family of nonintersecting reversible adiabatic surfaces,

$$\sigma(t, X, X') = \text{const.},$$

where the constant can take on various values. Any point in this space may be determined by specifying the value of σ along with X and X', so that we may regard the internal energy function U as a function of σ, X, and X'. Then,

$$dU = \left(\frac{\partial U}{\partial \sigma}\right)_{X,\,X'} d\sigma + \left(\frac{\partial U}{\partial X}\right)_{\sigma,\,X'} dX + \left(\frac{\partial U}{\partial X'}\right)_{\sigma,\,X} dX',$$

and

$$đQ = \left(\frac{\partial U}{\partial \sigma}\right)_{X,\,X'} d\sigma + \left[-Y + \left(\frac{\partial U}{\partial X}\right)_{\sigma,\,X'}\right] dX + \left[-Y' + \left(\frac{\partial U}{\partial X'}\right)_{\sigma,\,X}\right] dX'. \tag{7-4}$$

Since the coordinates σ, X, and X' are independent variables, this equation must be true for all values of $d\sigma$, dX, and dX'. Suppose that two of the differentials, $d\sigma$ and dX, are zero and that dX' is not. The provision that $d\sigma = 0$ (or $\sigma = \text{const.}$) is the condition for a reversible adiabatic process in which $đQ = 0$, and therefore the coefficient of dX' must also vanish. Alternatively, if we take $d\sigma$ and dX' to be zero, then by the same reasoning the coefficient of dX must vanish. It follows therefore that, in order for the coordinates σ, X, X' to be independent, and also for $đQ$ to be zero whenever $d\sigma$ is zero, the equation for $đQ$ must reduce to the form

$$đQ = \left(\frac{\partial U}{\partial \sigma}\right)_{X,\,X'} d\sigma. \tag{7-5}$$

If we *define* a function λ by the equation

$$\lambda = \left(\frac{\partial U}{\partial \sigma}\right)_{X,\,X'}, \tag{7-6}$$

we get the result

$$\boxed{đQ = \lambda\, d\sigma.} \tag{7-7}$$

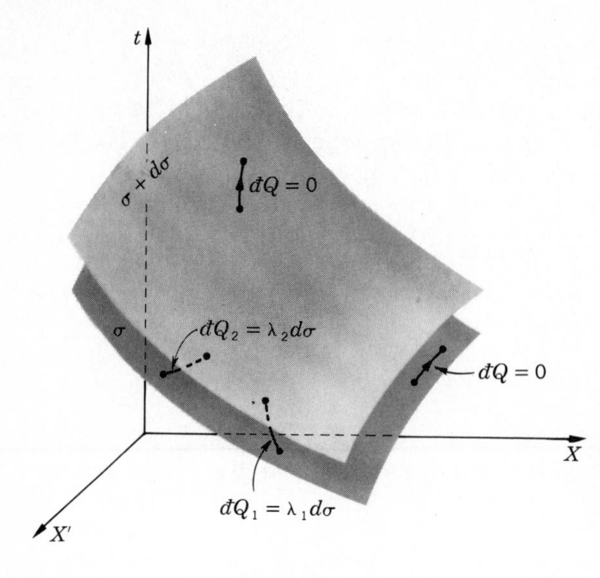

Figure 7-6 Two reversible adiabatic surfaces, infinitesimally close. When the process is represented by a curve connecting the two surfaces, heat $đQ = \lambda\, d\sigma$ is transferred.

According to its definition, given in Eq. (7-6), λ is a function of σ, X, and X'. Since σ is a function of t, X, and X', however, we may imagine X' to be eliminated, with the result that λ is a function of t, σ, and X.

It is seen from Eq. (7-7) that the function $1/\lambda$ is an integrating factor, such that when $đQ$ is multiplied by $1/\lambda$ the result is an exact differential $d\sigma$. Now, an infinitesimal of the type

$$P\,dx + Q\,dy + R\,dz + \cdots,$$

known as a *linear differential form* or a *Pfaffian expression*, when it involves three or more independent variables, does *not* admit, in general, of an integrating factor. *It is only because of the existence of the second law that the differential form for đQ referring to a physical system of any number of independent coordinates possesses an integrating factor.*

Two infinitesimally neighboring reversible adiabatic surfaces are shown in Fig. 7-6. One surface is characterized by a constant value of the function σ, and the other by a slightly different value $\sigma + d\sigma$. In any process represented by a curve *on* either of the two surfaces $đQ = 0$. When a reversible process is represented by a curve *connecting* the two surfaces, however, heat $đQ = \lambda\, d\sigma$ is transferred. All curves joining the two surfaces represent processes with the same $d\sigma$, *but the values of λ are different.*

7-9 PHYSICAL SIGNIFICANCE OF λ

The various infinitesimal processes that may be chosen to connect the two neighboring reversible adiabatic surfaces shown in Fig. 7-6 involve the same change of σ but take place at different values of λ, because λ is a function of t, σ,

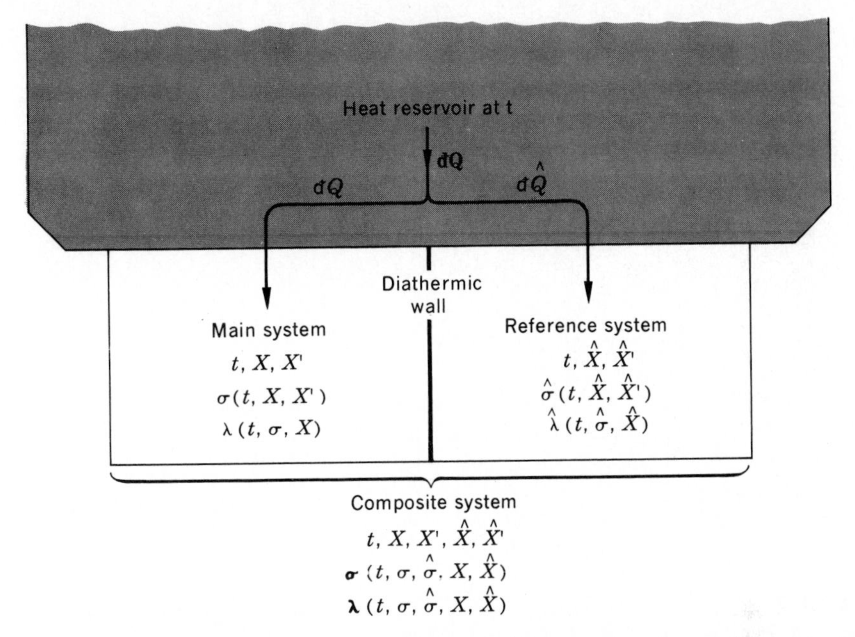

Figure 7-7 Two systems in thermal equilibrium, constituting a composite system receiving heat from a reservoir.

and X. To find the temperature dependence of λ, we go back to the fundamental concept of temperature as the property of a system determining thermal equilibrium between it and another system. Let us therefore consider two systems, each of three independent coordinates (for mathematical generality), in contact through a diathermic wall, as depicted schematically in Fig. 7-7. The two systems are assumed to be at all times in thermal equilibrium having a common temperature t, and together they constitute a composite system with five independent coordinates.

1. *Main system.* The three independent coordinates are t, X, and X', and the reversible adiabatic surfaces are specified by different values of the function σ of t, X, and X'. When heat $đQ$ is transferred, σ changes by $d\sigma$, and $đQ = \lambda\, d\sigma$ where λ is a function of t, σ, and X.
2. *Reference system.*† The three independent coordinates are t, $\hat{X}$, and $\hat{X}'$, and the reversible adiabatic surfaces are specified by different values of the function $\hat{\sigma}$ of t, $\hat{X}$, and $\hat{X}'$. When heat $đ\hat{Q}$ is transferred, $\hat{\sigma}$ changes by $d\hat{\sigma}$, and $d\hat{Q} = \hat{\lambda}\, d\hat{\sigma}$ where $\hat{\lambda}$ is a function of t, $\hat{\sigma}$, and $\hat{X}$.
3. *Composite system.* The five independent coordinates are t, X, X', $\hat{X}$, and $\hat{X}'$, and the reversible adiabatic hypersurfaces are specified by different values of the function $\boldsymbol{\sigma}$ of these independent variables.

† The diacritical mark or accent over the symbols referring to the reference system is called a *circumflex*.

Using the equation for σ of the main system, we may express X' in terms of t, σ, and X. Similarly, using the equation for $\hat{\sigma}$ of the reference system, $\hat{X}'$ may be expressed in terms of t, $\hat{\sigma}$, and $\hat{X}$. The primed quantities X' and $\hat{X}'$ may therefore be eliminated from the expression for $\boldsymbol{\sigma}$ of the composite system, and $\boldsymbol{\sigma}$ may be regarded as a function of t, σ, $\hat{\sigma}$, X and $\hat{X}$. For an infinitesimal process between two neighboring reversible adiabatic hypersurfaces specified by $\boldsymbol{\sigma}$ and $\boldsymbol{\sigma} + d\boldsymbol{\sigma}$, the heat transferred is $\text{đ}\mathbf{Q} = \boldsymbol{\lambda}\, d\boldsymbol{\sigma}$, where $\boldsymbol{\lambda}$ is also a function of t, σ, $\hat{\sigma}$, X, and $\hat{X}$. We have

$$d\boldsymbol{\sigma} = \frac{\partial \boldsymbol{\sigma}}{\partial t}\, dt + \frac{\partial \boldsymbol{\sigma}}{\partial \sigma}\, d\sigma + \frac{\partial \boldsymbol{\sigma}}{\partial \hat{\sigma}}\, d\hat{\sigma} + \frac{\partial \boldsymbol{\sigma}}{\partial X}\, dX + \frac{\partial \boldsymbol{\sigma}}{\partial \hat{X}}\, d\hat{X}. \tag{7-8}$$

Now suppose that, in a reversible process, there is a transfer of heat $\text{đ}\mathbf{Q}$ between the composite system and an external reservoir, as shown in Fig. 7-7, with heats $\text{đ}Q$ and $\text{đ}\hat{Q}$ being transferred, respectively, to the main and to the reference systems. Then,

$$\text{đ}\mathbf{Q} = \text{đ}Q + \text{đ}\hat{Q},$$

and

$$\boldsymbol{\lambda}\, d\boldsymbol{\sigma} = \lambda\, d\sigma + \hat{\lambda}\, d\hat{\sigma},$$

or

$$d\boldsymbol{\sigma} = \frac{\lambda}{\boldsymbol{\lambda}}\, d\sigma + \frac{\hat{\lambda}}{\boldsymbol{\lambda}}\, d\hat{\sigma}. \tag{7-9}$$

Comparing the two expressions for $d\boldsymbol{\sigma}$ given by Eqs. (7-8) and (7-9), we get

$$\frac{\partial \boldsymbol{\sigma}}{\partial t} = 0, \qquad \frac{\partial \boldsymbol{\sigma}}{\partial X} = 0, \qquad \text{and} \qquad \frac{\partial \boldsymbol{\sigma}}{\partial X} = 0;$$

therefore $\boldsymbol{\sigma}$ does *not* depend on t, X, or $\hat{X}$ but only on σ and $\hat{\sigma}$. That is,

$$\boldsymbol{\sigma} = \boldsymbol{\sigma}(\sigma, \hat{\sigma}). \tag{7-10}$$

Again comparing the two expressions for $d\boldsymbol{\sigma}$, we see that

$$\frac{\lambda}{\boldsymbol{\lambda}} = \frac{\partial \boldsymbol{\sigma}}{\partial \sigma} \qquad \text{and} \qquad \frac{\hat{\lambda}}{\boldsymbol{\lambda}} = \frac{\partial \boldsymbol{\sigma}}{\partial \hat{\sigma}}; \tag{7-11}$$

therefore, the two ratios $\lambda/\boldsymbol{\lambda}$ and $\hat{\lambda}/\boldsymbol{\lambda}$ are also independent of t, X, and $\hat{X}$. These two ratios depend only on the σ's, but each *separate* λ *must depend on temperature as well.* [For example, if λ depended only on σ and on nothing else, then since $\text{đ}Q = \lambda\, d\sigma$, $\text{đ}Q$ would equal $f(\sigma)\, d\sigma$, which is an exact differential!] In order, therefore, for each λ to depend on temperature, and at the same time for the ratios of the λ's to depend only on the σ's, the λ's *must* have the following structure:

$$\lambda = \phi(t) f(\sigma),$$

$$\hat{\lambda} = \phi(t) \hat{f}(\hat{\sigma}), \tag{7-12}$$

and

$$\boldsymbol{\lambda} = \phi(t) g(\sigma, \hat{\sigma})$$

where $\phi(t)$ is an arbitrary function of the empirical temperature t. (The quantity λ cannot contain X, nor can $\hat{\lambda}$ contain $\hat{X}$, since $\lambda/\boldsymbol{\lambda}$ and $\hat{\lambda}/\boldsymbol{\lambda}$ must be functions of the σ's only.)

Referring now only to our main system as representative of any system of any number of independent coordinates, we have, from the top line of Eq. (7-12),

$$\boxed{đQ = \phi(t) f(\sigma)\, d\sigma.} \tag{7-13}$$

Since $f(\sigma)\, d\sigma$ is an exact differential, the quantity $1/\phi(t)$ *is an integrating factor for* $đQ$. It is an extraordinary circumstance that not only does an integrating factor exist for the $đQ$ of any system, but *this integrating factor is a function of temperature only and is the* same *function for all systems!* This universal character of $\phi(t)$ enables us to define an *absolute temperature.*

The fact that a system of *two* independent variables has a $đQ$ which always admits an integrating factor regardless of the second law is interesting, of course; but *its importance in physics* is not established until it is shown that the integrating factor is a function of temperature *only* and that it is the *same* function for all systems.

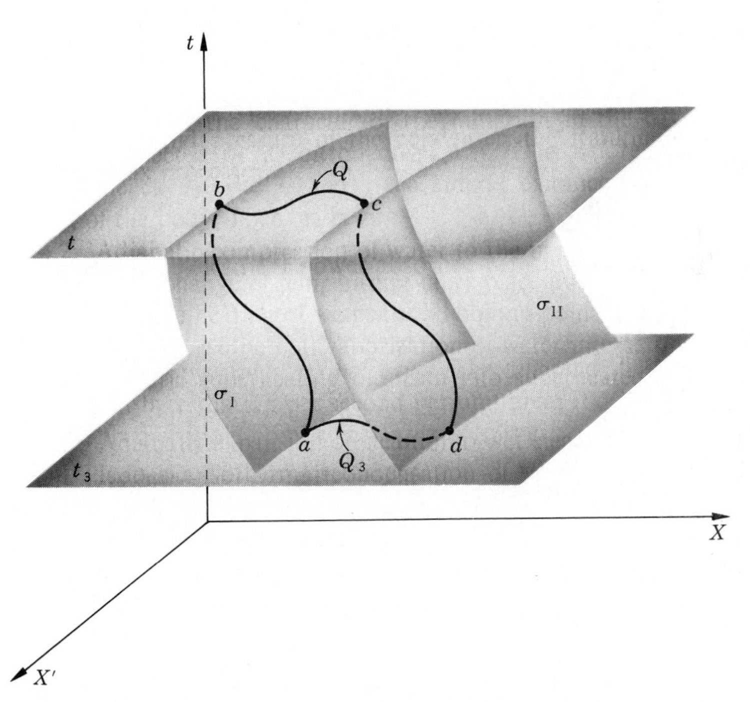

Figure 7-8 Two isothermal heat transfers, Q at t from b to c and Q_3 at t_3 from a to d, *between the same two reversible adiabatic surfaces* σ_{I} and σ_{II}. The cycle *abcda* is a Carnot cycle.

7-10 KELVIN TEMPERATURE SCALE

Consider a system of three independent variables t, X, and X', for which two isothermal surfaces and reversible adiabatic surfaces are drawn in Fig. 7-8. Suppose that there is a reversible isothermal transfer of heat Q between the system and a reservoir at the temperature t, so that the system proceeds from a state b, lying on a reversible adiabatic surface characterized by the value σ_{I}, to another state c, lying on another reversible adiabatic surface specified by σ_{II}. Then, since Eq. (7-13) tells us that $đQ = \phi(t) f(\sigma)\, d\sigma$, we have

$$Q = \phi(t) \int_{\sigma_{\mathrm{I}}}^{\sigma_{\mathrm{II}}} f(\sigma)\, d\sigma \qquad \text{(at const. } t\text{).}$$

For any reversible isothermal process $a \to d$ at the temperature t_3 *between the same two reversible adiabatic surfaces*, the heat Q_3 is

$$Q_3 = \phi(t_3) \int_{\sigma_{\mathrm{I}}}^{\sigma_{\mathrm{II}}} f(\sigma)\, d\sigma \qquad \text{(at const. } t_3\text{).}$$

Taking the ratio of Q to Q_3, we get

$$\frac{Q}{Q_3} = \frac{\phi(t)}{\phi(t_3)} = \frac{\text{a function of the temp. at which } Q \text{ is transferred}}{\text{the } \textit{same} \text{ function of temp. at which } Q_3 \text{ is transferred}};$$

therefore, we define the ratio of two Kelvin temperatures T/T_3 by the relation

$$\boxed{\frac{Q \text{ (between } \sigma_{\mathrm{I}} \text{ and } \sigma_{\mathrm{II}} \text{ at } T)}{Q_3 \text{ (between } \sigma_{\mathrm{I}} \text{ and } \sigma_{\mathrm{II}} \text{ at } T_3)} = \frac{T}{T_3}.} \tag{7-14}$$

Thus, *two temperatures on the Kelvin scale are to each other as the heats transferred between the same two reversible adiabatic surfaces at these two temperatures.* It is seen that the Kelvin temperature scale is independent of the peculiar characteristics of any particular substance. It therefore supplies exactly what is lacking in the ideal-gas scale.

If the temperature T_3 is taken arbitrarily to be the triple point of water (the standard fixed point) and T_3 is chosen to have the value 273.16 K, then the Kelvin temperature is defined to be

$$\boxed{T = 273.16 \text{ K} \frac{Q}{Q_{TP}}} \qquad \text{(between the same two reversible adiabatic surfaces).} \tag{7-15}$$

To measure a Kelvin temperature, we must therefore measure or calculate the heats transferred at the unknown temperature and at the triple point of water during reversible isothermal processes between the same two reversible adiabatic surfaces. Comparing this equation with the corresponding equation for the ideal-gas temperature (see Art. 5-1)

$$\theta = 273.16 \frac{\lim (PV)}{\lim (PV)_{TP}},$$

it is seen that, in the Kelvin scale, Q plays the role of a "thermometric property." This does not have the objection attached to a coordinate of an arbitrarily chosen thermometer, however, inasmuch as Q/Q_3 is independent of the nature of the system.

It follows from Eq. (7-15) that the heat transferred isothermally between two given reversible adiabatic surfaces decreases as the temperature decreases. Conversely, the smaller the value of Q, the lower the corresponding T. The smallest possible value of Q is zero, and the corresponding T is absolute zero. Thus, *if a system undergoes a reversible isothermal process between two reversible adiabatic surfaces without transfer of heat, the temperature at which this process takes place is called absolute zero.*

It should be noticed that the definition of absolute zero holds for all substances and is therefore independent of the peculiar properties of any one arbitrarily chosen substance. Furthermore, the definition is in terms of purely macroscopic concepts. No reference is made to molecules or to molecular energy. Whether absolute zero may be achieved experimentally is a question that we shall defer until a later chapter.

7-11 EQUALITY OF IDEAL-GAS TEMPERATURE AND KELVIN TEMPERATURE

For the sake of generality, systems with three or more independent coordinates have been used in most of the discussions in this chapter. The systems encountered most frequently in practical applications of thermodynamics, however, usually have no more than two independent variables. In such cases, isothermal and reversible adiabatic surfaces degenerate into plane curves such as those shown on the θV diagram of an ideal gas in Fig. 7-9.

For any infinitesimal reversible process of an ideal gas, the first law may be written

$$đQ = C_V \, d\theta + P \, dV.$$

When this equation is applied to the isothermal process $b \rightarrow c$, the heat transferred is found to be

$$Q = \int_{V_b}^{V_c} P \, dV = nR\theta \ln \frac{V_c}{V_b}.$$

Similarly, for the isothermal process $a \rightarrow d$, the heat transferred is

$$Q_3 = nR\theta_3 \ln \frac{V_d}{V_a}.$$

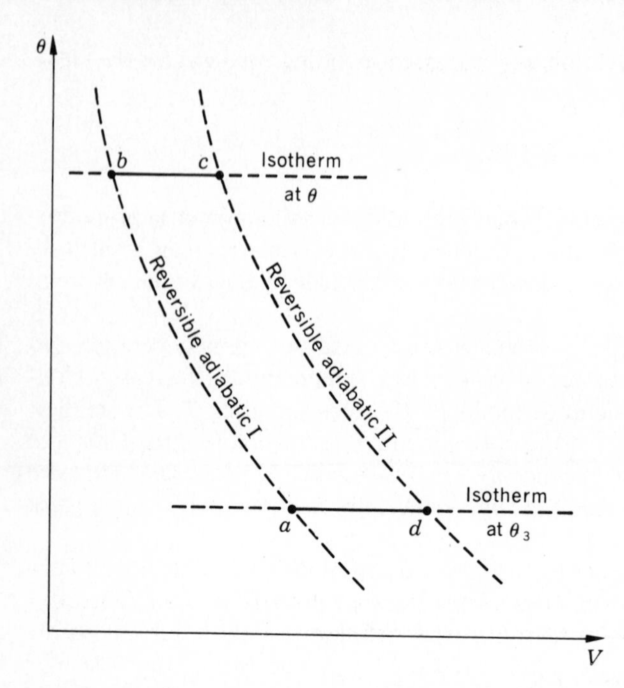

Figure 7-9 Two isotherms, at θ and θ_3, between two reversible adiabatics of an ideal gas. The cycle *abcda* is a Carnot cycle.

Therefore,

$$\frac{Q}{Q_3} = \frac{\theta \ln (V_c/V_b)}{\theta_3 \ln (V_d/V_a)}.$$

Since the process $a \to b$ is adiabatic, we may write for any infinitesimal portion

$$C_V \, d\theta = -P \, dV = -\frac{nR\theta}{V} \, dV.$$

Integrating from a to b, we get

$$\frac{1}{nR} \int_{\theta_3}^{\theta} C_V \frac{d\theta}{\theta} = \ln \frac{V_a}{V_b}.$$

Similarly, for the adiabatic process $d \to c$,

$$\frac{1}{nR} \int_{\theta_3}^{\theta} C_V \frac{d\theta}{\theta} = \ln \frac{V_d}{V_c}.$$

Therefore,
$$\ln \frac{V_a}{V_b} = \ln \frac{V_d}{V_c} \quad \text{and} \quad \ln \frac{V_c}{V_b} = \ln \frac{V_d}{V_a},$$

and we get, finally,

$$\frac{Q}{Q_3} = \frac{\theta}{\theta_3}.$$

Since, however, the Kelvin temperature scale is defined by the same sort of equation, we have

$$\frac{\theta}{\theta_3} = \frac{T}{T_3}.$$

If θ and T refer to any temperature and if θ_3 and T_3 refer to the triple point of water,

$$\theta_{TP} = T_{TP} = 273.16 \text{ K},$$

and

$$\boxed{\theta = T.} \tag{7-14}$$

The Kelvin temperature is therefore numerically equal to the ideal-gas temperature and, in the proper range, may be measured with a gas thermometer.

PROBLEMS

7-1 A gas is contained within a cylinder-piston combination. In the following five sets of conditions, tell (1) whether $đW = -P\,dV$ or not and (2) whether the process is reversible, quasi-static, or irreversible:

(*a*) There is no external pressure on the piston and no friction between the piston and the cylinder wall.

(*b*) There is no external pressure, and friction is small.

(*c*) The piston is jerked out faster than the average molecular speed.

(*d*) The friction is adjusted to allow the gas to expand slowly.

(*e*) There is no friction, but the external pressure is adjusted to allow the gas to expand slowly.

7-2 Continuing the reasoning at the beginning of Art. 7-8 concerning a system of two independent variables, show that the expression for $đQ$ admits an integrating factor.

7-3 Consider the differential (or Pfaffian) expression

$$yz\,dx + dy + dz.$$

To determine whether an integrating factor exists, we investigate the possible solutions of the Pfaffian equation

$$yz\,dx + dy + dz = 0. \tag{P7-1}$$

(*a*) Holding x constant, show that the resulting equation has a solution

$$y + z = F(x),$$

but that this cannot be a solution of Eq. (P7-1).

(*b*) Holding z constant, show that the resulting equation has a solution

$$y = G(z)e^{-zx},$$

but that this cannot be a solution of Eq. (P7-1).

(*c*) Do the two "cuts" (x = const. and z = const.) produce a smooth surface?

(*d*) Does an integrating factor exist?

7-4 Determine whether the Pfaffian equation

$$y\,dx + x\,dy + 2z\,dz = 0$$

has a solution; and if so, find the equation of the family of surfaces.

7-5 Consider the Pfaffian expression

$$a^2y^2z^2\,dx + b^2z^2x^2\,dy + c^2x^2y^2\,dz.$$

(*a*) By inspection and guessing, find an integrating factor.

(*b*) Find the equation of the family of surfaces satisfying the Pfaffian equation obtained by setting the expression equal to zero.

7-6 The expression for $\hat{\lambda}$ given in Eq. (7-12) enables one to write $đ\hat{Q} = \varphi(t)\hat{f}(\hat{\sigma})\,d\hat{\sigma}$, which is of the same form as Eq. (7-13). The expression for $\boldsymbol{\lambda}$, however, would give an equation for $đQ$ of the same form as Eq. (7-13), *provided that* $g(\sigma, \hat{\sigma}) = f(\boldsymbol{\sigma})$. The purpose of this problem is to lead the way toward a proof of the functional dependence of $g(\sigma, \hat{\sigma})$ on $\boldsymbol{\sigma}$.

(*a*) Show that

$$f = g\frac{\partial \boldsymbol{\sigma}}{\partial \sigma} \quad \text{and} \quad \hat{f} = g\frac{\partial \boldsymbol{\sigma}}{\partial \hat{\sigma}}.$$

(*b*) Differentiate f with respect to $\hat{\sigma}$ and equate to zero. Differentiate $\hat{f}$ with respect to σ and equate to zero.

(*c*) Subtract the two equations of part (*b*).

(*d*) Rewrite part (*c*) in the form of a determinant, and interpret the result.

7-7 A Carnot cycle, as shown in Figs. 7-8 and 7-9, consists of a reversible adiabatic process from a lower temperature T_C to a higher temperature T_H, then a reversible isothermal process at T_H in which heat Q_H is transferred, then a reversible adiabatic process from T_H to T_C, and finally a reversible isothermal process at T_C in which heat Q_C is transferred. Draw qualitatively a Carnot cycle for the following:

(*a*) An ideal gas on a PV diagram.

(*b*) A liquid in equilibrium with its vapor on a PV diagram.

(*c*) A reversible electric cell whose emf is a function of temperature only, on an $\mathcal{E}Z$ diagram, assuming reversible adiabatics to have a positive slope.

(*d*) A paramagnetic substance obeying Curie's law on an $\mathcal{H}M$ diagram, assuming $\mathcal{H}/T$ to be practically constant during reversible adiabatic processes.

7-8 Apply the definition of the Kelvin scale to *any* Carnot cycle and calculate the following:

(*a*) The efficiency of a Carnot engine.

(*b*) The coefficient of performance of a Carnot refrigerator.

7-9 Which is the more effective way to increase the efficiency of a Carnot engine: to increase T_H while keeping T_C constant; or to decrease T_C while keeping T_H constant?

7-10 Cause a gas whose equation of state is $P(v - b) = R\theta$ and whose molar energy is a function of θ only to undergo a Carnot cycle, and prove that $\theta = T$.

7-11 In a cylinder, 1 mol of ideal gas with heat capacity c_P is separated by a movable adiabatic frictionless piston from 1 mol of another different ideal gas with heat capacity c'_P. If the first gas receives heat $đq$ and the second $đq'$ from separate reservoirs at different temperatures, the total heat received by this nonhomogeneous composite system is

$$đQ = đq + đq'.$$

Under what circumstances will $đQ$ have an integrating factor?

CHAPTER

EIGHT

ENTROPY

8-1 THE CONCEPT OF ENTROPY

In a system of any number of independent thermodynamic coordinates, all states accessible from a given initial state by reversible adiabatic processes lie on a surface (or hypersurface) $\sigma(t, X, X', \ldots) = \text{const}$. The entire $t, X, X', \ldots$ space may be conceived as being crossed by many nonintersecting surfaces of this kind, each corresponding to a different value of σ. In a *reversible nonadiabatic* process involving a transfer of heat $đQ$, a system in a state represented by a point lying on a surface σ will change until its state point lies on another surface $\sigma + d\sigma$. We have seen that

$$đQ = \lambda\, d\sigma,$$

where $1/\lambda$, the integrating factor of $đQ$, is given by

$$\lambda = \phi(t) f(\sigma)$$

and therefore

$$đQ = \phi(t) f(\sigma)\, d\sigma.$$

Since the Kelvin temperature T is defined so that $T/T' = đQ/đQ'$, with $d\sigma$ being the same for both heat transfers, it follows that

$$T = k\phi(t),$$

where k is an arbitrary constant. Therefore,

$$\frac{đQ}{T} = \frac{1}{k} f(\sigma)\, d\sigma.$$

Since σ is an actual function of $t, X, X', \ldots$, the right-hand member is an exact differential, which we may designate by dS; whence

$$dS = \frac{đQ_R}{T}, \tag{8-1}$$

where the subscript R is written to emphasize that $đQ$ *must* be transferred reversibly. The quantity S is called the *entropy* of the system, and dS is an infinitesimal *entropy change* of the system. In a finite change of state from i to f, the entropy change is $S_f - S_i$, where

$$S_f - S_i = \int_{R\ i}^{f} \frac{đQ}{T}. \tag{8-2}$$

The entropy of a system is a function of the thermodynamic coordinates whose change *is equal to the integral of* $đQ_R/T$ *between the terminal states, integrated along* any *reversible path connecting the two states.* It is important to understand that only an entropy *change* is defined, not an absolute entropy—just as in the case of the internal energy function, whose *change* is defined as the adiabatic work but whose absolute value is undefined.

A third relation may be obtained by integrating Eq. (8-1) around a reversible cycle, so that the initial and final entropies are the same. For a reversible cycle, we get

$$\oint_R \frac{đQ}{T} = 0, \tag{8-3}$$

an equation known as the *Clausius theorem.*

The concept of entropy was first introduced into theoretical physics by R. J. Clausius about the middle of the nineteenth century. Until this time there had been much confusion concerning the relation between heat and work and their roles in the operation of a heat engine. The great French engineers Carnot, Petit, Clément, and Désormes had little knowledge of the first law of thermodynamics. Carnot believed that the work output of an engine was the result of an amount of heat leaving a hot reservoir and the same amount of heat entering a cold reservoir. Petit and Clément computed the efficiency of a heat engine by calculating the work done *only* in the power stroke without considering, as Carnot insisted one must do, the entire cycle. In the words of Mendoza, "In the hands of Clapeyron, Kelvin, and Clausius, thermodynamics began to make headway only when it was divorced from engine design."

Clausius proved the existence of an entropy function by first deriving his theorem [Eq. (8-3); see Prob. 8-1] and then applying it to a cycle consisting of a reversible path R_1 between two equilibrium states i and f, followed by another reversible path R_2 bringing the system back to i. For this cycle,

$$\oint_R \frac{đQ}{T} = \int_{R_1\, i}^{f} \frac{đQ}{T} + \int_{R_2\, f}^{i} \frac{đQ}{T} = 0,$$

or

$$\int_{R_1\, i}^{f} \frac{đQ}{T} = \int_{R_2\, i}^{f} \frac{đQ}{T} = \text{independent of path.}$$

It follows that there exists a function S whose change is

$$S_f - S_i = \int_{R\, i}^{f} \frac{đQ}{T}.$$

The derivation of the Clausius theorem, the properties of Carnot engines on which the theorem is based, and Clausius' derivation of the existence of an entropy function are in every way equivalent to and as general as the methods of Caratheodory. The only superiority of the Caratheodory approach is that it focuses the attention on the system, its coordinates, its states, etc., whereas these are apt to be overlooked in the engineering approach. Physicists and engineers should appreciate both points of view.

8-2 ENTROPY OF AN IDEAL GAS

If a system absorbs an infinitesimal amount of heat $đQ_R$ during a reversible process, the entropy change of the system is equal to

$$dS = \frac{đQ_R}{T}.$$

If $đQ_R$ is expressed as a sum of differentials involving thermodynamic coordinates, then, upon dividing by T, the expression may be integrated and the entropy of the system obtained. As an example of this procedure, consider one of the expressions for $đQ_R$ of an ideal gas, namely,

$$đQ_R = C_P\, dT - V\, dP.$$

Dividing by T, we get

$$\frac{đQ_R}{T} = C_P \frac{dT}{T} - \frac{V}{T}\, dP,$$

or

$$dS = C_P \frac{dT}{T} - nR \frac{dP}{P}.$$

Let us now calculate the entropy change ΔS of the gas between an arbitrarily chosen *reference state* with coordinates T_r, P_r and any other state with coordinates T, P. Integrating between these two states, we get

$$\Delta S = \int_{T_r}^{T} C_P \frac{dT}{T} - nR \ln \frac{P}{P_r}.$$

Suppose that we ascribe to the reference state an entropy S_r and choose *any arbitrary numerical value* for this quantity. Then an entropy S may be associated with the other state where $S - S_r = \Delta S$. To make the discussion simpler, let C_P be constant. Then,

$$S - S_r = C_P \ln \frac{T}{T_r} - nR \ln \frac{P}{P_r},$$

which may be written thus:

$$S = C_P \ln T - nR \ln P + (S_r - C_P \ln T_r + nR \ln P_r).$$

Denoting the quantity in parentheses by the *constant* S_0, we get finally

$$S = C_P \ln T - nR \ln P + S_0.$$

Substituting for T and P thousands of different values, we may calculate thousands of corresponding values of S which, after tabulation, constitute an *entropy table*. Any one value from this table, taken alone, will have no meaning. The difference between two values, however, will be an actual entropy change.

Let us now return to the original differential equation,

$$dS = C_P \frac{dT}{T} - nR \frac{dP}{P}.$$

Again, for simplicity, assuming C_P to be constant, we may take the indefinite integral and obtain

$$S = C_P \ln T - nR \ln P + S_0,$$

where S_0 is the constant of integration. Since this is precisely the equation obtained previously, we see that, in taking the indefinite integral of dS, we do not obtain an "absolute entropy," but merely an entropy referred to a nonspecified reference state whose coordinates are contained within the constant of integration. Thus, for an ideal gas,

$$\boxed{S = \int C_P \frac{dT}{T} - nR \ln P + S_0.} \tag{8-4}$$

To calculate the entropy of an ideal gas as a function of T and V, we use the other expression for $đQ_R$ of an ideal gas. Thus,

$$\frac{đQ_R}{T} = C_V \frac{dT}{T} + \frac{P}{T} dV,$$

and

$$dS = C_V \frac{dT}{T} + nR \frac{dV}{V}.$$

Proceeding in the same way as before, we get for the entropy, referred to an unspecified reference state, the expression

$$\boxed{S = \int C_V \frac{dT}{T} + nR \ln V + S_0,} \tag{8-5}$$

which becomes, if C_V is constant,

$$S = C_V \ln T + nR \ln V + S_0.$$

8-3 *TS* DIAGRAM

For each infinitesimal amount of heat that enters a system during an infinitesimal portion of a reversible process, there is an equation

$$đQ_R = T\,dS.$$

It follows therefore that the total amount of heat transferred in a reversible process is given by

$$Q_R = \int_i^f T\,dS.$$

This integral can be interpreted graphically as the area under a curve on a diagram in which T is plotted along the Y axis and S along the X axis. The nature of the curve on the TS diagram is determined by the kind of reversible process that the system undergoes. Obviously, an isothermal process is a horizontal line.

In the case of a reversible adiabatic process, we have

$$dS = \frac{đQ_R}{T},$$

and

$$đQ_R = 0;$$

whence, if T is not zero,

$$dS = 0,$$

and S is constant. Therefore, during a reversible adiabatic process, the entropy of a system remains constant; or in other words, the system undergoes an *isentropic process*. An isentropic process on a TS diagram is obviously a vertical line.

If two equilibrium states are infinitesimally near, then

$$đQ = T\,dS,$$

and

$$\frac{đQ}{dT} = T\frac{dS}{dT}.$$

At constant volume,

$$\boxed{\left(\frac{đQ}{dT}\right)_V = C_V = T\left(\frac{\partial S}{\partial T}\right)_V;} \tag{8-6}$$

and at constant pressure,

$$\boxed{\left(\frac{đQ}{dT}\right)_P = C_P = T\left(\frac{\partial S}{\partial T}\right)_P.} \tag{8-7}$$

If the temperature variation of C_V is known, the entropy change during an isochoric process may be calculated from the equation

$$S_f - S_i = \int_i^f \frac{C_V}{T}\, dT \qquad \text{(isochoric)}. \tag{8-8}$$

Similarly, for an isobaric process,

$$S_f - S_i = \int_i^f \frac{C_P}{T}\, dT \qquad \text{(isobaric)}. \tag{8-9}$$

The foregoing equations provide a general method for calculating an entropy change but no way of calculating the absolute entropy of a system in a given state. If a set of tables is required that is to be used to obtain entropy differences and not absolute entropy, then it is a convenient procedure to choose an arbitrary standard state and calculate the entropy change of the system from

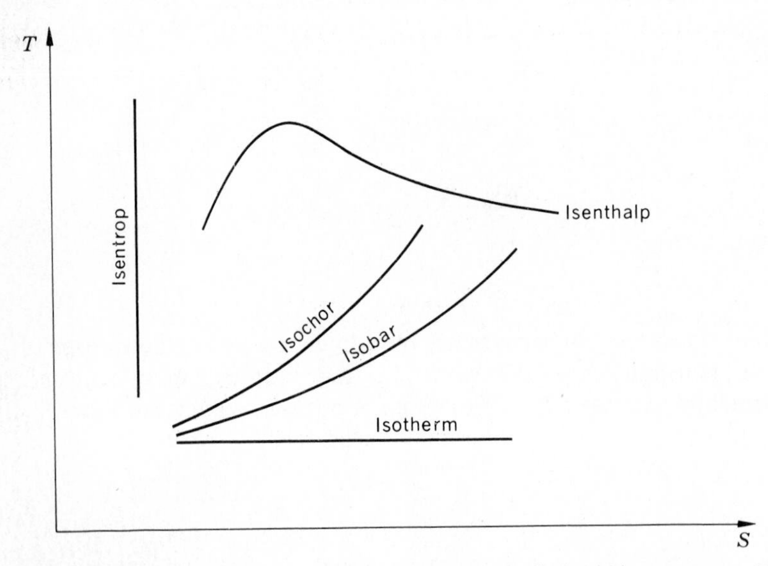

Figure 8-1 Curves representing reversible processes of a hydrostatic system on a TS diagram.

this standard state to all other states. Thus, in the case of water, the standard state is chosen to be that of saturated water at 0.01°C and its own vapor pressure 611 Pa, and all entropies are referred to this state.

The slope of a curve on a *TS* diagram representing a reversible isocheric process, from Eq. (8-6), is

$$\left(\frac{\partial T}{\partial S}\right)_V = \frac{T}{C_V};$$

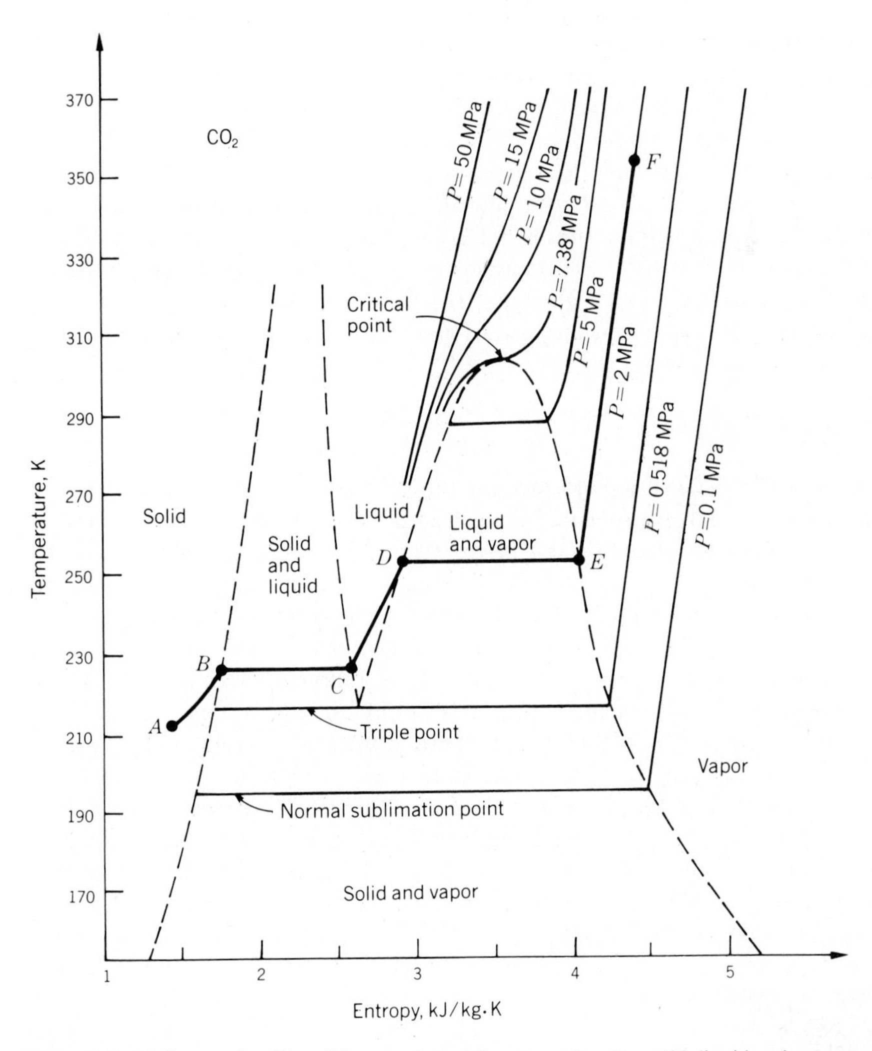

Figure 8-2 *TS* diagram for CO_2. (The two dashed lines bounding the solid–liquid region are a guess.)

and from Eq. (8-7), the slope of a reversible isobar is

$$\left(\frac{\partial T}{\partial S}\right)_P = \frac{T}{C_P}.$$

Curves representing various types of processes of a hydrostatic system are shown on a *TS* diagram in Fig. 8-1.

The *TS* diagram for a substance such as CO_2 is shown in Fig. 8-2. The curve from *A* to *F* is a typical isobar representing a series of reversible isobaric processes in which solid is transformed finally into vapor. Thus,

AB = isobaric heating of solid to its melting point.
BC = isobaric isothermal melting.
CD = isobaric heating of liquid to its boiling point.
DE = isobaric isothermal vaporization.
EF = isobaric heating of vapor (superheating).

The area under the line *BC* represents the heat of fusion at the particular temperature, and the area under the line *DE* represents the heat of vaporization. Similarly, the heat of sublimation is represented by the area under any sublimation line. It is obvious from the diagram that the heat of vaporization becomes zero at the critical point and, also, that the heat of sublimation is equal to the sum of the heat of fusion and the heat of vaporization at the triple point.

8-4 CARNOT CYCLE

During a part of the cycle performed by the system in an engine, some heat is absorbed from a hot reservoir; during another part of the cycle, a smaller amount of heat is rejected to a cooler reservoir. The engine is therefore said to operate between these two reservoirs. Since it is a fact of experience that some heat is always rejected to the cooler reservoir, the efficiency of an actual engine is never 100 percent. If we assume that we have at our disposal two reservoirs at given temperatures, it is important to answer the following questions: (1) What is the maximum efficiency that can be achieved by an engine operating between these two reservoirs? (2) What are the characteristics of such an engine? (3) Of what effect is the nature of the substance undergoing the cycle?

The importance of these questions was recognized by Nicolas Léonard Sadi Carnot, a brilliant young French engineer who in the year 1824, before the first law of thermodynamics was firmly established, described in a paper entitled "Réflexions sur la puissance motrice du feu" an ideal engine operating in a particularly simple cycle known today as the *Carnot cycle.* A general Carnot cycle is depicted in Fig. 7-8, and one executed by an ideal gas with only two independent variables in Fig. 7-9.

An engine operating in a Carnot cycle is called a *Carnot engine.* A Carnot engine operates between two reservoirs in a particularly simple way. *All the heat that is absorbed is absorbed at a constant high temperature, namely, that of the hot reservoir. Also, all the heat that is rejected is rejected at a constant lower temperature, that of the cold reservoir.* The processes connecting the high- and low-temperature isotherms are reversible and adiabatic. Since all four processes are reversible, the Carnot cycle is a reversible cycle. (See Prob. 6-1.)

If an engine is to operate between only two reservoirs and still operate in a reversible cycle, then it must be a Carnot engine. For example, if an Otto cycle were performed between only two reservoirs, the heat transfers in the two isochoric processes would involve finite temperature differences and, therefore, could not be reversible. Conversely, if the Otto cycle were performed reversibly, it would require a series of reservoirs, not merely two. The expression "Carnot engine," therefore, means "a reversible engine operating between only two reservoirs."

A Carnot engine absorbing heat Q_H from a hot reservoir at T_H and rejecting heat Q_C to a cooler reservoir at T_C has an efficiency η equal to $1 - |Q_C|/|Q_H|$. Since, between the same two isentropic surfaces,

$$\frac{|Q_C|}{|Q_H|} = \frac{T_C}{T_H},$$

$$\boxed{\eta \text{ (Carnot)} = 1 - \frac{T_C}{T_H}.} \qquad (8\text{-}10)$$

For a Carnot engine to have an efficiency of 100 percent, T_C must be zero. Since nature does not provide us with a reservoir at absolute zero, a heat engine with 100 percent efficiency is a practical impossibility.

A temperature-entropy diagram is particularly suited to display the characteristics of a Carnot cycle. The two reversible adiabatic processes are vertical lines, and the two reversible isothermal processes are horizontal lines lying between the two vertical lines, so that the Carnot cycle is represented by a rectangle, as shown in Fig. 8-3. This is true regardless of the nature of the system and of the number of independent thermodynamic coordinates.

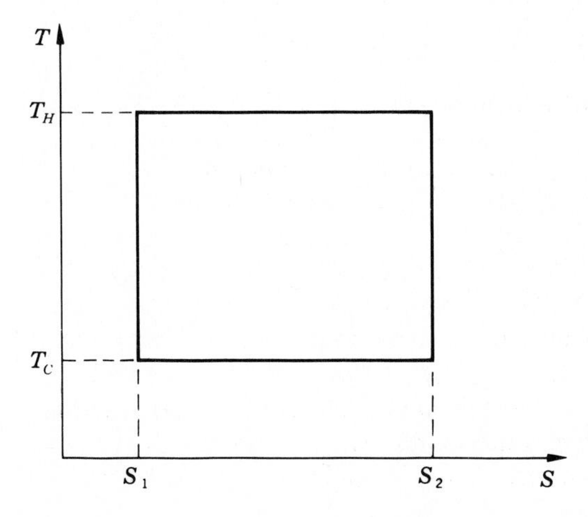

Figure 8-3 A Carnot cycle of any system of any number of independent coordinates when represented on a *TS* diagram is a rectangle.

8-5 ENTROPY AND REVERSIBILITY

In order to understand the physical meaning of entropy and its significance in the world of science, it is necessary to study all the entropy changes that take place when a system undergoes a process. If we calculate the entropy change of the system and add to this the entropy change of the local surroundings, we obtain a quantity that is the sum of all the entropy changes brought about by this particular process. We may call this the *entropy change of the universe* due to the process in question.

When a finite amount of heat is absorbed or rejected by a reservoir, extremely small changes in the coordinates take place in every unit of mass. The entropy change of a unit of mass is therefore very small. Since the total mass of a reservoir is large, however, the total entropy change is finite. Suppose that a reservoir is in contact with a system and that heat Q is absorbed by the reservoir at the temperature T. The reservoir undergoes nondissipative changes determined entirely by the quantity of heat absorbed. Exactly the same changes *in the reservoir* would take place if the same amount of heat Q were transferred reversibly. Hence, the entropy change of the reservoir is Q/T. Therefore, *whenever a reservoir absorbs heat Q at the temperature T from any system during any kind of process, the entropy change of the* reservoir *is Q/T.*

Consider now the entropy change of the universe that is brought about by the performance of any reversible process. The process will, in general, be accompanied by a flow of heat between a system and a set of reservoirs ranging in temperature from T_i to T_f. During *any* infinitesimal portion of the process, an amount of heat $đQ_R$ is transferred between the system and one of the reservoirs at the temperature T. Let $đQ_R$ be a positive number. If $đQ_R$ is absorbed by the system, then

$$dS \text{ of the system} = +\frac{đQ_R}{T},$$

$$dS \text{ of the reservoir} = -\frac{đQ_R}{T},$$

and the entropy change of the universe $\sum dS$ is zero. If $đQ_R$ is rejected by the system, then obviously

$$dS \text{ of the system} = -\frac{đQ_R}{T},$$

$$dS \text{ of the reservoir} = +\frac{đQ_R}{T},$$

and the entropy change of the universe $\sum dS$ is again zero. If $đQ_R$ is zero, then neither the system nor the reservoir will have an entropy change, and the entropy change of the universe is still zero. Since this is true for any infinitesimal portion of the reversible process, it is true for all such portions; therefore we may conclude that, *when a reversible process is performed, the entropy of the universe remains unchanged.* However, all natural processes are irreversible.

8-6 ENTROPY AND IRREVERSIBILITY

When a system undergoes an irreversible process between an initial equilibrium state and a final equilibrium state, the entropy change of the system is equal to

$$\Delta S \text{ (system)} = S_f - S_i = \int_{R, i}^{f} \frac{đQ}{T},$$

where R indicates *any reversible process arbitrarily chosen* by which the system may be brought from the given initial state to the given final state. No integration is performed over the original irreversible path. The irreversible process is replaced by a reversible one. This can easily be done when the initial and the final state of the system are equilibrium states. When either the initial or the final state is a nonequilibrium state, special methods must be used. At first, we shall limit ourselves to irreversible processes all of which involve initial and final states of equilibrium.

Processes Exhibiting External Mechanical Irreversibility

(*a*) Those involving the isothermal dissipation of work through a system (which remains unchanged) into internal energy of a reservoir, such as:

1. Irregular stirring of a viscous liquid in contact with a reservoir.
2. Coming to rest of a rotating or vibrating liquid in contact with a reservoir.
3. Inelastic deformation of a solid in contact with a reservoir.
4. Transfer of electricity through a resistor in contact with a reservoir.
5. Magnetic hysteresis of a material in contact with a reservoir.

In the case of any process involving the isothermal transformation of work W through a system into internal energy of a reservoir, there is no entropy change of the system because the thermodynamic coordinates do not change. There is a flow of heat Q into the reservoir where $Q = W$. Since the reservoir absorbs Q units of heat at the temperature T, its entropy change is $+Q/T$ or $+W/T$. The entropy change of the universe is therefore W/T, which is a positive quantity.

(*b*) Those involving the adiabatic dissipation of work into internal energy of a system, such as:

1. Irregular stirring of a viscous thermally insulated liquid.
2. Coming to rest of a rotating or vibrating thermally insulated liquid.
3. Inelastic deformation of a thermally insulated solid.
4. Transfer of electricity through a thermally insulated resistor.
5. Magnetic hysteresis of a thermally insulated material.

In the case of any process involving the adiabatic transformation of work W into internal energy of a system whose temperature rises from T_i to T_f at constant pressure, there is no flow of heat to or from the surroundings, and therefore the entropy change of the local surroundings is zero. To calculate the entropy change of the system, the original irreversible process must be replaced by a reversible one that will take the system from the given initial state (temperature T_i, pressure P) to the final state (temperature T_f, pressure P). Let us replace the irreversible performance of work by a reversible isobaric flow of heat from a series of reservoirs ranging in temperature from T_i to T_f. The entropy change of the system will then be

$$\Delta S\ (\text{system}) = \int_{R\ T_i}^{T_f} \frac{đQ}{T}.$$

For an isobaric process,

$$đQ_R = C_P\, dT,$$

and

$$\Delta S\ (\text{system}) = \int_{T_i}^{T_f} C_P \frac{dT}{T}.$$

Finally, if C_P is assumed constant,

$$\Delta S\ (\text{system}) = C_P \ln \frac{T_f}{T_i},$$

and the entropy change of the universe is $C_P \ln (T_f/T_i)$, which is a positive quantity.

Processes Exhibiting Internal Mechanical Irreversibility

Those involving the transformation of internal energy of a system into mechanical energy and then back into internal energy again, such as:

1. Ideal gas rushing into a vacuum (free expansion).
2. Gas seeping through a porous plug (throttling process).
3. Snapping of a stretched wire after it is cut.
4. Collapse of a soap film after it is pricked.

In the case of a free expansion of an ideal gas, the entropy change of the local surroundings is zero. To calculate the entropy change of the system, the free expansion must be replaced by a reversible process that will take the gas from its original state (volume V_i, temperature T) to the final state (volume V_j, temperature T). Evidently, the most convenient reversible process is a reversible isothermal expansion at the temperature T from a volume V_i to the volume V_f. The entropy change of the system is then

$$\Delta S\ (\text{system}) = \int_{R\ V_i}^{V_f} \frac{đQ}{T}.$$

For an isothermal process of an ideal gas,

$$đQ_R = P\,dV,$$

and

$$\frac{đQ_R}{T} = nR\frac{dV}{V};$$

whence

$$\Delta S\ (\text{system}) = nR \ln \frac{V_f}{V_i}.$$

The entropy change of the universe is therefore $nR \ln (V_f/V_i)$, which is a positive number.

Processes Exhibiting External Thermal Irreversibility

Those involving a transfer of heat by virtue of a finite temperature difference, such as:

1. Conduction or radiation of heat from a system to its cooler surroundings.
2. Conduction or radiation of heat through a system (which remains unchanged) from a hot reservoir to a cooler one.

In the case of the conduction of Q units of heat through a system (which remains unchanged) from a hot reservoir at T_1 to a cooler reservoir at T_2, the following steps are obvious:

$$\Delta S\ (\text{system}) = 0,$$

$$\Delta S\ (\text{hot reservoir}) = -\frac{Q}{T_1},$$

$$\Delta S\ (\text{cold reservoir}) = +\frac{Q}{T_2},$$

and

$$\sum \Delta S = \Delta S\ (\text{universe}) = \frac{Q}{T_2} - \frac{Q}{T_1}.$$

Processes Exhibiting Chemical Irreversibility

Those involving a spontaneous change of internal structure, chemical composition, density, etc., such as:

1. A chemical reaction.
2. Diffusion of two dissimilar inert ideal gases.
3. Mixing of alcohol and water.
4. Freezing of supercooled liquid.
5. Condensation of a supersaturated vapor.
6. Solution of a solid in water.
7. Osmosis.

Table 8-1 Entropy change of the universe due to natural processes

Type of irreversibility	Irreversible process	Entropy change of the system ΔS (syst.)	Entropy change of the local surroundings ΔS (surr.)	Entropy change of the universe $\sum \Delta S$
External mechanical irreversibility	Isothermal dissipation of work through a system into internal energy of a reservoir	0	$\frac{W}{T}$	$\frac{W}{T}$
	Adiabatic dissipation of work into internal energy of a system	$C_P \ln \frac{T_f}{T_i}$	0	$C_P \ln \frac{T_f}{T_i}$
Internal mechanical irreversibility	Free expansion of an ideal gas	$nR \ln \frac{V_f}{V_i}$	0	$nR \ln \frac{V_f}{V_i}$
External thermal irreversibility	Transfer of heat through a medium from a hot to a cooler reservoir	0	$\frac{Q}{T_2} - \frac{Q}{T_1}$	$\frac{Q}{T_2} - \frac{Q}{T_1}$
Chemical irreversibility	Diffusion of two dissimilar inert ideal gases	$2R \ln 2$	0	$2R \ln 2$

Assuming the diffusion of two dissimilar inert ideal gases to be equivalent to two separate free expansions, for one of which

$$\Delta S = nR \ln \frac{V_f}{V_i},$$

and taking a mole of each gas with $V_i = v$ and $V_f = 2v$, we obtain

$$\sum \Delta S = 2R \ln 2,$$

which is a positive number. All the results of this article are summarized in Table 8-1.

8-7 ENTROPY AND NONEQUILIBRIUM STATES

The calculation of the entropy changes associated with the irreversible processes discussed in Art. 8-6 presented no special difficulties because, in all cases, the system either did not change at all (in which case only the entropy changes of reservoirs had to be calculated) or the terminal states of a system were equilibrium states that could be connected by a suitable reversible process. Consider, however, the following process involving internal thermal irreversibility. A thermally conducting bar, brought to a nonuniform temperature distribution by contact at one end with a hot reservoir and at the other end with a cold reservoir, is removed from the reservoirs and then thermally insulated and kept at constant pressure. An internal flow of heat will finally bring the bar to a uniform temperature, but the transition will be from an initial nonequilibrium state to a final equilibrium state. It is obviously impossible to find one reversible process by which the system may be brought from the same initial to the same final state. What meaning, therefore, may be attached to the entropy change associated with this process?

Let us consider the bar to be composed of an infinite number of infinitesimally thin sections, each of which has a different initial temperature but all of which have the same final temperature. Suppose that we imagine all the sections to be insulated from one another and kept at the same pressure and then each section to be put in contact successively with a series of reservoirs ranging in temperature from the initial temperature of the particular section to the common final temperature. This defines an infinite number of reversible isobaric processes, which may be used to take the system from its initial non-equilibrium state to its final equilibrium state. We shall now define the entropy change as the result of integrating $đQ/T$ over all these reversible processes. In other words, in the absence of one reversible process to take the system from i to f, we conceive of an infinite number of reversible processes—one for each volume element.

As an example, consider the uniform bar of length L depicted in Fig. 8-4. A typical volume element at x has a mass

$$dm = \rho A\, dx,$$

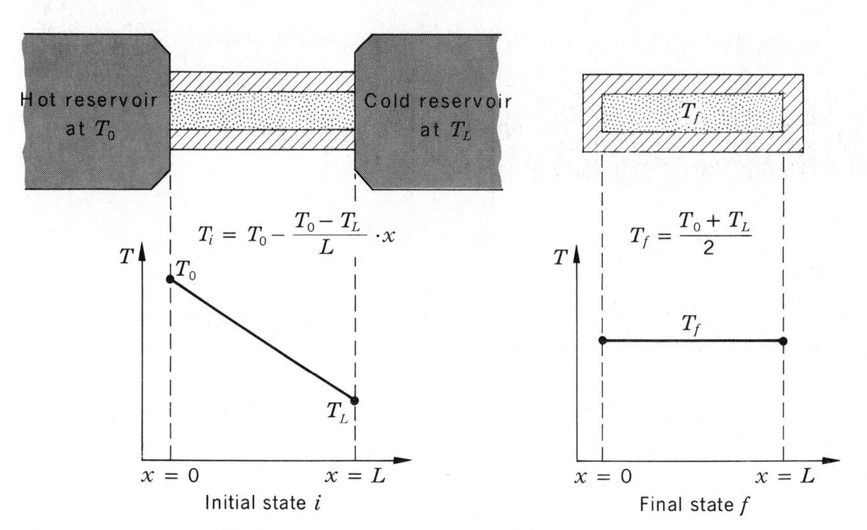

Figure 8-4 Process exhibiting internal thermal irreversibility.

where ρ is the density and A the cross-sectional area. The heat capacity of the section is

$$c_P\,dm = c_P \rho A\,dx.$$

Let us suppose that the initial temperature distribution is linear, so that the section at x has an initial temperature

$$T_i + T_0 - \frac{T_0 - T_L}{L} x.$$

If no heat is lost and if we assume for the sake of simplicity that the thermal conductivity, density, and heat capacity of all sections remain constant, then the final temperature will be

$$T_f = \frac{T_0 + T_L}{2}.$$

Integrating $đQ/T$ over a reversible isobaric transfer of heat between the volume element and a series of reservoirs ranging in temperature from T_i to T_f, we get, for the entropy change of *this one volume element*,

$$c_P \rho A\,dx \int_{T_i}^{T_f} \frac{dT}{T} = c_P \rho A\,dx \ln \frac{T_f}{T_i}$$

$$= c_P \rho A\,dx \ln \frac{T_f}{T_0 - \dfrac{T_0 - T_L}{L} x}$$

$$= -c_P \rho A\,dx \ln \left(\frac{T_0}{T_f} - \frac{T_0 - T_L}{L T_f} x \right).$$

Upon integrating over the whole bar, the total entropy change is

$$\sum \Delta S = -c_P \rho A \int_0^L \ln \left(\frac{T_0}{T_f} - \frac{T_0 - T_L}{L T_f} x \right) dx,$$

which, after integration† and simplification, becomes

$$\sum \Delta S = C_P \left(1 + \ln T_f + \frac{T_L}{T_0 - T_L} \ln T_L - \frac{T_0}{T_0 - T_L} \ln T_0 \right).$$

To show that the entropy change is positive, let us take a convenient numerical case such as $T_0 = 400$ K, $T_L = 200$ K; whence $T_f = 300$ K. Then,

$$\sum \Delta S = 2.30 C_P \left(\frac{1}{2.30} + 2.477 + 2.301 - 2 \times 2.602 \right)$$

$$= 0.019 C_P.$$

† $\int \ln (a + bx)\,dx = \frac{1}{b}(a + bx) \ln (a + bx) - x.$

The same method may be used to compute the entropy change of a system during a process from an initial nonequilibrium state characterized by a nonuniform pressure distribution to a final equilibrium state where the pressure is uniform. Examples of such processes are given in the problems at the end of this chapter.

8-8 PRINCIPLE OF THE INCREASE OF ENTROPY

The entropy change of the universe associated with each of the irreversible processes treated up to now was found to be positive. We are led to believe, therefore, that whenever an irreversible process takes place the entropy of the universe increases. To establish this proposition, known as the *entropy principle*, in a general manner, it is sufficient to confine our attention to adiabatic processes only, since we have already seen that the entropy principle is true for all processes involving the irreversible transfer of heat. We start the proof by considering the special case of an adiabatic irreversible process between two equilibrium states of a system.

1. Let the system, as usual, have three independent coordinates T, X, and X' and let the initial state be represented by point i on the diagram shown in Fig. 8-5. Suppose that the system undergoes an *irreversible adiabatic process* to the state f; then the entropy change is

$$\Delta S = S_f - S_i.$$

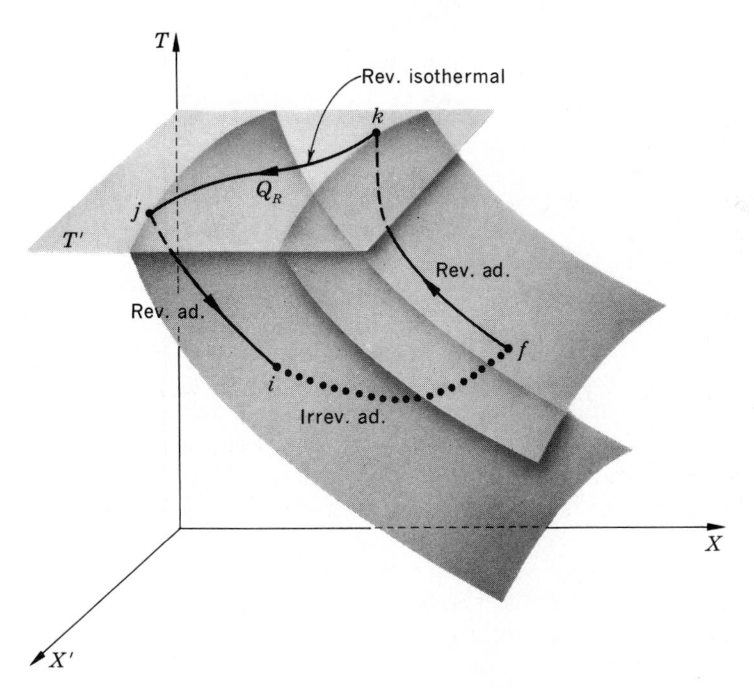

Figure 8-5 The process $i \to f$ is irreversible and adiabatic. The cycle *ifkji* contradicts the second law unless $S_f > S_i$.

A temperature change may or may not have taken place. Whether or not, let us cause the system to undergo a *reversible adiabatic process* $f \to k$ in such a direction as to bring its temperature to that of any arbitrarily chosen reservoir, say, at T'. Now suppose that the system is brought into contact with the reservoir and caused to undergo a *reversible isothermal process* $k \to j$ until its entropy is the same as at the beginning. A final *reversible adiabatic process* $j \to i$ will now bring the system back to its original state. The net entropy change for the cycle is zero, and entropy changes take place only during the two processes $i \to f$ and $k \to j$. Consequently,

$$(S_f - S_i) + (S_j - S_k) = 0.$$

If ΔS denotes the entropy change associated with the irreversible part of the cycle ($\Delta S = S_f - S_i$), it follows that

$$\Delta S = S_k - S_j.$$

The only heat transfer Q_R that has taken place in the cycle is during the isothermal process $k \to j$, where

$$Q_R = T'(S_j - S_k).$$

A net amount of work W (net) has been done in the cycle, where

$$W\ (\text{net}) = Q_R.$$

It is clear from the second law of thermodynamics that the heat Q_R cannot have entered the system—that is, Q_R cannot be positive—for then we would have a cyclic process in which no effect has been produced other than the extraction of heat from a reservoir and the performance of an equivalent amount of work. Therefore, $Q_R \leq 0$, and

$$T'(S_j - S_k) \leq 0,$$

and finally,

$$\Delta S \geq 0.$$

2. If we assume that the original irreversible adiabatic process took place without any change in entropy, then it would be possible to bring the system back to i by means of one reversible adiabatic process. Moreover, since the net heat transferred in this cycle is zero, the net work would also be zero. Therefore, under these circumstances, the system and its surroundings would have been restored to their initial states without producing changes elsewhere, which implies that the original process was reversible. Since this is contrary to our original assertion, the entropy of the system cannot remain unchanged. Therefore,

$$\Delta S > 0. \tag{8-11}$$

3. Let us now suppose that the system is not homogeneous and not of uniform temperature and pressure and that it undergoes an irreversible adiabatic process in which mixing and chemical reaction may take place. If we assume that the system may be subdivided into parts (each one infinitesimal, if necessary)

and that it is possible to ascribe a definite temperature, pressure, composition, etc., to each part, so that each part shall have a definite entropy depending on its coordinates, then we may define the entropy of the whole system as the sum of the entropies of its parts. If we now assume that it is possible to take *each part* back to its initial state by means of the reversible processes described above in (1), using the same reservoir for each part, then it follows that ΔS of the whole system is positive.

It should be emphasized that we have had to make two assumptions, namely, (1) that the entropy of a system may be defined by subdividing the system into parts and summing the entropies of these parts and (2) that reversible processes may be found or imagined by which mixtures may be unmixed and reactions may be caused to proceed in the opposite direction. The justification for these assumptions rests to a small extent on experimental grounds. Thus, in a later chapter, there will be described a device involving semipermeable membranes whereby a mixture of two different inert ideal gases may be separated reversibly. A similar device through which a chemical reaction may be caused to proceed reversibly in any desired direction may also be conceived. Nevertheless, the main justification for these assumptions, and therefore for the entropy principle, lies in the fact that they lead to results in complete agreement with experiment; for the experimental physicist, this is sufficient.

4. As the last step in our argument, let us consider an assemblage of systems *and* reservoirs in an adiabatic enclosure. All heat transfers involving finite temperature differences involve net entropy increases, and all adiabatic processes involving irreversible state changes, mixing, chemical reactions, etc., are also attended by net entropy increases. The adiabatic enclosure constitutes the "universe" since it includes all systems and reservoirs that interact during the process under consideration. It follows, therefore, that the behavior of the *entropy of the universe* as a result of *any* kind of process may now be represented in the following succinct manner:

$$\boxed{\sum \Delta S \geq 0,} \tag{8-12}$$

where the equality sign refers to reversible processes and the inequality sign to irreversible processes.

8-9 ENGINEERING APPLICATIONS OF THE ENTROPY PRINCIPLE

Whenever irreversible processes take place, the entropy of the universe increases. In the actual operation of a device such as an engine or a refrigerator, it is often possible to calculate the sum of all the entropy changes. The fact that this sum is positive enables us to draw useful conclusions concerning the behavior of the device. Two important examples from the field of mechanical engineering will illustrate the power and simplicity of the entropy principle.

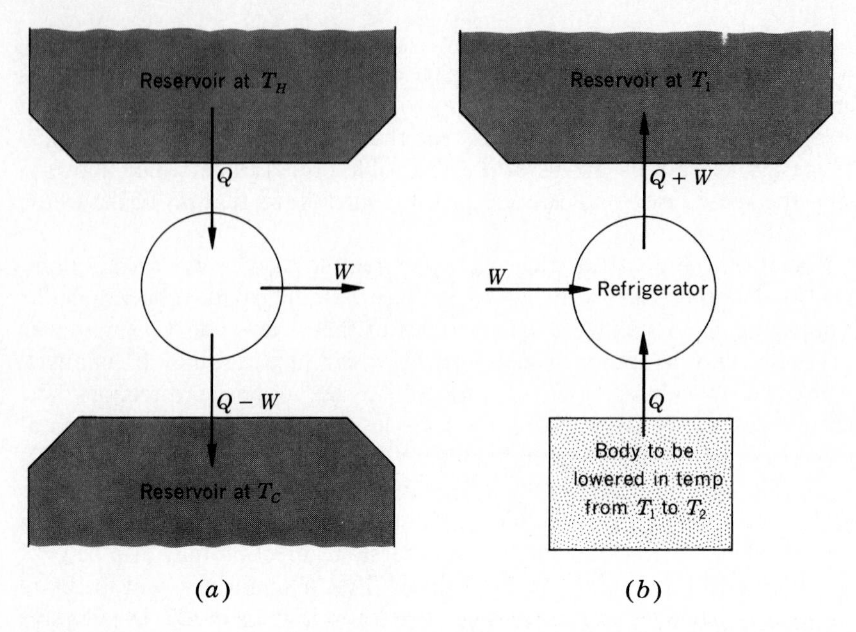

Figure 8-6 (*a*) Engine operating between reservoirs at T_H and T_C. (*b*) Refrigerator operating between a reservoir at T_1 and a finite body and lowering the temperature of the body from T_1 to T_2.

1. Consider a heat engine undergoing *any arbitrary cycle*, as shown in Fig. 8-6*a*, extracting heat Q from a reservoir at T_H delivering an amount of work W, and rejecting heat $Q - W$ to a colder reservoir at T_C. According to the entropy principle,

$$\sum \Delta S \text{ (universe)} = \frac{Q - W}{T_C} - \frac{Q}{T_H} \geq 0,$$

or

$$W \leq Q - \frac{T_C}{T_H} Q;$$

whence

$$W_{\max} = Q\left(1 - \frac{T_C}{T_H}\right).$$

Since $W_{\max}/Q$ is the maximum efficiency of an engine extracting Q from a reservoir at T_H and rejecting heat to a reservoir at T_C, and since $1 - T_C/T_H$ was shown, in Eq. (8-10), to be the efficiency of a Carnot engine, we have the result that *the maximum efficiency of any engine operating between two reservoirs is that of a Carnot engine operating between the same two reservoirs.*

2. Suppose that it is desired to freeze water or to liquefy air, i.e., to lower the temperature of a body of finite mass from the temperature T_1 of its surroundings to any desired temperature T_2. A refrigerator operating in a cycle between a reservoir at T_1 and the body itself is utilized, and after a finite number of complete cycles has been traversed, a quantity of heat Q has been removed from

the body, a quantity of work W has been supplied to the refrigerator, and a quantity of heat $Q + W$ has been rejected to the reservoir, as shown in Fig. 8-6(*b*). Listing the entropy changes, we have

$$\Delta S \text{ of the body} = S_2 - S_1,$$

$$\Delta S \text{ of the refrigerant} = 0,$$

and

$$\Delta S \text{ of the reservoir} = \frac{Q + W}{T_1}.$$

Applying the entropy principle,

$$S_2 - S_1 + \frac{Q + W}{T_1} \geq 0;$$

whence

$$W \geq T_1(S_1 - S_2) - Q.$$

It follows that the smallest value for W is

$$W\text{ (min)} = T_1(S_1 - S_2) - Q.$$

If tables of the thermodynamic properties of the material are available, a knowledge of the initial and final states is all that is needed to read from the tables the values of $S_1 - S_2$ and, if the body undergoes an isobaric process, of Q. The calculated value of W (min) is used to provide an estimate of the minimum cost of operation of the refrigeration plant.

8-10 ENTROPY AND UNAVAILABLE ENERGY

Suppose that a quantity of heat Q may be extracted from a reservoir at the temperature T and that it is desired to convert as much as possible of this heat into work. If T_0 is the temperature of the *coldest* reservoir at hand, then

$$W\text{ (max)} = Q\left(1 - \frac{T_0}{T}\right),$$

which represents the maximum amount of energy available for work when Q units of heat are extracted from a reservoir at T. It is obvious, therefore, that any energy which resides within the reservoir at T_0 and which may be extracted only in the form of heat is in a form in which it is completely unavailable for work. The potential energy, however, of a frictionless mechanical device (as measured from the position of lowest potential energy) is in a form in which it is completely available for work. It is desired to establish the proposition that, *whenever an irreversible process takes place, the effect on the universe is the same as that which would be produced if a certain quantity of energy were converted from a form in which it is completely available for work into a form in which it is completely unavailable for work. This amount of energy E is T_0 times the entropy change of the universe brought about by the irreversible process.*

Since the general proof of this proposition is somewhat abstract, let us consider first a special case, namely, the irreversible conduction of heat under a finite temperature gradient. Suppose that heat Q is conducted along a bar from a region at temperature T_1 to a region at temperature T_2. After conduction has taken place, we have heat Q available at the lower temperature T_2, of which the following amount is available for work:

$$\text{Max. work after conduction} = Q\left(1 - \frac{T_0}{T_2}\right).$$

If conduction had not taken place, heat Q would have been available at the higher temperature T_1, and the maximum amount of work that could have been obtained from this is

$$\text{Max. work before conduction} = Q\left(1 - \frac{T_0}{T_1}\right).$$

Evidently, the amount of energy E that has become unavailable for work is the difference

$$\begin{aligned} E &= Q\left(1 - \frac{T_0}{T_1}\right) - Q\left(1 - \frac{T_0}{T_2}\right) \\ &= T_0\left(\frac{Q}{T_2} - \frac{Q}{T_1}\right) \\ &= T_0\,\Delta S \qquad \text{(universe).} \end{aligned}$$

The proposition is therefore seen to be true for the special case of heat conduction. Since it is not possible to handle all irreversible processes in this simple manner, we shall have to adopt a more abstract point of view in order to establish the proposition generally.

Consider a mechanical device such as a suspended object or a compressed spring capable of doing work on a system. Suppose that the system is in contact with a reservoir at the temperature T. The mechanical device and the reservoir at T constitute the *local surroundings* of the system. Suppose that an irreversible process takes place in which the mechanical device does work W on the system, the internal energy of the system changes from U_i to U_f, and heat Q is transferred between the system and the reservoir. Then the first law demands that

$$Q = U_f - U_i - W,$$

and the second law that

$$S_f - S_i \text{ (system and local surroundings)} > 0.$$

Now suppose that it is desired to produce exactly the same changes in the system and the local surroundings which resulted from the performance of the irreversible process, *but by reversible processes only*. This would require, in general, the services of Carnot engines and refrigerators, which, in turn, would have to be operated in conjunction with an auxiliary mechanical device and an auxiliary reservoir. The auxiliary mechanical device may be considered, as usual, to be either a suspended object or a compressed spring. For the auxiliary reservoir let us choose the one whose temperature is the lowest at hand, say, T_0. These constitute the *auxiliary surroundings.* With the aid of suitable Carnot engines and refrigerators all operating in cycles, in conjunction with the auxiliary surroundings, it is now possible to produce in the system and the local surroundings, by reversible processes only, the same changes that were formerly brought about by the irreversible process. If this is done, the entropy change of the system and the local surroundings is the same as before, since they have gone from the same initial states to the same final states. The auxiliary surroundings, however, must undergo an equal and opposite entropy change, because the net entropy change of the universe during reversible processes is zero.

Since the entropy change of the system and local surroundings is positive, the entropy change of the auxiliary surroundings is negative. Therefore the reservoir at T_0 must have rejected a certain amount of heat, say, E. Since no extra energy has appeared in the system and local surroundings, the energy E must have been transformed into work on the auxiliary mechanical device. We have the result therefore that, *when the same changes which were formerly produced in a system and local surroundings by an irreversible process are brought about reversibly, an amount of energy E leaves an auxiliary reservoir at T_0 in the form of heat and appears in the form of work on an auxiliary mechanical device.* In other words, energy E is converted from a form in which it was completely unavailable for work into a form in which it is completely available for work. Since the original process was not performed reversibly, the energy E was not converted into work, and therefore *E is the energy that is rendered unavailable for work* because of the performance of the irreversible process.

It is a simple matter to calculate the energy that becomes unavailable during an irreversible process. If the same changes are brought about reversibly, the entropy change of the system and local surroundings is the same as before, namely, $S_f - S_i$. The entropy change of the auxiliary surroundings is merely the entropy change of the auxiliary reservoir due to the rejection of E units of heat at the temperature T_0, that is, $-E/T_0$. Since the sum of the entropy changes of the system, local surroundings, and auxiliary surroundings is zero, we have

$$S_f - S_i - \frac{E}{T_0} = 0;$$

whence

$$\boxed{E = T_0(S_f - S_i).} \tag{8-13}$$

Therefore, *the energy that becomes unavailable for work during an irreversible process is T_0 times the entropy change of the universe that is brought about by the irreversible process.* Since no energy becomes unavailable during a reversible process, it follows that *the maximum amount of work is obtained when a process takes place reversibly.*

Since irreversible processes are continually going on in nature, energy is continually becoming unavailable for work. This conclusion, known as the *principle of the degradation of energy* and first developed by Kelvin, provides an important physical interpretation of the entropy change of the universe. It must be understood that energy which becomes unavailable for work is not energy which is lost. The first law is obeyed at all times. Energy is merely transformed from one form into another. In picturesque language, one may say that energy is "running downhill."

8-11 ENTROPY AND DISORDER

It has been emphasized that work, as it is used in thermodynamics, is a macroscopic concept. There must be changes that are describable by macroscopic coordinates. Haphazard motions of individual molecules against intermolecular forces do not constitute work. Work involves order or orderly motion. Whenever work is dissipated into internal energy, the disorderly motion of molecules is increased. Thus, during either the isothermal or adiabatic dissipation of work into internal energy, the disorderly motion of the molecules of either a reservoir or a system is increased. Such processes therefore involve a transition from order to disorder. Similarly, two gases that are mixed represent a higher degree of disorder than when they are separated. It is possible to regard all natural processes from this point of view, and in all cases the result obtained is that *there is a tendency on the part of nature to proceed toward a state of greater disorder.*

The increase of entropy of the universe during natural processes is an expression of this transition. In other words, we may state roughly that *the entropy of a system or of a reservoir is a measure of the degree of molecular disorder existing in the system or reservoir.* To put these ideas on a firm foundation, the concept of disorder must be properly defined. It will be shown in Chap. 11 that the disorder of a system may be calculated by the theory of probability and expressed by a quantity Ω known as the *thermodynamic probability*. The relation between entropy and disorder is then shown to be

$$S = \text{const.} \ln \Omega. \tag{8-14}$$

By means of this equation a meaning may be given to the entropy of a system in a nonequilibrium state. That is, a nonequilibrium state corresponds to a certain degree of disorder and, therefore, to a definite entropy.

8-12 ENTROPY AND DIRECTION; ABSOLUTE ENTROPY

The second law of thermodynamics provides an answer to the question that is not contained within the scope of the first law: In what direction does a process take place? The answer is that a process always takes place in such a direction as to cause an increase in the entropy of the universe. In the case of an isolated system, it is the entropy of the system that tends to increase. To find out, therefore, the equilibrium state of an isolated system, it is necessary merely to express the entropy as a function of certain coordinates and to apply the usual rules of calculus to render the function a maximum. When the system is not isolated but instead, let us say, is maintained at constant temperature and pressure, there are other entropy changes to be taken into account. It will be shown later, however, that there exists another function, referring to the system alone and known as the *Gibbs function*, whose behavior determines equilibrium under these conditions.

In practical applications of thermodynamics, one is interested only in the amount by which the entropy of a system changes in going from an initial to a final state. In cases where it is necessary to perform many such calculations with the minimum of effort—for example, in steam engineering, in problems in refrigeration and gas liquefaction, etc.—it is found expedient to set up an *entropy table* in which the "entropy" of the system in thousands of different states is represented by appropriate numbers. This is done by assigning the value zero to the entropy of the system in an arbitrarily chosen standard state and calculating the entropy change from this standard state to all other states. When this is done, it is understood that one value of what is listed as "the entropy" has no meaning but that the difference between two values is actually the entropy change.

It is a very interesting and also a very important question in physics as to whether there exists an absolute standard state of a system in which the entropy is really zero, so that the number obtained by calculating the entropy change from the zero state to any other represents the "absolute entropy" of the system. It was first suggested by Planck that the entropy of a single crystal of a pure element at the absolute zero of temperature should be taken to be zero. Zero entropy, however, has statistical implications implying, in a rough way, the absence of all molecular, atomic, electronic, and nuclear disorder. Before any meaning can be attached to the idea of zero entropy, one must know all the factors that contribute to the disorder of a system. An adequate discussion requires the application of quantum ideas to statistical mechanics.

Fowler and Guggenheim, who have considered the subject exhaustively, summarize the situation as follows:

> We may assign, if we please, the value zero to the entropy of all perfect crystals of a single pure isotope of a single element in its idealized state at the absolute zero of temperature, but even this has no theoretical significance on account of nuclear spin weights. For the purpose of calculating experimental results, some conventional zero must be chosen, and the above choice or a similar one is thus often convenient. But its conventional character will no longer be so likely to be overlooked that any importance will in the future be attached to absolute entropy, an idea which has caused much confusion and been of very little assistance in the development of the subject.

8-13 ENTROPY FLOW AND ENTROPY PRODUCTION

Consider the conduction of heat along a copper wire that lies between a hot reservoir at temperature T_1 and a cooler reservoir at T_2. Suppose that the heat current or rate of flow of heat is represented by the symbol I_Q. In unit time, the hot reservoir undergoes a decrease of entropy I_Q/T_1, the copper wire suffers no entropy change because, once in the steady state, its coordinates do not change, and the cooler reservoir undergoes an entropy increase I_Q/T_2. The entropy change of the universe per unit time is $I_Q/T_2 - I_Q/T_1$, which is of course positive.

This process may be considered, however, from a point of view in which the attention is focused on the wire, rather than on the universe. Since the hot reservoir underwent an entropy decrease, we may say that it lost entropy to the wire, or that *there was a flow of entropy into the wire* equal to I_Q/T_1 per unit time. Since the cooler reservoir underwent an entropy increase, we may say that the reservoir gained entropy from the wire, or that *there was a flow of entropy out of the wire* equal to I_Q/T_2 per unit time. But I_Q/T_2 is greater than I_Q/T_1, and hence this point of view leads us to a situation in which the *flow of entropy out of the wire exceeds the flow in.* If entropy is to be regarded as a quantity that can flow, it is necessary to assume that entropy is produced or generated inside the wire at a rate sufficient to compensate for the difference between the rate of outflow and the rate of inflow. If the rate of production of entropy within the wire is written $dS/d\tau$, we have

$$\frac{dS}{d\tau} = \frac{I_Q}{T_2} - \frac{I_Q}{T_1} = I_Q \frac{T_1 - T_2}{T_1 T_2},$$

and if the temperatures of the reservoirs are $T + \Delta T$ and T, so that only a small temperature difference exists across the wire,

$$\frac{dS}{d\tau} = I_Q \frac{\Delta T}{T^2} = \frac{I_Q}{T} \frac{\Delta T}{T}.$$

Since I_Q stands for a heat current, we may interpret I_Q/T as an entropy current I_S, or

$$I_S = \frac{I_Q}{T}.$$

We have therefore the result that, when heat is conducted along a wire across which there is a temperature difference ΔT, *entropy flows through the wire at a rate I_S and entropy is produced within the wire at a rate*

$$\frac{dS}{d\tau} = I_S \frac{\Delta T}{T}.$$

Suppose now that an electric current I is maintained in this same copper wire by virtue of a difference of potential $\Delta\mathscr{E}$ across its ends, while the wire is in contact with a reservoir at the temperature T. Electrical energy of amount $I\,\Delta\mathscr{E}$

is dissipated in the wire per unit time, and heat flows out of the wire at the same rate $I\,\Delta\mathcal{E}$, since the wire itself undergoes no energy change. The reservoir undergoes an increase of entropy $I\,\Delta\mathcal{E}/T$ per unit time, and there is no entropy change of the wire. Hence, the entropy change of the universe per unit time is $I\,\Delta\mathcal{E}/T$, which is positive. Changing our point of view, as before, to a consideration of the wire, we may say that there was no flow of entropy into the wire, but the entropy flowed out at the rate $I\,\Delta\mathcal{E}/T$. To provide for this outflow of entropy, we assume an entropy production inside the wire at the rate

$$\frac{dS}{d\tau} = I\frac{\Delta\mathcal{E}}{T}.$$

If, now, *both* a heat current and an electric current exist in the wire simultaneously, we may say that entropy is being generated within the wire by virtue of *both processes* at a rate given by

$$\boxed{\frac{dS}{d\tau} = I_S\frac{\Delta T}{T} + I\frac{\Delta\mathcal{E}}{T}.} \tag{8-15}$$

It is an interesting fact of experimental physics that, in the absence of a potential difference, a heat current depends only on the temperature difference; but when there is a potential difference as well, the heat current (also the entropy current) depends on *both* the temperature difference and the potential difference. Similarly, when both temperature and potential differences exist across a wire, the electric current also depends on *both* of these differences. The heat flow (and entropy flow) and the electricity flow are irreversible *coupled flows*, which exist by virtue of a departure from equilibrium conditions in the wire. If the departure from equilibrium is not too great, it may be assumed that both I_S and I are *linear* functions of the temperature and potential differences. Thus,

$$I_S = L_{11}\frac{\Delta T}{T} + L_{12}\frac{\Delta\mathcal{E}}{T} \tag{8-16}$$

and

$$I = L_{21}\frac{\Delta T}{T} + L_{22}\frac{\Delta\mathcal{E}}{T} \tag{8-17}$$

are the famous *Onsager equations*, which express the linearity between the flows (or currents) and the *generalized forces* $\Delta T/T$ and $\Delta\mathcal{E}/T$. The L's are coefficients connected with electric resistance, thermal conductivity, and the thermoelectric properties of the wire. Only three of the four L's are independent, for it can be proved rigorously by means of statistical mechanics that, if the departure from equilibrium is small,

$$\boxed{L_{12} = L_{21},} \tag{8-18}$$

which is known as *Onsager's reciprocal relation.*

By means of this strange point of view involving entropy flow and entropy production, and with the aid of Onsager's equations and reciprocal relation, the famous equations for a thermocouple will be derived in Chap. 14.

PROBLEMS

8-1 Figure P8-1 represents schematically a system undergoing a reversible cycle during which heat transfers Q_1, Q_2, ..., take place between it and a set of reservoirs at T_1, T_2, ..., where any T is the temperature of the system at the moment it is exchanging heat with the reservoir at T. (If the cycle were not reversible, the temperature of a reservoir and that of the system would not necessarily be

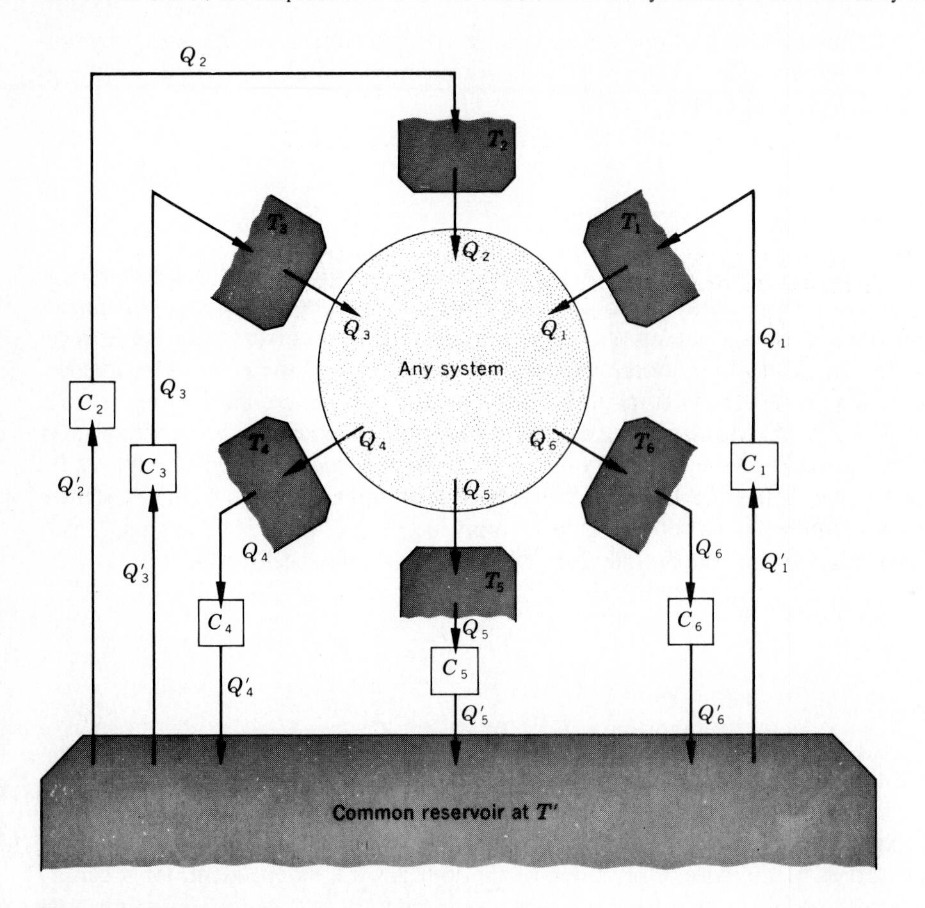

Figure P8-1 System undergoing a reversible cycle.

the same.) Some Q's are positive and some are negative. Let C_1, C_2, ... represent devices each of which operates in a Carnot cycle either as an engine or as a refrigerator, between one of the reservoirs and a common reservoir at T'. Suppose each device is arranged so that in one or more complete cycles it exchanges the same amount of heat with its reservoir that the reservoir exchanges with the system, so that *each reservoir is left unchanged*. Let Q'_1, Q'_2, ... be the heats exchanged between the Carnot devices and the common reservoir. By invoking the second law of thermodynamics, derive Clausius' theorem,

$$\oint_R \frac{đQ}{T} = 0.$$

8-2 (*a*) Derive the expression for the efficiency of a Carnot engine directly from a *TS* diagram.

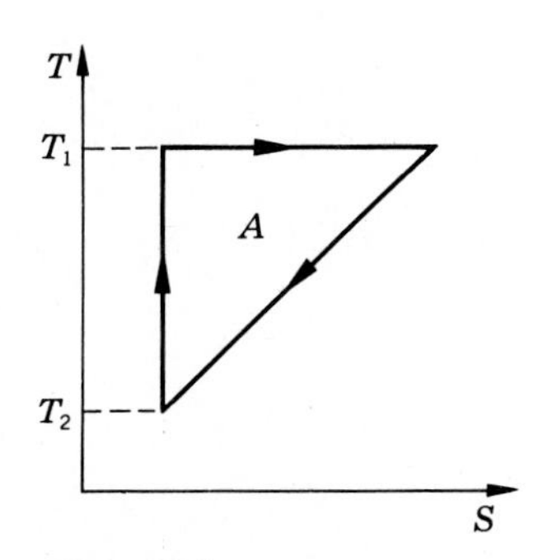

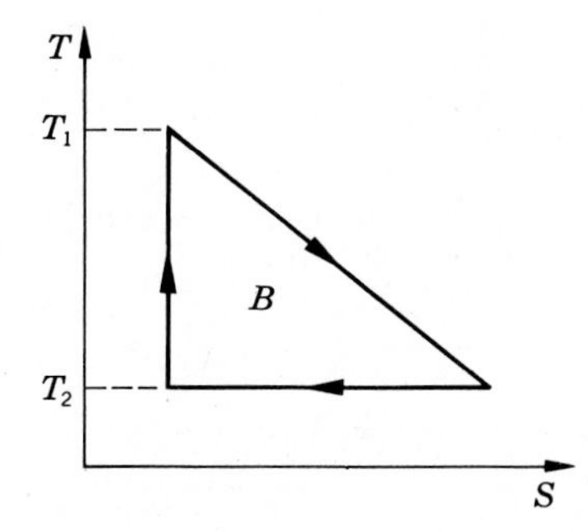

Figure P8-2

(*b*) Compare the efficiencies of cycles *A* and *B* of Fig. P8-2.

8-3 Draw rough *TS* diagrams for the following ideal-gas cycles: Stirling, Otto, and Diesel; a rectangle on a *PV* diagram; a "right triangle" on a *PV* diagram in which the base is an isobaric, the altitude an isochoric, and the "hypotenuse" an adiabatic.

8-4 Given a system whose coordinates are the temperature *T*, any number of generalized forces *Y*, *Y'*, ..., and their corresponding generalized displacements *X*, *X'*, ...

(*a*) Prove that

$$\frac{1}{T} = \left(\frac{\partial S}{\partial U}\right)_{X, X', \ldots}.$$

(*b*) What is the expression for $1/T$ of a fluid?

(*c*) Of a paramagnetic gas?

8-5 For an ideal gas with constant heat capacities, show that:

(*a*) The entropy is given by

$$S = C_V \ln P + C_P \ln V + \text{const.}$$

(*b*) The adiabatic compressibility is

$$\kappa_S = -\frac{1}{V}\left(\frac{\partial V}{\partial P}\right)_S = \frac{1}{\gamma P}.$$

(*c*) If a gas is both ideal and paramagnetic, obeying Curie's law, show that the entropy is given by

$$S = C_{V,M} \ln T + nR \ln V - \frac{M^2}{2C_C'} + \text{const.},$$

where $C_{V,M}$ is the heat capacity at constant volume and magnetization, assumed constant, and C_C' is the Curie constant.

8-6 An electric current of 10 A is maintained for 1 s in a resistor of 25 Ω while the temperature of the resistor is kept constant at 27°C.

(*a*) What is the entropy change of the resistor?

(*b*) What is the entropy change of the universe?

The same current is maintained for the same resistor, but now thermally insulated, whose initial temperature is 27°C. If the resistor has a mass of 0.01 kg and $c_P = 0.84$ kJ/kg · K:

(*c*) What is the entropy change of the resistor?

(*d*) What is the entropy change of the universe?

8-7 (*a*) A kilogram of water at 273 K is brought into contact with a heat reservoir at 373 K. When the water has reached 373 K, what is the entropy change of the water? Of the heat reservoir? Of the universe?

(*b*) If the water had been heated from 273 K by first bringing it in contact with a reservoir at 323 K and then with a reservoir at 373 K, what would have been the entropy change of the universe?

(*c*) Explain how the water might be heated from 273 to 373 K with almost no change of entropy of the universe.

8-8 A body of constant heat capacity C_P and at a temperature T_i is put in contact with a reservoir at a higher temperature T_f. The pressure remains constant while the body comes to equilibrium with the reservoir. Show that the entropy change of the universe is equal to

$$C_P[x - \ln(1 + x)],$$

where $x = -(T_f - T_i)/T_f$. Prove that this entropy change is positive.

8-9 According to Debye's law, the molar heat capacity at constant volume of a diamond varies with the temperature as follows:

$$c_V = 3R\frac{4\pi^4}{5}\left(\frac{T}{\Theta}\right)^3.$$

What is the entropy change in units of R of a diamond of 1.2 g mass when it is heated at constant volume from 10 to 350 K? Atomic weight of carbon is 12, and Θ is 2230 K.

8-10 Calculate the entropy change of the universe as a result of each of the following processes:

(*a*) A copper block of 0.4 kg mass and with a total heat capacity at constant pressure of 150 J/K at 100°C is placed in a lake at 10°C.

(*b*) The same block, at 10°C, is dropped from a height of 100 m into the lake.

(*c*) Two such blocks, at 100 and 0°C, are joined together.

8-11 What is the entropy change of the universe as a result of each of the following processes?

(*a*) A capacitor of capacitance 1 μF is connected to a 100-V reversible battery at 0°C.

(*b*) The same capacitor, after being charged to 100 V, is discharged through a resistor kept at 0°C.

8-12 Thirty-six grams of water at a temperature of 20°C is converted into steam at 250°C at constant atmospheric pressure. Assuming the heat capacity per gram of liquid water to remain practically constant at 4.2 J/g · K and the heat of vaporization at 100°C to be 2260 J/g, and using Table 5-2, calculate the entropy change of the system.

8-13 Ten grams of water at 20°C is converted into ice at −10°C at constant atmospheric pressure. Assuming the heat capacity per gram of liquid water to remain constant at 4.2 J/g · K and that of ice to be one-half of this value, and taking the heat of fusion of ice at 0°C to be 335 J/g, calculate the total entropy change of the system.

8-14 A thermally insulated cylinder closed at both ends is fitted with a frictionless heat-conducting piston which divides the cylinder into two parts. Initially, the piston is clamped in the center with 10^{-3} m^3 of air at 300 K and 2×10^5 Pa pressure on one side and 10^{-3} m^3 of air at 300 K at 1×10^5 Pa pressure on the other side. The piston is released and reaches equilibrium in pressure and temperature at a new position. Compute the final pressure and temperature and the total increase of entropy. What irreversible process has taken place?

8-15 A cylinder closed at both ends, with adiabatic walls, is divided into two parts by a movable frictionless *adiabatic* piston. Originally, the pressure, volume, and temperature are the same on both sides of the piston (P_0, V_0, T_0). The gas is ideal, with C_V independent of T and $\gamma = 1.5$. By means of a heating coil in the gas on the left-hand side, heat is slowly supplied to the gas on the left until the pressure reaches $27P_0/8$. In terms of nR, V_0, and T_0:

(*a*) What is the final right-hand volume?
(*b*) What is the final right-hand temperature?
(*c*) What is the final left-hand temperature?
(*d*) How much heat must be supplied to the *gas* on the left? (Ignore the coil!)
(*e*) How much work is done on the gas on the right?
(*f*) What is the entropy change of the gas on the right?
(*g*) What is the entropy change of the gas on the left?
(*h*) What is the entropy change of the universe?

8-16 According to Eq. (8-5), the entropy of n mol of ideal gas of constant heat capacity C_V at a temperature T and volume V is equal to

$$S = C_V \ln T + nR \ln V + S_0.$$

Imagine a box divided by a partition into two equal compartments of volume V, each containing 1 mol of the same gas at the same temperature and pressure.
(*a*) Calculate the entropy of the two portions of gas while the partition is in place.
(*b*) Calculate the entropy of the entire system after the partition has been removed.
(*c*) Has any process taken place? If so, was it reversible or irreversible?
(*d*) Has any entropy change taken place? If not, why not?

8-17 A mass m of water at T_1 is isobarically and adiabatically mixed with an equal mass of water at T_2. Show that the entropy change of the universe is

$$2mc_P \ln \frac{(T_1 + T_2)/2}{\sqrt{T_1 T_2}},$$

and prove that this is positive by drawing a semicircle of diameter $T + T_2$.

8-18 According to Eq. (8-9), the entropy change of a solid undergoing an isobaric process at pressure P from a standard state ($T = 0$, $S = S_0$) to another state of the solid (T, S) is equal to

$$S = S_0 + \int_0^T \frac{C_P}{T} \, dT.$$

Is it possible to draw conclusions concerning the behavior of the C_P of a solid *as the temperature approaches absolute zero?* Try these possibilities:
(*a*) C_P remains constant.
(*b*) C_P varies inversely as T, inversely as T^2, etc.
(*c*) C_P varies directly as T, as T^2, etc.

If S does not approach infinity as T approaches zero, what conclusion can you make about the low-temperature behavior of C_P?

8-19 Using the entropy principle, given in Eq. (8-12), prove:
(*a*) The Kelvin-Planck statement of the second law.
(*b*) The Clausius statement of the second law.

8-20 A body of finite mass is originally at a temperature T_1, which is higher than that of a reservoir at temperature T_2. Suppose that an engine operates in a cycle between the body and the reservoir until it lowers the temperature of the body from T_1 to T_2, thus extracting heat Q from the body. If the engine does work W, it will reject heat $Q - W$ to the reservoir at T_2. Applying the entropy principle, prove that the maximum work obtainable from the engine is

$$W\,(\max) = Q - T_2(S_1 - S_2),$$

where $S_1 - S_2$ is the entropy decrease of the body.

8-21 Two identical bodies of constant heat capacity at temperatures T_1 and T_2, respectively, are used as reservoirs for a heat engine. If the bodies remain at constant pressure and undergo no change of phase, show that the amount of work obtainable is

$$W = C_P(T_1 + T_2 - 2T_f),$$

where T_f is the final temperature attained by both bodies. Show that, when W is a maximum,

$$T_f = \sqrt{T_1 T_2}.$$

8-22 Two identical bodies of constant heat capacity are at the same initial temperature T_i. A refrigerator operates between these two bodies until one body is cooled to temperature T_2. If the bodies remain at constant pressure and undergo no change of phase, show that the minimum amount of work needed to do this is

$$W\,(\text{min}) = C_P\left(\frac{T_i^2}{T_2} + T_2 - 2T_i\right).$$

8-23 An ideal gas cycle suggested by A. S. Arrott of British Columbia, Canada, is shown in Fig. P8-3, where there are shown on a PV diagram two isotherms intersected by an adiabat, referring to 1 mol of an ideal monatomic gas. The gas starts at the upper intersection point A and expands isothermally at 600 K to a very special state B. It then is put in contact with a cold reservoir at 300 K so that it cools isochorically to state C. Then there is a further isothermal expansion from C to the lower intersection point D. The remainder of the zilch cycle is accomplished by an adiabatic compression from D back to A. The isochor BC is chosen to satisfy the condition that the net work in the cycle is zero.

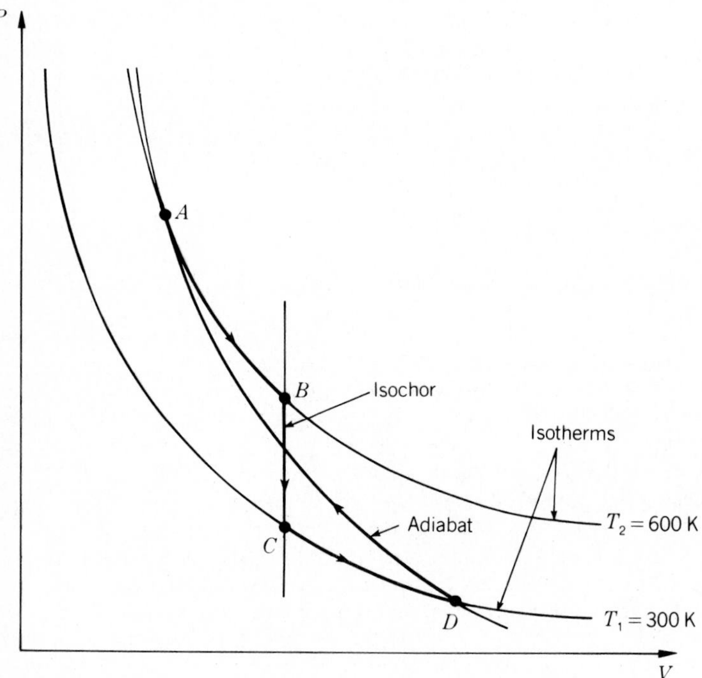

Figure P8-3 Arrott's zilch cycle for 1 mol of an ideal monatomic gas.

(*a*) Calculate W_{DA}.
(*b*) Calculate Q_{BC}.
(*c*) Calculate the net entropy change of the gas (*not the reservoirs*) and obtain the relation

$$\frac{Q_{AB}}{600\text{ K}} + \frac{Q_{CD}}{300\text{ K}} = 8.64\frac{\text{J}}{\text{mol}\cdot\text{K}}.$$

(*d*) Calculate W_{AB}.
(*e*) Calculate W_{CD}.
(*f*) Calculate the net entropy change of the reservoirs.
(*g*) Draw the TS diagram.

8-24 When both a heat current and an electric current are maintained in the same wire simultaneously by a temperature difference ΔT and a potential difference $\Delta\mathcal{E}$ prove that

(*a*)
$$\left(\frac{\partial I_s}{\partial\,\Delta\mathcal{E}}\right)_{\Delta T} = \left(\frac{\partial I}{\partial\,\Delta T}\right)_{\Delta\mathcal{E}}$$

(*b*) $L_{11} = KA/\Delta x$, where K is the thermal conductivity and A and Δx are the area and length, respectively, of the wire.

(*c*) $L_{22} = T/R'$, where R' is the electric resistance of the wire.

8-25 Show that, in the case of irreversible coupled flows of heat and electricity,

(*a*)
$$T^2\frac{dS}{d\tau} = L_{11}(\Delta T)^2 + (L_{12} + L_{21})\,\Delta T\,\Delta\mathcal{E} + L_{22}(\Delta\mathcal{E})^2.$$

(*b*)
$$\frac{\partial}{\partial\,\Delta\mathcal{E}}\left(T\frac{dS}{d\tau}\right)_{\Delta T} = 2I \qquad \text{and} \qquad \frac{\partial}{\partial\,\Delta T}\left(T\frac{dS}{d\tau}\right)_{\Delta\mathcal{E}} = 2I_s.$$

(*c*) Show that, with ΔT fixed, the equilibrium state obtained when $I = 0$ involves a minimum rate of entropy production.

(*d*) Show that, with $\Delta\mathcal{E}$ fixed, the equilibrium state obtained when $I_s = 0$ involves a minimum rate of entropy production.

8-26 Three identical finite bodies of constant heat capacity are at temperatures 300, 300, and 100 K. If no work or heat is supplied from the outside, what is the highest temperature to which any one of the bodies can be raised by the operation of heat engines or refrigerators?

CHAPTER

NINE

PURE SUBSTANCES

9-1 ENTHALPY

The laws of thermodynamics were stated and their consequences were developed in a sufficiently general manner to apply to systems of any number of coordinates. When there are three or more independent coordinates, one speaks of isothermal surfaces and isentropic (adiabatic reversible) surfaces. If, as is often the case, there are only two independent coordinates, these surfaces reduce to simple plane curves. The most important system of two independent coordinates is a hydrostatic one, consisting of a single pure substance of constant mass. Once the thermodynamic equations are developed for this system, we shall see how simple it is to write down the analogous equations for any other two-coordinate system.

In discussing some of the properties of gases in Chap. 4, the sum of U and PV appeared several times (see Probs. 4-8, 4-9, and 4-11). It has been found very useful to define a new function H, called the *enthalpy*,† by the relation

$$\boxed{H = U + PV.} \tag{9-1}$$

In order to study the properties of this function, consider the change in enthalpy

† Pronounced *en-thal′-pi.*

that takes place when a system undergoes an infinitesimal process from an initial equilibrium state to a final equilibrium state. We have

$$dH = dU + P\,dV + V\,dP;$$

but

$$đQ = dU + P\,dV.$$

Therefore,

$$dH = đQ + V\,dP. \tag{9-2}$$

Dividing both sides by dT, we obtain

$$\frac{dH}{dT} = \frac{đQ}{dT} + V\frac{dP}{dT},$$

and, at constant P,

$$\left(\frac{\partial H}{\partial T}\right)_P = \left(\frac{đQ}{dT}\right)_P = C_P. \tag{9-3}$$

Since

$$dH = đQ + V\,dP,$$

the change in enthalpy during an isobaric process is equal to the heat that is transferred. The so-called *latent heat* measured during a phase transition at constant pressure (e.g., melting, boiling, sublimation) is simply the change of enthalpy. That is,

$$\left.\begin{aligned} H_f - H_i &= Q, \\ \text{or} \qquad H_f - H_i &= \int_i^f C_P\,dT. \end{aligned}\right\} \quad \text{(isobaric)} \tag{9-4}$$

Since isobaric processes are much more important in engineering and chemistry than isochoric processes, the enthalpy is of greatest use in these branches of science.

If a pure substance undergoes an infinitesimal reversible process, Eq. (9-2) may be written

$$dH = T\,dS + V\,dP,$$

showing that

$$\left(\frac{\partial H}{\partial S}\right)_P = T \quad \text{and} \quad \left(\frac{\partial H}{\partial P}\right)_S = V. \tag{9-5}$$

The relations given in Eqs. (9-5) suggest that the properties of a pure substance could be displayed to advantage on a diagram in which H is plotted as a function of S and P. This three-dimensional graph would be a surface, and T and V would be indicated at any point by the two slopes that determine a plane tangent to the surface at the point.

One of the most interesting properties of the enthalpy function is in connection with a *throttling process*. Imagine a cylinder thermally insulated and equipped with two nonconducting pistons on opposite sides of a porous wall, as shown in Fig. 9-1*i*. The wall, shaded in horizontal lines, is a porous plug, a

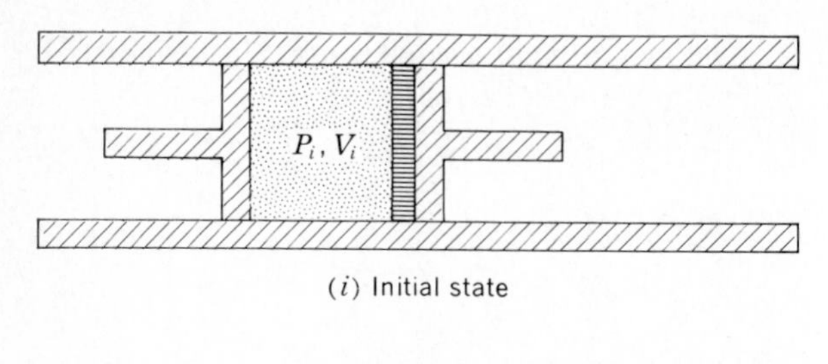

(i) Initial state

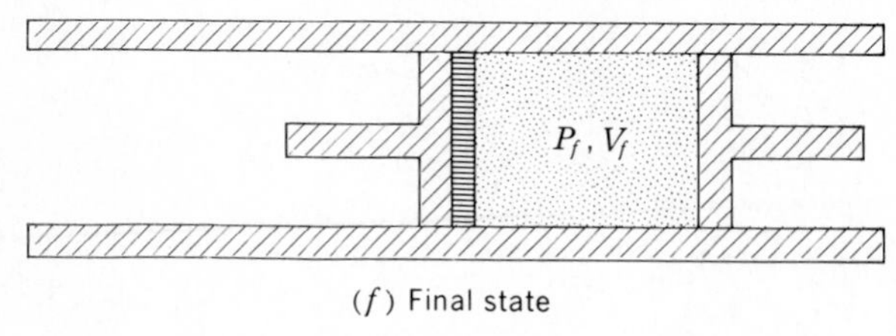

(f) Final state

Figure 9-1 Throttling process.

narrow constriction, or a series of small holes. Between the left-hand piston and the wall there is a gas at a pressure P_i and a volume V_i; and since the right-hand piston is against the wall, with any gas being thus prevented from seeping through, the initial state of the gas is an equilibrium state. Now imagine moving both pistons simultaneously in such a way that a *constant pressure* P_i is maintained on the left-hand side of the wall and a *constant lower pressure* P_f is maintained on the right-hand side. After all the gas has seeped through the porous wall, the final equilibrium state of the system will be as shown in Fig. 9-1*f*. Such a process is a throttling process.

A throttling process is obviously an irreversible one, since the gas passes through nonequilibrium states on its way from the initial equilibrium state to its final equilibrium state. These nonequilibrium states cannot be described by thermodynamic coordinates, but an interesting conclusion can be drawn about the initial and final equilibrium states. Applying the first law to the throttling process

$$Q = U_f - U_i - W,$$

we have

$$Q = 0,$$

and

$$W = -\int_0^{V_f} P_f \, dV - \int_{V_i}^{0} P_i \, dV.$$

Since both pressures remain constant,

$$W = -(P_f V_f - P_i V_i).$$

The negative of the foregoing expression is known in engineering as *flow work*, since it represents the work necessary to keep the gas flowing. Therefore,

$$0 = U_f - U_i + P_f V_f - P_i V_i,$$

or

$$U_i + P_i V_i = U_f + P_f V_f.$$

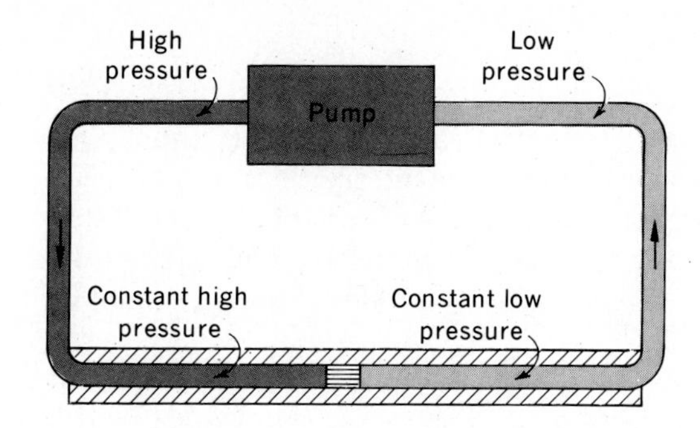

Figure 9-2 Apparatus for performing a continuous throttling process.

Table 9-1 Comparison of U and H

Internal energy U	Enthalpy H
In general $dU = đQ - P\,dV$ $\left(\frac{dU}{\partial T}\right)_V = C_V$	In general $dH = đQ + V\,dP$ $\left(\frac{\partial H}{\partial T}\right)_P = C_P$
Isochoric process $U_f - U_i = Q$ $U_f - U_i = \int_i^f C_V\,dT$	Isobaric process $H_f - H_i = Q$ $H_f - H_i = \int_i^f C_P\,dT$
Adiabatic process $U_f - U_i = -\int_i^f P\,dV$	Adiabatic process $H_f - H_i = \int_i^f V\,dP$
Free expansion $U_i = U_f$	Throttling process $H_i = H_f$
For an ideal gas $U = \int C_V\,dT + \text{const.}$	For an ideal gas $H = \int C_P\,dT + \text{const.}$
Nearby equil. states $dU = T\,dS - P\,dV$ $T = \left(\frac{\partial U}{\partial S}\right)_V$ $-P = \left(\frac{\partial U}{\partial V}\right)_S$	Nearby equil. states $dH = T\,dS + V\,dP$ $T = \left(\frac{\partial H}{\partial S}\right)_P$ $V = \left(\frac{\partial H}{\partial P}\right)_S$

And, finally,

$$H_i = H_f \qquad \text{(throttling process).} \tag{9-6}$$

In a throttling process, therefore, the initial and final enthalpies are equal. One is not entitled to say that the enthalpy remains constant, since one cannot speak of the enthalpy of a system that is passing through such nonequilibrium states. In plotting a throttling process on any diagram, the initial and final equilibrium states may be represented by points. The intermediate states, however, cannot be plotted.

A continuous throttling process may be achieved by a pump that maintains a constant high pressure on one side of a constriction or porous wall and a constant lower pressure on the other side, as shown in Fig. 9-2 (see page 215). For every mole of fluid that undergoes the throttling process, we may write

$$h_i = h_f,$$

where the lowercase letters indicate *molar enthalpy*.

The properties of the enthalpy function must be clearly understood by the student, for they will be used continually throughout the remainder of this book. The comparison of the internal energy and the enthalpy given in Table 9-1 (see page 215) will help the student to remember these properties.

9-2 THE HELMHOLTZ AND GIBBS FUNCTIONS

The *Helmholtz function* (sometimes called the *Helmholtz free energy*) is defined as

$$\boxed{F = U - TS.} \tag{9-7}$$

For an infinitesimal reversible process,

$$dF = dU - T\,dS - S\,dT,$$

and

$$T\,dS = dU + P\,dV.$$

Hence,

$$dF = -S\,dT - P\,dV. \tag{9-8}$$

From this it follows: (1) for a *reversible isothermal process*,

$$dF = -P\,dV,$$

or

$$F_f - F_i = -\int_i^f P\,dV. \tag{9-9}$$

Hence the change of the Helmholtz function during a reversible isothermal process equals the work done *on* the system. (2) For a *reversible isothermal and isochoric process*,

$$dF = 0,$$

and

$$F = \text{const.} \tag{9-10}$$

These properties are of interest in chemistry and are useful in considering chemical reactions that take place isothermally and isochorically. The main importance of the Helmholtz function, however, is in statistical mechanics, where it is closely associated with the partition function Z defined by the equation

$$Z = \sum g_i e^{-\epsilon_i/kT},$$

where ϵ_i and g_i represent the energy values and degeneracies, respectively, of the various energy levels of the system of particles.

Since

$$dF = -S\,dT - P\,dV,$$

the entropy and the pressure may then be calculated by performing the simple differentiations:

$$S = -\left(\frac{\partial F}{\partial T}\right)_V \quad \text{and} \quad P = -\left(\frac{\partial F}{\partial V}\right)_T.$$

The *Gibbs function* (also called the *Gibbs free energy*) is defined as

$$\boxed{G = H - TS.} \tag{9-11}$$

For an infinitesimal reversible process,

$$dG = dH - T\,ds - S\,dT.$$

It will be recalled, however, that

$$dH = T\,dS + V\,dP;$$

whence

$$dG = -S\,dT + V\,dP. \tag{9-12}$$

In the case of a *reversible isothermal and isobaric process,*

$$dG = 0,$$

and

$$G = \text{const.}$$

This is a particularly important result in connection with processes involving a change of phase. Sublimation, fusion, and vaporization take place isothermally and isobarically and can be conceived of as occurring reversibly. Hence, during such processes, the Gibbs function of the system remains constant. If we denote by the symbols g', g'', and g''' the molar Gibbs functions of a saturated solid, saturated liquid, and saturated vapor, respectively, then the equation of the fusion curve is

$$g' = g'',$$

the equation of the vaporization curve is

$$g'' = g''',$$

and the equation of the sublimation curve is

$$g' = g'''.$$

At the triple point two equations hold simultaneously, namely,

$$g' = g'' = g'''.$$

All the g's can be regarded as functions of T and P only, and hence the two equations above serve to determine the T and P of the triple point uniquely.

The Gibbs function is of the utmost importance in chemistry, since chemical reactions can be conceived of as taking place at constant T and P. It is also of some use in engineering.

9-3 TWO MATHEMATICAL THEOREMS

Theorem 1 If a relation exists among x, y, and z, we may imagine z expressed as a function of x and y; whence

$$dz = \left(\frac{\partial z}{\partial x}\right)_y dx + \left(\frac{\partial z}{\partial y}\right)_x dy.$$

If we let

$$M = \left(\frac{\partial z}{\partial x}\right)_y \quad \text{and} \quad N = \left(\frac{\partial z}{\partial y}\right)_x,$$

then

$$dz = M\,dx + N\,dy,$$

where z, M, and N are all functions of x and y. Partially differentiating M with respect to y, and N with respect to x, we get

$$\left(\frac{\partial M}{\partial y}\right)_x = \frac{\partial^2 z}{\partial x\,\partial y} \quad \text{and} \quad \left(\frac{\partial N}{\partial x}\right)_y = \frac{\partial^2 z}{\partial y\,\partial x}.$$

Since the two second derivatives of the right-hand terms are equal, it follows that

$$\boxed{\left(\frac{\partial M}{\partial y}\right)_x = \left(\frac{\partial N}{\partial x}\right)_y.} \tag{9-13}$$

This is known as the *condition for an exact differential*.

Theorem 2 If a quantity f is a function of x, y, and z, and a relation exists among x, y, and z, then f may be regarded as a function of *any two* of x, y, and z. Similarly, any one of x, y, and z may be considered to be a function of f and one other of x, y, and z. Thus, regarding x to be a function of f and y,

$$dx = \left(\frac{\partial x}{\partial f}\right)_y df + \left(\frac{\partial x}{\partial y}\right)_f dy.$$

Considering y to be a function of f and z,

$$dy = \left(\frac{\partial y}{\partial f}\right)_z df + \left(\frac{\partial y}{\partial z}\right)_f dz.$$

Substituting this expression for dy in the preceding equation, we get

$$dx = \left[\left(\frac{\partial x}{\partial f}\right)_y + \left(\frac{\partial x}{\partial y}\right)_f \left(\frac{\partial y}{\partial f}\right)_z\right] df + \left[\left(\frac{\partial x}{\partial y}\right)_f \left(\frac{\partial y}{\partial z}\right)_f\right] dz.$$

But
$$dx = \left(\frac{\partial x}{\partial f}\right)_z df + \left(\frac{\partial x}{\partial z}\right)_f dz.$$

Equating the dz terms of the last two equations, we get

$$\boxed{\begin{gathered}\left(\frac{\partial x}{\partial y}\right)_f \left(\frac{\partial y}{\partial z}\right)_f = \left(\frac{\partial x}{\partial z}\right)_f, \\ \left(\frac{\partial x}{\partial y}\right)_f \left(\frac{\partial y}{\partial z}\right)_f \left(\frac{\partial z}{\partial x}\right)_f = 1.\end{gathered}} \tag{9-14}$$

9-4 MAXWELL'S RELATIONS

We have seen that the properties of a pure substance are conveniently represented in terms of these four functions:

Internal energy U,
Enthalpy $H = U + PV$,
Helmholtz function $F = U - TS$,
Gibbs function $G = H - TS$.

Any one of these may be regarded as a function of *any two* of P, V, and T. Suppose, for example, that both U and S are expressed as functions of V and T, thus:

$$U = \text{function of } (V, T)$$

and
$$S = \text{function of } (V, T).$$

The second equation may be imagined to be solved for T in terms of S and V; substituting this value of T in the first equation, we should then have

$$U = \text{function of } (S, V).$$

Consequently, we may go further and say that any one of the eight quantities P, V, T, S, U, H, F, and G may be expressed as a function of *any two others*.

Now imagine a hydrostatic system undergoing an infinitesimal reversible process from one equilibrium state to another:

1. The internal energy changes by an amount
$$dU = đQ - P\,dV$$
$$= T\,dS - P\,dV,$$
where U, T, and P are all imagined to be functions of S and V.
2. The enthalpy changes by an amount
$$dH = dU + P\,dV + V\,dP$$
$$= T\,dS + V\,dP,$$
where H, T, and V are all imagined to be functions of S and P.
3. The Helmholtz function changes by an amount
$$dF = dU - T\,dS - S\,dT$$
$$= -S\,dT - P\,dV,$$
where F, S, and P are all imagined to be functions of T and V.
4. The Gibbs function changes by an amount
$$dG = dH - T\,dS - S\,dT$$
$$= -S\,dT + V\,dP,$$
where G, S, and V are all imagined to be functions of T and P.

Since U, H, F, and G are actual functions, their differentials are exact differentials of the type
$$dz = M\,dx + N\,dy,$$
where z, M, and N are all functions of x and y. Therefore,
$$\left(\frac{\partial M}{\partial y}\right)_x = \left(\frac{\partial N}{\partial x}\right)_y.$$
Applying this result to the four exact differentials dU, dH, dF, and dG, we obtain:

$$\begin{aligned}
&1.\ dU = T\,dS - P\,dV;\ \text{hence}\ \left(\frac{\partial T}{\partial V}\right)_S = -\left(\frac{\partial P}{\partial S}\right)_V.\\
&2.\ dH = T\,dS + V\,dP;\ \text{hence}\ \left(\frac{\partial T}{\partial P}\right)_S = \left(\frac{\partial V}{\partial S}\right)_P.\\
&3.\ dF = -S\,dT - P\,dV;\ \text{hence}\ \left(\frac{\partial S}{\partial V}\right)_T = \left(\frac{\partial P}{\partial T}\right)_V.\\
&4.\ dG = -S\,dT + V\,dP;\ \text{hence}\ \left(\frac{\partial S}{\partial P}\right)_T = -\left(\frac{\partial V}{\partial T}\right)_P.
\end{aligned} \tag{9-15}$$

The four equations on the right are known as *Maxwell's relations*. These do not refer to a process but express relations which hold at any equilibrium state of a hydrostatic system.

Maxwell's relations are enormously useful, because they provide relationships between measurable quantities and those which either cannot be measured or are difficult to measure. For example, the fourth Maxwell relation,

$$\left(\frac{\partial S}{\partial P}\right)_T = -\left(\frac{\partial V}{\partial T}\right)_P,$$

may be combined with the statistical interpretation of entropy in order to provide information concerning the volume expansivity β of a pure substance in the following way. If a substance is compressed isothermally and if no unusual molecular rearrangements take place (such as association or dissociation), the molecules merely occupy a smaller volume and are therefore in a more orderly state. In the language of information theory, our knowledge about these molecules is increased. The entropy is therefore decreased, and the derivative $(\partial S/\partial P)_T$ is negative. It follows that $(\partial V/\partial T)_P$ is positive and that the substance must have a positive expansivity.

9-5 THE $T\,dS$ EQUATIONS

The entropy of a pure substance can be imagined as a function of T and V; whence

$$dS = \left(\frac{\partial S}{dT}\right)_V dT + \left(\frac{\partial S}{\partial V}\right)_T dV,$$

and

$$T\,dS = T\left(\frac{\partial S}{\partial T}\right)_V dT + T\left(\frac{\partial S}{\partial V}\right)_T dV.$$

Since $T\,dS = đQ$ for a reversible process, it follows that

$$T\left(\frac{\partial S}{\partial T}\right)_V = C_V.$$

And, from Maxwell's third relation,

$$\left(\frac{\partial S}{\partial V}\right)_T = \left(\frac{\partial P}{\partial T}\right)_V;$$

whence

$$\boxed{T\,dS = C_V\,dT + T\left(\frac{\partial P}{\partial T}\right)_V dV.} \qquad (9\text{-}16)$$

We shall call Eq. (9-16) the *first $T\,dS$ equation*. It is useful in a variety of ways. For example, 1 mol of a van der Waals gas undergoes a reversible isothermal expansion from a volume v_i to a volume v_f. How much heat has been transferred?

For 1 mol,

$$T\,ds = c_V\,dT + T\left(\frac{\partial P}{\partial T}\right)_V dv.$$

Using the van der Waals equation of state,

$$P = \frac{RT}{v-b} - \frac{a}{v^2},$$

and $$\left(\frac{\partial P}{\partial T}\right)_V = \frac{R}{v-b};$$

hence $$T\,ds = c_V\,dT + RT\frac{dv}{v-b}.$$

Since T is constant, $c_V\,dT = 0$; and since the process is reversible, $q = \int T\,ds$. Therefore,

$$q = RT\int_{v_i}^{v_f}\frac{dv}{v-b},$$

and finally

$$q = RT\ln\frac{v_f - b}{v_i - b}.$$

If the entropy of a pure substance is regarded as a function of T and P, then

$$dS = \left(\frac{\partial S}{\partial T}\right)_P dT + \left(\frac{\partial S}{\partial P}\right)_T dP,$$

and $$T\,dS = T\left(\frac{\partial S}{\partial T}\right)_P dT + T\left(\frac{\partial S}{\partial P}\right)_T dP.$$

But $$T\left(\frac{\partial S}{\partial T}\right)_P = C_P.$$

And, from Maxwell's fourth relation,

$$\left(\frac{\partial S}{\partial P}\right)_T = -\left(\frac{\partial V}{\partial T}\right)_P;$$

whence $$\boxed{T\,dS = C_P\,dT - T\left(\frac{\partial V}{\partial T}\right)_P dP.} \qquad (9\text{-}17)$$

This is the *second T dS equation.* A third will be found among the problems at the end of the chapter. Two important applications of the second $T\,dS$ equation follow.

1. *Reversible isothermal change of pressure* When T is constant,

$$T\,dS = -T\left(\frac{\partial V}{\partial T}\right)_P dP,$$

and

$$Q = -T\int\left(\frac{\partial V}{\partial T}\right)_P dP.$$

Remembering that the volume expansivity is

$$\beta = \frac{1}{V}\left(\frac{\partial V}{\partial T}\right)_P,$$

we obtain

$$Q = -T\int V\beta\,dP,$$

which can be integrated when the dependence of V and β on the pressure is known. In the case of a solid or liquid, neither V nor β is very sensitive to a change in pressure. For example, in the case of mercury, Bridgman found that as the pressure was increased from zero to 1000 atm (1.013×10^8 Pa) at 0°C the volume of 1 mol of mercury changed from 14.72 to 14.67 cm^3, a change of only $\frac{1}{3}$ percent; and the volume expansivity changed from $181 \times 10^{-6}\ K^{-1}$ to $174 \times 10^{-6}\ K^{-1}$, a 4 percent change. The volume and the expansivity of most solids and liquids behave similarly; therefore V and β may be taken out from under the integral sign and replaced by average values, $\bar{V}$ and $\bar{\beta}$. (A bar over a quantity indicates an average value.) We then have

$$Q = -T\bar{V}\bar{\beta}\int_{P_i}^{P_f} dP,$$

or

$$Q = -T\bar{V}\bar{\beta}(P_f - P_i).$$

It is seen from this result that, as the pressure is increased isothermally, heat will flow *out* if $\bar{\beta}$ is positive but that, for a substance with a negative expansivity (such as water between 0 and 4°C), an isothermal increase of pressure causes an absorption of heat.

If the pressure on 15 cm^3 of mercury at 0°C is increased reversibly and isothermally from zero to 1000 atm, the heat transferred will be

$$Q = -T\bar{V}\bar{\beta}(P_f - P_i),$$

where $T = 273$ K, $\bar{V} = 1.5 \times 10^{-5}\ m^3$, $\bar{\beta} = 178 \times 10^{-6}\ K^{-1}$, $P_i = 0$, and $P_f = 1.013 \times 10^8$ Pa. Hence,

$$\begin{aligned} Q &= -273\ K \times 1.5 \times 10^{-5}\ m^3 \times 178 \times 10^{-6}\ K^{-1} \times 1.013 \times 10^8\ Pa \\ &= -73.8\ Pa \cdot m^3 \\ &= -73.8\ J. \end{aligned}$$

It is interesting to compare the heat liberated with the work done during the compression.

$$W = -\int P\,dV;$$

and at constant temperature,

$$W = -\int \left(\frac{\partial V}{\partial P}\right)_T P\,dP.$$

Recalling the isothermal compressibility, $\kappa = -(1/V)(\partial V/\partial P)_T$, we get

$$W = \int_{P_i}^{P_f} V\kappa P\,dP.$$

The isothermal compressibility is also fairly insensitive to a change of pressure. Bridgman showed that the compressibility of mercury at 0°C changed from 3.92×10^{-6} to $3.83 \times 10^{-6}\ \text{atm}^{-1}$ (a 2 percent change) as the pressure was increased from zero to 1000 atm. We may therefore again replace V and κ by average values and obtain

$$W = \frac{\bar{V}\bar{\kappa}}{2}(P_f^2 - P_i^2);$$

and taking $\bar{\kappa} = 3.88 \times 10^{-6}\ \text{atm}^{-1}$ $(3.83 \times 10^{-11}\ \text{Pa}^{-1})$ for mercury, we get

$$\begin{aligned} W &= \tfrac{1}{2}(1.5 \times 10^{-5}\ \text{m}^3 \times 3.83 \times 10^{-11}\ \text{Pa}^{-1} \times 1.027 \times 10^{16}\ \text{Pa}^2) \\ &= 2.95\ \text{Pa} \cdot \text{m}^3 \\ &= 2.95\ \text{J}. \end{aligned}$$

Therefore, it is seen that, when the pressure on 15 cm^3 of mercury at 0°C is increased from zero to 1000 atm, 73.8 J of heat is liberated but only 2.95 J of work done! The extra amount of heat comes, of course, from the store of internal energy, which has changed by an amount

$$\begin{aligned} \Delta U &= Q + W \\ &= -73.8 + 3.0 = -70.8\ \text{J}. \end{aligned}$$

A similar result is obtained in the case of any substance with a positive expansivity. For a substance with a negative expansivity, heat is absorbed and the internal energy is increased.

2. *Reversible adiabatic change of pressure* Since the entropy remains constant,

$$T\,dS = 0 = C_P\,dT - T\left(\frac{\partial V}{\partial T}\right)_P dP,$$

or

$$dT = \frac{T}{C_P}\left(\frac{\partial V}{\partial T}\right)_P dP = \frac{TV\beta}{C_P}\,dP.$$

In the case of a solid or liquid, an increase of pressure of as much as 1000 atm produces only a small temperature change. Also, experiment shows that C_P hardly changes even for an increase of 10,000 atm. The equation above, when applied to a solid or a liquid, may therefore be written

$$\Delta T = \frac{T\bar{V}\beta}{C_P}(P_f - P_i).$$

It is clear from the discussion above that an adiabatic increase of pressure will produce an increase of temperature in any substance with a positive expansivity and a decrease in temperature in a substance with a negative expansivity.

If the pressure on 15 cm^3 of mercury ($\bar{C}_P = 28.6$ J/K) at 0°C is increased isentropically from zero to 1000 atm, the temperature change will be

$$\Delta T = \frac{273 \text{ K} \times 1.5 \times 10^{-5} \text{ m}^3 \times 178 \times 10^{-6} \text{ K}^{-1} \times 1.013 \times 10^8 \text{ Pa}}{28.6 \text{ J/K}}$$

$$= 2.58 \text{ K}.$$

9-6 ENERGY EQUATIONS

If a pure substance undergoes an infinitesimal reversible process between two equilibrium states, the change of internal energy is

$$dU = T\,dS - P\,dV.$$

Dividing by dV, we get

$$\frac{dU}{dV} = T\frac{dS}{dV} - P,$$

where U, S, and P are regarded as functions of T and V. If T is held constant, then the derivatives become partial derivatives, and

$$\left(\frac{\partial U}{\partial V}\right)_T = T\left(\frac{\partial S}{\partial V}\right)_T - P.$$

Using Maxwell's third relation, $(\partial S/\partial V)_T = (\partial P/\partial T)_V$, we get

$$\boxed{\left(\frac{\partial U}{\partial V}\right)_T = T\left(\frac{\partial P}{\partial T}\right)_V - P.} \qquad (9\text{-}18)$$

We shall call this equation the *first energy equation*. Two examples of its usefulness follow.

1. *Ideal gas*

$$P = \frac{nRT}{V},$$

$$\left(\frac{\partial P}{\partial T}\right)_V = \frac{nR}{V},$$

and

$$\left(\frac{\partial U}{\partial V}\right)_T = T\frac{nR}{V} - P = 0.$$

Therefore, U does not depend on V but is a function of T only.

2. *Van der Waals gas (1 mol)*

$$P = \frac{RT}{v - b} - \frac{a}{v^2},$$

$$\left(\frac{\partial P}{\partial T}\right)_V = \frac{R}{v - b},$$

and

$$\left(\frac{\partial u}{\partial v}\right)_T = T\frac{R}{v - b} - \frac{RT}{v - b} + \frac{a}{v^2} = \frac{a}{v^2}.$$

Consequently,

$$du = c_V\, dT + \frac{a}{v^2}\, dv$$

and

$$u = \int c_V\, dT - \frac{a}{v} + \text{const.}$$

It follows, therefore, that the internal energy of a van der Waals gas increases as the volume increases, with the temperature remaining constant.

The second energy equation shows the dependence of energy on pressure. We start, as usual, with the equation

$$dU = T\, dS - P\, dV,$$

and divide by dP. Then,

$$\frac{dU}{dP} = T\frac{dS}{dP} - P\frac{dV}{dP},$$

where U, S, and V are imagined to be functions of T and P. If T is held constant, then the derivatives become partial derivatives, and

$$\left(\frac{\partial U}{\partial P}\right)_T = T\left(\frac{\partial S}{\partial P}\right)_T - P\left(\frac{\partial V}{\partial P}\right)_T.$$

Using Maxwell's fourth relation, $(\partial S/\partial P)_T = -(\partial V/\partial T)_P$, we get

$$\boxed{\left(\frac{\partial U}{\partial P}\right)_T = -T\left(\frac{\partial V}{\partial T}\right)_P - P\left(\frac{\partial V}{\partial P}\right)_T,} \qquad \text{(9-19)}$$

which is the *second energy equation.*

9-7 HEAT-CAPACITY EQUATIONS

Equating the first and second $T\,dS$ equations,

$$C_P\,dT - T\left(\frac{\partial V}{\partial T}\right)_P dP = C_V\,dT + T\left(\frac{\partial P}{\partial T}\right)_V dV;$$

solving for dT, we obtain

$$dT = \frac{T\left(\frac{\partial P}{\partial T}\right)_V}{C_P - C_V}\,dV + \frac{T\left(\frac{\partial V}{\partial T}\right)_P}{C_P - C_V}\,dP.$$

But

$$dT = \left(\frac{\partial T}{\partial V}\right)_P dV + \left(\frac{\partial T}{\partial P}\right)_V dP.$$

Therefore,

$$\left(\frac{\partial T}{\partial V}\right)_P = \frac{T\left(\frac{\partial P}{\partial T}\right)_V}{C_P - C_V},$$

and

$$\left(\frac{\partial T}{\partial P}\right)_V = \frac{T\left(\frac{\partial V}{\partial T}\right)_P}{C_P - C_V}.$$

Both of the foregoing equations yield the result that

$$C_P - C_V = T\left(\frac{\partial V}{\partial T}\right)_P\left(\frac{\partial P}{\partial T}\right)_V.$$

It was shown in Art. 2-7 that

$$\left(\frac{\partial P}{\partial T}\right)_V = -\left(\frac{\partial V}{\partial T}\right)_P\left(\frac{\partial P}{\partial V}\right)_T,$$

and therefore

$$\boxed{C_P - C_V = -T\left(\frac{\partial V}{\partial T}\right)_P^2\left(\frac{\partial P}{\partial V}\right)_T.} \tag{9-20}$$

This is one of the most important equations of thermodynamics, and it shows that:

1. Since $(\partial P/\partial V)_T$ is always negative for all known substances and $(\partial V/\partial T)_P^2$ must be positive, then $C_P - C_V$ can never be negative; or *C_P can never be less than C_V.*
2. As $T \to 0$, $C_P \to C_V$; or *at the absolute zero the two heat capacities are equal.*
3. $C_P = C_V$ when $(\partial V/\partial T)_P = 0$. For example, at 4°C, at which the density of water is a maximum, $C_P = C_V$.

Laboratory measurements of the heat capacity of solids and liquids usually take place at constant pressure and therefore yield values of C_P. It would be extremely difficult to measure with any degree of accuracy the C_V of a solid or liquid. Values of C_V, however, must be known for purposes of comparison with theory. The equation for the difference in the heat capacities is very useful in calculating C_V in terms of C_P and other measurable quantities. Remembering that

$$\beta = \frac{1}{V}\left(\frac{\partial V}{\partial T}\right)_P,$$

and

$$\kappa = -\frac{1}{V}\left(\frac{\partial V}{\partial P}\right)_T,$$

we may write the equation in the form

$$C_P - C_V = \frac{TV\left[\frac{1}{V}\left(\frac{\partial V}{\partial T}\right)_P\right]^2}{-\frac{1}{V}\left(\frac{\partial V}{\partial P}\right)_T},$$

$$\boxed{C_P - C_V = \frac{TV\beta^2}{\kappa}.} \tag{9-21}$$

As an example, let us calculate the molar heat capacity at constant volume of mercury at 0°C and 1 atm pressure. From experiment we have $c_P = 28.0$ J/mol · K, $T = 273$ K, $v = 1.47 \times 10^{-5}$ m^3/mol, $\beta = 181 \times 10^{-6}$ K^{-1}, and $\kappa = 3.94 \times 10^{-6}$ atm^{-1} (3.89×10^{-11} Pa^{-1}). Hence,

$$c_P - c_V = \frac{273 \text{ K} \times 1.47 \times 10^{-5} \text{ m}^3/\text{mol} \times (181)^2 \times 10^{-12} \text{ K}^{-2}}{3.89 \times 10^{-11} \text{ Pa}^{-1}}$$

$$= 3.40 \text{ J/mol} \cdot \text{K},$$

and

$$c_V = 28.0 - 3.4 = 24.6 \text{ J/mol} \cdot \text{K}.$$

Finally,

$$\gamma = \frac{c_P}{c_V} = \frac{28.0}{24.6} = 1.14.$$

The two $T\,dS$ equations are

$$T\,ds = C_P\,dT - T\left(\frac{\partial V}{\partial T}\right)_P dP,$$

and

$$T\,dS = C_V\,dT + T\left(\frac{\partial P}{\partial T}\right)_V dV.$$

At constant S,

$$C_P\, dT_S = T\left(\frac{\partial V}{\partial T}\right)_P dP_S,$$

and

$$C_V\, dT_S = -T\left(\frac{\partial P}{\partial T}\right)_V dV_S.$$

Dividing, we obtain

$$\frac{C_P}{C_V} = -\left[\frac{(\partial V/\partial T)_P}{(\partial P/\partial T)_V}\right]\left(\frac{\partial P}{\partial V}\right)_S.$$

But the quantity in brackets is equal to $-(\partial V/\partial P)_T$. Therefore,

$$\boxed{\frac{C_P}{C_V} = \frac{(\partial P/\partial V)_S}{(\partial P/\partial V)_T}.} \tag{9-22}$$

The *adiabatic compressibility* is defined as

$$\kappa_S = -\frac{1}{V}\left(\frac{\partial V}{\partial P}\right)_S;$$

and, as usual,

$$\kappa = -\frac{1}{V}\left(\frac{\partial V}{\partial P}\right)_T.$$

We have, therefore,

$$\frac{C_P}{C_V} = \gamma = \frac{\kappa}{\kappa_S}. \tag{9-23}$$

9-8 HEAT CAPACITY AT CONSTANT PRESSURE

The experimental measurement of c_P has already been discussed in Art. 4-7, and the general features of one type of calorimeter suitable for such measurements were described, as well as some details of technique. Data on the heat capacities of elements, alloys, compounds, plastics, etc., taken over as wide a temperature interval as possible, are of great importance in pure science and in engineering. To the physicist the most important temperature region is from absolute zero to about room temperature (300 K), so that we shall emphasize this region more than any other. In this range, most materials are in the solid phase. We shall limit ourselves to solids in the form of crystals, either a single crystal or a rod or powder consisting of a large number of small crystals.

The behavior of three different crystalline nonmetals is shown in Fig. 9-3. (Metals exhibit a special behavior because of the effect of free electrons; such behavior will be discussed later in this chapter.) *The c_P of all materials approaches zero as T approaches zero.* In the first 20 to 40 degrees, c_P rises

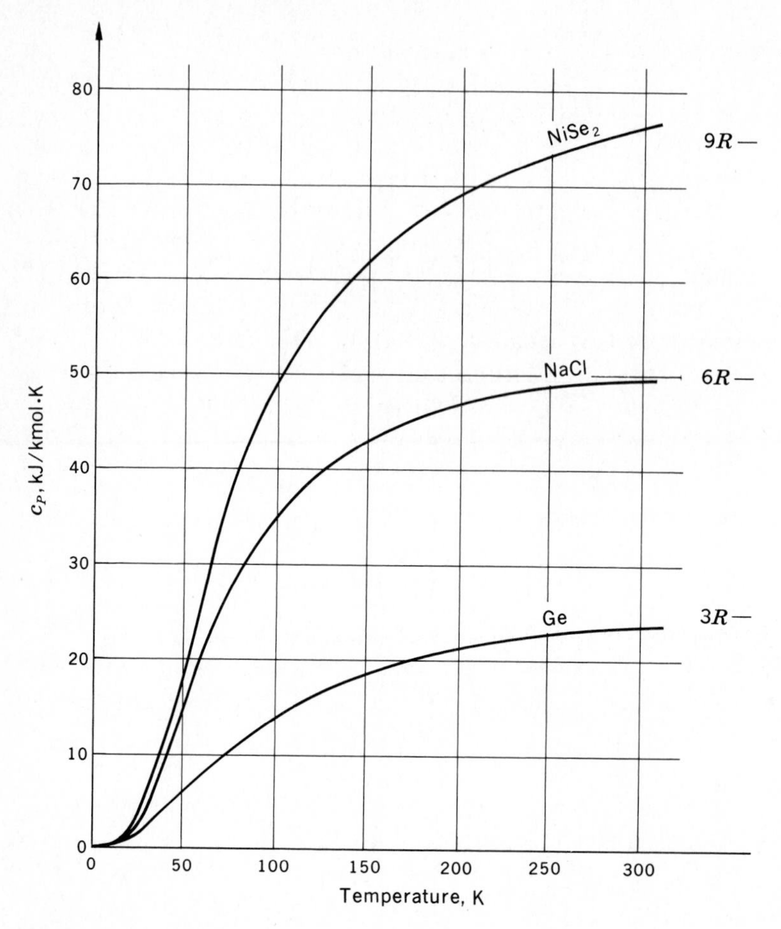

Figure 9-3 Heat capacity at constant pressure of crystalline nonmetals. *(Ge, Keesom and Seidel; NaCl, Clusius, Goldmann, and Perlick; $NiSe_2$, Gronvold and Westrum.)* ($R = 8.31$ kJ/kmol · K.)

rapidly but then bends and begins to flatten out in the neighborhood of room temperature. In none of the three crystals, however, does the c_P curve actually become horizontal.

In the Ge crystal, lattice sites are occupied by single germanium atoms, and therefore 1 mol of Ge consists of N_A vibrating particles, where N_A is Avogadro's number. The value of c_P at room temperature is very nearly equal to $3R$. In NaCl, the lattice sites are occupied by sodium ions and by chlorine ions in a face-centered cubic arrangement. Therefore, 1 mol of NaCl consists of N_A sodium ions in addition to N_A chlorine ions, so that altogether there are $2N_A$ vibrating particles. The value of c_P at room temperature is very nearly equal to $6R$. In $NiSe_2$, the lattice sites are occupied by nickel atoms and by selenium atoms, with the center of the line joining the two selenium atoms and the nickel atom forming a face-centered cubic arrangement similar to that of NaCl. In 1 mol of $NiSe_2$, there are $3N_A$ vibrating particles, and the room temperature value of c_P is very nearly equal to $9R$. In all cases, the c_P at room temperature of *exactly* N_A *atoms or ions* is in the neighborhood of $3R$, about 25 J/mol · K.

The curves in Fig. 9-3 correspond to crystals that were deliberately chosen to illustrate a regularity at room temperature. There is no magic in the temperature 300 K, however. Not all crystals have values of c_P whose rapid increase is tapering off at 300 K. The c_P of diamond, for example, rises so slowly that at 300 K it is still quite far from the value $3R$. Furthermore, c_P never approaches any value asymptotically but continues to rise at all temperatures. The laws governing the temperature variation of heat capacity cannot be stated simply in terms of c_P. To express experimental results in a neat form and also to appreciate the relation between experiment and theory, it is necessary to study the temperature variation of c_V. For this purpose, we use the thermodynamic formula

$$c_P - c_V = \frac{Tv\beta^2}{\kappa}, \tag{9-21}$$

which requires a knowledge of the complete temperature dependence of β, κ, and v, as given, for example, for NaCl in Table 9-2, compiled by Meincke and Graham. Before we use this equation, let us first learn something about thermal expansivity and compressibility.

Table 9-2 Thermal properties of NaCl

(Compiled by P. P. M. Meincke and G. M. Graham)

T, K	c_P, kJ/kmol · K	β, (MK)$^{-1}$	κ, (TPa)$^{-1}$	v, l/kmol	c_V, kJ/kmol · K	κ_S, (TPa)$^{-1}$
10	0.151	0.171	38.9	26.4	0.151	38.9
20	1.30	1.72	38.9	26.4	1.30	38.9
30	4.76	7.44	38.9	26.4	4.76	38.9
40	9.98	17.2	38.9	26.4	9.97	38.9
50	15.7	29.3	39.0	26.4	15.7	38.9
60	21.0	41.4	39.2	26.4	20.9	39.1
70	25.5	52.2	39.4	26.4	25.3	39.2
80	29.3	61.5	39.6	26.5	29.1	39.3
90	32.3	69.5	39.8	26.5	32.0	39.4
100	35.0	75.8	40.0	26.5	34.7	39.6
125	40.1	88.2	40.4	26.6	39.5	39.8
150	43.3	96.3	40.7	26.6	42.4	39.9
175	45.4	103	41.1	26.7	44.2	40.0
250	48.6	114	42.3	26.9	46.6	40.6
290	49.2	118	43.0	27.0	46.7	40.8

9-9 THERMAL EXPANSIVITY

In modern experiments on the expansion of solids, the linear expansivity α is usually measured. If the three rectangular dimensions of a solid are L_1, L_2, and L_3, then

$$V = L_1 L_2 L_3,$$

$$\frac{\partial V}{\partial T} = L_2 L_3 \frac{\partial L_1}{\partial T} + L_1 L_3 \frac{\partial L_2}{\partial T} + L_1 L_2 \frac{\partial L_3}{\partial T},$$

$$\frac{1}{V}\frac{\partial V}{\partial T} = \frac{1}{L_1}\frac{\partial L_1}{\partial T} + \frac{1}{L_2}\frac{\partial L_2}{\partial T} + \frac{1}{L_3}\frac{\partial L_3}{\partial T},$$

and
$$\beta = \alpha_1 + \alpha_2 + \alpha_3,$$

where α_1, α_2, and α_3 are the linear expansivities along the three directions. In the case of quartz, the two linear coefficients perpendicular to the z axis are equal, so that $\beta = 2\alpha_\perp + \alpha_\parallel$. If the solid is isotropic, as in the case of a cubic crystal, $\alpha_1 = \alpha_2 = \alpha_3 = \alpha$ and $\beta = 3\alpha$.

There are four absolute methods of measuring the linear expansivity of solids: X-ray diffraction determinations, interference fringes of visible light, variation of electric capacitance, and variation of intensity of light. Modern methods, particularly those which are used to make measurements in the 0 to 50 K range, involve a lot of auxiliary cryogenic equipment, but the physical principles are easily understood.

In the X-ray diffraction method, the lattice parameter of the specimen crystal is measured as a function of the temperature, using an X-ray beam of known wavelength.

In Fig. 9-4 there is a rough schematic diagram of an interferometer which is

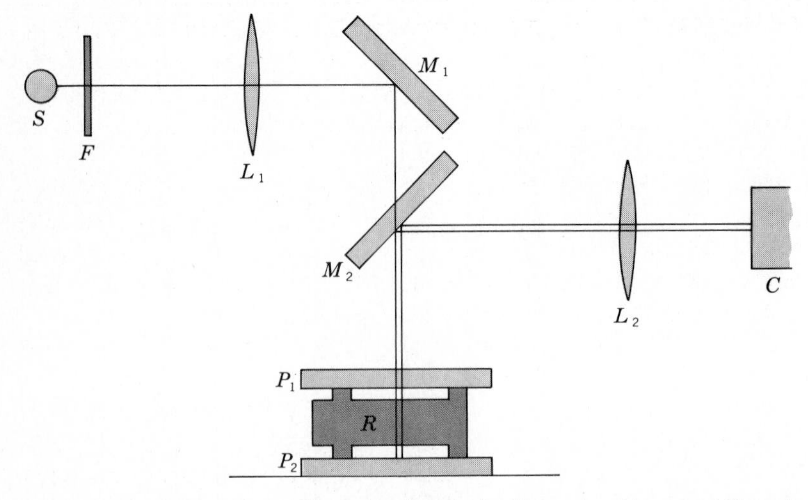

Figure 9-4 Modern version of Fizeau interferometric dilatometer. *(James and Yates.)*

a modification due to James and Yates of a device originally designed by Abbe and Pulfrich. Highly monochromatic light, such as the red light from a low-pressure cadmium lamp, after it has passed through a filter F is concentrated by lens L_1 and reflected from mirror M_1 onto plates P_1 and P_2, which are separated by a ring or cylinder R made of the material whose expansivity is to be studied. The ring and plates are placed at the bottom of a cryostat, where liquid hydrogen or liquid helium is used to provide the low temperatures at which measurements are often made. In Fig. 9-4 all details of the cryostat, heater, thermometer, venting tubes, electric leads, etc., have been omitted. Interference takes place between the rays of light reflected from the bottom of P_1 and the top of P_2, and a camera C is used to photograph the interference fringes. The temperature is varied very slowly from, let us say, 4 K up to room temperature, and the fringe system is photographed at regular intervals.

If N fringes travel across the field of view while the temperature changes from T_0 to T, then the optical path difference has changed by $N\lambda$, where λ is the wavelength of the light, and the thickness of the air space has changed by $N\lambda/2$. If L_0 is the length of the specimen at temperature T_0 and L is the length at T, then

$$\frac{L - L_0}{L_0} = \frac{N\lambda}{2L_0}.$$

If, therefore, $N\lambda/2L_0$ is plotted against T and the slope of the resulting curve is taken at various temperatures, the linear expansivity is obtained. Thus,

$$\alpha = \frac{d}{dT}\left(\frac{N\lambda}{2L_0}\right).$$

In order to avoid the delay involved in photographic processing, R. K. Kirby and his staff at the U.S. National Bureau of Standards have developed a photoelectric interferometer in which the movement of interference fringes is detected by a photomultiplier tube and the number of fringes is automatically plotted on a recorder against the measured temperature of the specimen. Thus, all hand operations are eliminated, and the data are presented on a chart in a form suitable for immediate determination of expansivities.

In the electric method, the expansion of the specimen is communicated to one of the plates of a capacitor, whose other plate is fixed nearby. The change in capacitance is measured by an extremely sensitive bridge; and in the hands of the Australian physicists headed by G. K. White, this method has proved capable of detecting length changes of the order of 10^{-10} m.

The optical grid method of R. V. Jones, which has been used with great success by Andres in Switzerland, is extraordinarily sensitive. The specimen is connected to one of the grids shown in Fig. 9-5, the other remaining fixed. The left-hand grid is regular, and the one on the right has an irregularity in the middle. In position (a), no light can pass the grids in the lower half; whereas in position (b), in which the right-hand grid has been lowered with respect to the

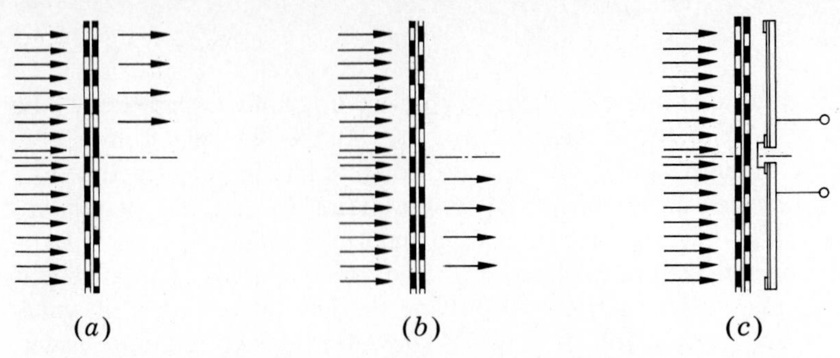

Figure 9-5 Optical grid method of measuring thermal expansion. *(R. V. Jones.)*

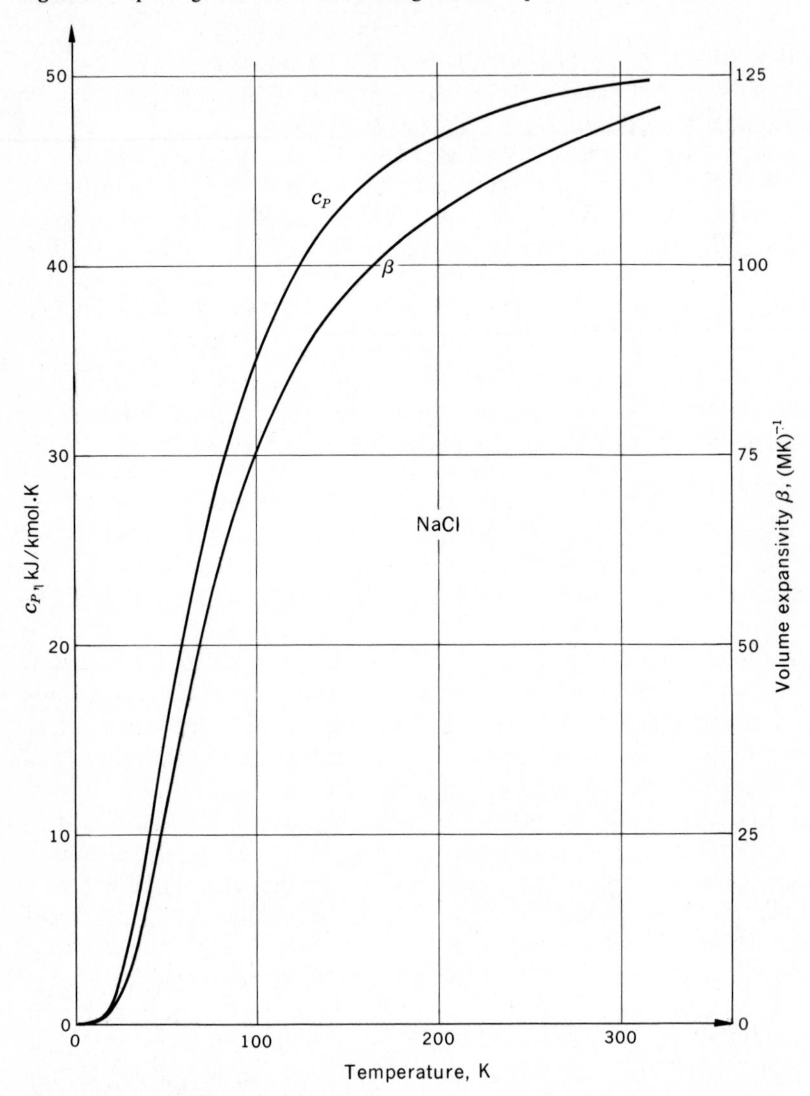

Figure 9-6 Temperature dependence of β and c_P are almost identical. *(β, Meincke and Graham; c_P, Clusius, Goldmann, and Perlick.)*

other by one line width, no light can pass the grids in the upper half. In position (*c*), the two halves are equally transparent, and as a result, the net output of two oppositely connected photocells (one behind the upper half, and the other behind the lower half) is zero. Thus, a change in length of the specimen produces a change in the net output of the two oppositely connected photocells.

The temperature dependence of the volume expansivity of many substances is the same as that of NaCl, shown in Fig. 9-6; namely, β is zero at absolute zero, rises rapidly in the interval from 0 to 50 K, then bends and flattens out without actually becoming horizontal. Thus, the temperature behavior of β is almost identical with that of c_P, as shown in Fig. 9-6. Another similarity between β and c_P is the insensitivity of both quantities to changes of pressure.

9-10 COMPRESSIBILITY

Compressibility measurements are made in two ways and for two different reasons. Installations capable of subjecting solids to enormous hydrostatic pressures at constant temperature and capable of providing numerical values of the *isothermal compressibility* at pressures up to about a million atmospheres are used to study phase transitions, changes of crystal structure, and other internal changes of solids and liquids. These are *static measurements*. Measurements of the speed of longitudinal waves in liquids and both longitudinal and transverse waves in solids, at atmospheric or moderate pressures, are *dynamic* in character and provide numerical values of the *adiabatic compressibility* κ_S, where

$$\kappa_S = -\frac{1}{V}\left(\frac{\partial V}{\partial P}\right)_S.$$

It was shown in Chap. 5 that the speed of a longitudinal wave w in a fluid is given by

$$w = \sqrt{\frac{1}{\rho \kappa_S}},$$

where ρ is the density. Measurements of w and ρ are sufficient to provide κ_S of a fluid, but the measurement of κ_S of a crystalline solid is more difficult. It is necessary to measure the speed of transverse waves as well as that of longitudinal waves and, from the two measurements, to calculate two different elastic constants. In the case of NaCl, these quantities are designated c_{11} and c_{12}, and it is shown in books on elasticity that

$$\kappa_S = \frac{3}{c_{11} + 2c_{12}}.$$

Once κ_S is obtained, the isothermal compressibility may be calculated by using the two thermodynamic equations derived in this chapter, namely,

$$c_P - c_V = \frac{Tv\beta^2}{\kappa}, \tag{9-21}$$

and

$$c_V\kappa = c_P\kappa_S. \tag{9-23}$$

Substituting for c_V in the first equation the value found from the second, we get

$$c_P - \frac{c_P\kappa_S}{\kappa} = \frac{Tv\beta^2}{\kappa},$$

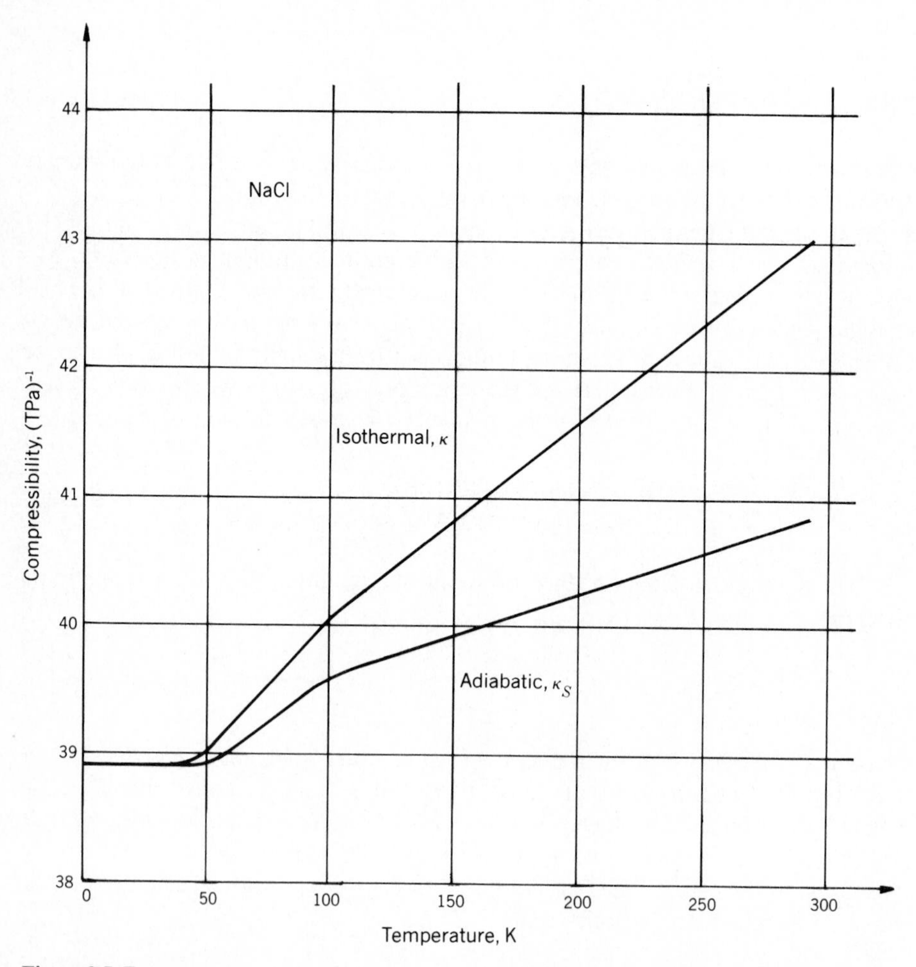

Figure 9-7 Temperature variation of isothermal and adiabatic compressibilities of NaCl. *(Overton and Swim.)*

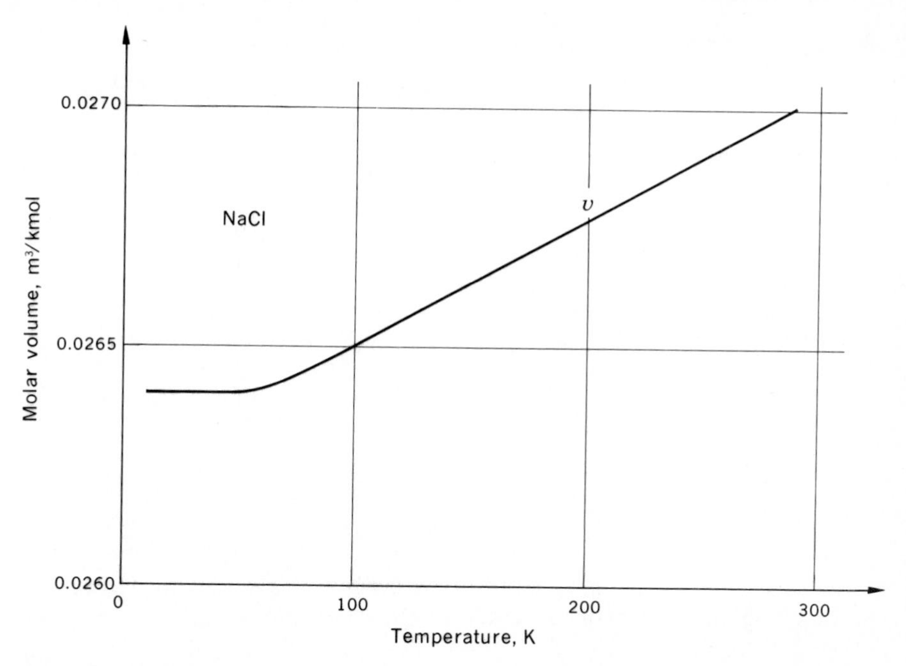

Figure 9-8 Temperature variation of molar volume of NaCl, similar to that of isothermal compressibility. *(Heinglen.)*

which reduces to

$$\kappa - \kappa_S = \frac{Tv\beta^2}{c_P}. \tag{9-24}$$

The method of measuring wave velocities that is most suitable for crystalline solids is one which incorporates the familiar radar technique. Short ultrasonic pulses of about 1 μs duration are sent through the crystal by a quartz crystal transducer. After reflection from an end of the crystal, the quartz emitter is used as a receiver. The pulse and the echo are observed on an oscilloscope, and the wave speed is calculated from the dimensions of the crystal and the time delay between pulse and echo. This is the method used by Overton and Swim to obtain the values of κ_S listed in Table 9-2.

The temperature variation of κ_S and κ of NaCl are shown in Fig. 9-7, where it may be seen that both κ_S and κ—unlike c_P and β—do *not* approach zero as T approaches zero. From zero to 40 K, the adiabatic and isothermal compressibilities are nearly equal. At higher temperatures, κ is larger than κ_S, as required by Eq. (9-24).

The molar volume of NaCl is shown in Fig. 9-8, where the temperature variation may be seen to be very similar to that of κ. In a rough sort of way, c_P and β vary in the same manner, while v and κ are also similar, but with an entirely different kind of temperature variation.

Table 9-3 Thermal properties of water

T, °C	w, m/s	β, $(\mathrm{MK})^{-1}$	c_P, kJ/kg · K	ρ, kg/m^3	κ, $(\mathrm{GPa})^{-1}$	κ_S, $(\mathrm{GPa})^{-1}$
0	1402.4	−67.89	4.2177	999.84	0.50885	0.50855
10	1447.2	87.96	4.1922	999.70	0.47810	0.47758
20	1482.3	206.80	4.1819	998.21	0.45891	0.45591
30	1509.1	303.23	4.1785	995.65	0.44770	0.44100
40	1528.8	385.30	4.1786	992.22	0.44240	0.43119
50	1542.5	457.60	4.1807	998.04	0.44174	0.42536
60	1550.9	523.07	4.1844	983.20	0.44496	0.42281
70	1554.7	583.74	4.1896	977.76	0.45161	0.42307
80	1554.4	641.11	4.1964	971.79	0.46143	0.42584
90	1550.4	696.24	4.2051	965.31	0.47430	0.43093
100	1543.2	750.30	4.2160	958.35	0.49018	0.43819

The speed of longitudinal waves in a liquid is usually measured with the aid of an acoustic interferometer such as the one described in Art. 5-8, which was used with helium gas to serve as an absolute thermometer in the range from 4 to 10 K. Results of measurements on water are listed in Table 9-3 and are shown in Fig. 9-9. The minimum achieved by the isothermal compressibility of water at about 50°C is quite anomalous. As a rule, the isothermal compressibility of most liquids increases as the temperature is raised and follows a simple exponential equation quite well:

$$\kappa = \kappa_0 e^{bT} \tag{9-25}$$

where κ_0 and b are constants. The constant b for mercury is $1.37 \times 10^{-3}\ \mathrm{K}^{-1}$.

All liquids, including water, become less compressible the more they are

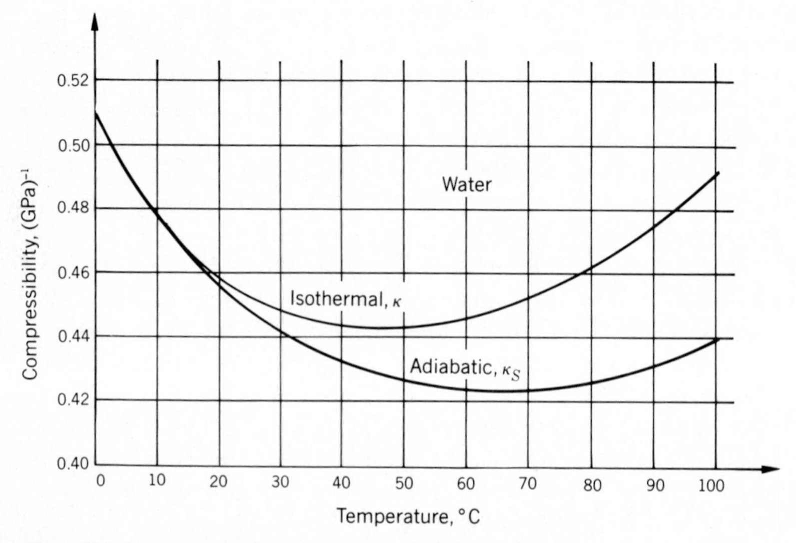

Figure 9-9 Isothermal and adiabatic compressibilities of water.

compressed; the reciprocal of the isothermal compressibility increases linearly with respect to the pressure, where

$$\frac{1}{\kappa} - \frac{1}{\kappa_0} = cP, \tag{9-26}$$

where κ_0 is the compressibility at zero pressure and c is 6.7 for water and 8.2 for mercury.

9-11 HEAT CAPACITY AT CONSTANT VOLUME

The measurement of c_P, β, and κ of crystalline solids both metallic and nonmetallic, particularly at low temperatures, is a constant challenge to experimental physicists throughout the world. This field is still a very active one, for each quantity is of theoretical interest in its own right. Our purpose at present is to use these measurements in conjunction with the thermodynamic equation,

$$c_P - c_V = \frac{Tv\beta^2}{\kappa}, \tag{9-21}$$

to find the complete temperature dependence of c_V. All the measurements on NaCl are listed in Table 9-2, along with the calculated values of c_V, and both c_P

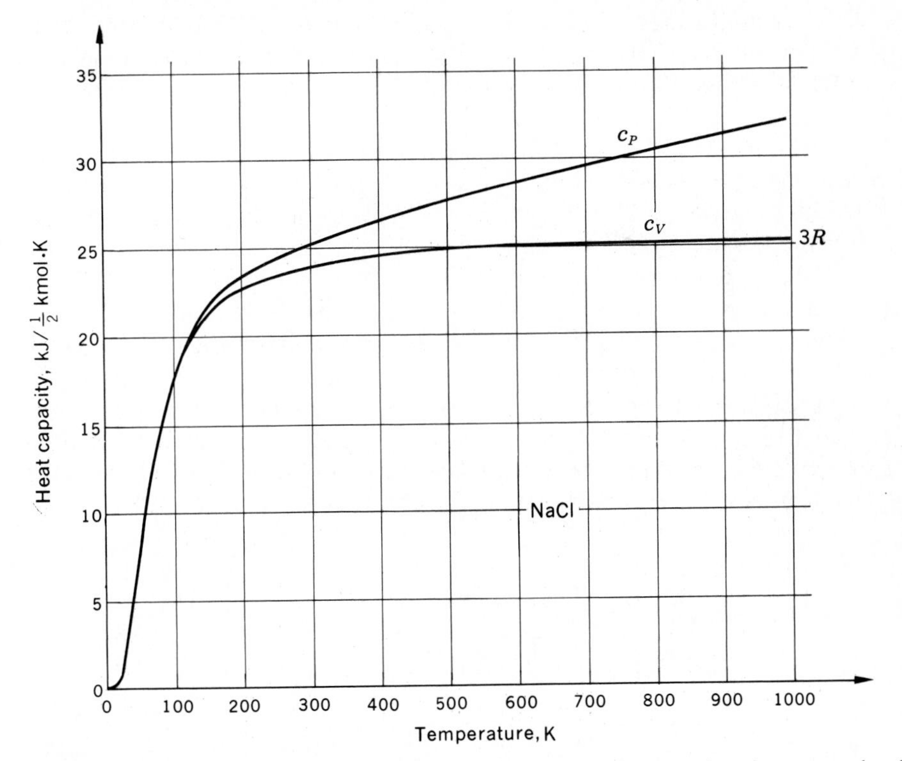

Figure 9-10 Temperature variation of c_P and c_V of $\frac{1}{2}$ mol of NaCl. The value of c_V approaches $3R$ as a limit.

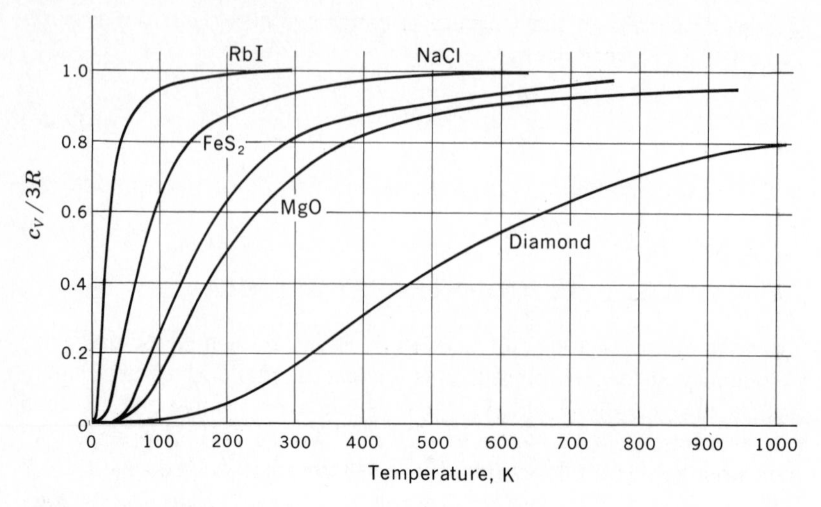

Figure 9-11 Temperature variation of $c_V/3R$ of nonmetals. (1 mol of diamond; $\frac{1}{2}$ mol of RbI, NaCl, MgO; and $\frac{1}{3}$ mol of FeS_2.)

and c_V are plotted against T up to 1000 K in Fig. 9-10. Since 1 mol of NaCl consists of $2N_A$ ions, the heat capacities refer to $\frac{1}{2}$ mol, or N_A ions.

At low temperatures, below 100 K, c_P and c_V are practically the same. At all higher temperatures, while c_P continues to increase, c_V approaches a constant value $3R$, which is called the *Dulong and Petit value*, named after the physicists who first observed that c_P came near this value at about room temperature. We see now that this value is actually approached by c_V and is exceeded only in special situations. The temperature dependence of $c_V/3R$ of five representative nonmetals is shown in Fig. 9-11, where it may be seen that RbI practically reaches the Dulong and Petit value even before room temperature, whereas diamond has reached only one-fifth this value. As a matter of fact, it requires a temperature greater than 2000 K to bring the c_V of diamond near $3R$.

Although the five curves in Fig. 9-11 differ markedly in the temperature at which $c_V \to 3R$, the curves are still very similar in shape. An experienced experimenter would be led to suspect that there existed a parameter—Θ, let us say—small for RBI and large for diamond, such that c_V is a *universal function* of the ratio T/Θ. Such a variation is called a *law of corresponding states*, and we shall see in Chap. 12 that an approximate law of this sort is provided by theory.

PROBLEMS

9-1 Show that, for an ideal gas:

(*a*)
$$F = \int C_V\,dT - T\int \frac{C_V}{T}\,dT - nRT\ln V - \text{const.}\;T + \text{const.}$$

(*b*)
$$G = \int C_P\,dT - T\int \frac{C_P}{T}\,dT + nRT\ln P - \text{const.}\;T + \text{const.}$$

(*c*) Apply the above equations to 1 mol of an ideal monatomic gas.

9-2 Defining the *Massieu function* F_M by the equation

(*a*)
$$F_M = -\frac{U}{T} + S,$$

show that

$$dF_M = \frac{U}{T^2}\,dT + \frac{P}{T}\,dV.$$

(*b*) Defining the *Planck function* F_P by the equation

$$F_P = -\frac{H}{T} + S,$$

show that

$$dF_P = \frac{H}{T^2}\,dT - \frac{V}{T}\,dP.$$

9-3 From the fact that dV/V is an exact differential, derive the relation

$$\left(\frac{\partial \beta}{\partial P}\right)_T = -\left(\frac{\partial \kappa}{\partial T}\right)_P.$$

9-4 Derive the following equations:

(*a*) $$U = F - T\left(\frac{\partial F}{\partial T}\right)_V = -T^2\left(\frac{\partial F/T}{\partial T}\right)_V.$$

(*b*) $$C_V = -T\left(\frac{\partial^2 F}{\partial T^2}\right)_V.$$

(*c*) $$H = G - T\left(\frac{\partial G}{\partial T}\right)_P = -T^2\left(\frac{\partial G/T}{\partial T}\right)_P$$ (Gibbs-Helmholtz equation).

(*d*) $$C_P = -T\left(\frac{\partial^2 G}{\partial T^2}\right)_P.$$

9-5 Derive the third $T\,dS$ equation,

$$T\,dS = C_V\left(\frac{\partial T}{\partial P}\right)_V dP + C_P\left(\frac{\partial T}{\partial V}\right)_P dV,$$

and show that the three $T\,ds$ equations may be written as follows:

(*a*)
$$T\,dS = C_V\,dT + \frac{\beta T}{\kappa}\,dV,$$

(*b*)
$$T\,dS = C_P\,dT - V\beta T\,dP,$$

(*c*)
$$T\,dS = \frac{C_V \kappa}{\beta}\,dP + \frac{C_P}{\beta V}\,dV.$$

9-6 The pressure on 500 g of copper is increased reversibly and isothermally from zero to 500 atm at 100 K. (Take the density $\rho = 8.93 \times 10^3$ kg/m^3, volume expansivity $\beta = 31.5 \times 10^{-6}$ K^{-1}, isothermal compressibility $\mathrm{K} = 7.21 \times 10^{-12}$ Pa^{-1}, and heat capacity $C_P = 0.254$ kJ/kg · K to be constant.)

(*a*) How much heat is transferred during the compression?

(*b*) How much work is done during the compression?

(*c*) Determine the change of internal energy.

(*d*) What would have been the rise of temperature if the copper had been subjected to a reversible adiabatic compression?

9-7 The pressure on 0.2 kg of water is increased reversibly and isothermally from 1 to 3×10^8 Pa at 0°C. (Numerical values are given in Table 9-3.)

(*a*) How much heat is transferred?

(*b*) How much work is done?

(*c*) Calculate the change in internal energy.

9-8 The pressure on 1 g of water is increased from 0 to 10^8 Pa reversibly and adiabatically. Calculate the temperature change when the initial temperature has the three different values given below:

Temperature, °C	Specific volume v, 10^{-3} m³/kg	β, 10^{-6} K^{-1}	c_P, kJ/kg · K
0	1.000	− 68	4.220
5	1.000	+ 16	4.200
50	1.012	+458	4.180

9-9 A gas obeys the equation $P(v - b) = RT$, where b is constant and has a constant c_V. Show that:

(*a*) u is a function of T only.

(*b*) γ is constant.

(*c*) A relation that holds during an adiabatic process is

$$P(v - b)^\gamma = \text{const.}$$

9-10 Show that for a gas obeying the van der Waals equation $(P + a/v^2)(v - b) = RT$, with c_V a function of T only, an equation for an adiabatic process is

$$T(v - b)^{R/c_V} = \text{const.}$$

9-11 (*a*) Using the virial expansion

$$Pv = RT\left(1 + \frac{B}{v} + \frac{C}{v^2} + \cdots\right),$$

calculate $(\partial u/\partial v)_T$ and its limit as $v \to \infty$.

(*b*) Using the same expansion, calculate $(\partial P/\partial v)_T$ and its limit as $v \to \infty$.

(*c*) Using (*a*) and (*b*), calculate $(\partial u/\partial P)_T$ and its limit as $v \to \infty$. (Compare the solution with the results of Rossini and Frandsen given in Art. 5-2.)

(*d*) Using the virial expansion

$$Pv = RT + B'P + C'P^2 + \cdots$$

and bearing in mind that $B' = B$, calculate $(\partial u/\partial P)_T$ directly from Eq. (9-19).

9-12 Show that the differentials of the three thermodynamic functions U, H, and F may be written

$$dU = (C_P - PV\beta)\, dT + V(\kappa P - \beta T)\, dP.$$

$$dH = C_P\, dT + V(1 - \beta T)\, dP.$$

$$dF = -(PV\beta + S)\, dT + PV\kappa\, dP.$$

9-13 (*a*) Derive the equation

$$\left(\frac{\partial C_V}{\partial V}\right)_T = T\left(\frac{\partial^2 P}{\partial T^2}\right)_V.$$

(*b*) Prove that C_V of an ideal gas is a function of T only.

(*c*) In the case of a gas obeying the equation of state

$$\frac{Pv}{RT} = 1 + \frac{B}{v},$$

where B is a function of T only, show that

$$c_V = -\frac{RT}{v}\frac{d^2}{dT^2}(BT) + (c_V)_0,$$

where $(c_V)_0$ is the value at very large volumes.

9-14 (*a*) Derive the equation

$$\left(\frac{\partial C_P}{\partial P}\right)_T = -T\left(\frac{\partial^2 V}{\partial T^2}\right)_P.$$

(*b*) Prove that C_P of an ideal gas is a function of T only.
(*c*) In the case of a gas obeying the equation of state

$$Pv = RT + BP,$$

where B is a function of T only, show that

$$c_P = -T\frac{d^2B}{dT^2}P + (c_P)_0,$$

where $(c_P)_0$ is the value at very low pressures.

9-15 In the accompanying table are listed the thermal properties of liquid neon, compiled by Gladun. Calculate and plot against temperature (*a*) c_V, (*b*) κ_S, and (*c*) γ.

T, K	ρ, kmol/m^3	β, 10^{-2} K^{-1}	κ, 10^{-8} Pa^{-1}	c_P, kJ/kmol · K
25	61.5	1.33	0.43	36.6
27	59.9	1.46	0.50	37.6
29	58.1	1.63	0.62	39.2
31	56.2	1.84	0.79	41.2
33	54.0	2.12	1.03	43.9
35	51.7	2.52	1.40	47.7
37	49.1	3.14	2.04	53
39	46.1	4.24	3.4	62
41	42.3	6.8	6.9	82
42	40.0	10	11	100
43	36.8	18	26	160

9-16 Derive the following equations:

(*a*)
$$C_V = -T\left(\frac{\partial P}{\partial T}\right)_V\left(\frac{\partial V}{\partial T}\right)_S.$$

(*b*)
$$\left(\frac{\partial V}{\partial T}\right)_S = -\frac{C_V\kappa}{\beta T}.$$

(*c*)
$$\frac{(\partial V/\partial T)_S}{(\partial V/\partial T)_P} = \frac{1}{1-\gamma}.$$

9-17 Derive the following equations:

(*a*)
$$C_P = T\left(\frac{\partial V}{\partial T}\right)_P\left(\frac{\partial P}{\partial T}\right)_S.$$

(*b*)
$$\left(\frac{\partial P}{\partial T}\right)_S = \frac{C_P}{V\beta T}.$$

(*c*)
$$\frac{(\partial P/\partial T)_S}{(\partial P/\partial T)_V} = \frac{\gamma}{\gamma - 1}.$$

9-18 (*a*) A measure of the result of a Joule free expansion is provided by the *Joule coefficient* $\eta = (\partial T/\partial V)_U$. Show that

$$\eta = -\frac{1}{C_V}\left(\frac{\beta T}{\kappa} - P\right).$$

(*b*) A measure of the result of a Joule-Kelvin expansion (throttling process) is provided by the *Joule-Kelvin coefficient* $\mu = (\partial T/\partial P)_H$. Show that

$$\mu = \frac{V}{C_P}(\beta T - 1).$$

CHAPTER

TEN

PHASE TRANSITIONS: MELTING, VAPORIZATION, AND SUBLIMATION

10-1 FIRST-ORDER TRANSITION; CLAPEYRON'S EQUATION

In the familiar phase transitions—melting, vaporization, and sublimation—as well as in some less familiar transitions, such as from one crystal modification to another, the temperature and pressure remain constant while the entropy and volume change. Consider n_0 moles of material in phase i with molar entropy $s^{(i)}$ and molar volume $v^{(i)}$. Both $s^{(i)}$ and $v^{(i)}$ are functions of T and P and hence remain constant during the phase transition which ends with the material in phase f with molar entropy $s^{(f)}$ and molar volume $v^{(f)}$. (The different phases are indicated by superscripts in order to reserve subscripts for specifying different states of the same phase or different substances.) Let x equal the fraction of the initial phase that has been transformed into the final phase at any moment. Then the entropy and volume of the mixture at any moment—S and V, respectively—are given by

$$S = n_0(1 - x)s^{(i)} + n_0 x s^{(f)}$$

and

$$V = n_0(1 - x)v^{(i)} + n_0 x v^{(f)},$$

and S and V are seen to be linear functions of x.

If the phase transition takes place reversibly, the heat (commonly known as a *latent heat*) transferred per mole is given by

$$l = T(s^{(f)} - s^{(i)}).$$

The existence of a latent heat, therefore, means that there is a change of entropy. Since

$$dg = -s\,dT + v\,dP,$$

$$s = -\left(\frac{\partial g}{\partial T}\right)_P,$$

and

$$v = \left(\frac{\partial g}{\partial P}\right)_T,$$

we may characterize the familiar phase transitions by either of the following equivalent statements:

1. There are changes of entropy and of volume.
2. The first-order derivatives of the Gibbs function change discontinuously.

Any phase change that satisfies these requirements is known as a *phase change of the first order*. For such a phase change, the temperature variations of G, S, V, and C_P are shown by four crude graphs in Fig. 10-1. The phase transi-

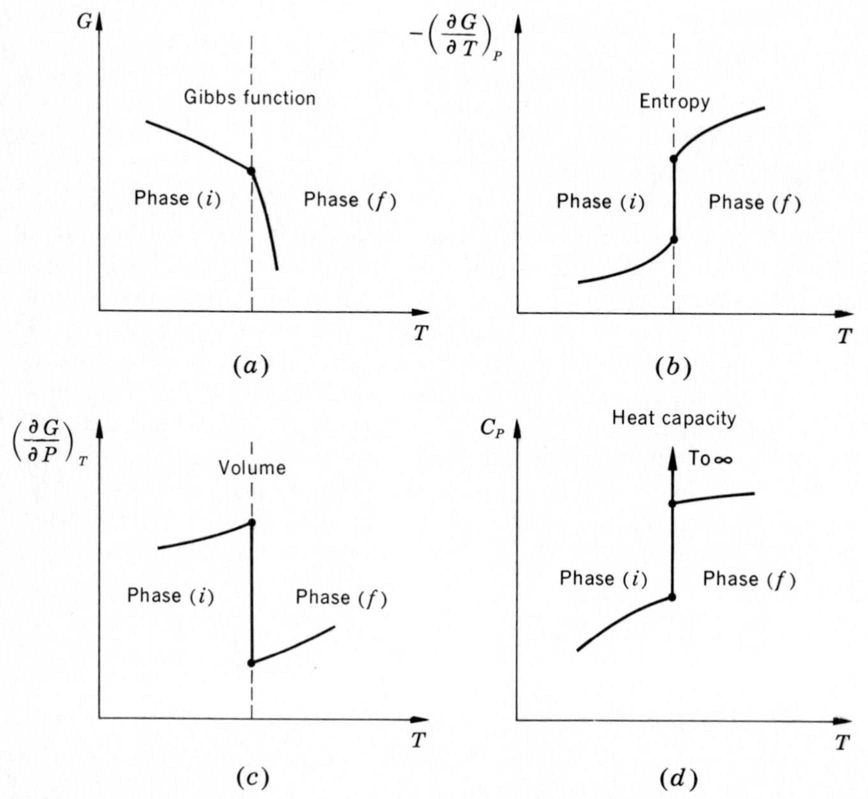

Figure 10-1 Characteristics of a first-order phase transition. (*a*) Gibbs function; (*b*) entropy; (*c*) volume; (*d*) heat capacity.

tion may be regarded as accomplished reversibly in either direction. The fourth graph showing the behavior of C_P is particularly significant in that the C_P of a *mixture of two phases during the phase transition* is infinite. This is true because the transition occurs at constant T and P. When P is constant, $dT = 0$; or when T is constant $dP = 0$. Therefore,

$$C_P = T\left(\frac{\partial S}{\partial T}\right)_P = \infty, \quad \beta = \frac{1}{V}\left(\frac{\partial V}{\partial T}\right)_P = \infty, \quad \kappa = -\frac{1}{V}\left(\frac{\partial V}{\partial P}\right)_T = \infty.$$

It should be noticed, however, that these statements are true only when both phases are present. As shown in Fig. 10-1*d*, the C_P of phase (i) remains finite right up to the transition temperature. It does not "anticipate" the onset of a phase transition by starting to rise before this temperature is reached. This is always true of a first-order transition, but not of all transitions, as will be shown in Chap. 13.

The second $T\,dS$ equation provides an indeterminate result when applied to a first-order phase transition. For any small portion,

$$T\,dS = C_P\,dT - TV\beta\,dP,$$

where $C_P = \infty$ and $dT = 0$; also, $\beta = \infty$ and $dP = 0$.

The first $T\,dS$ equation, however, may be integrated through the phase transition. When 1 mole of substance is converted reversibly, isothermally, and isobarically from phase (i) to phase (f), the first $T\,dS$ equation,

$$T\,ds = c_V\,dT + T\left(\frac{\partial P}{\partial T}\right)_V dv,$$

may be integrated with the understanding that the various P's and T's at which a phase transition occurs obey a relation in which P is a function of T only, independent of V, so that $(\partial P/\partial T)_V = dP/dT$. Hence,

$$T(s^{(f)} - s^{(i)}) = T\frac{dP}{dT}(v^{(f)} - v^{(i)}).$$

The left-hand member of this equation is the latent heat per mole, and hence

$$\boxed{\frac{dP}{dT} = \frac{l}{T(v^{(f)} - v^{(i)})}.} \tag{10-1}$$

This equation, known as *Clapeyron's equation*, applies to any first-order change of phase or transition that takes place at constant T and P.

It is instructive to derive Clapeyron's equation in another way. It was shown in Chap. 9 that the Gibbs function remains constant during a reversible process taking place at constant temperature and pressure. Hence, for a change of phase at T and P,

$$g^{(i)} = g^{(f)};$$

and for a phase change at $T + dT$ and $P + dP$,

$$g^{(i)} + dg^{(i)} = g^{(f)} + dg^{(f)}.$$

Subtracting, we get

$$dg^{(i)} = dg^{(f)},$$

or

$$-s^{(i)}\, dT + v^{(i)}\, dP = -s^{(f)}\, dT + v^{(f)}\, dP.$$

Therefore,

$$\frac{dP}{dT} = \frac{s^{(f)} - s^{(i)}}{v^{(f)} - v^{(i)}},$$

and, finally,

$$\frac{dP}{dT} = \frac{l}{T(v^{(f)} - v^{(i)})}.$$

In dealing with phase transitions it is necessary to indicate in a simple way the initial and final phases and the corresponding heat of transition. The notation used in this book is as follows. A symbol representing any property of the solid phase will be distinguished with a prime; for the liquid phase it will have a double prime, and for the vapor phase a triple prime. Thus, v' stands for the molar volume of a solid, v'' of a liquid, and v''' of a vapor. The heat of fusion (melting) per mole will be l_F, the heat of vaporization (boiling) l_V, and the heat of sublimation l_S.

10-2 FUSION

The simplest method of measuring the heat of fusion of a solid is to supply electrical energy at a constant rate and to measure the temperature at convenient time intervals. Plotting the temperature against the time, a heating curve is obtained in which the phase transition appears as a straight line at constant temperature, of length $\Delta\tau$ as measured along the time axis. The apparatus, shielding, precautions, etc., are exactly the same as those required in the measurement of heat capacity, which were described in Chap. 4. If n moles of solid melt in time $\Delta\tau$, with electrical energy supplied at the rate $\mathcal{E}I$, then

$$l_F = \frac{\mathcal{E}I\, \Delta\tau}{n}.$$

If T_M is the normal melting point of a solid and l_{FM} is the latent heat of fusion at the normal melting point, the entropy change associated with melting at this temperature is l_{FM}/RT_M, expressed in units of R. This entropy change is listed in Table 10-1 for 15 nonmetallic solids and 15 metals, and it may be seen that metals show more of a regularity than nonmetals. Very roughly, l_{FM}/RT_M is about 1 for metals.

The various pressures and temperatures at which the solid and liquid phases coexist in equilibrium determine the melting curve, and one of the first tasks of the experimenter is to determine the equation of this curve. In the region of low

Table 10-1 Entropy change accompanying fusion at the normal melting point T_M

Nonmetal	T_M, K	l_{FM}, kJ/mol	$\frac{l_{FM}}{RT_M}$	Metal	T_M, K	l_{FM}, kJ/mol	$\frac{l_{FM}}{RT_M}$
O_2	54.8	0.445	0.98	Hg	234	2.33	1.20
N_2	63.3	0.721	1.37	Cs	302	2.09	0.83
CO	68.1	0.837	1.48	Rb	312	2.20	0.85
Ar	84.0	1.21	1.73	K	337	2.40	0.86
C_2H_6	89.9	2.86	3.83	Na	371	2.64	0.86
CH_4	90.7	0.941	1.25	Li	452	4.60	1.22
C_2H_4	104	3.35	3.87	Cd	594	6.11	1.24
Kr	117	1.63	1.68	Pb	601	5.12	1.02
Xe	161	3.10	2.32	Zn	693	6.67	1.16
Cl_2	172	6.41	4.48	Mg	922	9.04	1.18
CCl_4	250	2.47	1.19	Ag	1235	11.3	1.10
CH_4ON_2	408	14.5	4.27	Au	1338	12.7	1.14
$CaCl_2$	1055	25.5	2.91	Cu	1358	13.0	1.15
Ge	1211	34.7	3.45	Be	1551	11.7	0.91
Si	1683	39.6	2.83	Ni	1726	17.6	1.23

temperatures, the melting temperatures and pressures are often measured by the *blocked-capillary* method, illustrated in Fig. 10-2. The material in the gaseous phase is compressed to a high pressure and is forced into a steel capillary, part of which is immersed in a low-temperature bath whose temperature may be adjusted to any desired value by careful choice of the bath liquid and its pressure. Two manometers, M_1 before the bath and M_2 after, communicate with the capillary. The melting pressure associated with the temperature of the bath is the maximum reading of M_2. Four typical melting curves, those of neon, argon, krypton, and xenon, are shown in Fig. 10-3.

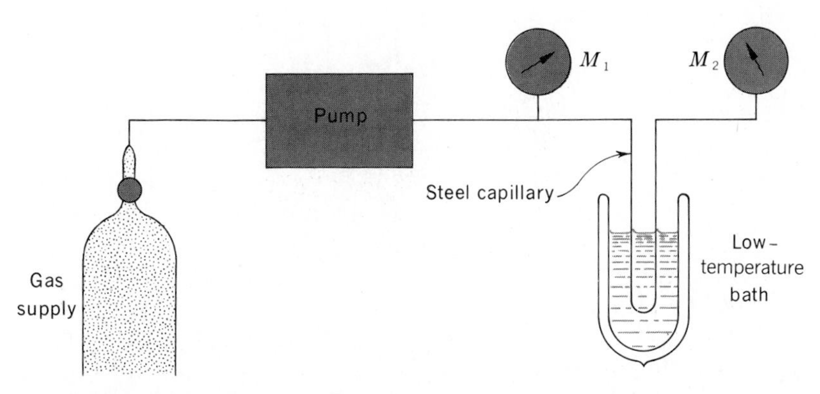

Figure 10-2 Blocked-capillary method of measuring melting pressure.

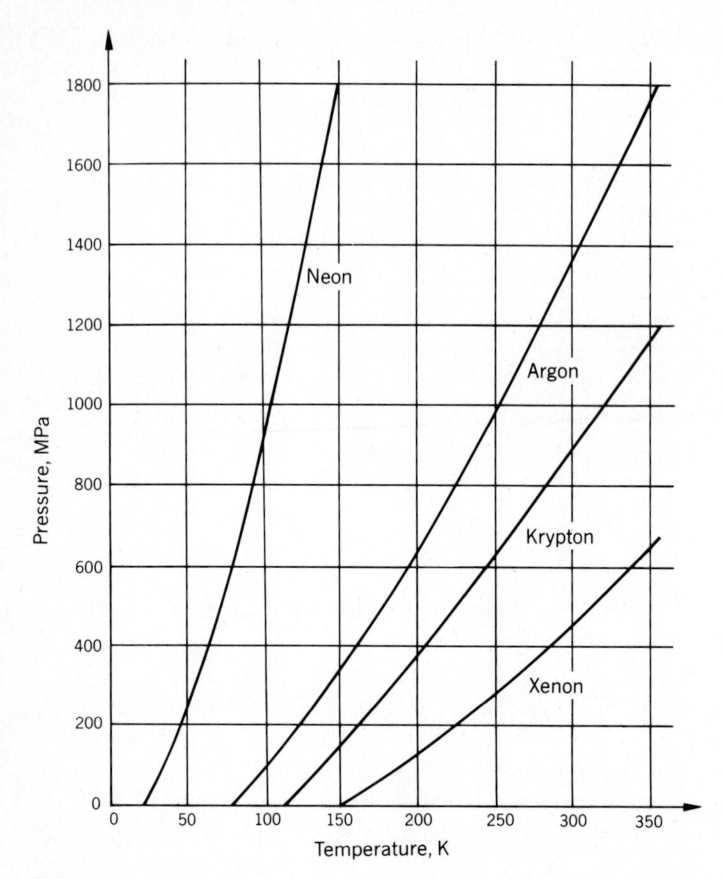

Figure 10-3 Melting curves of neon, argon, krypton, and xenon. *(S. E. Babb; G. L. Pollack.)*

In 1929, F. E. Simon and G. Glatzel suggested an equation that has been fairly successful in representing the data on melting curves, as follows:

$$P - P_{TP} = a\left[\left(\frac{T}{T_{TP}}\right)^c - 1\right] \tag{10-2}$$

where T_{TP} and P_{TP} are the coordinates of the triple point, and a and c are constants that also depend on the substance. At the high pressures, P_{TP} is negligible, so that the equation is usually used in the form

$$\frac{P}{a} = \left(\frac{T}{T_{TP}}\right)^c - 1.$$

The values of a and c for the four condensed inert gases shown in Fig. 10-3 are listed in Table 10-2, and those for many other solids have been given by S. E. Babb.

The slope of the melting curve is negative for those few substances like ice I which contract upon melting. This behavior is shown by Bi, Ge, Si, and Ga and

Table 10-2 Melting parameters of condensed inert gases

Solidified inert gas	T_{TP}, K	P_{TP}, kPa	a, MPa	c
Ne	24.6	43.2	103.6	1.6
Ar	83.8	69.0	227.0	1.5
Kr	116	73.3	305.0	1.4
Xe	161	81.7	345.5	1.31

requires values of T in Simon's equation less than T_{TP}. Consequently the values of a are negative. The various values of a and c for four of the forms of ice are shown in Fig. 10-4. In Fig. 10-5 are shown the enormous pressures and temperatures needed to produce gaseous and liquid carbon and also the solid crystalline form of diamond.

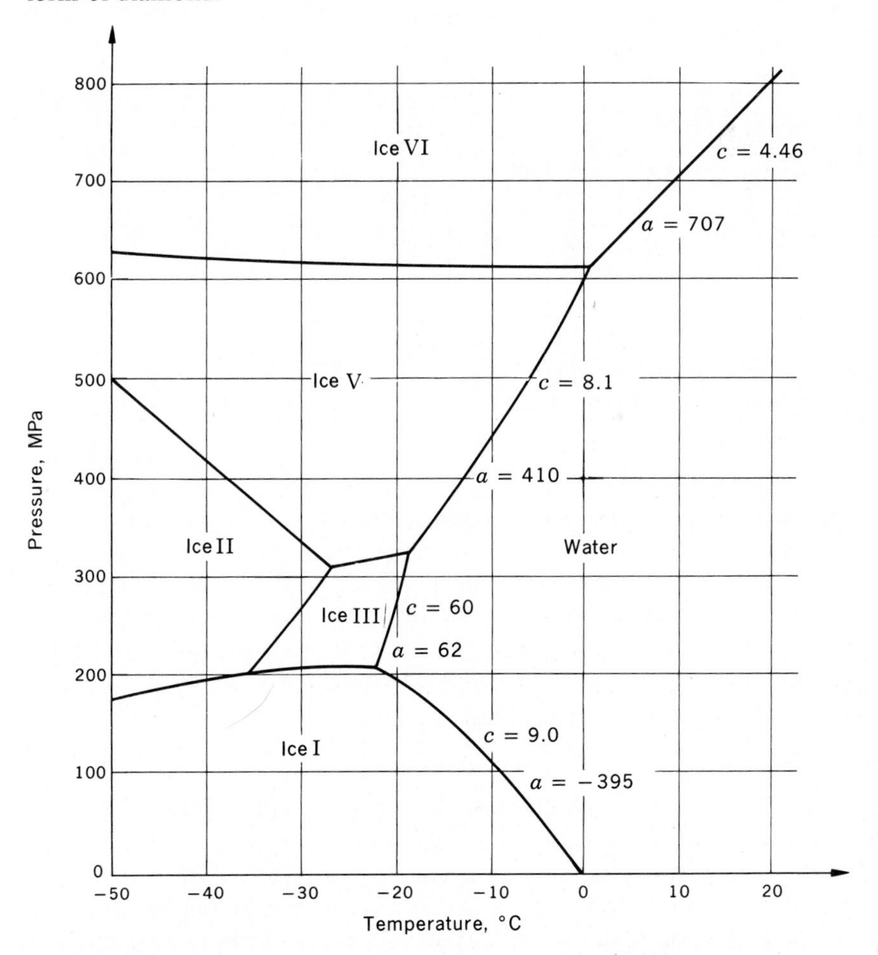

Figure 10-4 Phase diagram of water with melting parameters of ice I, III, V, and VI.

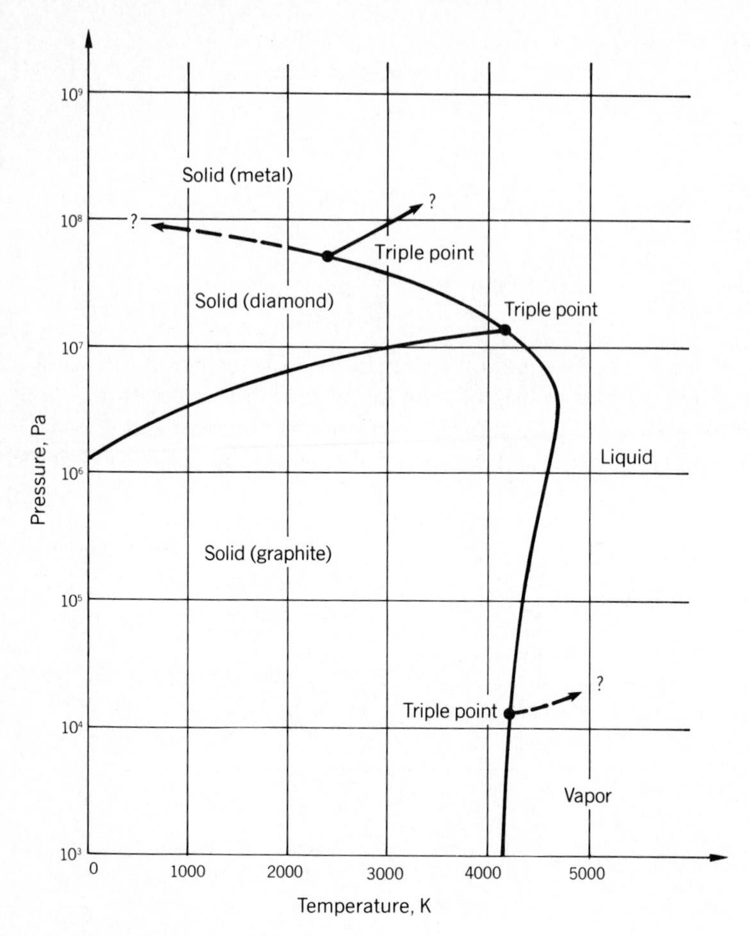

Figure 10-5 Phase diagram for carbon.

Theories concerning the exact processes that take place when a solid melts have engaged the attention of physicists for many years. It was first suggested by Lindemann that a solid melts when the amplitude of the lattice vibrations becomes large enough to break down the attractive forces holding the lattice together; in more picturesque language, "In melting, a solid shakes itself to pieces." With this point of view, Lindemann derived the formula

$$\frac{\mathcal{M}v^{2/3}\Theta^2}{T_M} = \text{const.,} \tag{10-3}$$

where $\mathcal{M}$ and v are the molecular weight and molar volume, Θ is the Debye characteristic temperature, and T_M is the melting temperature. This relation holds quite well for many metals and nonmetals, but there are a few of each that depart radically from the formula. This suggests that the melting process is not exclusively a question of lattice vibrations. Dislocations and vacancies in the crystal lattice, as well as the quantities specifying the law of force among the molecules of both solid and liquid, are all considered to play a role. On the basis of such ideas, it is possible to provide a slight theoretical foundation for the Simon equation.

10-3 VAPORIZATION

The heat of vaporization of liquids with normal boiling points from, let us say, 250 K to about 550 K is generally measured directly with a calorimeter resembling that shown in Fig. 10-6. The sample liquid L_2 is contained in a small vessel and has immersed in it a small heating coil R_2. Completely surrounding this vessel is a temperature bath consisting of a mixture of air and the vapor of another liquid L_1. By choosing a suitable liquid L_1 and keeping it at its boiling point by means of the heating coil R_1 in the presence of air at the proper pressure, the temperature bath may be maintained at any desired temperature. At this chosen temperature, the liquid L_2 is in equilibrium with its vapor. The small vessel containing L_2 communicates with another vessel on the outside (not shown in the figure), which may be maintained at any desired temperature by a separately controlled heating or cooling device.

If the temperature of the outside container is maintained at less than that of L_2, a pressure gradient is produced, and some of L_2 will distill over. By maintaining a small current I in the heating coil R_2, the temperature of L_2 is kept equal to that of its surroundings, and the energy necessary to vaporize it is thus supplied. There is therefore a steady distillation of L_2 into the outside container, with the heat of vaporization being supplied by the heating coil R_2 and the heat of condensation being withdrawn by the surroundings of the outside container.

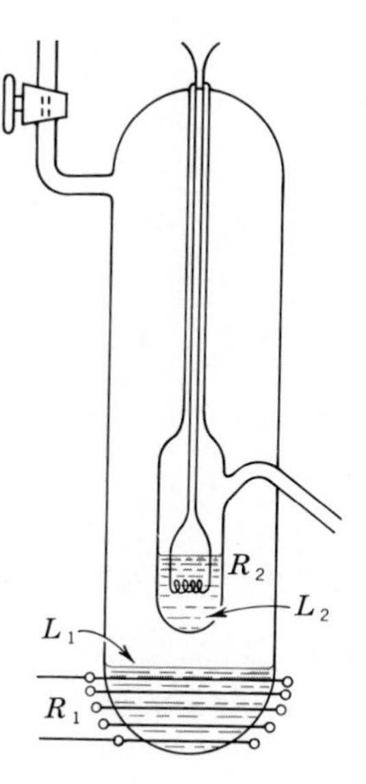

Figure 10-6 Apparatus for measuring heat of vaporization.

Moreover, all the energy supplied by the heater R_2 is used to vaporize L_2, since there is no heat loss between the inner tube and its surroundings. Consequently, if n moles are vaporized in a time τ, the heat of vaporization per mole is

$$l_V = \frac{\mathcal{E} I \tau}{n}.$$

Of much more interest are the cryogenic liquids with normal boiling points in the neighborhood of 100 K or less. For these liquids, one needs the sort of information found in engineering handbooks—that is, the pressure, entropy, enthalpy, and volume of both saturated liquid and saturated vapor at temperatures from the triple point to the critical point. A few such tables are now available, and the heat of vaporization may be obtained by performing the subtraction $h''' - h''$. In Table 10-3, vaporization data for some simple

Table 10-3 Vaporization data†

Substance	T, K	T/T_C	l_V, J/mol	l_V/T_C, J/mol · K	P, kPa	$v''' - v''$, 1/mol	$P(v''' - v'')/T$ J/mol · K
N_2	63.15	0.500	5956	47.18	12.53	41.35	8.20
$T_C = 126.25$ K	77.35	0.613	5536	43.85	101.3	6.042	7.91
$P_C = 3.396$ MPa	94	0.745	4869	38.57	499.5	1.317	7.00
	104	0.824	4292	34.00	1016	0.624	6.10
	111	0.879	3754	29.73	1554	0.374	5.24
	116	0.919	3244	26.70	2047	0.251	4.43
	120	0.950	2681	21.24	2515	0.172	3.60
	124	0.982	1818	14.40	3057	0.097	2.39
Ar	83.78	0.555	6463	42.84	68.75	9.834	8.07
$T_C = 150.86$ K	87.29	0.579	6375	42.26	101.3	6.882	7.99
$P_C = 4.898$ MPa	106	0.703	5760	38.18	507.4	1.523	7.29
	117	0.776	5245	34.77	1022	0.758	6.62
	124	0.822	4825	31.98	1499	0.501	6.06
	130	0.862	4390	29.10	2020	0.352	5.47
	135	0.895	3950	26.18	2545	0.259	4.88
	139	0.921	3539	23.46	3032	0.198	4.32
CO	72.4	0.516	6429	45.85	30.4	19.047	8.00
$T_C = 140.23$ K	81.63	0.582	6040	43.07	101.3	6.325	7.85
$P_C = 3.498$ MPa	99	0.706	5124	36.54	506.5	1.304	6.67
	109	0.777	4490	32.02	1012	0.650	6.04
	115	0.820	4131	29.46	1418	0.450	5.55
	121	0.863	3522	25.12	2026	0.279	4.67
	126	0.899	2802	19.98	2532	0.185	3.72
	130	0.927	1990	14.19	3039	0.109	2.55

† Taken from N. B. Vargaftik, *Tables on the Thermophysical Properties of Liquids and Gases*, Hemisphere Publishing Company, Washington, D.C., 1975.

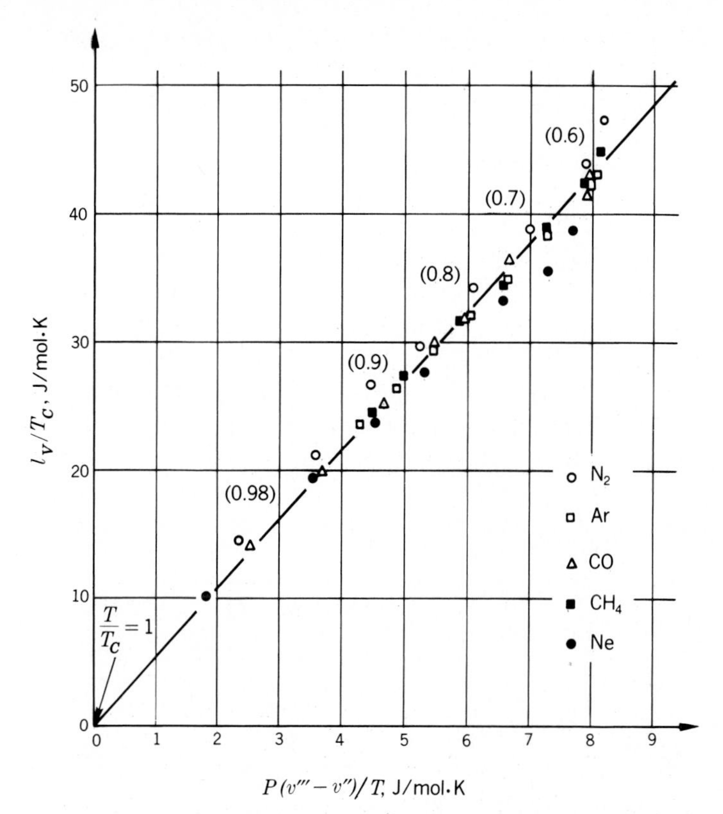

Figure 10-7 A law of corresponding states valid at reduced temperatures between about 0.5 and 1.

liquids, obtained from thermodynamic tables prepared by Vargaftik, are listed. In Fig. 10-7, the heat of vaporization l_V divided by the critical temperature T_C has been plotted against the quantity $P(v''' - v'')/T$ from about $0.5T_C$ to $0.98T_C$. It may be seen that the points for five gases lie on a common straight line, and therefore one is led to assume that similar points for other simple liquids would lie on the same straight line. By the term "simple" we mean liquids such as Kr, Xe, and O_2 whose molecules have no dipole moment (or at least only a small one) and do not associate in the liquid or the vapor phase. Determining the slope of the line in Fig. 10-7 to be 5.4, we may write

$$\frac{l_V/T_C}{P(v''' - v'')/T} = 5.4 \qquad \left(\text{for } 0.5 < \frac{T}{T_C} < 1\right) \tag{10-4}$$

This relation may be regarded as a law of corresponding states. In the present form it is of limited usefulness, however, since it involves a knowledge of so many quantities. It is therefore of interest to examine the consequences of this strange proportionality. We first write Clapeyron's equation in the form

$$\frac{dP/P}{dT/T^2} = \frac{l_V}{P(v''' - v'')/T} = \frac{(l_V/T_C)T_C}{P(v''' - v'')/T}.$$

Notice that the right-hand member is equal to $5.4T_C$. The resulting equation,

$$\frac{dP}{P} = 5.4T_C \frac{dT}{T^2},$$

may be integrated from T to T_C and from P to P_C, provided that T/T_C is not less than 0.5. Hence,

$$\ln \frac{P_C}{P} = 5.4T_C\left(\frac{1}{T} - \frac{1}{T_C}\right),$$

or

$$\ln \frac{P}{P_C} = 5.4\left(1 - \frac{T_C}{T}\right) \qquad \left(\text{for } 0.5 < \frac{T}{T_C} < 1\right). \tag{10-5}$$

Equation (10-5) is a genuine *law of corresponding states*, expressed in terms of reduced temperature and reduced pressure. It was first suggested by E. A. Guggenheim, who plotted the logarithm of the reduced vapor pressure against

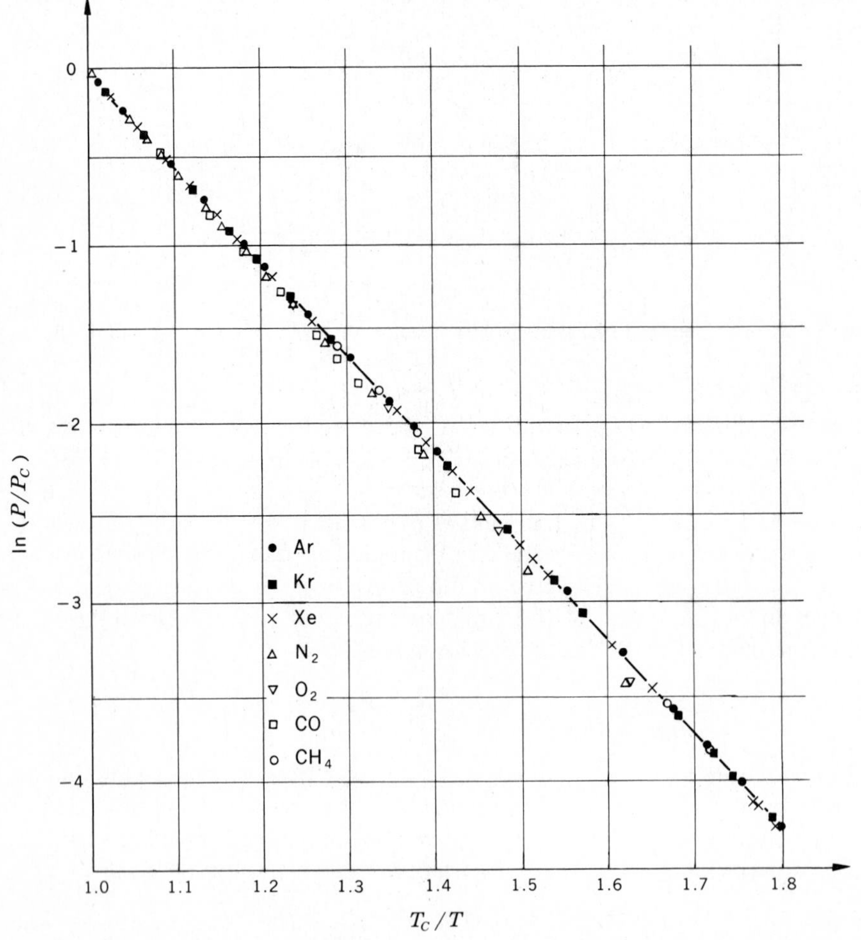

Figure 10-8 A law of corresponding states for simple liquids. *(E. A. Guggenheim, Thermodynamics, Interscience, 1967.)*

the reciprocal of the reduced temperature for seven simple liquids, as shown in Fig. 10-8. The points are seen to fall well along one straight line whose equation is

$$\ln \frac{P}{P_C} = 5.3\left(1 - \frac{T_C}{T}\right) \qquad \left(\text{for } 0.55 < \frac{T}{T_C} < 1\right),$$

and the numerical agreement with Eq. (10-5) is quite satisfactory. It is conceivable that liquids whose molecules have large electric dipole moments and exert unusual forces on one another will obey a law of corresponding states with a different numerical constant.

There is another interesting consequence of the law of corresponding states, as given in Eq. (10-5), which is brought to light if we restrict ourselves to a *small region of temperature* far enough from the critical point to allow us to regard l_V as constant, say, around the normal boiling point. In this region, v'' is negligible compared with v''', and the vapor pressure is small enough to approximate that of an ideal gas, or $v''' = RT/P$. Under these circumstances, Clapeyron's equation becomes

$$\frac{dP}{dT} = \frac{l_V}{RT^2/P},$$

or

$$\frac{l_V}{R} = -\frac{d \ln P}{d(1/T)} = -\frac{d \ln (P/P_C)}{d(1/T)}.$$

Table 10-4 Heat of vaporization at the normal boiling point, l_{VB}†

Liquid	T_B, K	T_C, K	T_B/T_C	l_{VB}, KJ/kmol	l_{VB}/R, K	l_{VB}/RT_B
Ne	27.1	44.5	0.609	2112	254	9.4
N_2	77.3	126	0.613	5583	671	8.7
CO	81.7	133	0.614	6051	728	8.9
F_2	85.2	144	0.592	6046	727	8.5
Ar	87.3	151	0.578	6288	757	8.7
O_2	90.2	154	0.586	6833	822	9.1
CH_4	111	191	0.581	8797	1058	9.5
Kr	120	209	0.574	9812	1180	9.8
Xe	165	290	0.569	12,644	1521	9.2
C_2H_4	175	283	0.601	14,680	1766	10.4
C_2H_6	185	308	0.601	16,241	1953	10.6
HCl	188	325	0.578	16,183	1946	10.4
HBr	206	363	0.567	17.618	2119	10.3
Cl_2	238	417	0.570	18,408	2214	9.3

† Taken from Kuzman Ražnjević, *Handbook of Thermodynamic Tables and Charts*, McGraw-Hill, New York, 1976.

If we integrate this equation over a very small temperature range around T_B, where l_V has the constant value l_{VB}, we get

$$\ln \frac{P}{P_C} = \text{const.} - \frac{l_{VB}}{RT}. \tag{10-6}$$

In Table 10-4, the normal boiling points of 14 simple liquids are listed along with their critical points, and the reduced normal boiling points, T_B/T_C, are seen to lie between 0.57 and 0.61, within the range of the law of corresponding states. Therefore, comparing Eqs. (10-5) and (10-6), we get

$$\frac{l_{VB}}{R} = 5.4T_C. \tag{10-7}$$

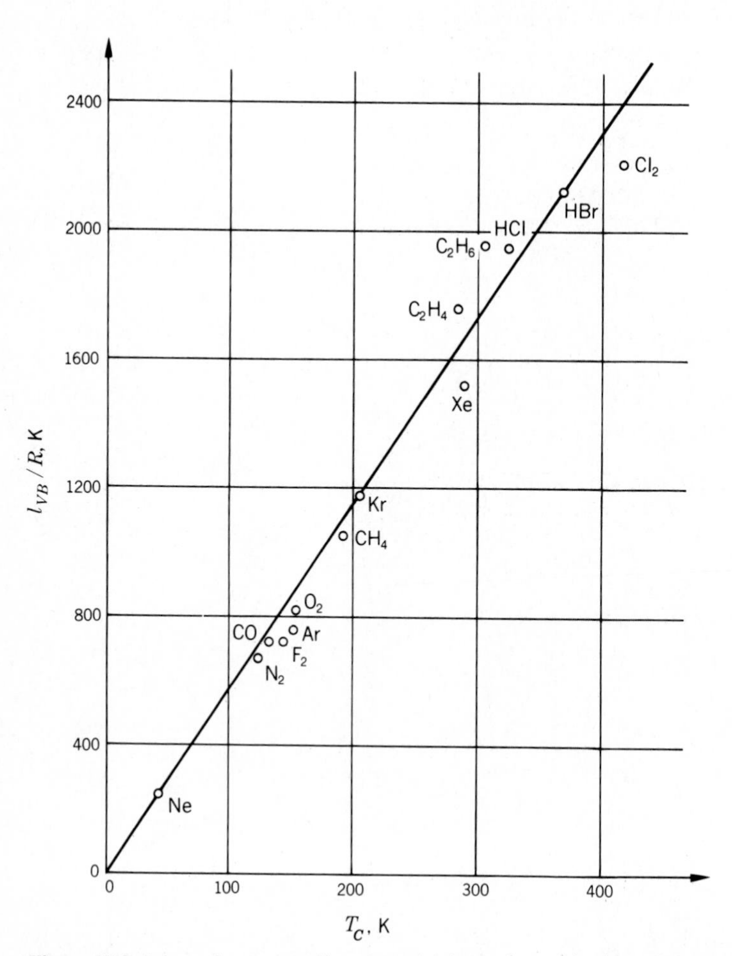

Figure 10-9 A law of corresponding states for heat of vaporization of simple liquids at a reduced temperature of about 0.6.

Using the data in Table 10-4, a graph of l_{VB}/R vs. T_C is shown in Fig. 10-9, and again a single straight line is found for 14 simple liquids. The slope of this line is found to be 5.8, in fair agreement with the expected value. (The agreement would be much better if only the first nine liquids were considered.)

Since the reduced boiling point of many liquids is in the neighborhood of 0.6, Eq. (10-7) may be regarded as a law of corresponding states, with the statement:

$$l_{VB}\left(\text{at } \frac{T}{T_C} \sim 0.6\right) = 5.4RT_C.$$

Referring to Table 10-4, note that the ratio given in the last column, constituting the entropy change due to vaporization at the normal boiling point, is *not* constant but increases with T_B. The increase is small, however, so that a rough approximation is provided by taking l_V/RT_B to be about 9—a working rule known as *Trouton's rule*, which is useful in cases where T_C is not known.

10-4 SUBLIMATION; KIRCHHOFF'S EQUATION

Clapeyron's equation for sublimation is

$$\frac{dP}{dT} = \frac{l_S}{T(v''' - v')},$$

where v''' stands for the molar volume of a vapor and v' of a solid. Sublimation usually takes place at a low pressure, where the vapor may be regarded as an ideal gas, so that

$$v''' \approx \frac{RT}{P}.$$

Since P is small, v''' is large—indeed, so much larger than the molar volume of the solid that v' may be neglected, or

$$v''' - v' \approx v'''.$$

Clapeyron's equation can then be written

$$\begin{aligned} l_S &= R\frac{dP/P}{dT/T^2} \\ &= -R\frac{d \ln P}{d(1/T)} \\ &= -2.30R\frac{d \log P}{d(1/T)}, \end{aligned}$$

from which it can be seen that l_S is equal to $-2.30R$ times the slope of the curve obtained when $\log P$ is plotted against $1/T$. Vapor pressures of solids are usually measured over only a small range of temperature. Within this range, the graph of log P against $1/T$ is practically a straight line, or

$$\log P = -\frac{\text{const.}}{T} + \text{const.}$$

For example, within the temperature interval from 700 to 739 K, the vapor pressure of magnesium satisfies with reasonable accuracy the equation

$$\log P = -\frac{7527}{T} + 8.589.$$

In the case of zinc between 575 and 630 K, the vapor pressure is given by

$$\log P = -\frac{6787}{T} + 8.972.$$

Therefore, from 700 to 739 K the heat of sublimation for magnesium is $2.30R \times 7527 = 144$ kJ/mol; for zinc between 575 and 630 K, $l_S = 2.30R \times 6787 = 130$ kJ/mol. At other temperatures, the heat of sublimation is different. If reliable vapor-pressure data existed over other temperature ranges, the temperature variation of l_S could be obtained. As a rule, however, this is impossible because at low temperatures the vapor pressure of a solid is too small to measure. In the following pages we shall derive *Kirchhoff's equation* for the heat of sublimation at any desired temperature.

An infinitesimal change of molar enthalpy between two states of equilibrium of a chemical system is given by

$$dh = T\,ds + v\,dP.$$

Introducing the second $T\,ds$ equation, we get

$$dh = c_P\,dT + \left[v - T\left(\frac{\partial v}{\partial T}\right)_P\right] dP$$
$$= c_P\,dT + v(1 - \beta T)\,dP.$$

A finite change of enthalpy between the two states $P_i T_i$ and $P_f T_f$ is

$$h_f - h_i = \int_i^f c_P\,dT + \int_i^f v(1 - \beta T)\,dP.$$

Let us apply this equation to a solid whose initial state i' is at zero pressure and at a temperature of absolute zero and whose final state f' is that of a saturated solid (solid about to sublime) represented by a point on the solid-saturation curve below the triple point. These two states are shown on a PVT surface in Fig. 10-10. To calculate the enthalpy change from i' of f', we may integrate along any reversible path from i' to f'. The most convenient is the path represented by the two steps $i' \to A$ and $A \to f'$, the first being isothermal at absolute zero and

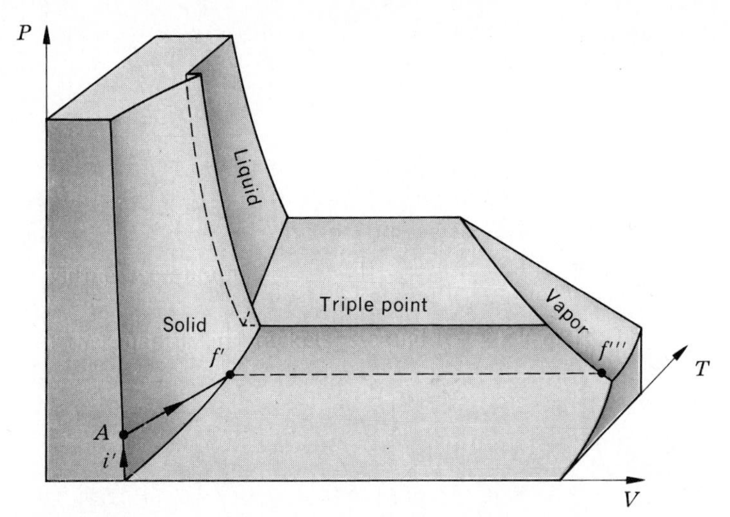

Figure 10-10 Portion of *PVT* surface below the triple point.

the second isobaric at pressure P. Denoting the final enthalpy by h' and the initial by h'_0,

$$h' - h'_0 = \int_{i'}^{A} v(1 - \beta T)\, dP + \int_{A}^{f'} c_P\, dT$$

$$= \int_0^P v'\, dP + \int_0^T c'_P\, dT,$$

where v' is the molar volume of the solid at absolute zero and c'_P the molar heat capacity at a constant pressure P. Now the pressure at all points on the sublimation curve is very small for most solids. For example, for ordinary ice it ranges from 0 to about 650 Pa; for cadmium, from 0 to 15 Pa. Therefore, if we limit the application of this formula to solids at temperatures where the vapor pressure is very small, we may ignore $\int_0^P v'\, dP$, and

$$h' = \int_0^T c'_P\, dT + h'_0. \tag{10-8}$$

Since the c'_P of a solid does not vary appreciably with the pressure, the value of c'_P at atmospheric pressure may be used in the integral above.

The enthalpy of the saturated vapor indicated by point f''' in Fig. 10-10 may be calculated on the basis of the assumption that the saturated vapor at such a low pressure behaves like an ideal gas. Going back to the general equation $c_P = (\partial h/\partial T)_P$ and remembering that the enthalpy of an ideal gas is a function of the temperature only, we have

$$dh''' = c'''_P\, dT.$$

Integrating from absolute zero to T, we have

$$h''' = \int_0^T c_P''' \, dT + h_0''', \tag{10-9}$$

where h_0''' is the molar enthalpy of a saturated vapor at absolute zero.

Consider now the reversible sublimation of 1 mole of a solid at temperature T and pressure P corresponding to the transition from f' to f''' in Fig. 10-10. We have

$$\begin{aligned} l_S &= h''' - h' \\ &= \int_0^T c_P''' \, dT - \int_0^T c_P' \, dT + h_0''' - h_0'. \end{aligned}$$

Since the two integrals approach zero as T approaches zero, it follows that

$$l_S \to h_0''' - h_0' \qquad \text{as } T \to 0,$$

and $h_0''' - h_0'$ *is the heat of sublimation at absolute zero*, which is denoted by l_0. Hence,

$$\boxed{l_S = \int_0^T c_P''' \, dT - \int_0^T c_P' \, dT + l_0.} \tag{10-10}$$

This equation is known as *Kirchhoff's equation.* It is only an approximate equation, being subject to the restrictions that the pressure is low and that the saturated vapor behaves like an ideal gas.

10-5 VAPOR-PRESSURE CONSTANT

If the vapor in equilibrium with a solid is assumed to behave like an ideal gas, and if the volume of the solid is neglected in comparison with that of the vapor, Clapeyron's equation becomes

$$\frac{dP}{P} = \frac{l_S}{RT^2} \, dT.$$

If, in addition to these assumptions, we suppose that the vapor pressure is very small, Kirchhoff's equation may be used. Thus,

$$l_S = l_0 + \int_0^T c_P''' \, dT - \int_0^T c_P' \, dT.$$

The molar heat capacity of an ideal gas can be represented as the sum of a constant term and a term that is a function of the temperature. Thus,

$$c_P''' = c_0''' + c_i''', \tag{10-11}$$

where c_0''' is equal to $\frac{5}{2}R$ for all monatomic gases and to $\frac{7}{2}R$ for all diatomic gases except hydrogen. The factor c_i''' is due to internal degrees of freedom of the vapor; it has the property that it approaches zero rapidly as T approaches zero or when the gas is monatomic. Kirchhoff's equation may therefore be written

$$l_S = l_0 + c_0'''T + \int_0^T c_i''' \, dT - \int_0^T c_P' \, dT;$$

and after substitution in Clapeyron's equation, we get

$$\frac{dP}{P} = \frac{l_0}{RT^2} dT + \frac{c_0'''}{RT} dT + \frac{\int_0^T c_i''' \, dT}{RT^2} dT - \frac{\int_0^T c_P' \, dT}{RT^2} dT.$$

Integrating this equation, we get finally

$$\boxed{\ln P = -\frac{l_0}{RT} + \frac{c_0'''}{R} \ln T + \frac{1}{R}\int_0^T \frac{\int_0^T c_i''' \, dT}{T^2} dT - \frac{1}{R}\int_0^T \frac{\int_0^T c_P' \, dT}{T^2} dT + i,} \tag{10-12}$$

where i is a constant of integration. This relation is not rigorous, but it is accurate enough to be used in conjunction with experimental measurements of the vapor pressure of solids. Such measurements are usually attended by errors that are much greater than those brought about by the simplifying assumptions introduced in the derivation.

If the vapor in equilibrium with a solid is monatomic, c_0''' has the value of $\frac{5}{2}R$ and c_i''' is zero. The equation of the sublimation curve then becomes

$$\ln P = -\frac{l_0}{RT} + \tfrac{5}{2} \ln T - \frac{1}{R}\int_0^T \frac{\int_0^T c_P' \, dT}{T^2} dT + i. \tag{10-13}$$

Changing to common logarithms and expressing the pressure in atmospheres, we get

$$\log P = -\frac{l_0}{2.30RT} + \tfrac{5}{2} \log T - \frac{1}{2.30R}\int_0^T \frac{\int_0^T c_P' \, dT}{T^2} dT + \frac{i}{2.30} - \log 1{,}013{,}250.$$

The last two terms are known as the *practical vapor-pressure constant* i'. Thus,

$$i' = \frac{i}{2.30} - \log 1{,}013{,}250$$

$$= \frac{i}{2.30} - 6.0052.$$

Finally, expressing the pressure in millimeters, introducing the numerical values

$$2.30R = 19.1 \text{ J/mol} \cdot \text{deg}$$

and

$$\log 760 = 2.881,$$

and calling

$$B = \frac{1}{2.30R} \int_0^T \frac{\int_0^T c'_P \, dT}{T^2} \, dT,$$

the equation becomes

$$\log P = -\frac{l_0}{19.1T} + \tfrac{5}{2} \log T - B + i' + 2.881.$$

This is the form in which the equation is most useful to the physicist or chemist in the laboratory.

The sublimation equation is used in two ways: (1) to obtain experimental measurements of the vapor-pressure constant i', which are to be compared with theoretical calculations of i'; and (2) to calculate the vapor pressure of a substance at temperatures at which P is too small to measure. In both cases, the integral B must be evaluated on the basis either of experimental measurements of c'_P or of theoretical values for c'_P. In order to do this, c'_P is plotted against T from absolute zero to as high a temperature as is necessary. The area under the curve at various values of T is obtained by numerical integration, and the temperature variation of $\int_0^T c'_P \, dT$ is thus obtained. These values are now divided by T^2 and plotted on another graph against T. The area under this new curve at various values of T provides, finally, the temperature variation of B.

If measurements of the vapor pressure exist over a wide temperature range, the numerical value of $\log P - \frac{5}{2} \log T + B$ is plotted against $1/T$. Since

$$\log P - \tfrac{5}{2} \log T + B = -\frac{l_0}{19.1} \frac{1}{T} + i' + 2.881,$$

the resulting graph is a straight line whose

$$\text{Slope} = -\frac{l_0}{19.1},$$

and

$$\text{Intercept} = i' + 2.881.$$

Table 10-5 Data for the determination of i' of cadmium

T, K	$\log P$	$\frac{5}{2} \log T$	B	$\log P - \frac{5}{2} \log T + B$	$1/T$
360	−7.44	6.38	1.82	−12.00	0.00278
380	−6.57	6.45	1.88	−11.14	0.00263
400	−5.80	6.50	1.94	−10.36	0.00250
450	−4.17	6.63	2.08	− 8.72	0.00222
500	−2.86	6.75	2.20	− 7.41	0.00200
550	−1.77	6.85	2.32	− 6.30	0.00182
594	−0.99	6.94	2.41	− 5.52	0.00168

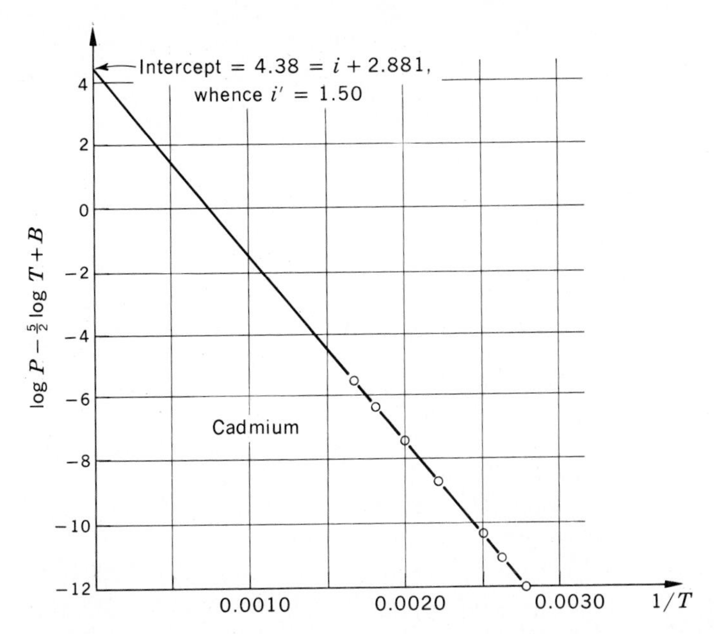

Figure 10-11 Graph for determination of vapor-pressure constant of cadmium.

The data for cadmium are listed in Table 10-5. The vapor-pressure measurements were made by Egerton and Raleigh, and the measurements of B by Lange and Simon. From the graph shown in Fig. 10-11, l_0 is found to be 112 kJ/mol, and i' to be 1.50.

10-6 MEASUREMENT OF VAPOR PRESSURE

The determination of the heat of sublimation at absolute zero l_0 and the vapor pressure constant i' require accurate measurements of the vapor pressure of solids. It was seen in Art. 10-3 that the most convenient method of determining the heat of vaporization of a liquid also requires measuring the vapor pressure. It is no wonder that such measurements have engaged the attention of physicists and chemists for many years and still constitute a very active branch of modern thermophysics. Let us consider just a few methods that are commonly used:

1. *Static method.* When the vapor pressure is within the range from about 10^{-3} to 10^3 mm Hg (10^{-1} to 10^5 Pa), the vessel containing the solid or liquid is connected to a liquid column manometer, and the pressure is obtained directly. There are at least a dozen manometers that are more sensitive (e.g., McLeod gauge, Bourdon gauge, hot-wire, ionization), which enable smaller pressures down to about 10^{-8} mm Hg (10^{-6} Pa) to be measured. The static

method usually is quite adequate for liquids but often is of little value in measuring the vapor pressure for high-melting-point solids.

2. *Langmuir's evaporation method.* The solid whose vapor pressure is to be measured is carefully weighed, and its surface area A is determined. It is then placed in an evacuated enclosure and raised to the desired temperature. It evaporates at a constant rate, with the vapor being drawn off. The assumption is made that the rate of evaporation is equal to the rate at which vapor molecules would strike the solid if there were equilibrium between the solid and the vapor. This rate was shown in Chap. 5 (Prob. 5-5e) to be equal to $P/\sqrt{2\pi mkT}$; hence, the rate of loss of mass per unit area of surface $\dot{M}/A$ is

$$\frac{\dot{M}}{A} = P\sqrt{\frac{\mathcal{M}}{2\pi RT}} \qquad \text{or} \qquad P = \frac{\dot{M}}{A}\sqrt{\frac{2\pi RT}{\mathcal{M}}},$$

where $\mathcal{M}$ is the molecular weight. This method is very useful when the substance is a high-melting-point wire.

3. *Knudsen's effusion method.* This is a variation of the Langmuir method in which the weight and area of the solid do not have to be measured. Instead, the evaporating vapor is allowed to pass through an opening of known area and is then condensed in a cold trap. A measurement of the mass of condensed vapor after a time interval provides the quantity $\dot{M}$.

There are many very ingenious methods involving optical absorption, rotation of the direction of vibration of linearly polarized light, measurement of radioactivity, isotope exchange, etc., but these are of interest only to research workers in this field. None of the methods is as simple as it may sound, and all are attended by errors that may go as high as 5 or 10 percent. A critical account of all this work, along with extensive tables, is given in a fine book by A. N. Nesmeyanov (*Vapor Pressure of the Chemical Elements*, Academic Press, New York, 1963).

The complete temperature dependence of vapor pressure requires a formula with four adjustable constants. Many formulas have been suggested, but the one found most satisfactory by Nesmeyanov is

$$\log P = A - \frac{B}{T} + CT + D \log T. \tag{10-14}$$

Values of the constants and of l_0 are given in his book.

A curious regularity has been found between l_0 and the limit (as $T \to 0$) of the ratio c_V/β. In Table 10-6, values of these two quantities are given for 20 metallic elements. The graph drawn in Fig. 10-12 shows that

$$l_0 \approx \frac{3}{5}\left(\frac{c_V}{\beta}\right)_0. \tag{10-15}$$

Table 10-6 Ratio of heat capacity to expansivity and heat of sublimation of metals

[$(c_V/\beta)_0$ is due to R. K. Kirby; l_0 is due to A. N. Nesmeyanov]

Metal	$(c_V/\beta)_0$, kJ/mol	l_0, kJ/mol	Metal	$(c_V/\beta)_0$, kJ/mol	l_0, kJ/mol
Ag	470	284	Na	120	108
Al	374	312	Nb	1150	722
Au	632	367	Pb	332	197
Be	513	321	Pd	725	382
Cd	302	112	Pt	964	555
Cu	500	338	Sn	507	302
In	343	238	Ta	1280	780
Fe	700	436	Th	835	470
Li	177	160	Ti	943	472
Mg	340	146	Zn	288	131

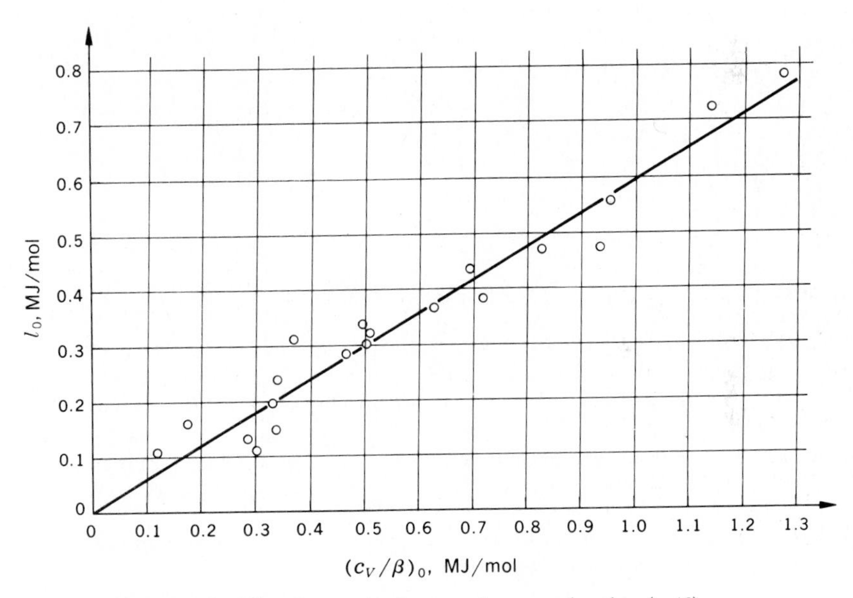

Figure 10-12 Heat of sublimation at absolute zero is proportional to $(c_V/\beta)_0$.

PROBLEMS

10-1 Saturated liquid carbon dioxide at a temperature of 293 K and a pressure of 5.72×10^6 Pa undergoes throttling to a pressure of 1.01×10^5 Pa. The temperature of the resulting mixture of solid and vapor is 195 K. What fraction is vaporized? (The enthalpy of saturated liquid at the initial state is 24,200 J/mol, and the enthalpy of saturated solid at the final state is 6750 J/mol. The heat of sublimation at the final state is 25,100 J/mol.)

10-2 Prove that, during a phase transition of the first order, (*a*) the entropy of the entire system is a linear function of the total volume and (*b*) the energy change is given by

$$\Delta U = L\left(1 - \frac{d \ln T}{d \ln P}\right).$$

10-3 When lead is melted at atmospheric pressure, the melting point is 600 K, the density decreases from 11.01 to 10.65 g/cm^3, and the latent heat of fusion is 24.5 J/g. What is the melting point at a pressure of 1.01×10^7 Pa?

10-4 Water at its freezing point (T_i, P_i) completely fills a strong steel container. The temperature is reduced to T_f at constant volume, with the pressure rising to P_f.

(*a*) Show that the fraction y of water that freezes is given by

$$y = \frac{v''_f - v''_i}{v''_f - v'_f}.$$

(*b*) State explicitly the simplifying assumptions that must be made in order that y may be written

$$y = \frac{v''[\beta''(T_f - T_i) - \kappa''(P_f - P_i)]}{v''_f - v'_f}.$$

(*c*) Calculate y for $i = 0°\text{C}, 1.01 \times 10^5$ Pa; $f = -5°\text{C}, 5.98 \times 10^7$ Pa; $\beta'' = -67 \times 10^{-6}\ \text{K}^{-1}$; $\kappa'' = 12.04 \times 10^{-11}\ \text{Pa}^{-1}$; $v''_f - v'_f = -1.02 \times 10^{-4}\ \text{m}^3/\text{kg}$.

10-5 (*a*) Prove that, for a single phase,

$$\left(\frac{\partial P}{\partial T}\right)_S = \frac{c_P}{Tv\beta}.$$

(*b*) Calculate $(\partial P/\partial T)_S$ for ice at $-3°$C, where $c_P = 2.01$ kJ/kg · K, $v = 1.09 \times 10^{-3}$ m^3/kg and $\beta = 1.58 \times 10^{-4}\ \text{K}^{-1}$.

(*c*) Ice is originally at $-3°$C and 1.01×10^5 Pa. The pressure is increased adiabatically until the ice reaches the melting point. At what temperature and pressure is this melting point? [*Hint:* At what point does a line whose slope is $(\partial P/\partial T)_S$ cut a line whose slope is that of fusion curve, -1.35×10^7 Pa/K?]

10-6 Figure P10-1 shows a thermodynamic surface for water viewed from the high-temperature end. Consider 1 kg of ice in the state i ($P_i = 1.01 \times 10^5$ Pa, $T_i = 273$ K). If the ice undergoes an isentropic compression to a state f:

(*a*) Why is the state f in the mixture region? In other words, why does some of the ice melt?

(*b*) Show that the fraction x of ice that is melted is given by

$$x = \frac{s'_f - s'_i}{s''_f - s'_f}.$$

(*c*) State explicitly the simplifying assumptions that must be made in order that x may be written

$$x = -\frac{c'_P(T_f - T_i) - T_f v'\beta'(P_f - P_i)}{(l_F)_f}.$$

(*d*) Calculate x when $T_f = 272$ K, $P_f = 1.35 \times 10^7$ Pa, $c'_P = 2.01$ kJ/kg · K, $v' = 1.09 \times 10^{-3}$ m^3/kg, $\beta' = 1.58 \times 10^{-4}\ \text{K}^{-1}$, $(l_F)_f = 331$ kJ/kg.

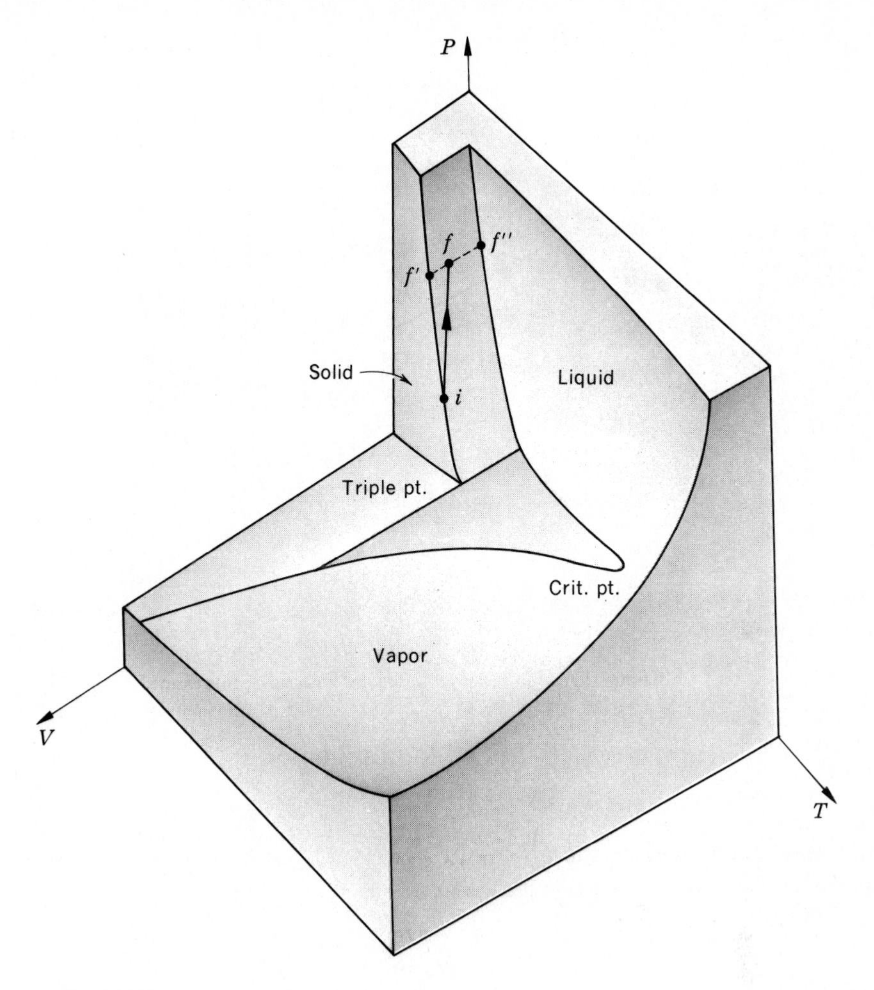

Figure P10-1

10-7 A steel bar in the form of a rectangular parallelepiped of height a and breadth b is embedded in a cake of ice as shown in Fig. P10-2. With the aid of an external magnetic field, a constant force $\mathcal{F}$ is exerted downward on the bar. The whole system is at 0°C.

(*a*) Show that the *decrease* in temperature of the ice directly under the bar is

$$\Delta T = \frac{\mathcal{F}T(v' - v'')}{bcl_F}.$$

(*b*) Ice melts (see Prob. 10-6) under the bar, and all the water thus formed is forced to the top of the bar, where it refreezes. This phenomenon is known as *regelation*. Heat is therefore liberated above the bar, is conducted through the metal and a layer of water under the metal, and is absorbed by the ice under the layer of water. Show that the speed with which the bar sinks through the ice is

$$\frac{dy}{d\tau} = \frac{U'T(v' - v'')\mathcal{F}}{\rho l_F^2 bc},$$

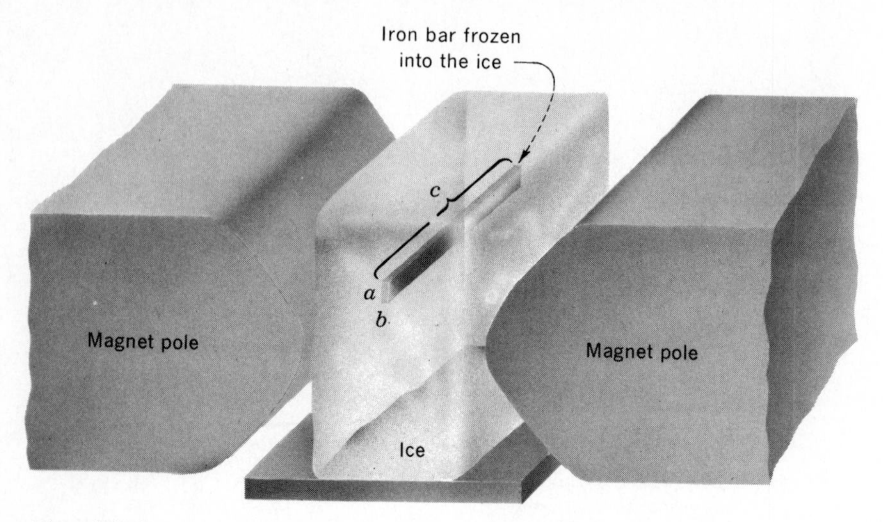

Figure P10-2

where U' is the overall heat-transfer coefficient of the composite heat-conducting path consisting of the metal and the water layer. U' is given by

$$\frac{1}{U'} = \frac{x_m}{K_m} + \frac{x_w}{K_w},$$

where x_m and x_w are the thicknesses of metal and water layer, respectively, and K_m and K_w are their respective thermal conductivities.

(*c*) Assuming the water layer to have a thickness of about 10^{-5} m and a thermal conductivity of about 0.6 W/m · K, and the bar to be 0.1 m long, with a and b each equal to 10^{-3} m, with what speed will the bar descend when $\mathcal{F} = 10^2$ N? (Thermal conductivity of steel is 60 W/m · K.)

10-8 Some of the properties of saturated water and steam near the critical point are given in the accompanying table:

T, K	P, 10^5 Pa	v'' 10^3 m^3/kg	v''', 10^3 m^3/kg	h'', kJ/kg	h''', kJ/kg
473	15.551	1.1565	127.14	852.4	2791.4
523	39.776	1.2513	50.02	1085.8	2799.5
573	85.917	1.4041	21.62	1345.4	2748.4
623	165.37	1.7407	8.822	1672.9	2566.1
647	221.145	3.147	3.147	2095.2	2095.2

Plot l_V/T_C against $(Pv''' - v'')/T$, determine the slope, and compare it with that of Fig. 10-7.

10-9 In Fig. P10-3 several examples of two neighboring phase transitions—one at (T, P) and the other at $(T + dT, P + dP)$—are shown on an ST diagram. The heat capacity at constant saturation is defined as

$$c_{\text{sat}}^{(i)} = T\frac{ds^{(i)}}{dT},$$

and

$$c_{\text{sat}}^{(f)} = T\frac{ds^{(f)}}{dT}.$$

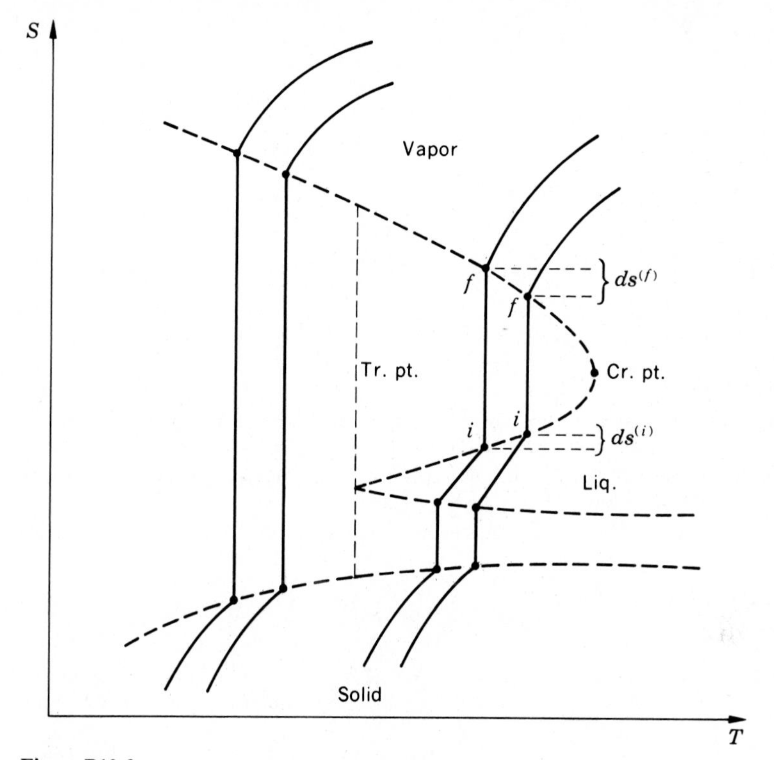

Figure P10-3

Show that:

(a) $$\frac{dl}{dT} - \frac{l}{T} = c_{\text{sat}}^{(f)} - c_{\text{sat}}^{(i)}.$$

(b) $$c_{\text{sat}} = c_P - Tv\beta\frac{dP}{dT}.$$

(c) $$\frac{dl}{dT} = c_P^{(f)} - c_P^{(i)} + l\left(\frac{1}{T} - \frac{(v\beta)^{(f)} - (v\beta)^{(i)}}{v^{(f)} - v^{(i)}}\right).$$

10-10 Consider two neighboring vaporizations—one at (T, P) and the other at $(T + dT, P + dP)$. Assuming that $v''' \gg v''$ and $v''' = RT/P$, show that the *expansivity at constant saturation* of the vapor is

$$\frac{1}{v'''}\left(\frac{dv'''}{dT}\right)_{\text{sat}} = \frac{1}{T}\left(1 - \frac{l_V}{RT}\right).$$

10-11 Using the van der Waals equation of state

$$P = \frac{RT}{v - b} - \frac{a}{v^2},$$

show that

$$P_C = \frac{a}{27b^2}, \qquad v_C = 3b, \qquad T_C = \frac{8a}{27bR},$$

and calculate $RT_C/P_C v_C$.

10-12 Using the Dieterici equation of state

$$P = \frac{RT}{v - b} e^{-a/RTv},$$

show that

$$P_C = \frac{a}{4e^2b^2}, \qquad v_C = 2b, \qquad T_C = \frac{a}{4Rb},$$

and calculate $RT_C/P_C v_C$.

10-13 Iodine crystals have an atomic weight of 127 kg/kmol and a specific heat of 0.197 kJ/kg · K. Iodine vapor may be assumed to be an ideal diatomic gas with a constant c_P. At 301 K, the vapor pressure is 51.5 N/m^2; at 299 K, it is 43.5 N/m^2. Compute the latent heat of sublimation (*a*) at 300 K, (*b*) at absolute zero, and (*c*) at 200 K.

10-14 The heat of sublimation of zinc at 600 K is found to be 130 kJ/mol. The heat capacity of solid zinc is such that

$$\int_0^{600} c'_P \, dT = 13{,}800 \text{ J/mol},$$

and B at 600 K is 1.96. The vapor pressure of zinc at 600 K is 4.67×10^{-3} mm Hg. Assuming zinc vapor to be an ideal gas, calculate (*a*) the heat of sublimation at absolute zero and (*b*) the vapor-pressure constant.

10-15 The vapor pressure in Pa of solid ammonia is given by $\ln P = 27.92 - 3754/T$, and that of liquid ammonia by $\ln P = 24.38 - 3063/T$.

(*a*) What is the temperature of the triple point?

(*b*) What are the three latent heats at the triple point?

10-16 The triple point of hydrogen is at $T_{TP} = 14$ K. The density of solid H_2 at this point is 81 kg/m^3; that of liquid H_2 is 71 kg/m^3. The vapor pressure of the liquid is given by

$$\ln P(\text{atm}) \approx 6.8 - \frac{122}{T} - 0.3 \ln T,$$

and the melting temperature is given by

$$T_M = 14 + \frac{P(\text{atm})}{33}.$$

(*a*) Compute the three latent heats at the triple point in terms of R, to within 5 percent.

(*b*) Compute the slope of the vapor pressure curve of the solid at T_{TP}.

10-17 The intensity I of the atomic beam emerging from a narrow slit in an oven containing a solid in equilibrium with its vapor at the temperature T will be shown in Prob. 11-10

$$I = \frac{P}{\sqrt{2\pi mkT}},$$

where m is the atomic mass, k is Boltzmann's constant, and P is the vapor pressure. If the molar latent heat is l, the molar volume of the solid is negligible, and the vapor behaves like an ideal gas, show that

$$\frac{1}{I}\frac{dI}{dT} = \frac{1}{T}\left(\frac{l}{RT} - \frac{1}{2}\right).$$

PART TWO

APPLICATIONS OF FUNDAMENTAL CONCEPTS

CHAPTER

ELEVEN

STATISTICAL MECHANICS

11-1 FUNDAMENTAL PRINCIPLES

In the treatment of kinetic theory given in Art. 5-10, the molecules of an ideal gas could not be regarded as completely independent of one another, for then they could not arrive at an equilibrium distribution of velocities. It was therefore assumed that interaction *did* take place, but only during collisions with other molecules and with the walls. To describe this limited form of interaction we refer to the molecules as "weakly interacting" or "quasi-independent." The treatment of strongly interacting particles is beyond the scope of the present discussion.

The molecules of an ideal gas have another characteristic besides their quasi-independence. They are *indistinguishable*, because they are not localized in space. It was emphasized in Art. 5-10 that the molecules have neither a preferred location nor a preferred velocity. The particles occupying regular lattice sites in a crystal are distinguishable, however, because they are constrained to oscillate about fixed positions; therefore one particle can be distinguished from its neighbors by its location. The statistical treatment of an ideal crystal as a number of distinguishable, quasi-independent particles will be given in the next chapter. In this chapter we shall confine our attention to the indistinguishable, quasi-independent particles of an ideal gas.

Suppose that a monatomic ideal gas consists of N particles, where N, as usual, is an enormous number—say, about 10^{20}. Let the gas be contained in a cubical enclosure whose edge has a length L, and let the energy ϵ of any particle, as a first step, be entirely kinetic energy of translation. In the x direction,

$$\epsilon_x = \tfrac{1}{2}m\dot{x}^2 = \frac{(m\dot{x})^2}{2m} = \frac{p_x^2}{2m},$$

where p_x is the x component of the momentum. If the particle is assumed to move freely back and forth between two planes a distance L apart, the simplest form of quantum mechanics provides that, in a complete cycle (from one wall to the other and back again, or a total distance of $2L$), the constant momentum p_x multiplied by the total path $2L$ is an integer n_x times Planck's constant h. Thus,

$$p_x \cdot 2L = n_x h.$$

Substituting this result into the previous equation, we get

$$\epsilon_x = n_x^2 \frac{h^2}{8mL^2},$$

or

$$n_x = \frac{L}{h}\sqrt{8m\epsilon_x}.$$

The allowed values of kinetic energy ϵ_x are *discrete*, corresponding to integer values of n_x; but when n_x changes by unity, the corresponding change in ϵ_x is very small, because n_x itself is exceedingly large. To see that a typical value of n_x is very large, consider a cubical box containing gaseous helium at 300 K, whose edge is, say, 10 cm. It was shown at the end of Chap. 5 [Eq. (5-22)] that the average energy of an ideal gas is $\frac{3}{2}kT$. Since a molecule has 3 degrees of freedom, and there is no preferred direction, it follows that the average energy associated with each translational degree of freedom is $\frac{1}{2}kT$. Then,

$$\begin{aligned}\epsilon_x = \tfrac{1}{2}kT &= \tfrac{1}{2} \times 1.4 \times 10^{-23}\,\frac{\text{J}}{\text{K}} \times 300\ \text{K}\\ &= 2.1 \times 10^{-21}\ \text{J},\end{aligned}$$

and

$$\begin{aligned}n_x &= \frac{0.1\ \text{m}}{6.6 \times 10^{-34}\ \text{J}\cdot\text{s}}\sqrt{8 \times 6.6 \times 10^{-27}\ \text{kg} \times 2.1 \times 10^{-21}\ \text{J}}\\ &= \frac{0.1 \times 10.5 \times 10^{-24}}{6.6 \times 10^{-34}}\\ &\approx 10^9.\end{aligned}$$

Therefore, the change of energy when n_x changes by unity is so small that, for most practical purposes, the energy may be assumed to vary continuously. This will be of advantage later on, when it will be found useful to replace a sum by an integral.

Table 11-1 $n_x^2 + n_y^2 + n_z^2 = 66$

	1	2	3	4	5	6	7	8	9	10	11	12
n_x	8	1	1	7	1	4	7	4	1	5	5	4
n_y	1	8	1	4	7	1	1	7	4	5	4	5
n_z	1	1	8	1	4	7	4	1	7	4	5	5

Taking into account the three components of momentum, we get for the total kinetic energy of a particle

$$\epsilon = \frac{p_x^2 + p_y^2 + p_z^2}{2m} = \frac{h^2}{8mL^2}(n_x^2 + n_y^2 + n_z^2). \tag{11-1}$$

The specification of an integer for each n_x, n_y, and n_z is a specification of a *quantum state* of a particle. All states characterized by values of the n's such that $n_x^2 + n_y^2 + n_z^2 = \text{const.}$ will have the same energy. To use an example given by Guggenheim,† the states corresponding to the values of n_x, n_y, and n_z in Table 11-1 all have the energy $\epsilon = 66h^2/8mL^2$. There are twelve quantum states associated with the same *energy level*, and we therefore refer to this energy level as having a *degeneracy* of 12. In any actual case, $n_x^2 + n_y^2 + n_z^2$ is an enormous number, so that the degeneracy of an actual energy level is extremely large.

However close they may be, there is still only a discrete number of energy levels for the molecules of an ideal gas. It is the fundamental problem of statistical mechanics to determine, at equilibrium, the *populations* of these energy levels—that is, the number of particles N_1 having energy ϵ_1, the number N_2 having energy ϵ_2, and so on. It is a simple matter to show that (see Prob. 11-2) the number of quantum states g_i corresponding to an energy level i (the degeneracy of the level) is very much larger than the number of particles occupying that level. Thus,

$$g_i \gg N_i. \tag{11-2}$$

It is very unlikely, therefore, that more than one particle will occupy the same quantum state at any one time.

At any one moment, some particles are moving rapidly and some slowly, so that the particles are distributed among a large number of different quantum states. As time goes on, the particles collide with one another and with the walls, or emit and absorb photons, so that each particle undergoes many changes from one quantum state to another. The fundamental assumption of statistical mechanics is that *all quantum states have equal likelihood of being occupied*. The

† E. A. Guggenheim, *Boltzmann's Distribution Law*, Interscience Publishers, Inc., New York, 1955.

1	2	3	4	5	6	7	8	9	10	11	12	13	14	15	16
	A					*B*						*C*			
	A					*C*						*B*			
	B					*A*						*C*			
	B					*C*						*A*			
	C					*A*						*B*			
	C					*B*						*A*			

Figure 11-1 There are six ways in which three *distinguishable* particles *A*, *B*, and *C* can occupy three given quantum states.

probability that a particle may find itself in a given quantum state is the same for all states.

Now consider the N_i particles in any of the g_i quantum states associated with the energy ϵ_i. Any one particle would have g_i choices in occupying g_i different quantum states. A second particle would have the same g_i choices, and so on. The total number of ways in which N_i *distinguishable* particles could be distributed among g_i quantum states would therefore be $g_i^{N_i}$. But the quantity $g_i^{N_i}$ is much too large, since it holds for distinguishable particles such as *A*, *B*, and *C* in Fig. 11-1. This figure shows six different ways in which three distinguishable particles can occupy quantum states 2, 7, and 13. If the particles had no identity, there would be only one way to occupy these particular quantum states. That is, one must divide by 6, which is 3! The number of permutations of N_i distinguishable objects is N_i! If the quantity $g_i^{N_i}$ is divided by this factor, the resulting expression will then hold for indistinguishable particles. Therefore,

$$\left\{\begin{array}{l}\text{No. of ways that } N_i \textit{ indistinguishable} \\ \text{particles can be distributed among } g_i \\ \text{quantum states}\end{array}\right\} = \frac{g_i^{N_i}}{N_i!}. \tag{11-3}$$

It should be pointed out that the N indistinguishable, quasi-independent particles were assumed to be contained within a cubical box only for the sake of simplicity. A rectangular box with three different dimensions could easily have been chosen, in which case Eq. (11-3) would be unchanged.

11-2 EQUILIBRIUM DISTRIBUTION

We have seen that, in the case of an ideal gas, there are many quantum states corresponding to the same energy level and that the degeneracy of each level is much larger than the number of particles which would be found in any one level at any one time. The specification, at any one moment, that there are

N_1 particles in energy level ϵ_1 with degeneracy g_1
N_2 particles in energy level ϵ_2 with degeneracy g_2

$\vdots$

N_i particles in energy level ϵ_i with degeneracy g_i

$\vdots$

in a container of volume V when the gas has a total number of particles N and an energy U is a description of a *macrostate* of the gas. The number of ways Ω in which this macrostate may be achieved is given by a product of terms of the type of Eq. (11-3), or

$$\boxed{\Omega = \frac{g_1^{N_1}}{N_1!}\frac{g_2^{N_2}}{N_2!}\cdots.} \tag{11-4}$$

The quantity Ω is called the *thermodynamic probability* of the particular macrostate. Other names for this quantity are the *number of microstates* and the *number of complexions*. Whatever its name, the larger Ω is, the greater the probability of finding the system of N particles in this state. It is assumed that, if V, N, and U are kept constant, *the equilibrium state of the gas will correspond to that macrostate in which Ω is a maximum*. To find the equilibrium populations of the energy levels, therefore, we look for the values of the individual N's that render Ω—or more simply, $\ln \Omega$—a maximum.

Since $\ln \Omega$ contains factorials of large numbers, it is convenient to use Stirling's approximation, which may be derived in the following way: the natural logarithm of factorial x is

$$\ln (x!) = \ln 2 + \ln 3 + \cdots + \ln x.$$

If we draw a series of steps on a diagram, as shown in Fig. 11-2, where the integers are plotted along the x axis and $\ln x$ along the y axis, the area under each step is exactly equal to the natural logarithm, since the width of each step equals unity. The area under the steps from $x = 1$ to $x = x$ is therefore $\ln (x!)$. When x is large, we may replace the steps by a smooth curve, shown as a dashed curve in Fig. 11-2; therefore, approximately, when x is large,

$$\ln (x!) \approx \int_1^x \ln x \, dx.$$

Integrating by parts, we get

$$\ln (x!) \approx x \ln x - x + 1.$$

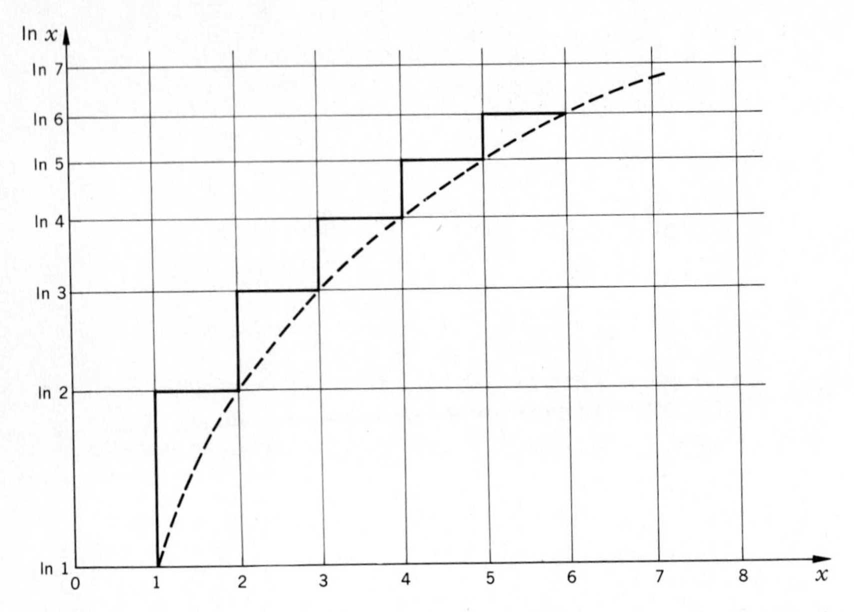

Figure 11-2 The area under the dashed curve approximates the area under the steps (ln 2 + ln 3 + ln 4 + ⋯ + ln x) when x is large.

Then, if we neglect 1 compared with x,

$$\boxed{\ln (x!) \approx x \ln x - x.} \tag{11-5}$$

This formula is *Stirling's approximation*.

Using Stirling's approximation in Eq. (11-4), we get

$$\begin{aligned}\ln \Omega &= N_1 \ln g_1 - N_1 \ln N_1 + N_1 + N_2 \ln g_2 - N_2 \ln N_2 + N_2 + \cdots \\ &= \sum N_i \ln g_i - \sum N_i \ln N_i + \sum N_i,\end{aligned}$$

or

$$\ln \Omega = \sum N_i \ln \frac{g_i}{N_i} + N, \tag{11-6}$$

where we have used the fact that $\sum N_i = N$. Our problem now is to render ln Ω a maximum subject to the conditions that

$$\sum N_i = N = \text{const.}, \tag{11-7}$$

$$\sum N_i \epsilon_i = U = \text{const.} \tag{11-8}$$

Before we proceed to solve this problem by the method of Lagrange multipliers, it is important to bear in mind that the ϵ's and g's are constants. The only variables are the populations of the energy levels, and their sum N is constant.

Since $dN = 0$, the differential of $\ln \Omega$ is

$$d \ln \Omega = \sum d\left(N_i \ln \frac{g_i}{N_i}\right) = \sum \ln \frac{g_i}{N_i}\, dN_i + \sum \frac{N_i^2}{g_i}\left(-\frac{g_i}{N_i^2}\right) dN_i$$

$$= \sum \ln \frac{g_i}{N_i}\, dN_i - d \sum N_i$$

$$= \sum \ln \frac{g_i}{N_i}\, dN_i. \qquad (11\text{-}9)$$

Setting the differential of $\ln \Omega$ equal to zero and taking the differential of Eqs. (11-7) and (11-8), we get

$$\ln \frac{g_1}{N_1}\, dN_1 + \ln \frac{g_2}{N_2}\, dN_2 + \cdots + \ln \frac{g_i}{N_i}\, dN_i + \cdots = 0,$$

$$dN_1 + dN_2 + \cdots + dN_i + \cdots = 0,$$

and

$$\epsilon_1\, dN_1 + \epsilon_2\, dN_2 + \cdots + \epsilon_i\, dN_i + \cdots = 0.$$

Multiplying the second equation by $\ln A$ and the third by $-\beta$, where $\ln A$ and $-\beta$ are Lagrange multipliers (see Appendix B), we get

$$\ln \frac{g_1}{N_1}\, dN_1 + \ln \frac{g_2}{N_2}\, dN_2 + \cdots + \ln \frac{g_i}{N_i}\, dN_i + \cdots = 0,$$

$$\ln A\, dN_1 + \ln A\, dN_2 + \cdots + \ln A\, dN_i + \cdots = 0,$$

and

$$-\beta\epsilon_1\, dN_1 - \beta\epsilon_2\, dN_2 - \cdots - \beta\epsilon_i\, dN_i - \cdots = 0.$$

If we add these equations, the coëfficient of each dN may be set equal to zero. Taking the ith term,

$$\ln \frac{g_i}{N_i} + \ln A - \beta\epsilon_i = 0,$$

or

$$\ln \frac{N_i}{g_i} - \ln A = -\beta\epsilon_i,$$

or

$$N_i = A g_i e^{-\beta\epsilon_i}. \qquad (11\text{-}10)$$

The population of any energy level at equilibrium is therefore seen to be proportional to the degeneracy of the level and to vary exponentially with the energy of the level.

The next step is to determine the physical significance of the Lagrange multipliers A and β.

11-3 SIGNIFICANCE OF A AND β

The population N_i of the ith energy level is given by

$$N_i = Ag_i e^{-\beta\epsilon_i}.$$

Summing over all the energy levels, we get

$$\sum N_i = A \sum g_i e^{-\beta\epsilon_i},$$

and

$$A = \frac{N}{\sum g_i e^{-\beta\epsilon_i}}. \tag{11-11}$$

The sum in the denominator plays a fundamental role in statistical mechanics. It was first introduced by Boltzmann, who called it the *Zustandsumme*, or "sum over states." We retain the first letter of *Zustandsumme* as a mathematical symbol, but the accepted English expression for this sum is the *partition function.* Thus,

$$Z = \sum g_i e^{-\beta\epsilon_i}, \tag{11-12}$$

and

$$A = \frac{N}{Z}. \tag{11-13}$$

Substituting this result into Eq. (10-10), we get

$$N_i = N\frac{g_i e^{-\beta\epsilon_i}}{Z}. \tag{11-14}$$

It will be shown later that Z is proportional to the volume of the container. Since the properties of a gas depend on temperature as well as on volume, one would expect a relation between β and the temperature. To introduce the concept of temperature into statistical mechanics, we must go back to the fundamental idea of thermal equilibrium between two systems, just like the procedure in Chap. 7 for relating the quantity λ to temperature. Consequently, let us consider an isolated composite system consisting of two samples of ideal gas separated by a diathermic wall, as shown in Fig. 11-3. For the second sample of gas, the symbols expressing energy levels, populations, etc., are distinguished

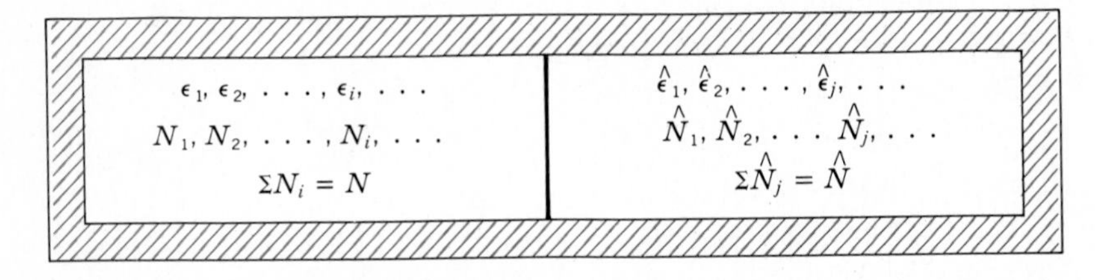

Figure 11-3 An isolated composite system of two samples of ideal gas separated by a diathermic wall. The total energy is constant.

with a circumflex. The thermodynamic probability of the composite system $\boldsymbol{\Omega}$ is the product of the separate thermodynamic probabilities, so that the logarithm is

$$\ln \boldsymbol{\Omega} = \sum N_i \ln \frac{g_i}{N_i} + N + \sum \hat{N}_j \ln \frac{\hat{g}_j}{\hat{N}_j} + \hat{N}.$$

Each sample has a constant number of molecules, so that

$$\sum N_i = N = \text{const.}$$

and

$$\sum \hat{N}_j = \hat{N} = \text{const.},$$

but the energy of each sample is *not constant*. Only the total energy of the composite system is constant; thus,

$$\sum N_i \epsilon_i + \sum \hat{N}_j \hat{\epsilon}_j = U = \text{const.}$$

To find equilibrium conditions, we proceed as before and get

$$\begin{aligned}
\sum \ln \frac{g_i}{N_i}\, dN_i + \sum \ln \frac{\hat{g}_j}{\hat{N}_j}\, d\hat{N}_j &= 0,\\
\ln A \sum dN_i \qquad\qquad\qquad &= 0,\\
\ln \hat{A} \sum d\hat{N}_j &= 0,\\
-\beta \sum \epsilon_i\, dN_i \quad - \beta \sum \hat{\epsilon}_j\, d\hat{N}_j &= 0.
\end{aligned}$$

Adding and setting each coefficient of dN and $d\hat{N}$ equal to zero, we get *two* sets of equations,

$$N_i = A g_i e^{-\beta\epsilon_i}$$

and

$$\hat{N}_j = \hat{A}\hat{g}_j e^{-\beta\hat{\epsilon}_j},$$

where all quantities are different, *except* β. When two systems separated by a diathermic wall come to equilibrium, the temperatures are the same and the β's are the same. The conclusion that β is connected with the temperature is inescapable.

It was shown in Chap. 8 that the entropy of an isolated system increases when the system undergoes a spontaneous, irreversible process. At the conclusion of such a process, when equilibrium is reached, the entropy has the maximum value consistent with its energy and volume. The thermodynamic probability also increases and approaches a maximum as equilibrium is approached. We therefore look for some correlation between S and Ω. Consider two similar systems A and B in thermal contact—one with entropy S_A and thermodynamic probability Ω_A, the other with values S_B and Ω_B. Since entropy is an extensive variable, the total entropy of the composite system is

$$S = S_A + S_B.$$

The thermodynamic probability, however, is the product, or

$$\Omega = \Omega_A \cdot \Omega_B.$$

If we let

$$S = f(\Omega),$$

then,

$$f(\Omega_A \Omega_B) = f(\Omega_A) + f(\Omega_B).$$

The only function that satisfies this relation is the logarithm. Introducing an arbitrary constant k', we may write

$$\boxed{S = k' \ln \Omega} \tag{11-15}$$

for the relation between entropy and thermodynamic probability.

The first law of thermodynamics applied to any infinitesimal process of any hydrostatic system is

$$đQ = dU + P\,dV.$$

If the process takes place between two neighboring equilibrium states, it may be performed reversibly, in which case $đQ = T\,dS$, and

$$dU = T\,dS - P\,dV.$$

If we now specify that the reversible process take place at constant V, we have the important *link between thermodynamics and statistical mechanics:*

$$\boxed{\frac{1}{T} = \left(\frac{\partial S}{\partial U}\right)_V.} \tag{11-16}$$

Since both S and U may be calculated by statistical mechanics, *the derivative* $(\partial S/\partial U)_V$ *gives the reciprocal of the Kelvin temperature. This is the way in which the macroscopic concept of temperature is injected into statistical mechanics.*

In employing the Lagrange method to find the equilibrium values of the energy level populations [Eqs. (11-9) and (11-10)], we found that

$$d \ln \Omega = \sum \ln \frac{g_i}{N_i}\, dN_i,$$

and

$$\ln \frac{g_i}{N_i} = \beta\epsilon_i - \ln A.$$

Therefore,

$$\begin{aligned} d \ln \Omega &= \sum \beta\epsilon_i\, dN_i - \ln A \sum dN_i \\ &= \beta\, d \sum \epsilon_i N_i - \ln A\, d \sum N_i \\ &= \beta\, dU, \end{aligned}$$

where U is the total energy of the system. Therefore,

$$\beta = \frac{d \ln \Omega}{dU} = \frac{1}{k'} \frac{d}{dU} k' \ln \Omega = \frac{1}{k'} \left(\frac{\partial S}{\partial U}\right)_V.$$

Since $(\partial S/\partial U)_V = 1/T$, we get the beautiful result

$$\boxed{\beta = \frac{1}{k'T}.} \tag{11-17}$$

When the actual values of the ϵ's appropriate to an ideal gas are introduced, it will be seen that k' is none other than Boltzmann's constant k.

11-4 PARTITION FUNCTION

We have seen that the population N_i of the ith energy level is

$$N_i = Ag_i e^{-\beta\epsilon_i}.$$

Substituting $1/k'T$ for β and N/Z for A, we get

$$\boxed{N_i = \frac{N}{Z} g_i e^{-\epsilon_i/k'T},} \tag{11-18}$$

where

$$\boxed{Z = \sum g_i e^{-\epsilon_i/k'T}.} \tag{11-19}$$

The partition function Z contains the heart of the statistical information about the particles of the system, so that it is worthwhile to express other properties of the system, such as U, S, and P, in terms of Z. If we differentiate Z with respect to T, holding V constant, we get

$$\begin{aligned}\left(\frac{\partial Z}{\partial T}\right)_V &= \sum g_i \left(\frac{\epsilon_i}{k'T^2}\right) e^{-\epsilon_i/k'T} \\ &= \frac{1}{k'T^2} \sum \epsilon_i g_i e^{-\epsilon_i/k'T} \\ &= \frac{Z}{Nk'T^2} \sum \epsilon_i N_i \\ &= \frac{ZU}{Nk'T^2}.\end{aligned}$$

It follows that

$$\boxed{U = Nk'T^2 \left(\frac{\partial \ln Z}{\partial T}\right)_V,} \tag{11-20}$$

and U may be calculated once $\ln Z$ is known as a function of T and V. Also,

$$S = k' \ln \Omega,$$

where, according to Eq. (11-6),

$$\ln \Omega = \sum N_i \ln \frac{g_i}{N_i} + N.$$

Hence,

$$S = -k' \sum N_i \ln \frac{N_i}{g_i} + k'N.$$

Substituting for N_i/g_i the value given in Eq. (11-18), we get

$$S = -k' \sum N_i \left(\ln \frac{N}{Z} - \frac{\epsilon_i}{k'T} \right) + k'N,$$

and, finally,

$$\boxed{S = Nk' \ln \frac{Z}{N} + \frac{U}{T} + Nk',} \tag{11-21}$$

which provides us with a method of calculating S once $\ln Z$ is known.

The Helmholtz function can also be evaluated in terms of the partition function using S from Eq. (11-21). Since

$$F = U - TS,$$

then

$$F = U - T\left(Nk' \ln \frac{Z}{N} + \frac{U}{I} + Nk' \right)$$

$$= -k'T(N \ln Z - N \ln N + N);$$

therefore,

$$\boxed{F = -k'T(N \ln Z - \ln N!).} \tag{11-22}$$

One more equation that is of value is the relation between pressure and the partition function. Since

$$dF = dU - T\,dS - S\,dT$$

$$= -P\,dV - S\,dT,$$

it follows that

$$P = -\left(\frac{\partial F}{\partial V} \right)_T.$$

Substituting the value for F in Eq. (11-22), we obtain

$$\boxed{P = Nk'T\left(\frac{\partial \ln Z}{\partial V} \right)_T,} \tag{11-23}$$

so that again the pressure may be calculated once ln Z is known as a function of T and V.

This is the advantage of statistical mechanics. It provides us with a simple set of rules for obtaining the properties of a system of weakly interacting particles:

1. Use quantum mechanics to find the ϵ values of the quantum states.
2. Find the partition function Z in terms of T and V.
3. Calculate the energy by differentiating ln Z with respect to T.
4. Calculate the pressure by differentiating ln Z with respect to V.
5. Calculate the entropy from Z and U.
6. Calculate the Helmholtz function directly from ln Z.

11-5 PARTITION FUNCTION OF AN IDEAL MONATOMIC GAS

To apply the rules laid down in the preceding article to an ideal gas, we must first calculate the appropriate partition function. This was defined to be

$$Z = \sum_{\text{levels}} g_i e^{-\epsilon_i/k'T},$$

where the summation was over all the energy levels. Exactly the same result is obtained if we sum the expression

$$Z = \sum_{\text{states}} e^{-\epsilon_j/k'T}$$

over all quantum states. We at first take into account only the kinetic energy of translation of the particles confined to a rectangular box whose x, y, and z sides are, respectively, a, b, and c. The energy of any quantum state j is given by Eq. (11-1) as

$$\epsilon_j = \frac{h^2}{8m}\left(\frac{n_x^2}{a^2} + \frac{n_y^2}{b^2} + \frac{n_z^2}{c^2}\right),$$

where n_x, n_y, and n_z are quantum numbers specifying the various quantum states. The partition function is therefore a threefold sum: thus,

$$Z = \sum_{n_x=1}^{\infty} \sum_{n_y=1}^{\infty} \sum_{n_z=1}^{\infty} e^{-(h^2/8mk'T)(n_x^2/a^2 + n_y^2/b^2 + n_z^2/c^2)}$$

or

$$Z = \sum e^{-(h^2/8mk'T)(n_x^2/a^2)} \sum e^{-(h^2/8mk'T)(n_y^2/b^2)} \sum e^{-(h^2/8mk'T)(n_z^2/c^2)}.$$

Since the values of n_x, n_y, and n_z that give rise to appreciable values of the energy are very large and since a change of n_x or n_y or n_z by unity produces a change of energy that is exceedingly small, no error is introduced by replacing each sum with an integral and by writing

$$Z = \left[\int_0^\infty e^{-(h^2/8mk'T)(n_x^2/a^2)}\, dn_x\right]\left[\int_0^\infty e^{-(h^2/8mk'T)(n_y^2/b^2)}\, dn_y\right] \times \left[\int_0^\infty e^{-(h^2/8mk'T)(n_z^2/c^2)}\, dn_z\right].$$

Each integral is of the type (see Appendix C)

$$\int_0^\infty e^{-ax^2}\, dx = \frac{1}{2}\sqrt{\frac{\pi}{a}}.$$

Therefore,

$$Z = \left[\frac{a}{2}\sqrt{\frac{8\pi mk'T}{h^2}}\right]\left[\frac{b}{2}\sqrt{\frac{8\pi mk'T}{h^2}}\right]\left[\frac{c}{2}\sqrt{\frac{8\pi mk'T}{h^2}}\right],$$

and since $abc = V$,

$$Z = V\left(\frac{2\pi mk'T}{h^2}\right)^{3/2}, \tag{11-24}$$

and

$$\boxed{\ln Z = \ln V + \tfrac{3}{2}\ln T + \tfrac{3}{2}\ln\left(\frac{2\pi mk'}{h^2}\right).} \tag{11-25}$$

1, *Pressure of an ideal monatomic gas*

$$P = Nk'T\left(\frac{\partial \ln Z}{\partial V}\right)_T$$

$$= Nk'T\left(\frac{1}{V}\right)$$

$$= \frac{N}{V}k'T. \tag{11-26}$$

Comparing this result with the expression for the pressure $P = NkT/V$ obtained with the kinetic theory of gases, given in Eq. (5-23), where k is Boltzmann's constant, we see that the arbitrary constant k' introduced in the equation $S = k' \ln \Omega$ is none other than the Boltzmann constant, or

$$\boxed{k' = k = \frac{R}{N_A}.} \tag{11-27}$$

2. *Energy of an ideal monatomic gas*

$$U = NkT^2\left(\frac{\partial \ln Z}{\partial T}\right)_V$$

$$= NkT^2\left(\frac{3}{2T}\right)$$

$$= \tfrac{3}{2}NkT. \tag{11-28}$$

This is exactly the same result which was obtained by the kinetic theory of gases for a monatomic ideal gas and shows that, when particles each having three translational degrees of freedom come to statistical equilibrium, the energy per particle equals $\frac{3}{2}kT$.

3. *Entropy of an ideal monatomic gas*

$$S = Nk \ln \frac{Z}{N} + \frac{U}{T} + Nk$$

$$= Nk\left[\ln \frac{V}{N} + \tfrac{3}{2}\ln T + \ln\left(\frac{2\pi mk}{h^2}\right)^{3/2}\right] + \tfrac{3}{2}Nk + Nk$$

$$= Nk\left[\tfrac{3}{2}\ln T + \ln \frac{V}{N} + \ln\left(\frac{2\pi mk}{h^2}\right)^{3/2} + \tfrac{5}{2}\right].$$

If we take 1 mol of gas, $N = N_A$ and $N_A k = R$. Therefore,

$$s = c_V \ln T + R \ln v + R \ln \frac{(2\pi mk/h^2)^{3/2}}{N_A} + \tfrac{5}{2}R. \tag{11-29}$$

The expression is to be compared with Eq. (8-5), namely,

$$s = c_V \ln T + R \ln v + s_0,$$

and we see that not only were we able to arrive at this equation by the methods of statistical mechanics but also we were able to calculate the constant s_0. Equation (11-29), which was first obtained by Sackur and Tetrode, usually bears their names.

11-6 EQUIPARTITION OF ENERGY

Both kinetic theory and statistical mechanics, when applied to the molecules of an ideal gas (each having three translational degrees of freedom), yield the result that at equilibrium the energy per particle associated with each degree of translational freedom is $\frac{1}{2}kT$. The methods of kinetic theory could not be applied to rotational and vibrational degrees of freedom, but the simple statistical method just developed is capable of dealing with all types of molecular energy, not just translational kinetic energy.

The property of the partition function which makes it so useful is that, whenever the energy of a molecule is expressed as a sum of independent terms each referring to a different degree of freedom,

$$\epsilon = \epsilon' + \epsilon'' + \epsilon''' + \cdots;$$

then,

$$\begin{aligned} Z &= \sum e^{-\epsilon/kT} = \sum e^{-(\epsilon' + \epsilon'' + \epsilon''' + \cdots)/kT} \\ &= \sum e^{-\epsilon'/kT} \sum e^{-\epsilon''/kT} \sum e^{-\epsilon'''/kT} \cdots \\ &= Z'Z''Z''' \cdots. \end{aligned} \tag{11-30}$$

If the various types of energy are calculated with classical physics, it is a simple matter to derive the *classical* principle of the equipartition of energy. We take Eq. (11-20), namely,

$$U = NkT^2 \left(\frac{\partial \ln Z}{\partial T} \right)_V,$$

and rewrite it thus:

$$\langle \epsilon \rangle = \frac{U}{N} = -\frac{\partial \ln Z}{\partial (1/kT)},$$

$$\langle \epsilon \rangle = -\frac{\partial \ln Z}{\partial \beta}.$$

Suppose ϵ to consist of terms representing translational kinetic energy of the type $\frac{1}{2}mw^2$, those representing rotational kinetic energy of the type $\frac{1}{2}I\omega^2$, those representing vibrational energy $\frac{1}{2}m\dot{\xi}^2 + \frac{1}{2}k\xi^2$, etc. All these forms of energy are expressed as squared terms of the type $b_i p_i^2$. Let there be f such terms, or

$$\epsilon = b_1 p_1^2 + b_2 p_2^2 + \cdots + b_f p_f^2.$$

Then, since the partition function is the product of the separate partition functions,

$$Z = \int_0^\infty e^{-\beta b_1 p_1^2}\, dp_1 \int_0^\infty e^{-\beta b_2 p_2^2}\, dp_2 \cdots \int_0^\infty e^{-b_f p_f^2}\, dp_f.$$

Let $\qquad y_i = \beta^{1/2} p_i \qquad$ and $\qquad dy_i = \beta^{1/2}\, dp_i;$

then

$$\begin{aligned} \int_0^\infty e^{-\beta b_i p_i^2}\, dp_i &= \int_0^\infty e^{-\beta b_i y_i^2/\beta}\, dy_i/\beta^{1/2} \\ &= \beta^{-1/2} \int_0^\infty e^{-b_i y_i^2}\, dy_i \\ &= \beta^{-1/2} K_i, \end{aligned}$$

where K_i does *not* contain β. The partition function now becomes

$$\begin{aligned} Z &= \beta^{-1/2} K_1 \cdot \beta^{-1/2} K_2 \cdots \beta^{-1/2} K_f, \\ &= \beta^{-f/2} K_1 K_2 \cdots K_f, \end{aligned}$$

where *none of the K's contains β*. Since $\langle\epsilon\rangle = -\partial(\ln Z)/\partial\beta$,

$$\langle\epsilon\rangle = -\frac{\partial}{\partial\beta}\left(-\frac{f}{2}\ln\beta + \ln K_1 + \ln K_2 + \cdots + \ln K_f\right)$$

$$= \frac{f}{2\beta};$$

and since $\beta = 1/kT$,

$$\boxed{\langle\epsilon\rangle = \frac{f}{2}kT.} \tag{11-31}$$

It has therefore been proposed that, when a large number of nondistinguishable, quasi-independent particles whose energy is expressed as the sum of f squared terms come to equilibrium, the average energy per particle is f times $\frac{1}{2}kT$. This is the famous principle of the equipartition of energy.

A monatomic ideal gas has only three translational degrees of freedom; therefore, N_A atoms (1 mol) will have an energy u equal to

$$u = N_A(\tfrac{3}{2}kT) = \tfrac{3}{2}RT$$

and

$$c_V = \frac{du}{dT} = \frac{3}{2}R.$$

Since $c_P = c_V + R$, $c_P = \frac{5}{2}R$ and $\gamma = \frac{5}{3}$. These calculations are in agreement with the experimental results listed in Art. 5-4.

The diatomic gases H_2, D_2, O_2, N_2, NO, and CO, in the neighborhood of room temperature—where rotation takes place, but not vibration—are dumbbell-shaped with two rotational degrees of freedom. Hence, the molar energy is

$$u = \tfrac{5}{2}RT,$$

and

$$c_V = \tfrac{5}{2}R, \qquad c_P = \tfrac{7}{2}R, \quad \text{and} \qquad \gamma = \tfrac{7}{5},$$

also in agreement with the results of Art. 5-4.

When polyatomic molecules are "soft" and vibrate easily with many frequencies, say, q, the principle of the equipartition of energy would require that

$$u = (3 + q)RT,$$

and

$$c_V = (3 + q)R, \qquad c_P = (4 + q)R, \qquad \text{and} \qquad \gamma = \frac{4 + q}{3 + q}.$$

If q is large, γ is nearly 1. This result is *in complete disagreement* with experiment. The heat capacities of such molecules are not constant but vary markedly with the temperature even around room temperature, and γ is not in the neighborhood of unity.

When the principle of the equipartition of energy is applied to solids and liquids, the disagreement is still worse, and this principle must be abandoned in favor of quantum ideas.

11-7 DISTRIBUTION OF MOLECULAR SPEEDS

If w denotes the speed of a molecule in an ideal gas consisting of N molecules, it is often necessary to calculate the average speed $\langle w \rangle$ or the average square of the speed $\langle w^2 \rangle$, where

$$\langle w \rangle = \frac{1}{N} \int_0^\infty w \, dN_w,$$

and

$$\langle w^2 \rangle = \frac{1}{N} \int_0^\infty w^2 \, dN_w,$$

and dN_w is the number of molecules with speeds between w and $w + dw$. In order to evaluate these integrals, one must know the expression for dN_w in terms of w—a relation known as the *Maxwellian law of distribution of molecular speeds.* In Maxwell's original derivation of this law, it was not necessary to use any physical laws governing the behavior of molecules when colliding with one another or with the wall. These physical laws are hidden within the fundamental assumption of chaos: that, at or near equilibrium, the molecules have no preferred direction of velocity and no preferred location within the container. Since these characteristics of chaos are the same as those involved in the simple treatment of statistical mechanics given in this chapter, we shall derive the Maxwell distribution law from statistical mechanics.

We start with the equation expressing the equilibrium value of the number of molecules N_ϵ with energy ϵ (the equilibrium population of energy level ϵ) given by Eq. (11-18), namely,

$$N_\epsilon = \frac{N}{Z} g_\epsilon e^{-\epsilon/kT}, \tag{11-18}$$

where g_ϵ is the number of quantum states available for molecules of energy ϵ (the degeneracy of the energy level ϵ). These states correspond to positive integer values for each n_x, n_y, and n_z according to Eq. (11-1), namely,

$$n_x^2 + n_y^2 + n_z^2 = \frac{L^2}{h^2} 8m\epsilon = r^2. \tag{11-1}$$

In a three-dimensional Euclidian space with coordinates n_x, n_y, and n_z, each unit volume will contain one quantum state. Such a space is shown in Fig. 11-4, where the space containing unit volume elements corresponding to molecular energies between ϵ and $\epsilon + d\epsilon$ lies between the positive octant of the spherical surface of radius $r = (L/h)(8m\epsilon)^{1/2}$ and the spherical surface of radius $r + dr$. The

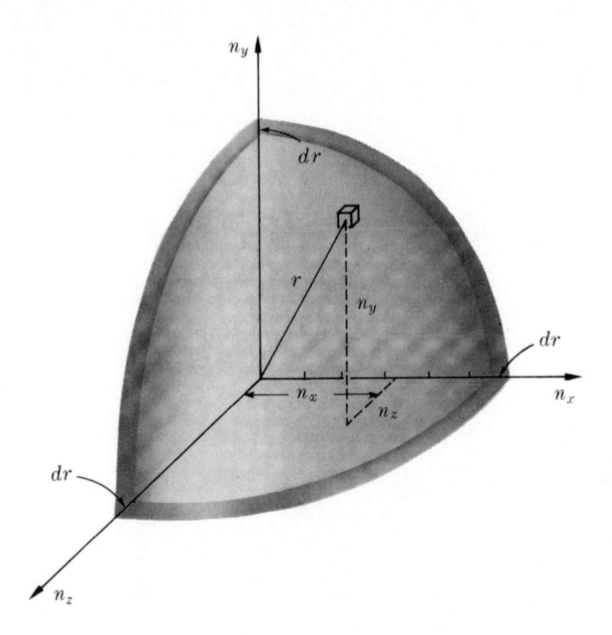

Figure 11-4 Quantum number space. All molecules, whose energy lies between ϵ and $\epsilon + d\epsilon$, have quantum numbers lying within the spherical surface of radius r and that of radius $r + dr$, where $r = (L/h)(8m\epsilon)^{1/2}$.

volume of this region is dg_ϵ, where

$$dg_\epsilon = \tfrac{1}{8}4\pi r^2\, dr$$

$$= \tfrac{1}{8}4\pi \cdot \frac{L^2}{h^2}\, 8m\epsilon\, d\left[\frac{L}{h}\,(8m\epsilon)^{1/2}\right].$$

Simplifying and setting $V = L^3$, we get

$$dg_\epsilon = \frac{2\pi V(2m)^{3/2}}{h^3}\,\epsilon^{1/2}\, d\epsilon. \tag{11-32}$$

The number of molecules dN_ϵ in the energy range between ϵ and $\epsilon + d\epsilon$ is clearly

$$dN_\epsilon = N\,\frac{dg_\epsilon}{Z}\,e^{-\epsilon/kT}. \tag{11-33}$$

We have shown in Eq. (11-24) that the partition function of our ideal gas is

$$Z = V\left(\frac{2\pi mkT}{h^2}\right)^{3/2}. \tag{11-24}$$

Hence,

$$\frac{dg_\epsilon}{Z} = \frac{2}{\pi^{1/2}(kT)^{3/2}}\,\epsilon^{1/2}\, d\epsilon.$$

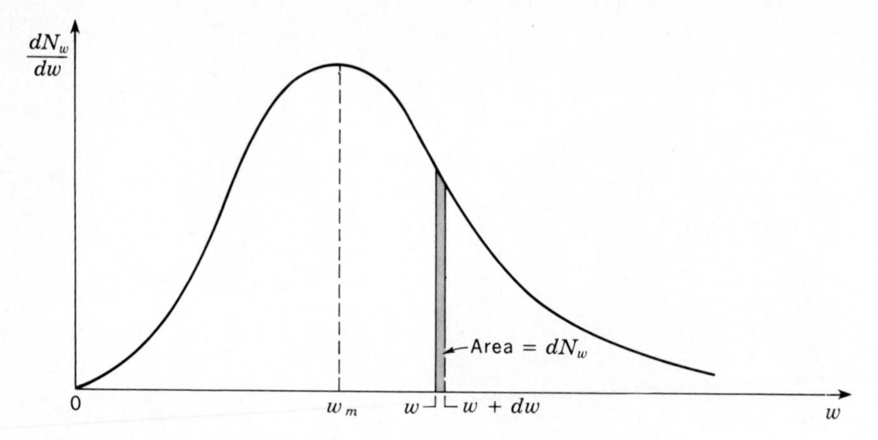

Figure 11-5 Graph of Maxwell speed-distribution function.

If the speed of a molecule is w, then

$$\epsilon = \tfrac{1}{2}mw^2,$$

and substituting dg_ϵ/Z and ϵ into Eq. (11-33), we get

$$dN_w = \frac{2N}{\pi^{1/2}(kT)^{3/2}}\left(\tfrac{1}{2}mw^2\right)^{1/2} mw\; d\left(we^{-(1/2)mw^2/kT}\right),$$

and finally

$$\boxed{\frac{dN_w}{dw} = \frac{2N}{\sqrt{2\pi}}\left(\frac{m}{kT}\right)^{3/2} w^2 e^{-(1/2)mw^2/kT}.} \tag{11-34}$$

The expression on the right is the famous *Maxwell speed-distribution* function, and is plotted in Fig. 11-5. Figure 11-6 shows the function for three different

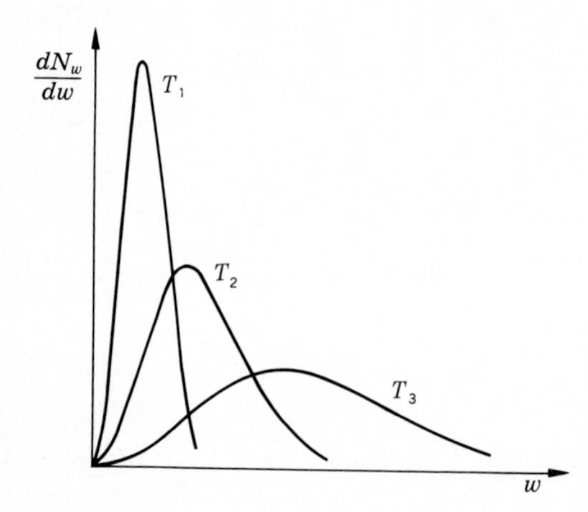

Figure 11-6 Maxwellian speed distribution at various temperatures, $T_3 > T_2 > T_1$.

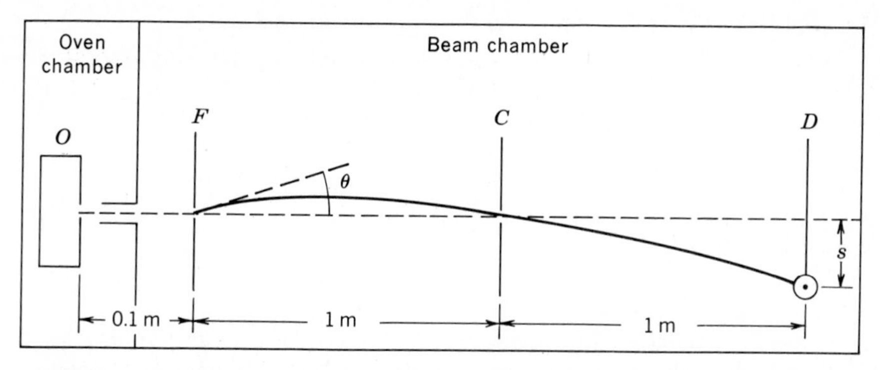

Figure 11-7 Apparatus of Estermann, Simpson, and Stern for the deflection of a molecular beam by gravity. *O*, oven; *F* and *C*, slits; *D*, detector.

temperatures. The higher the temperature, the wider is the spread of values of the speed.

Maxwell's equation for the distribution of molecular speeds has been experimentally verified directly as well as indirectly. One of the most convincing experiments was performed by Estermann, Simpson, and Stern in 1947, making use of the apparatus depicted schematically in Fig. 11-7, in which atoms are deflected *by gravity only*. Cesium atoms issue from a minute slit in an oven *O* at the left of a long, highly evacuated chamber. Most of these atoms are stopped by the diaphragm *F*, and those which go through the slit constitute a narrow, almost horizontal beam. The slit *C*, called the *collimating slit*, is halfway between *F* and the detector *D*. The atoms are detected by the method of surface ionization, in which nearly every cesium atom striking a hot tungsten wire leaves the wire as a positive ion and is collected by a negatively charged plate. The plate current is then a measure of the number of cesium atoms striking the detector wire per unit of time.

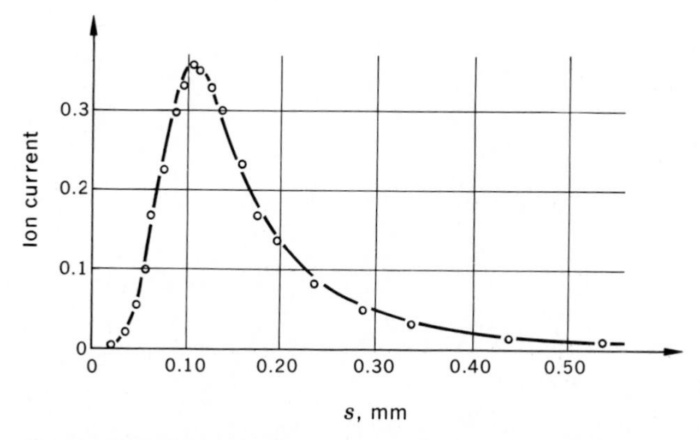

Figure 11-8 Distribution of speeds of cesium atoms in atomic beam deflected by gravity. *(Estermann, Simpson, and Stern.)*

A simple analysis of the parabolic paths of the atoms that pass through F and C provides a relation between the deflection s and the speed w. The departure angle θ is practically zero. The experimental results are shown in Fig. 11-8, where the smooth curve is a plot of Maxwell's function and the dots are experimental points.

11-8 STATISTICAL INTERPRETATION OF WORK AND HEAT

We have been considering the statistical equilibrium of a large number N of nondistinguishable, quasi-independent particles in a cubical container of volume V. The energy levels ϵ_i of individual particles undergoing translational motion only were given by

$$\epsilon_i = \frac{h^2}{8mL^2}(n_x^2 + n_y^2 + n_z^2).$$

Since $L^3 = V$, then $L^2 = V^{2/3}$, and letting B_i be the sum of the squares of the quantum numbers appropriate to the ith energy level, we get

$$\epsilon_i = \frac{h^2}{8m} B_i V^{-2/3}.$$

Given the set of quantum numbers that determines B_i, we may say that the *corresponding energy* ϵ_i *depends on volume only*. Taking the logarithm of ϵ_i,

$$\ln \epsilon_i = \ln \frac{h^2}{8m} + \ln B_i - \tfrac{2}{3} \ln V.$$

The effect of a small change of V on ϵ is given by taking the differential of this equation; whence

$$\frac{d\epsilon_i}{\epsilon_i} = -\frac{2}{3}\frac{dV}{V}.$$

Therefore,

$$d\epsilon_i = -\frac{2}{3}\frac{\epsilon_i}{V} dV,$$

$$N_i \, d\epsilon_i = -\frac{2}{3}\frac{N_i \epsilon_i}{V} dV,$$

and

$$\sum N_i \, d\epsilon_i = -\frac{2}{3}\frac{U}{V} dV. \tag{11-35}$$

Now, it has been shown both by kinetic theory and by statistical mechanics that the pressure of an ideal gas is given by

$$P = \frac{NkT}{V}.$$

Since the energy per particle is translational kinetic energy only, with three degrees of freedom,

$$U = \tfrac{3}{2}NkT.$$

It follows that

$$P = \frac{2}{3}\frac{U}{V}, \tag{11-36}$$

Substituting this result into Eq. (11-35), we get

$$\sum N_i \, d\epsilon_i = -P \, dV. \tag{11-37}$$

A change of volume, therefore, causes changes in the energy values of the energy levels, without producing changes in the populations of the levels. When the N_i change and the ϵ_i remain constant, we have from page 284,

$$d \ln \Omega = \beta \sum \epsilon_i \, dN_i .$$

Since $kd \ln \Omega = dS$, $k\beta \sum \epsilon_i \, dN_i = dS$, and setting $k\beta$ equal to $1/T$, we get, finally,

$$\sum \epsilon_i \, dN_i = T \, dS.$$

We see that a reversible heat transfer produces changes in the populations of the energy levels without changes in the energy values of the levels themselves. Thus, the equation $dU = \sum \epsilon_i \, dN_i + \sum N_i \, d\epsilon_i$ expresses the first and second laws of thermodynamics, with $\sum \epsilon_i \, dN_i = T \, dS$ and $\sum N_i \, d\epsilon_i = -P \, dV$.

11-9 DISORDER, ENTROPY, AND INFORMATION

Whenever work or kinetic energy is dissipated within a system because of friction, viscosity, inelasticity, electric resistance, or magnetic hysteresis, the disorderly motions of molecules are increased. Whenever different substances are mixed or dissolved or diffused with one another, the spatial positions of the molecules constitute a more disorderly arrangement. Rocks crumble, iron rusts, some metals corrode, wood rots, leather disintegrates, paint peels, and people age. All these processes involve the transition from some sort of "orderliness" to a greater disorder. This transition is expressed in the language of classical thermodynamics by the statement that the entropy of the universe increases. Molecular disorder and entropy go together, and if we measure disorder by the number of ways a particular macrostate may be achieved, the thermodynamic probability Ω is a measure of disorder. Then the equation $S = k \ln \Omega$ is the simple relation between entropy and disorder.

The number of ways in which a particular macrostate may be achieved can be given another interpretation. Suppose that you are called upon to guess a person's first name. The number of choices of names of men and women is staggeringly large. With no hint or clue, the number of ways in which one can arrive at a name is very large, and the information at one's disposal is small. Suppose, now, that we are told the person is a man. Immediately the number of choices of names is *reduced*, whereas the information is *increased*. Information is increased further if we are told that the man's name starts with H, for then the number of choices (or ways of picking a man's name) is reduced very greatly. It is clear that the *fewer* the number of ways a particular situation or a particular state of a system may be achieved, the greater is the information.

A convenient measure of the information conveyed when the number of choices is reduced from Ω_0 to Ω_1 is given by

$$I = k \ln \frac{\Omega_0}{\Omega_1}.$$

The bigger the reduction, the bigger the information. Since $k \ln \Omega$ is the entropy S, then

$$I = S_0 - S_1,$$

or

$$S_1 = S_0 - I,$$

which can be interpreted to mean that the entropy of a system is reduced by the amount of information about the state of a system. In the words of Brillouin, "Entropy measures the lack of information about the exact state of a system."

As an example of the connection between entropy and information, consider the isothermal compression of an ideal gas (N molecules) from a volume V_0 to a volume V_1. We know that the reduction of entropy is equal to

$$S_0 - S_1 = Nk \ln \frac{V_0}{V_1}.$$

But, when we decrease the volume of the gas, we decrease the number of ways of achieving this state, because there are fewer microstates with position coordinates in the smaller volume. Before the compression, each molecule is known to be in the volume V_0. The number of locations each molecule could occupy is $V_0/\Delta V$, where ΔV is some arbitrary small volume. After the compression, each molecule is to be found in volume V_1, with a smaller number of possible locations $V_1/\Delta V$. It follows that

$$I = k \ln \frac{\Omega_0}{\Omega_1} = k \ln \frac{V_0/\Delta V}{V_1/\Delta V} = k \ln \frac{V_0}{V_1},$$

and for the entire gas of N molecules,

$$I = Nk \ln \frac{V_0}{V_1},$$

in agreement with the result of classical thermodynamics. The increase of information as a result of the compression is seen to be identical with the corresponding entropy reduction.

The connection between entropy and information can be applied to the problem of *Maxwell's demon.* Maxwell imagined a small creature stationed near a trap door separating two compartments of a vessel containing a gas. Suppose that the demon opened the trap door only when fast molecules approached, thereby allowing the fast molecules to collect in one compartment and slow ones in the other. This would obviously result in a transition from disorder to order—thus violating the second law. According to Brillouin, the demon could not tell the difference between one kind of molecule and another because he and the molecules are in an enclosure at a uniform temperature and all are bathed in isotropic blackbody radiation. The demon could not see the individual atoms. However, suppose that we allow the demon, according to the analysis of Rodd, to use a flashlight whose radiation is not in equilibrium with the enclosure. Then the demon can get information about the molecules and thereby decrease the entropy of the system. But other phenomena come into the discussion: (1) the filament of the lamp in the flashlight undergoes an increase of entropy; (2) a photon scattered by a molecule is absorbed by the demon and serves to increase his entropy; (3) the action of the demon in opening the trap door reduces the number of microstates available to the molecules. (The entropy change of the battery of the flashlight can be ignored.) If all these processes are taken into account and the corresponding entropy changes are calculated from the standpoint of the increase or decrease of information, then Rodd was able to demonstrate that the total entropy change is positive. The second law is not violated.

Much has been written about reversibility and irreversibility, order and disorder, and supposed violations of the second law. It would be hard to find anything in any language to compare with the exposition given in *Feynman Lectures on Physics* (Addison-Wesley, Reading, Mass., 1963–1965, Chap. 46). This chapter is recommended wholeheartedly to all students—whether naïve or sophisticated in their level of attainment—for its brilliance, its depth, and its human warmth.

PROBLEMS

(Values of constants: $k = 1.38 \times 10^{-23}$ J/K and $h = 6.63 \times 10^{-34}$ J · s.)

11-1 A mercury atom moves in a cubical box whose edge is 1 m long. Its kinetic energy is equal to the average kinetic energy of a molecule of an ideal gas at 1000 K. If the quantum numbers n_x, n_y, and n_z are all equal to n, calculate n.

11-2 Show that, when N ideal gas atoms come to equilibrium,

$$\frac{g_i}{N_i} = \frac{Z}{N} e^{\epsilon_i/kT}$$

and

$$\frac{Z}{N} = \frac{(kT)^{5/2}}{P} \left(\frac{2\pi m}{h^2} \right)^{3/2}.$$

Taking $\epsilon_i = 3/2kT$, $T = 300$ K, $P = 10^3$ Pa, and $m = 10^{-26}$ kg, calculate g_i/N_i.

11-3 Consider a function f defined by the relation

$$f(\Omega_A \Omega_B) = f(\Omega_A) + f(\Omega_B).$$

First differentiate partially with respect to Ω_B, and then with respect to Ω_A. Integrate twice to show that

$$f(\Omega) = \text{const. } \ln \Omega + \text{const.}$$

11-4 In the case of N *distinguishable* particles, the number of ways Ω in which a macrostate defined by N_1 particles in g_1 quantum states with energy ϵ_1, N_2 particles in g_2 quantum states with energy ϵ_2, and so on, may be achieved is given by

$$\Omega = N! \frac{g_1^{N_1} g_2^{N_2}, \ldots}{N_1! N_2! \cdots},$$

when $g_i \gg N_i$.

(*a*) Using the Stirling approximation, calculate $\ln \Omega$.

(*b*) Render $\ln \Omega$ a maximum subject to the equations of constraint $\sum N_i = N = \text{const.}$ and $\sum N_i \epsilon_i = U = \text{const.}$, and explain why U and P should be the same as for indistinguishable particles but S should be different.

11-5 Given N indistinguishable, quasi-independent particles capable of existing in energy levels ϵ_1, ϵ_2, ..., with degeneracies g_1, g_2, ..., respectively. In any given macrostate in which there are N_1 particles in energy level ϵ_1, N_2 particles with energy ϵ_2, and so on, assume the thermodynamic probability to be given by the Bose-Einstein expression,

$$\Omega_{\text{BE}} = \frac{(g_1 + N_1)!\,(g_2 + N_2) \cdots}{g_1!\,N_1!\,g_2!\,N_2! \cdots}.$$

Using Stirling's approximation and the Lagrange method, render $\ln \Omega_{\text{BE}}$ a maximum subject to $\sum N_i = N = \text{const.}$ and $\sum N_i \epsilon_i = U = \text{const.}$, and show that

$$N_i = \frac{g_i}{Ae^{-\beta\epsilon_i} - 1}$$

11-6 Given the same system as in Prob. 11-5, except that the thermodynamic probability is given by the Fermi-Dirac expression

$$\Omega_{\text{FD}} = \frac{g_1!\,g_2! \cdots}{N_1!\,(g_1 - N_1)!\,N_2!\,(g_2 - N_2)! \cdots}.$$

Using Stirling's approximation and the Lagrange method, render $\ln \Omega_{\text{FD}}$ a maximum subject to $\sum N_i = N = \text{const.}$ and $\Sigma N_i \epsilon_i = U = \text{const.}$, and show

$$N_i = \frac{g_i}{Ae^{-\beta\epsilon_i} + 1}$$

11-7 Given a gaseous system of N_A indistinguishable, weakly interacting diatomic molecules:

(*a*) Each molecule may vibrate with the same frequency ν but with an energy ϵ_i, given by

$$\epsilon_i = (\tfrac{1}{2} + i)h\nu \qquad (i = 0, 1, 2, \ldots)$$

Show that the vibrational partition function Z_v is

$$Z_v = \frac{\cdot e^{-h\nu/2kT}}{1 - e^{-h\nu/kT}}.$$

(*b*) Each molecule may rotate, and the rotational partition function Z_r has the same form as that for translation, except that the volume V is replaced by the total solid angle 4π, the mass is replaced by the moment of inertia I, and the exponent $\frac{3}{2}$ (referring to three translational degrees of freedom) is replaced by $\frac{2}{2}$, since there are only two rotational degrees of freedom. Write the rotational partition function.

(*c*) Taking into account translation, vibration, and rotation, calculate the Helmholtz function.
(*d*) Calculate the pressure.
(*e*) Calculate the energy.
(*f*) Calculate the molar heat capacity at constant volume.

11-8 Defining the average speed $\langle w \rangle$ by the equation

$$\langle w \rangle = \frac{1}{N} \int_0^\infty w \, dN_w,$$

show that

$$\langle w \rangle = \sqrt{\frac{8kT}{\pi m}}.$$

11-9 (*a*) In Fig. 11-5, let w_m be the value of w at which dN_w/dw is a maximum. Calculate w_m.
(*b*) Choose a new variable $x = w/w_m$, and calculate $dN_x/N\,dx$. What is the maximum value of $dN_x/N\,dx$?

11-10 (*a*) Calculate $\langle 1/w \rangle$ and compare this with $1/\langle w \rangle$.
(*b*) Show that the number of molecules striking unit area of a wall per unit time is equal to

$$\frac{P}{\sqrt{2\pi mkT}}.$$

11-11 The Doppler broadening of a spectral line increases with the rms speed of the atoms in the source of light. Which should give narrower spectral lines: a mercury 198 lamp at 300 K or a krypton 86 lamp at 77 K?

11-12 At what temperature is the mean translational kinetic energy of a molecule equal to that of a singly charged ion of the same mass which has been accelerated from rest through a potential difference of (*a*) 1 V? (*b*) 1000 V? (*c*) 1,000,000 V? (Neglect relativistic effects.)

11-13 An oven contains cadmium vapor at a pressure of 2.28 Pa and at a temperature of 550 K. In one wall of the oven there is a slit with a length of 10^{-2} m and a width of 10^{-5} m. On the other side of the wall is a very high vacuum. If one assumes that all the atoms arriving at the slit pass through, what is the atomic beam current?

11-14 A vessel of volume V contains a gas that is kept at constant temperature. The gas slowly leaks out of a small hole of area A. The outside pressure is so low that no molecules leak back. Prove that the pressure P at any time τ is given by

$$P = P_0 e^{-k'\tau},$$

where P_0 is the initial pressure; and calculate k' in terms of V, A, and $\langle w \rangle$. (Assume that all the molecules arriving at the hole pass through.)

11-15 A spherical glass bulb 0.1 m in radius is maintained at 300 K, except for an appendix with a cross-sectional area of 10^{-4} m^2 immersed in liquid nitrogen, as shown in Fig. P11-1. The bulb

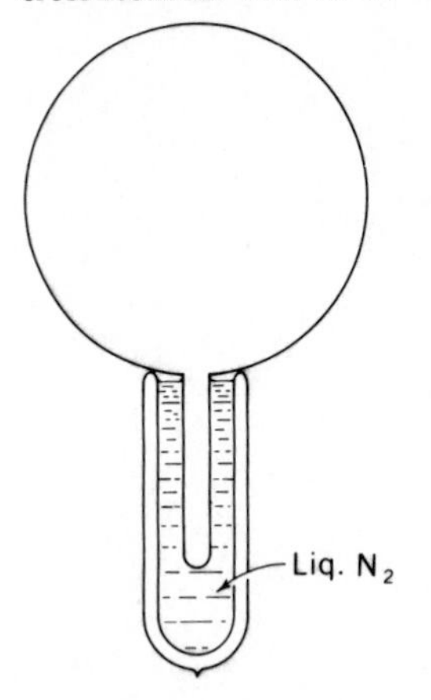

Figure P11-1

contains water vapor originally at a pressure of 13.3 Pa. Assuming that every water molecule which enters the appendix condenses on the wall and stays there, find the time required for the pressure to decrease to 1.33×10^{-4} Pa.

11-16 A vessel partially filled with mercury and closed except for a hole of area 10^{-7} m^2 above the liquid level is kept at 0°C in a continuously evacuated enclosure. After 30 days it is found that 2.4×10^{-5} kg of mercury has been lost. What is the vapor pressure of mercury at 0°C?

CHAPTER

TWELVE

THERMAL PROPERTIES OF SOLIDS

12-1 STATISTICAL MECHANICS OF A NONMETALLIC CRYSTAL

We concluded Chap. 9 with a discussion of the heat capacity at constant volume of nonmetallic crystalline solids (see Art. 9-11). Experimental values for c_P and c_V at low temperatures showed practically the same temperature variation. At higher temperatures, however, c_P continued to rise, while c_V approached a constant value of $3R$, known as the *rule of Dulong and Petit.*

Theoretical values of $c_V = (\partial u/\partial T)_V$, where u is the molar internal energy, may be calculated with the aid of statistical mechanics. In general, this calculation is extremely complicated, because many different phenomena contribute to the internal energy of the solid. Suppose, for example, that the solid is a crystal having a lattice composed of molecules, each of which consists of several atoms; and furthermore, suppose that there is about one free electron per molecule. Then the total internal energy may be due to:

1. Translational motions of the free electrons.
2. Vibrations of the molecules about their equilibrium positions, called briefly *lattice vibrations*.
3. Internal vibrations of atoms within each molecule.
4. Partial rotation of the molecules.
5. Excitation of upper energy levels of the molecules.
6. Anomalous effects.

It is fortunate that all these effects do not take place in all solids. For example, in the case of nonmetals, motions of free electrons do not exist, and in the case of metals the lattice consists of single atoms whose component parts do not rotate or vibrate. Furthermore, all effects do not take place in all temperature ranges. Thus, the motions of the free electrons of metals have an appreciable effect on the heat capacity only at very low temperatures, below about 20 K. Above this temperature, they may be ignored. Similarly, excitation of upper energy levels takes place only at very high temperatures and can therefore be ignored at moderate temperatures.

In this article, let us limit ourselves to a nonmetallic crystal in which the lattice sites are occupied either by a single atom or ion or by a rigid molecule whose internal vibrations, rotations, excitations, etc., may be ignored. We shall refer to the particles occupying lattice sites as *lattice points* and shall assume there are N of them. These lattice points are localized in space and are therefore distinguishable by their positions; furthermore, they are closely packed and interact very strongly with their neighbors. It would seem, at first glance, that the statistical methods described in Chap. 11 as appropriate to indistinguishable, weakly interacting particles are useless here. But this is not the case. Since each lattice point has three coordinates x, y, and z, the system of N lattice points has $3N$ coordinates. If each lattice point undergoes a displacement from its equilibrium position that is small compared with the space between lattice points, the change in potential energy of the crystal would involve many terms involving not only squares of these displacements but product terms as well. In the theory of small oscillations, it is shown that there *always* exists a new set of $3N$ coordinates (linear functions of the original coordinates) such that the potential energy is the sum of exactly $3N$ squared terms. Corresponding to these new coordinates, called *normal coordinates*, there $3N$ momenta, and the kinetic energy is the sum of $3N$ terms that each contains the square of a momentum.

A particle undergoing simple harmonic motion in the x direction has an energy equal to $\frac{1}{2}kx^2 + \frac{1}{2}mp_x^2$. The fact that the energy of N vibrating lattice points is given by $3N$ expressions of this type enables us to conclude that, when N lattice points undergo small displacements, *the motion of the crystal may be described as that of* 3N *independent simple harmonic oscillators*.

There harmonic oscillators (or *normal modes*) are *not associated with individual lattice points;* each one involves motion of the entire crystal. This conclusion is independent of the type of crystal lattice and is true only when displacements from equilibrium positions are small. When the vibrations get large enough, anharmonic effects take place and the oscillators are no longer independent. Actually, to determine the normal coordinates of a given crystal and to calculate the various frequencies of vibration of the $3N$ normal modes is a very complicated problem in mechanics. It is a fortunate circumstance that considerable information may be obtained by applying statistical mechanics to these normal modes, along with either of two simplifying assumptions—one due to Einstein and the other due to Debye.

Our problem in statistical mechanics has now been reduced to that of $3N$

independent (weakly interacting) but *distinguishable* simple harmonic oscillators. Suppose we have N_ν such oscillators *each vibrating with the same frequency* ν. According to the quantum theory, the energy ϵ of any such oscillator may take on only discrete values:

$$\epsilon_i = (i + \tfrac{1}{2})h\nu \qquad (i = 0, 1, 2, \ldots), \tag{12-1}$$

where h is Planck's constant, 6.63×10^{-34} J · s. Suppose that at any moment a macrostate of the crystal is specified by:

$$\begin{aligned} &N_0 \text{ oscillators with energy } \epsilon_0 = \tfrac{1}{2}h\nu \\ &N_1 \text{ oscillators with energy } \epsilon_1 = \tfrac{3}{2}h\nu \\ &N_2 \text{ oscillators with energy } \epsilon_2 = \tfrac{5}{2}h\nu \\ &\cdots\cdots\cdots\cdots\cdots\cdots \end{aligned}$$

These energy states are nondegenerate: that is, not more than one quantum state has the same energy. The number of ways in which N_ν vibrators may be distributed among the energy states according to the macrostate specified is the same as the number of ways in which N_ν *distinguishable* objects (colored balls, marked objects, etc.) can be distributed in boxes so that there are N_0 objects in box 0, N_1 objects in box 1, etc. To fix our ideas, suppose that we have only four objects a, b, c, and d to be distributed between two boxes so that one object will be in one box and three in the other. The number of different ways in which four lettered objects may be arranged in sequence is $4! = 24$, as shown in Fig. 12-1. Consider the arrangement (microstate) depicted in the upper left-hand corner,

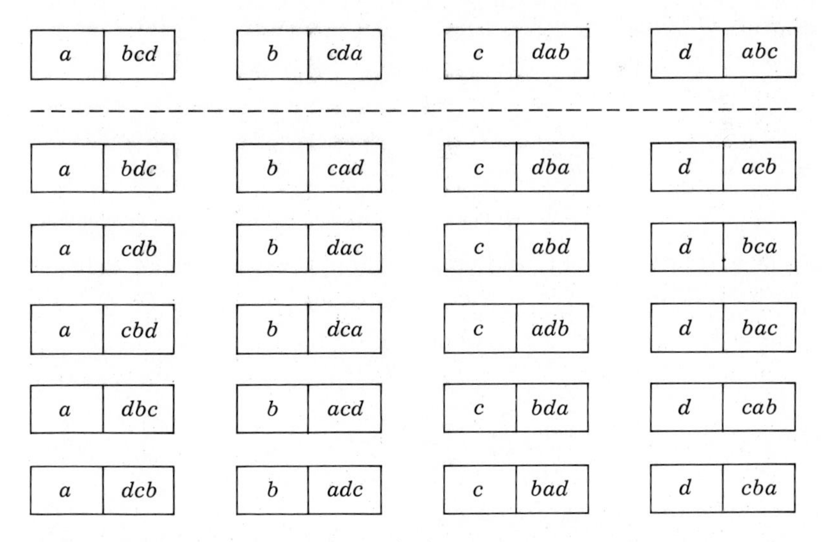

Figure 12-1 The number of different microstates corresponding to the macrostate in which one vibrator is in one energy state and three vibrators are in the other energy state is equal to $4!/3! = 4$. The four different microstates lie above the dashed line; those below are superfluous.

where a is in the left box and b, c, and d are in the right one. All the other microstates with a in the left box (depicted beneath the dashed line) are superfluous, since they involve merely a shift of position *within* a box. Of the six arrangements on the left, only one-sixth need be counted or $1/3!$. The same is true of the arrangements involving b in the left box, and so on. The total number of different arrangements is therefore $4!/3! = 4$.

In the general case where N_v vibrators are distributed among energy states with N_0 vibrators in the state with energy ϵ_0, N_1 in the state with energy ϵ_1, etc., the total number of microstates, or the thermodynamic probability Ω, is given by

$$\Omega = \frac{N_v!}{N_1!\,N_2!\cdots}, \tag{12-2}$$

which is the expression given in Prob. 11-5, where degeneracy was included. The entropy S is expressed as

$$\begin{aligned} S &= k \ln \Omega \\ &= k(\ln N_v! - \ln N_1! - \ln N_2! - \cdots) \\ &= k(N_v \ln N_v - N_v - N_1 \ln N_1 + N_1 - N_2 \ln N_2 + N_2 - \cdots) \\ &= k(N_v \ln N_v - \sum N_i \ln N_i), \end{aligned}$$

which must be maximized subject to the usual conditions,

$$\sum N_i = N_v = \text{const.},$$

and

$$\sum N_i \epsilon_i = U_v = \text{const.}$$

The details of this calculation are identical with those of Chap. 11, and only the results will be given. We get

$$N_i = N_v \frac{e^{-\epsilon_i/kT}}{Z_v}, \tag{12-3}$$

where Z_v is the partition function

$$Z_v = \sum e^{-\epsilon_i/kT}. \tag{12-4}$$

The expressions for energy U_v and pressure P are the same as for indistinguishable particles, namely,

$$U_v = N_v k T^2 \left(\frac{\partial \ln Z_v}{\partial T}\right)_V, \tag{12-5}$$

and

$$P = N_v k T \left(\frac{\partial \ln Z_v}{\partial V}\right)_T. \tag{12-6}$$

But the expression for the entropy is simpler:

$$S_v = N_v k \ln Z_v + \frac{U_v}{T}. \tag{12-7}$$

The evaluation of the partition function is particularly simple, since it is merely a geometric progression with a ratio less than 1 and an infinite number of terms:

$$Z_\nu = e^{-h\nu/2kT} + e^{-3h\nu/2kT} + e^{-5h\nu/2kT} + \cdots$$

$$= e^{-h\nu/2kT}(1 + e^{-h\nu/kT} + e^{-2h\nu/kT} + \cdots)$$

or

$$Z_\nu = \frac{e^{-h\nu/2kT}}{1 - e^{-h\nu/kT}} \tag{12-8}$$

and

$$\boxed{\ln Z_\nu = -\frac{1}{2}\frac{h\nu}{kT} - \ln(1 - e^{-h\nu/kT}).} \tag{12-9}$$

The energy U_ν of N_ν simple harmonic vibrators is

$$U_\nu = N_\nu kT^2 \frac{\partial}{\partial T}(\ln Z_\nu)$$

$$= N_\nu kT^2 \left[\frac{1}{2}\frac{h\nu}{kT^2} - \frac{-e^{-h\nu/kT}(h\nu/kT^2)}{1 - e^{-h\nu/kT}}\right]$$

$$= N_\nu \left(\tfrac{1}{2}h\nu + \frac{h\nu}{e^{h\nu/kT} - 1}\right);$$

therefore the average energy per vibrator is

$$\langle\epsilon\rangle = \frac{U_\nu}{N_\nu} = \frac{h\nu}{2} + \frac{h\nu}{e^{h\nu/kT} - 1}. \tag{12-10}$$

In general, all the $3N$ equivalent simple harmonic oscillators do *not* have the same frequency. Let dN_ν be the number of oscillators whose frequency lies between ν and $\nu + d\nu$. Then,

$$dN\nu = g(\nu)\,d\nu \tag{12-11}$$

where $g(\nu)$, the number of vibrators per unit frequency band, must be determined for a given crystal or class of crystals and must satisfy the condition

$$\int dN_\nu = \int g(\nu)\,d\nu = 3N. \tag{12-12}$$

The energy of N particles of the crystal is then

$$U = \int \langle\epsilon\rangle\,dN_\nu = \int \left(\frac{h\nu}{2} + \frac{h\nu}{e^{h\nu/kT} - 1}\right) g(\nu)\,d\nu, \tag{12-13}$$

and the heat capacity at constant volume of this amount of crystal is

$$C_V = \left(\frac{\partial U}{\partial T}\right)_V = \int \frac{(1/k)(h\nu/T)^2 e^{h\nu/kT}}{(e^{h\nu/kT} - 1)^2}\,g(\nu)\,d\nu. \tag{12-14}$$

12-2 FREQUENCY SPECTRUM OF CRYSTALS

The simplest assumption concerning the vibration characteristics of a crystal was that of Einstein, namely, that all the $3N$ equivalent harmonic oscillators had the same frequency ν_E (subscript E for Einstein), as depicted in Fig. 12-2*a*. Equation (12-14) then reduces to the following simple form:

$$C_V = 3N \frac{k(h\nu_E/kT)^2 e^{h\nu_E/kT}}{(e^{h\nu_E/kT} - 1)^2}.$$

If we define the *Einstein characteristic temperature* Θ_E by the expression

$$\Theta_E = \frac{h\nu_E}{k} \tag{12-15}$$

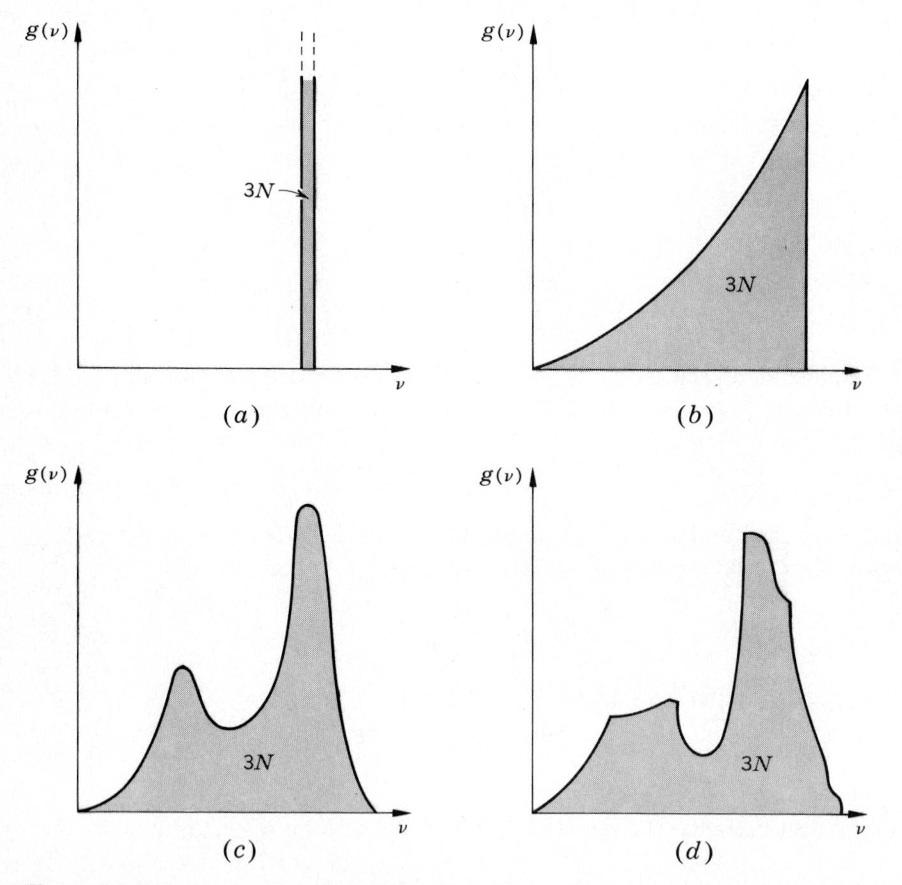

Figure 12-2 Frequency spectra of lattice vibrations. (*a*) Einstein approximation; (*b*) Debye approximation; (*c*) Blackman approximation; (*d*) more rigorous calculation.

and let $N = N_A$ (Avogadro's number)—remembering that $N_A k = R$ (the universal gas constant)—we get

$$\frac{c_V}{3R} = \left(\frac{\Theta_E}{T}\right)^2 \frac{e^{\Theta_E/T}}{(e^{\Theta_E/T} - 1)^2}. \tag{12-16}$$

This was the first attempt to apply quantum theory to the specific heat of solids, and although the assumption of equal frequencies for all equivalent harmonic oscillators is far from justified, the Einstein expression has the same general shape as the curves in Fig. 9-12. As $T \to \infty$, then $c_V/3R \to 1$, in agreement with the rule of Dulong and Petit. As $T \to 0$, then $c_V/3R$ approaches zero, in agreement with experiment; but it approaches zero exponentially, which is faster than experiment indicates.

The next approximation was made by Debye, who calculated a frequency distribution on the assumption that a crystal was a *continuous* medium supporting standing (or stationary) longitudinal and transverse waves. With this assumption it was a simple matter to show that a continuous spectrum of frequencies was present, starting with zero and terminating at a maximum frequency ν_m, according to the simple relation

$$g(\nu) = \frac{9N}{\nu_m^3} \nu^2. \tag{12-17}$$

It may be seen by inspection that $\int g(\nu)\, d\nu$ is equal to $3N$. The Debye approximation is depicted in Fig. 12-2*b*.

Substituting the Debye approximation into Eq. (12-14), we get

$$\frac{C_V}{3Nk} = \int_0^{\nu_m} \frac{(3\nu^2/\nu_m^3)(h\nu/kT)^2 e^{h\nu/kT}\, d\nu}{(e^{h\nu/kT} - 1)^2}.$$

It is convenient to define a new variable of integration x, where

$$x = \frac{h\nu}{kT} \quad \text{and} \quad x_m = \frac{h\nu_m}{kT} = \frac{\Theta}{T},$$

and the *Debye characteristic temperature*

$$\Theta = \frac{h\nu_m}{k}. \tag{12-18}$$

With these new quantities and with $N = N_A$,

$$\frac{c_V}{3R} = \frac{3}{(\Theta/T)^3} \int_0^{\Theta/T} \frac{x^4 e^x\, dx}{(e^x - 1)^2}. \tag{12-19}$$

At $T \to \infty$, the upper limit of the integral becomes small, and the integrand evaluated at small values of x reduces to x^2 and, when integrated, yields $\frac{1}{3}(\Theta/T)^3$. Therefore $c_V/3R \to 1$, as required by the rule of Dulong and Petit.

To find the limiting value of $c_V/3R$ as $T \to 0$, it is convenient to express Eq. (12-19) in a different form by integrating by parts.

$$\frac{3}{(\Theta/T)^3}\int_0^{\Theta/T}\frac{x^4 e^x\,dx}{(e^x-1)^2} = \frac{3}{(\Theta/T)^3}\int_0^{\Theta/T} x^4\, d\left(\frac{-1}{e^x-1}\right)$$

$$= \frac{3}{(\Theta/T)^3}\left[\int_0^{\Theta/T}\frac{4x^3\,dx}{e^x-1} - \frac{(\Theta/T)^4}{e^{\Theta/T}-1}\right];$$

therefore,

$$\frac{c_V}{3R} = 4\cdot\frac{3}{(\Theta/T)^3}\int_0^{\Theta/T}\frac{x^3\,dx}{e^x-1} - \frac{3(\Theta/T)}{e^{\Theta/T}-1}. \tag{12-20}$$

As $T \to 0$,

$$\frac{3(\Theta/T)}{e^{\Theta/T}-1} \to 0,$$

and

$$\int_0^{\theta/T}\frac{x^3\,dx}{e^x-1} \to \int_0^{\infty}\frac{x^3\,dx}{e^x-1} = 3!\,\zeta(4),$$

where $\zeta(4)$ is a *Riemann zeta function*,† equal to

$$\zeta(4) = 1 + \frac{1}{2^4} + \frac{1}{3^4} + \cdots = \frac{\pi^4}{90}.$$

Hence,

$$\frac{c_V}{3R} \to 4\cdot\frac{3}{(\Theta/T)^3}\,\frac{3!\,\pi^4}{90} = \frac{4\pi^4}{5}\left(\frac{T}{\Theta}\right)^3.$$

Since $4\pi^4/5 = 77.9$, and at low temperatures $c_V = c_P = c$, we may write *Debye's* T^3 *law* in either of these two forms:

$$\left.\begin{aligned}\frac{c}{3R} &= 77.9\left(\frac{T}{\Theta}\right)^3,\\ \text{or}\qquad c &= \frac{(125)^3\ \text{mJ}}{\text{mol}\cdot\text{K}}\left(\frac{T}{\Theta}\right)^3.\end{aligned}\right\}\quad\left(\text{for } \frac{T}{\Theta} < 0.04\right) \tag{12-21}$$

The entire course of $c_V/3R$ given by Eq. (12-19) or (12-20) cannot be reduced to a simple form but must be evaluated numerically. The results of this calculation are listed in Table 12-1 and are plotted in Fig. 12-3. A more extensive table will be found in the *AIP Handbook*.

† See Appendix D.

Table 12-1 Debye's heat capacity

$\frac{T}{\Theta}$	$\frac{c_V}{3R}$	$\frac{T}{\Theta}$	$\frac{c_V}{3R}$	$\frac{T}{\Theta}$	$\frac{c_V}{3R}$
0	0	0.175	0.293	0.7	0.916
0.025	0.00122	0.20	0.369	0.8	0.926
0.050	0.00974	0.25	0.503	0.9	0.942
0.075	0.0328	0.3	0.613	1.0	0.952
0.100	0.0758	0.4	0.746	1.2	0.963
0.125	0.138	0.5	0.825	1.5	0.980
0.150	0.213	0.6	0.869	2.0	0.988

12-3 THERMAL PROPERTIES OF NONMETALS

The fundamental assumption of the Debye theory, leading to a simple quadratic function for $g(\nu)$, as shown in Fig. 12-2*b*, is quite crude compared with theoretically derived functions such as that of Blackman shown in Fig. 12-2*c*. It is quite astonishing that the Debye curve in Fig. 12-3 is in such good agreement with experiment. Only when heat capacities are measured with great accuracy do

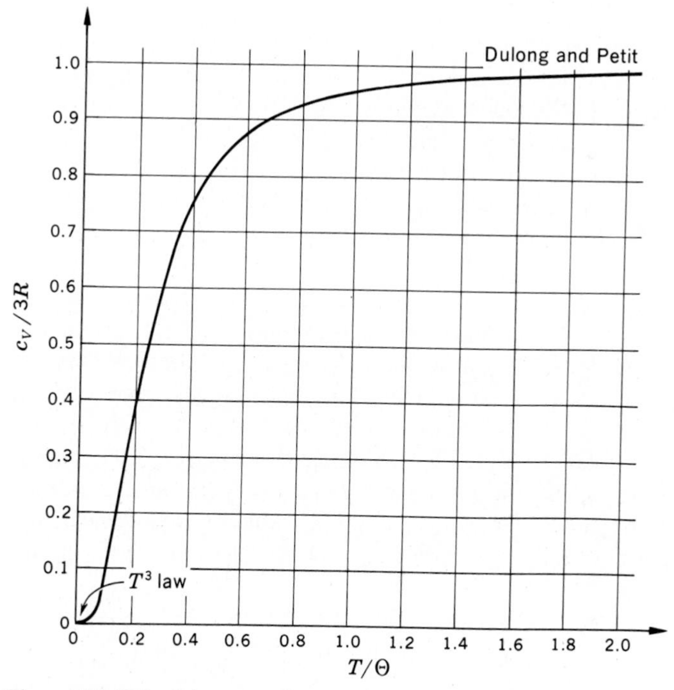

Figure 12-3 Debye's heat capacity.

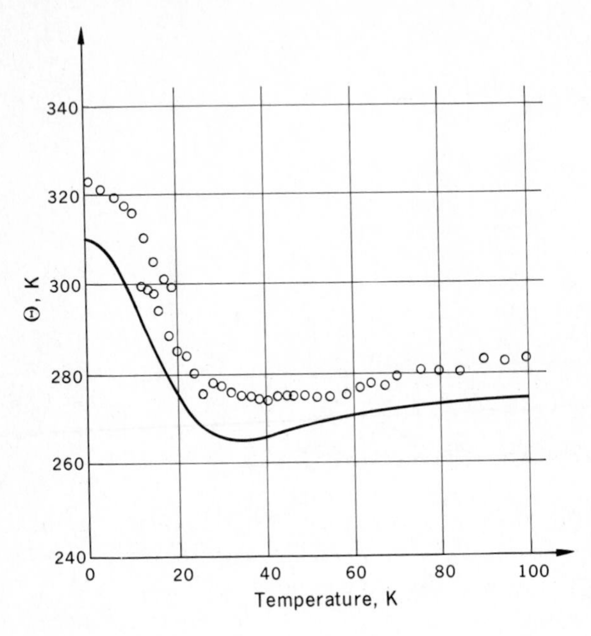

Figure 12-4 Variations of Θ with T for NaCl. *(Curve calculated by Lundquist, Lundstrom, Tenerz, and Waller.)*

departures from Debye's curve show up. These are obtained as follows. The accurate experimental value of $c_V/3R$ at a known T is compared with the Debye curve of Fig. 12-3, and the corresponding value of T/Θ is obtained; from this is derived the value of Θ. This is done at a number of temperatures. If all values of Θ so obtained were the same, the Debye theory would hold perfectly. Such is not the case, however, as the experimental points in Fig. 12-4 show. The exact calculation of the frequency spectrum of lattice vibrations is a very complicated and tedious task. The use of computers has helped a great deal in numerical calculations, and physicists are having reasonable success in explaining experimentally observed heat capacities.

Debye's model, although crude in the medium temperature range, is quite rigorous at very low temperatures, below Θ/100, well within the T^3 region. The reason may be seen in curves (*b*), (*c*), and (*d*) of Fig. 12-2. At very low temperatures, only the small frequencies of lattice vibration are excited, so that only the beginning segments of the curves play a role. No matter how rigorously the calculation of $g(\nu)$ is carried out, the beginning segment of $g(\nu)$ is always quadratic. One would expect therefore that, at very low temperatures, the Debye T^3 law should be strictly obeyed and also that the value of Θ obtained from such measurements should be the correct value of Θ at absolute zero. Using the T^3 law in the form

$$c_V = \left(\frac{125}{\Theta}\right)^3 T^3 \qquad \frac{\text{mJ}}{\text{mol} \cdot \text{K}},$$

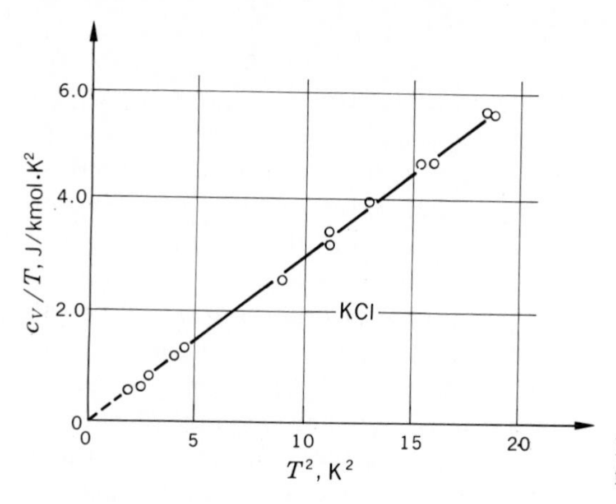

Figure 12-5 Verification of Debye T^3 law for KCl. *(Keesom and Pearlman.)*

and dividing both sides by T, we get

$$\frac{c_V}{T} = \left(\frac{125}{\Theta}\right)^3 T^2 \qquad \frac{\text{mJ}}{\text{mol} \cdot \text{K}^2}.$$

The graph in Fig. 12-5 shows how well the law is obeyed between 0 and 4.5 K for KCl. Value of Θ near absolute zero are given for a number of non-metals in Table 12-2.

Table 12-2 Debye temperatures for nonmetals

Nonmetal	Θ, K	Nonmetal	Θ, K
^{3}He (bcc)	16	KBr	174
^{4}He (hcp)	26	KCl	·235
Ne	75	KF	336
Ar	93	KI	132
Kr	72	RbBr	131
Xe	64	RbCl	165
Se	90	RbI	103
Te	153	InSb	200
As	282	CaF_2	510
Ge	370	SiO_2	470
Si	640	Fe_2O_3	660
C (graphite)	420	$FeSe_2$	366
C (diamond)	2230	FeS_2	637
LiCl	422	MgO	946
LiF	732	ZnS	315
NaBr	225	TiO_2	760
NaCl	321	$NiSe_2$	297
NaF	492	Bi_2Te_3	155
NaI	164	H_2O	192

12-4 THERMAL PROPERTIES OF METALS

In a metal crystal the lattice sites are occupied by single metal atoms whose vibrations obey the same laws that have been found to hold for nonmetals. In addition to the lattice vibrations, however, there are free electrons whose number is of the same order of magnitude as the number of atoms and whose motions resemble those of the molecules of a gas. If the classical theory of an ideal gas were valid for the electron gas in a metal, the molar heat capacity at constant volume should be augmented *at all temperatures* by the constant value $\frac{3}{2}R$. Hence at high temperatures, instead of reaching the Dulong and Petit value $3R$, it should reach the value $\frac{9}{2}R$, or about 37 J/mol · K; and at low temperatures, c_V should not approach zero as T approaches zero but should approach $\frac{3}{2}R$. The fact that the temperature variation of c_V of metals had the same general features as that of nonmetals remained a puzzle until Sommerfeld applied quantum statistics (Fermi-Dirac statistics) to the free electrons in a metal and showed that the contribution of the electrons to the heat capacity of the metal is linear in T and is appreciable only at low temperatures when the Debye T^3 term becomes small. At higher temperatures, the linear term is small compared with the effect of lattice vibrations. We shall return to this point later, after we have studied other thermal properties of metals.

The c_P and β of metals are insensitive to moderate changes of pressure and vary with temperature in the same way as nonmetals. The graphs for copper (as typical of all metals) in Fig. 12-6, which should be compared with those of NaCl

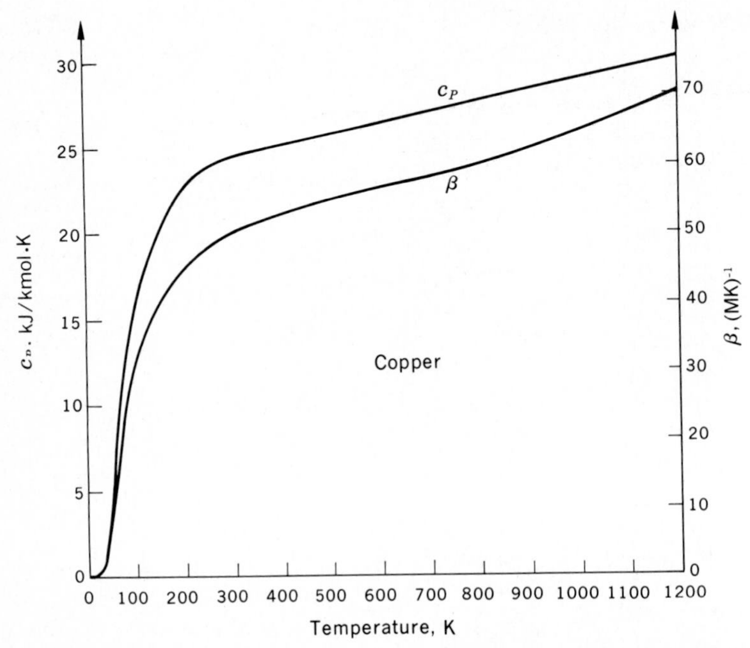

Figure 12-6 Temperature variation of c_P and β of copper is similar to that of nonmetals. (Compare with Fig. 9-6.)

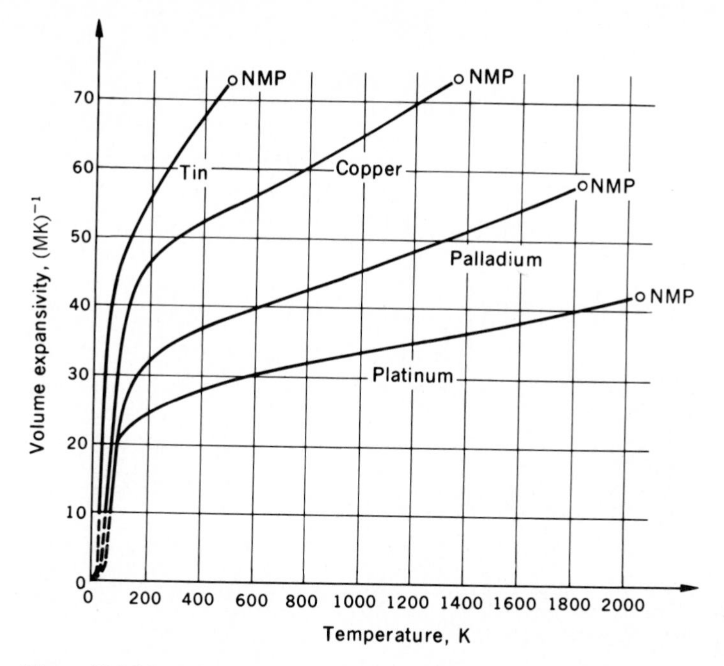

Figure 12-7 Normal temperature variation of volume expansivity of metals. (NMP, normal melting point.)

in Fig. 9-6, show that both c_P and β are zero at absolute zero and that, from 0 to 50 K, rise rapidly. At higher temperatures the curves flatten out but do not approach asymptotic values. The temperature dependence of the β of metals shows a curious regularity, as shown in Fig. 12-7, where it may be seen that the higher the melting point, the lower the volume expansivity. As a result, in the temperature interval from absolute zero to the melting point, all metals expand approximately the same fraction of their original volumes. It seems almost as if a metal like tin, realizing it is going to melt soon, expands rapidly with the temperature, whereas platinum, with a high melting point, slows down its rate.

In Fig. 12-8 it may be seen that the isothermal and adiabatic compressibilities of copper vary with the temperature like those of NaCl in Fig. 9-7. The complete temperature dependence of the thermal properties of copper is given in Table 12-3, and c_V is compared with c_P in Fig. 12-9. Notice that the c_V of copper goes somewhat beyond the Dulong and Petit value above 700 K. We shall see later that this additional heat capacity may be attributed to the free-electron gas inside the metal.

The free-electron gas in a metal differs from an ordinary gas in two main respects. In an ordinary gas, the number of quantum states accessible to the gas molecules is very much larger than the number of molecules ($g_i \gg N_i$), whereas in a metal the number of quantum states and the number of electrons are comparable. Let us see how this comes about. An electron moving more or less freely in a cubical metal of length L and volume $V = L^3$ has a kinetic energy ϵ

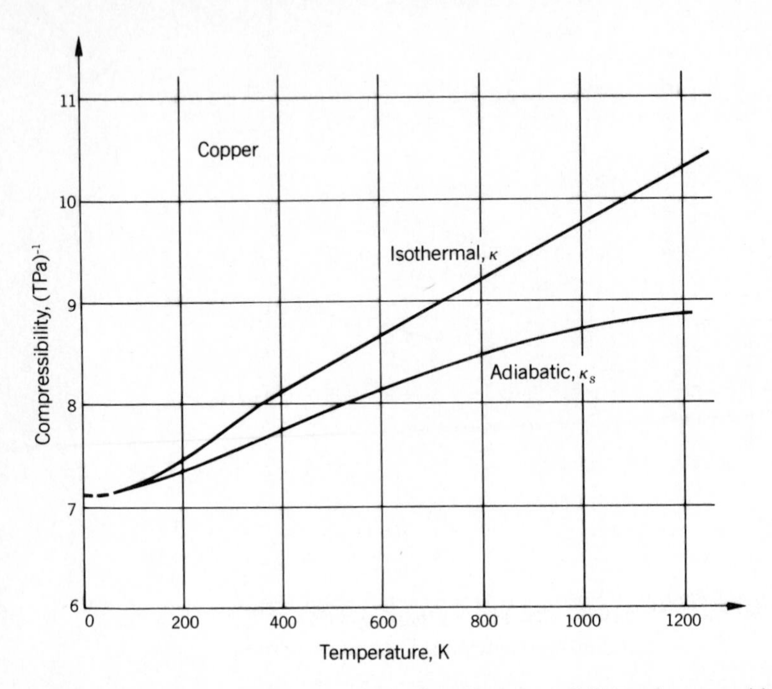

Figure 12-8 Variation of isothermal and adiabatic compressibilities of copper with temperature.

given by exactly the same expression as that developed in Chap. 11 for a molecule moving in a cubical box, namely,

$$\epsilon = \frac{h^2}{8mL^2}(n_x^2 + n_y^2 + n_z^2),$$

where m is the mass of an electron and n_x, n_y, and n_z are very large integers. Each triplet of values of n_x, n_y, and n_z refers to one quantum state, and all the quantum states referring to one energy ϵ lie on the surface of a sphere of radius r

Table 12-3 Thermal properties of copper

T, K	c_P kJ/mol K	β, (MK)$^{-1}$	κ (TPa)$^{-1}$	v liters/kmol	c_V kJ/kmol K	κ_S, (TPa)$^{-1}$
50	6.25	11.4	7.13	7.00	6.24	7.12
100	16.1	31.5	7.21	7.01	16.0	7.19
150	20.5	40.7	7.34	7.02	20.3	7.27
200	22.8	45.3	7.49	7.03	22.4	7.34
250	24.0	48.3	7.63	7.04	23.5	7.44
300	24.5	50.4	7.78	7.06	23.8	7.54
500	25.8	54.9	8.39	7.12	24.5	7.95
800	27.7	60.0	9.23	7.26	25.4	8.49
1200	30.2	70.2	10.31	7.45	26.0	8.89

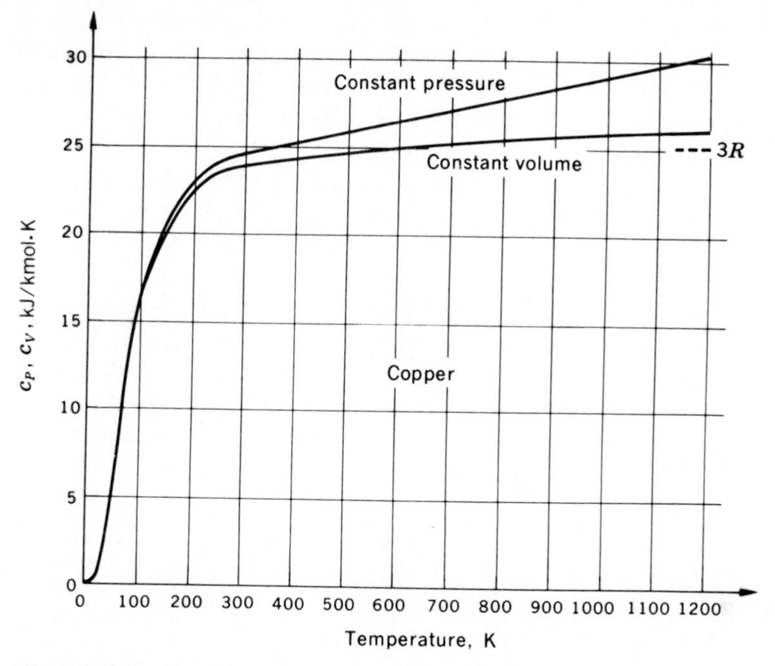

Figure 12-9 Temperature variation of c_P and c_V of copper.

in a space where n_x, n_y, and n_z are cartesian coordinates and

$$r^2 = n_x^2 + n_y^2 + n_z^2.$$

The quantum states referring to an energy interval between ϵ and $\epsilon + d\epsilon$ lie between the spheres of radius r and $r + dr$, and the number of these states is the volume of the positive part of this spherical shell. Since

$$r^2 = \frac{8mL^2}{h^2}\epsilon,$$

then

$$2r\,dr = \frac{8mL^2}{h^2}\,d\epsilon,$$

and the volume of the spherical shell between r and $r + dr$ is $4\pi r^2\,dr$. Therefore, the number of quantum states $g\,d\epsilon$ with energy between ϵ and $\epsilon + d\epsilon$ is one-eighth of this volume, or

$$g\,d\epsilon = \tfrac{1}{8}\cdot 4\pi r^2\,dr = \frac{\pi}{4}r\cdot 2r\,dr$$

$$= \frac{\pi}{4}\sqrt{\frac{8mL^2}{h^2}}\,\epsilon^{1/2}\frac{8mL^2}{h^2}\,d\epsilon,$$

or

$$g = 2\pi V\left(\frac{2m}{h^2}\right)^{3/2}\epsilon^{1/2}. \tag{12-22}$$

The number of quantum states per unit energy interval, g, is seen to depend on $m^{3/2}$. Since an electron has a mass only about 1/10,000 that of, say, a helium atom, it follows that, for a given energy, there may be $(10^{-4})^{3/2}$, or 10^{-6}, fewer quantum states for an electron gas than for an ordinary gas. Also, in a given volume (say, 1 cm^3), there may be 10^{19} atoms of helium, but in a metal of the same volume there might be about 10^{23} free electrons, assuming about one free electron per atom of metal. With 10^4 more electrons and about 10^{-6} fewer quantum states, it is clear that the number of electrons and the number of quantum states may be comparable.

Now, according to the Pauli exclusion principle, only two electrons (with opposite spin) can occupy the same quantum state. At a temperature of absolute zero, all the electrons cannot be in the lowest energy state; instead, all energy states are filled to an energy ϵ_F, known as the *Fermi energy.* Suppose that a volume V of metal contains N free electrons. At absolute zero, the number of states occupied by these electrons is $N/2$, where from Eq. (12-22),

$$\frac{N}{2} = 2\pi V\left(\frac{2m}{h^2}\right)^{1/2}\int_0^{\epsilon_F}\epsilon^{1/2}\,d\epsilon,$$

or

$$\frac{N}{V} = \frac{8\pi}{3}\left(\frac{2m}{h^2}\right)^{3/2}\epsilon_F^{3/2}. \tag{12-23}$$

To appreciate how large the Fermi energy is, it is instructive to calculate the temperature of an ordinary gas at which a molecule would have the energy ϵ_F. This temperature, called the *Fermi temperature*, is defined as

$$T_F = \frac{\epsilon_F}{k}; \tag{12-24}$$

where k is Boltzmann's constant. From Eq. (12-23), we get

$$T_F = \left(\frac{3}{8\pi}\right)^{2/3}\frac{h^2}{2mk}\left(\frac{N}{V}\right)^{2/3}. \tag{12-25}$$

The molar volume of copper is 7.1×10^{-3} m^3/kmol; and assuming one free electron for every two copper atoms, $N = 3.0 \times 10^{26}$ electrons per kilomole. Hence,

$$T_F = 0.24\frac{(6.6)^2 \times 10^{-68} \times 12 \times 10^{18}}{2 \times 9.1 \times 10^{-31} \times 1.4 \times 10^{-23}},$$

or

$$T_F \sim 50{,}000 \text{ K}.$$

This means that the closely packed electrons in a metal *at absolute zero* have the energy that an ordinary gas would have at 50,000 K! This energy, known as the *zero-point energy* of the electron gas, may be several hundred thousand joules per mole.

When the temperature of the metal is raised to a value T, only electrons whose energies are near ϵ_F can be raised to higher states, so that the increase of energy per unit temperature rise dU/dT, which is the heat capacity, is small. To calculate this electronic contribution to the molar heat capacity of the metal, c_e, it is necessary to develop a statistics of indistinguishable particles obeying the Pauli exclusion principle where $g_i \approx N_i$. First formulated by Fermi and Dirac, this is called the *Fermi-Dirac statistics* (see Prob. 11-6). When a system obeying this type of statistics comes to equilibrium, it is a simple matter to show that the number of particles N_i with energy ϵ_i is given by

$$N_i \frac{g_i}{e^{(\epsilon_i - \epsilon_F)/kT} + 1},$$

and the energy U of the entire system of N particles is

$$U = \frac{3N/V}{2\epsilon_F^{3/2}} \int_0^\infty \frac{\epsilon^{3/2}\, d\epsilon}{e^{(\epsilon - \epsilon_F)/kT} + 1}.$$

The evaluation of this integral is difficult. When T/T_F is small, the integrand may be expanded in a series and integrated term by term. When only the first two terms are retained,

$$U = \tfrac{3}{5} N\epsilon_F \left[1 + \frac{5\pi^2}{12}\left(\frac{kT}{\epsilon_F}\right)^2 - \cdots\right],$$

and

$$C_V = Nk \frac{\pi^2}{2}\frac{kT}{\epsilon_F} + \cdots .$$

Calling the constant γ', we may write for the molar heat capacity

$$\boxed{c_e = \gamma' T} \qquad (\text{for } T \ll T_F). \tag{12-26}$$

The entire course of c_e is shown in Fig. 12-10, where it may be seen that, when $T \approx T_F$, then $c_e \approx \frac{3}{2}R$. In the linear region of the curve,

$$c_e = \left(31 \frac{\text{J}}{\text{mole} \cdot \text{K}}\right)\frac{T}{T_F} \qquad (\text{for } T/T_F < 0.2).$$

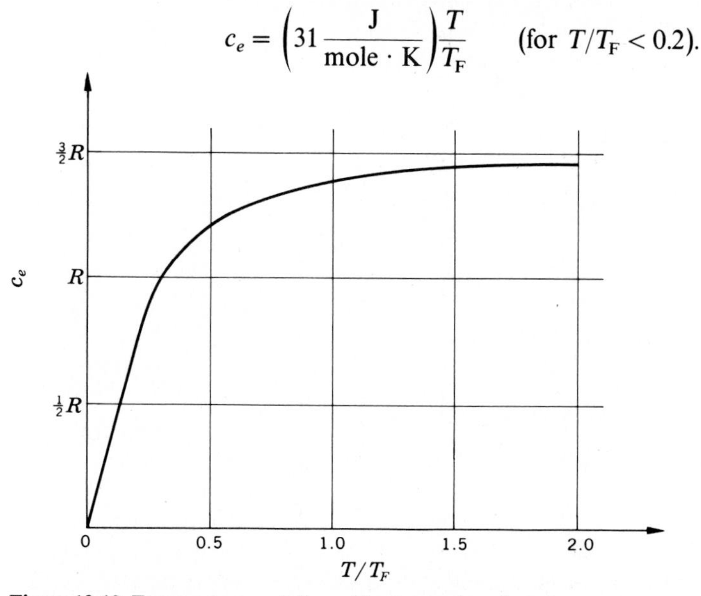

Figure 12-10 Temperature variation of heat capacity of an electron gas.

It was mentioned previously that the c_V of copper at 1200 K exceeds the Dulong and Petit value, as shown in Fig. 12-9. The electronic contribution at this temperature is

$$c_e = \left(31 \frac{\text{J}}{\text{mol} \cdot \text{K}}\right) \frac{1200}{50{,}000} = 0.75 \text{ J/mol} \cdot \text{K},$$

and when this is added to the lattice contribution of 25 J/mol · K, the result 25.8 J/mol · K is in good agreement with the value 26 J/mol · K shown in Fig. 12-9.

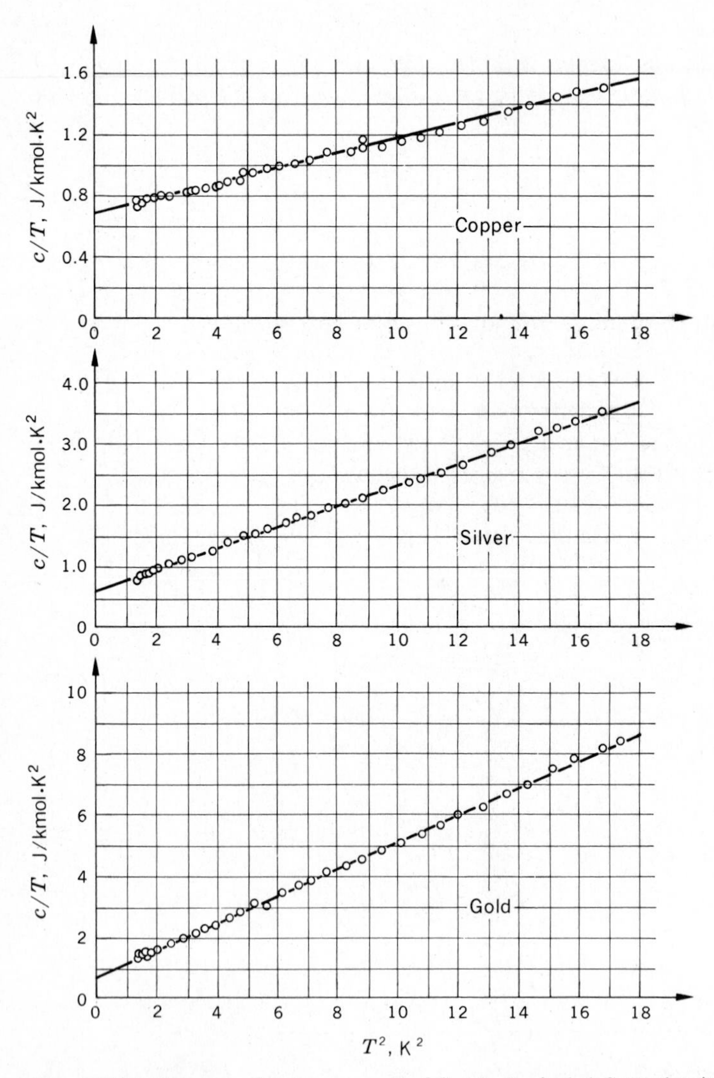

Figure 12-11 Heat-capacity measurements of Corak, Garfunkel, Satterthwaite, and Wexler.

Table 12-4 Debye temperatures and electronic constants for metals

Metal	Θ, K	γ', $\frac{\text{J}}{\text{kmol} \cdot \text{K}^2}$	Metal	Θ, K	γ', $\frac{\text{J}}{\text{kmol} \cdot \text{K}^2}$	Metal	Θ, K	γ', $\frac{\text{J}}{\text{kmol} \cdot \text{K}}$
Li	344	1.63	Rb	56	2.4	Yb	120	2.9
Be	1440	0.17	Sr	147	3.6	Lu	210	11.3
Na	158	1.4	Y	280	10.2	Hf	252	2.16
Mg	400	1.3	Zr	291	2.80	Ta	240	5.9
Al	428	1.35	Nb	275	7.79	W	400	1.3
K	91	2.1	Mo	450	2.0	Re	430	2.3
Ca	230	2.9	Ru	600	3.3	Os	500	2.4
Sc	360	10.7	Rh	480	4.9	Ir	420	3.1
Ti	420	3.5	Pd	274	9.42	Pt	240	6.8
V	380	9.8	Ag	225	0.650	Au	165	0.69
Cr	630	1.40	Cd	209	0.69	Hg	72	1.79
Mn	410	14	In	108	1.6	Tl	79	1.47
Fe	420	3.1	Sn	199	1.78	Pb	105	3.0
Co	445	4.7	Sb	211	0.112	Bi	119	0.021
Ni	450	7.1	Cs	38	3.2	Th	163	4.3
Cu	343	0.688	Ba	110	2.7	U	207	10.0
Zn	327	0.65	La	142	10	Pu	160	13
Ga	320	0.60	Dy	210				
As	282	0.19	Tm	200	10.5			

The region of greatest interest in the study of the electron gas in a metal is at low temperatures ($T < \Theta/25$), where Debye's T^3 law holds. In this region, the total heat capacity in mJ/mol · K is given by

$$c = \left(\frac{125}{\Theta}\right)^3 T^3 + \gamma' T,$$

a relation first derived by Sommerfeld. Dividing by T, we get

$$\frac{c}{T} = \left(\frac{125}{\Theta}\right)^3 T^2 + \gamma'; \tag{12-27}$$

therefore, a graph of c/T plotted against T^2 should yield a straight line with a slope equal to $(125/\Theta)^3$ and an intercept γ'. The experimental measurements on copper, silver, and gold made at the Westinghouse laboratories and plotted in Fig. 12-11 verify Eq. (12-27) very well and yield reliable values of Θ and γ'. Values of Θ and γ' of other metals are listed in Table 12-4.

It is an interesting fact that conduction electrons in a metal contribute to the thermal expansion as well as to the heat capacity and, according to the same relation,

$$\beta = aT + bT^3, \tag{12-28}$$

where the linear term represents the electronic contribution and the cubic term

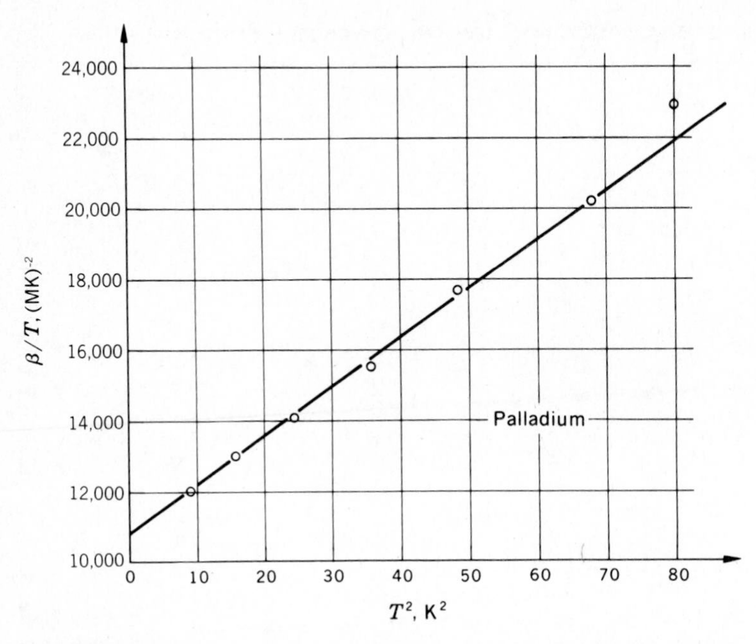

Figure 12-12 Thermal expansion of palladium to show the effect of conduction electrons. *(G. K. White, 1961.)*

Table 12-5 Expansion of metals at low temperatures

($\beta = aT + bT^3$)

Metal	a, $(\mathrm{MK})^{-2}$	b, $(\mathrm{kK})^{-4}$	Metal	a, $(\mathrm{MK})^{-2}$	b, $(\mathrm{kK})^{-4}$
Mo	1,290	11.4	Al	3,300	72
W	90	13.5	Pb	3,000	4,200
Re	2,700	20.4	Mg	3,600	126
Pt	6,600	177	Ta	3,100	96
Cd	600	1,140	Pd	10,800	129

that due to the lattice. A verification of this equation in the case of palladium is shown in Fig. 12-12, and numerical values of a and b are listed in Table 12-5.

PROBLEMS

12-1 It was shown in Art. 12-1 that the partition function of an Einstein crystal is

$$Z = \frac{e^{-h\nu/2kT}}{1 - e^{-h\nu/kT}}.$$

If the crystal consists of N_A lattice points, calculate (*a*) the Helmholtz function F, (*b*) the pressure P, and (*c*) the entropy S. (*d*) Express the zero-point energy in terms of Θ_E.

12.2 Using the Debye approximation, show that the total energy U of N_A lattice points ($3N_A$ independent harmonic oscillators) is

$$U = \tfrac{9}{8}R\Theta + \frac{9RT}{(\Theta/T)^3}\int_0^{\Theta/T} \frac{x^3\,dx}{e^x - 1}.$$

What is the interpretation of the term $\tfrac{9}{8}R\Theta$?

12-3 Plot the Einstein $c_V/3R$ curve against T/Θ_E on the same graph as the Debye $c_V/3R$ curve against T/Θ, and then compare the two curves.

12-4 A system consists of N_A distinguishable, independent particles, each of which is capable of existing in only two nondegenerate energy states 0 and ϵ.

(*a*) What is the partition function?
(b) Calculate the energy.
(*c*) Calculate c_V.
(*d*) Plot c_V/R as a function of kT/ϵ from $kT/\epsilon = 0$ to 1.

12-5 Given a crystal obeying Debye's approximation with Θ a function of V only, assume the entropy S to be a function of T/Θ. Γ is defined by the equation

$$\Gamma = -\frac{d \ln \Theta}{d \ln V}.$$

(*a*) Show that

$$\Gamma = \frac{\beta V}{C_V \kappa}.$$

[*Hint:* Use Maxwell's third relation.]
(*b*) Calculate Γ of NaCl at a few temperatures using Table 9-2, and plot Γ against T.
(*c*) Show that

$$\gamma = 1 + \Gamma\beta T.$$

12-6 The partition function of a Debye crystal is

$$\ln Z = -\frac{9}{(\Theta/T)^3}\int_0^{\Theta/T} x^2 \ln(1 - e^{-x})\,dx.$$

(*a*) Show that

$$\ln Z = -3\ln(1 - e^{-\Theta/T}) + \frac{9}{(\Theta/T)^3}\int_0^{\Theta/T} \frac{x^3\,dx}{e^x - 1}.$$

(*b*) Calculate the Helmholtz function.
(*c*) Show that the equation of state of the crystal is given by

$$PV + f(V) = \Gamma(U - U_0),$$

where U_0 is the zero-point energy.

CHAPTER

THIRTEEN

HIGHER-ORDER PHASE TRANSITIONS; CRITICAL PHENOMENA

13-1 JOULE-KELVIN EFFECT

In this chapter we shall study the behavior of pure substances during a higher-order transition from one phase to another. In the best-known first order phase transitions, namely, the melting of ice and the vaporization of water, the regions of temperature and pressure are easily accessible without special apparatus. Some of the most interesting materials, however, such as nitrogen, hydrogen, and helium, whose phase transitions contain still unsolved problems, exist only at low temperatures. It is important, therefore, to learn how these low temperatures are achieved and maintained. The first step is to liquefy air, and the most economical way to accomplish this is with the aid of the *Joule-Kelvin effect* or, as it was formerly called, the *porous-plug experiment*.

In the porous-plug experiment a gas is made to undergo a continuous throttling process. By means of a pump, a constant pressure is maintained on one side of a porous plug and a constant lower pressure on the other side. In the original experiments of Joule and Kelvin a cotton plug was used, and the gas flowed through it parallel to the axis of the pipe. In modern measurements a cup of a strong porous material capable of withstanding great force allows the gas to seep through in a radial direction. Rigid precautions are taken to provide adequate thermal insulation for the plug and the portion of the pipe near the plug. Suitable manometers and thermometers are used to measure the pressure and temperature of the gas on both sides of the plug.

The experiment is performed in the following way. The pressure P_i and temperature T_i on the high-pressure side of the plug are chosen arbitrarily. The

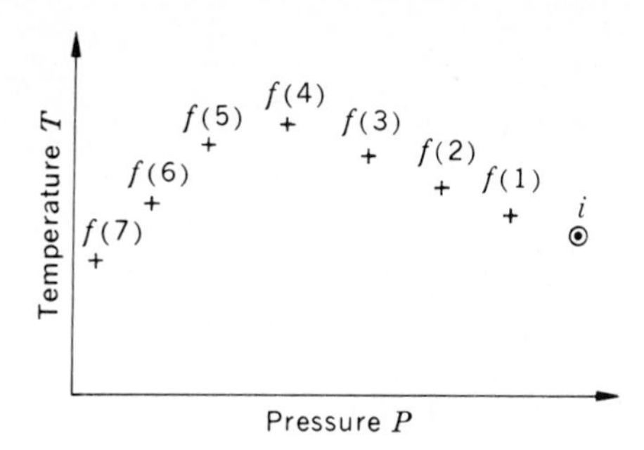

Figure 13-1 Isenthalpic states of a gas.

pressure P_f on the other side of the plug is then set at any value less than P_i, and the temperature of the gas T_f is measured. P_i and T_i are kept the same; P_f is changed to another value, and the corresponding T_f is measured. This is done for a number of different values of P_f, and the corresponding T_f is measured in each case. P_f is the independent variable of the experiment, and T_f is the dependent variable. The results provide a set of discrete points on a TP diagram, one point being $P_i T_i$ and the others being the various P_f's and T_f's indicated in Fig. 13-1 by numbers (1) to (7). Although the points shown in the figure do not refer to any particular gas, they are typical of most gases. It can be seen that, if a throttling process takes place between the states $P_i T_i$ and $P_f T_f$ (3), there is a rise of temperature. Between $P_i T_i$ and $P_f T_f$ (7), however, there is a drop of temperature. In general, the temperature change of a gas upon seeping through a porous plug depends on the three quantities P_i, T_i, and P_f and may be an increase or a decrease, or there may be no change whatever.

According to the principles developed in Art. 9-1, the eight points plotted in Fig. 13-1 represent equilibrium states of some constant mass of the gas (say, 1 kg) at which the gas has the same enthalpy. All equilibrium states of the gas corresponding to this enthalpy must lie on some curve, and it is reasonable to assume that this curve can be obtained by drawing a smooth curve through the discrete points. Such a curve is called an *isenthalpic curve*. The student must understand that *an isenthalpic curve is not the graph of a throttling process*. No such graph can be drawn because in any throttling process the intermediate states traversed by a gas cannot be described by means of thermodynamic coordinates. An isenthalpic curve is the locus of all points representing equilibrium states of the same enthalpy. The porous-plug experiment is performed to provide a few of these points, and the rest are obtained by interpolation.

The temperature T_i on the high-pressure side is now changed to another value, with P_i being kept the same. P_f is again varied, and the corresponding T_f's are measured. Upon plotting the new $P_i T_i$ and the new P_f's and T_f's, another discrete set of points is obtained, which determines another isenthalpic curve corresponding to a different enthalpy. In this way, a series of isenthalpic curves is obtained. Such a series is shown in Fig. 13-2 for nitrogen.

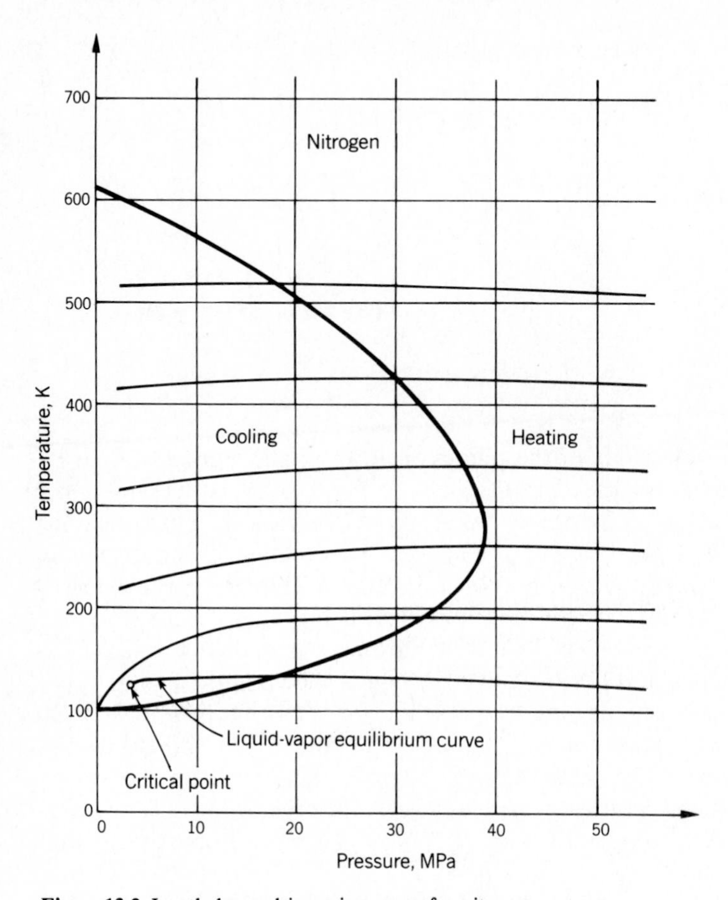

Figure 13-2 Isenthalps and inversion curve for nitrogen.

The numerical value of the slope of an isenthalpic curve on a *TP* diagram at any point is called the *Joule-Kelvin coefficient* and will be denoted by μ. Thus,

$$\mu = \left(\frac{\partial T}{\partial P}\right)_h. \tag{13-1}$$

The locus of all points at which the Joule-Kelvin coefficient is zero. i.e., the locus of the maxima of the isenthalpic curves, is known as the *inversion curve* and is shown for N_2 in Fig. 13-2 as a heavy closed curve. The region inside the inversion curve, where μ is positive, is called the *region of cooling;* whereas outside, where μ is negative, is the *region of heating*.

Since the Joule-Kelvin coefficient involves T, P, and h, we seek a relation among the differentials of T, P, and h. In general, the difference in molar enthalpy between two neighboring equilibrium states is

$$dh = T\,ds + v\,dP,$$

and, according to the second $T\,ds$ equation,

$$T\,ds = c_P\,dT - T\left(\frac{\partial v}{\partial T}\right)_P dP.$$

Substituting for $T\,ds$, we get

$$dh = c_P\,dT - \left[T\left(\frac{\partial v}{\partial T}\right)_P - v\right] dP,$$

or

$$dT = \frac{1}{c_P}\left[T\left(\frac{\partial v}{\partial T}\right)_P - v\right] dP + \frac{1}{c_P}\,dh.$$

Since $\mu = (\partial T/\partial P)_h$,

$$\boxed{\mu = \frac{1}{c_P}\left[T\left(\frac{\partial v}{\partial T}\right)_P - v\right].} \tag{13-2}$$

This is the thermodynamic equation for the Joule-Kelvin coefficient. It is evident that, for an ideal gas,

$$\mu = \frac{1}{c_P}\left(T\frac{R}{P} - v\right) = 0.$$

The most important application of the Joule-Kelvin effect is in the liquefaction of gases.

13-2 LIQUEFACTION OF GASES BY THE JOULE-KELVIN EFFECT

An inspection of the isenthalpic curves and the inversion curve of Fig. 13-2 shows that, for the Joule-Kelvin effect to give rise to cooling, the initial temperature of the gas must be below the point where the inversion curve cuts the temperature axis, i.e., below the maximum inversion temperature. For many gases, room temperature is already below the maximum inversion temperature, so that no precooling is necessary. Thus, if air is compressed to a pressure of 200 atm and a temperature of 52°C, then, after throttling to a pressure of 1 atm, it will be cooled to 23°C. On the other hand, if helium originally at 200 atm and 52°C is throttled to 1 atm, its temperature will rise to 64°C.

Figure 13-3 shows that, for the Joule-Kelvin effect to produce cooling in hydrogen, the hydrogen must be cooled below 200 K. Liquid nitrogen is used in most laboratories for this purpose. To produce Joule-Kelvin cooling in helium, the helium is first cooled with the aid of liquid hydrogen. Table 13-1 gives the maximum inversion temperatures of a few gases commonly used in low-temperature work. The inversion curve for helium is shown in Fig. 13-4.

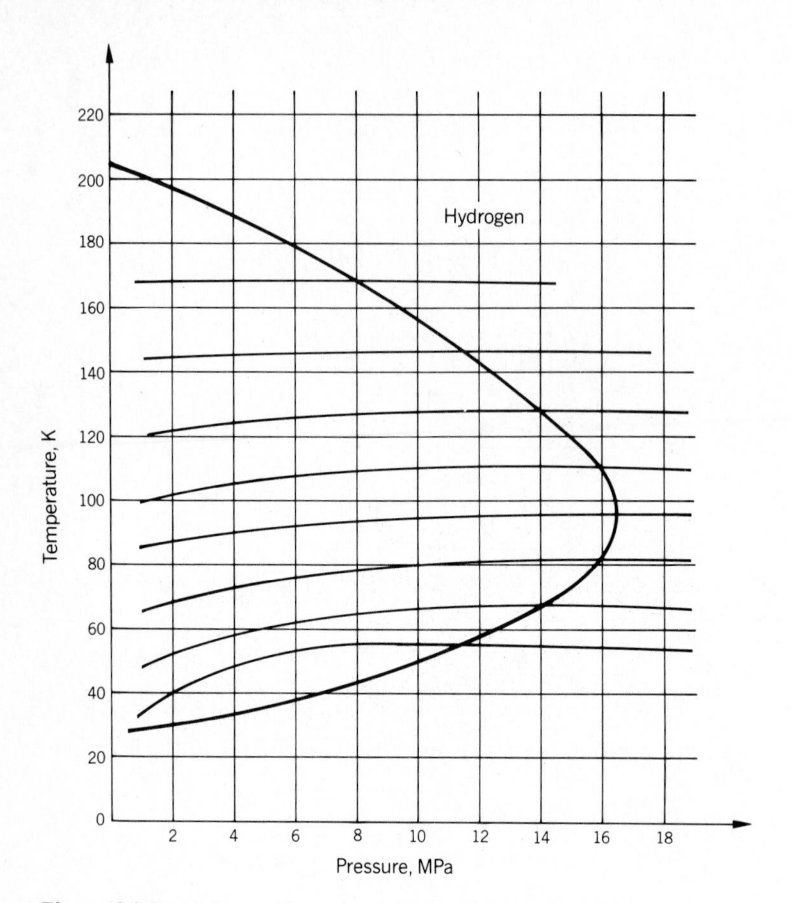

Figure 13-3 Isenthalps and inversion curve for hydrogen.

It is clear from Figs. 13-2, 13-3, and 13-4 that, once a gas has been precooled to a temperature lower than the maximum inversion temperature, the optimum pressure from which to start throttling corresponds to a point on the inversion curve. Starting at this pressure and ending at atmospheric pressure, the largest temperature drop is produced. This, however, is not large enough to produce liquefaction. Consequently, the gas that has been cooled by throttling is used to cool the incoming gas, which after throttling becomes still cooler. After many repetitions of these successive coolings, the gas is lowered to such a temperature that after throttling it becomes partly liquefied. The device used for this purpose, a *countercurrent heat exchanger*, is shown in Fig. 13-5.

The gas, after precooling, is sent through the middle tube of a long coil of double-walled pipe. After throttling, it flows back through the outer annular space surrounding the middle pipe. For the heat exchanger to be efficient, the temperature of the gas as it leaves must differ only slightly from the temperature

Table 13-1 Maximum inversion temperatures

Gas	Maximum inversion temperature, K
Xe	1486
CO_2	1275
Kr	1079
Ar	794
CO	644
N_2	607
Ne	228
H_2	204
^{4}He	43

at which it entered. To accomplish this, the heat exchanger must be quite long and well insulated, and the gas must flow through it with sufficient speed to cause turbulent flow, so that there is good thermal contact between the opposing streams of gas.

When the steady state is finally reached, liquid is formed at a constant rate: for every mass unit of gas supplied, a certain fraction y is liquefied, and the fraction $1 - y$ is returned to the pump. Considering only the heat exchanger and throttling valve completely insulated, as shown in Fig. 13-6, we have a process in

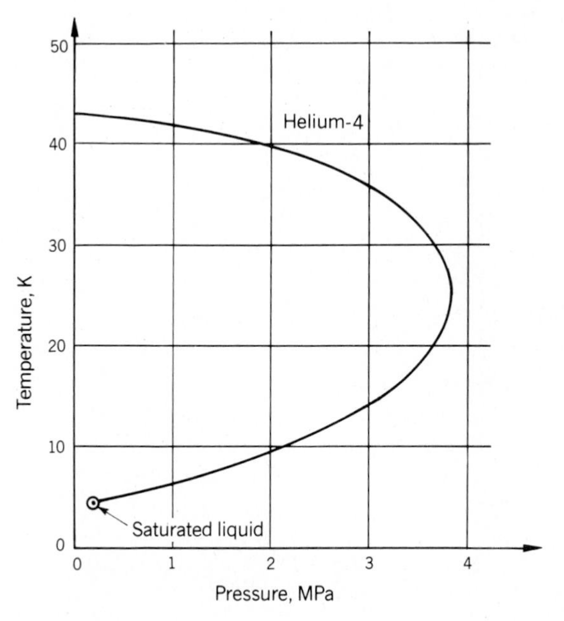

Figure 13-4 Isenthalps and inversion curve for helium.

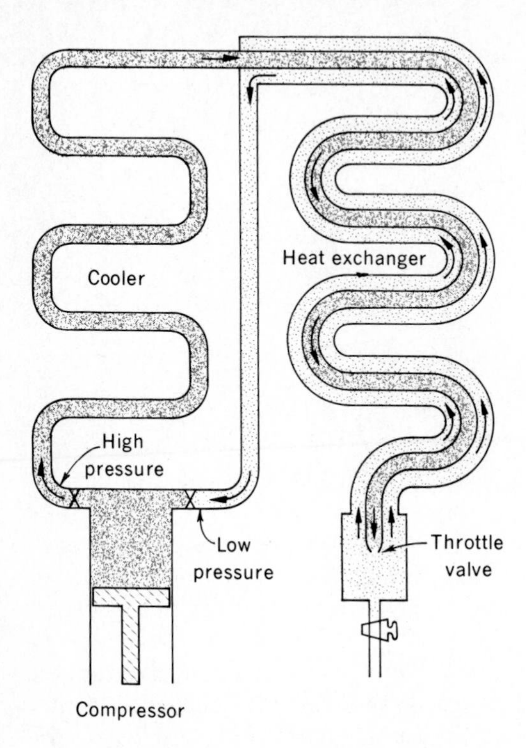

Figure 13-5 Liquefaction of a gas by means of Joule-Kelvin effect.

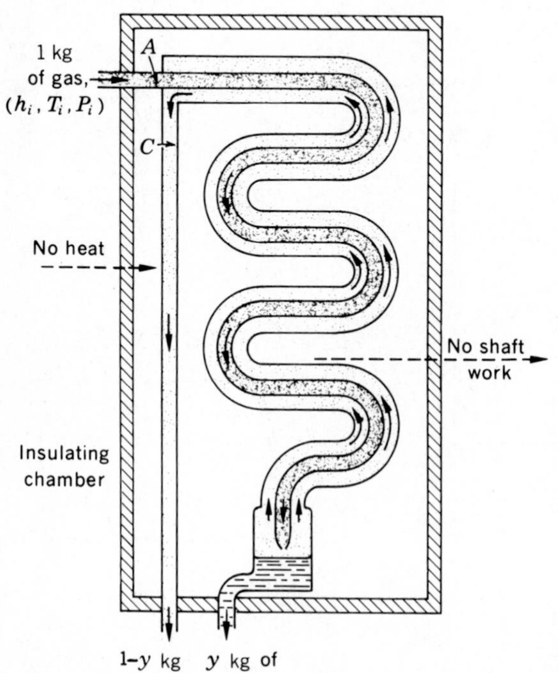

Figure 13-6 Throttling valve and heat exchanger in steady state.

which the enthalpy of the entering gas is equal to the enthalpy of y units of emerging liquid plus the enthalpy of $1 - y$ units of emerging gas. If

$$h_i = \text{enthalpy of entering gas at } (T_i, P_i),$$

$$h_L = \text{enthalpy of emerging liquid at } (T_L, P_L),$$

and

$$h_f = \text{enthalpy of emerging gas at } (T_f, P_f),$$

then

$$h_i = yh_L + (1 - y)h_f,$$

or

$$y = \frac{h_f - h_i}{h_f - h_L}. \tag{13-3}$$

Now, in the steady state, h_L is determined by the pressure on the liquid, which fixes the temperature, and hence is constant. h_f is determined by the pressure drop in the return tube and the temperature at C, which is only a little below that at A; hence, it remains constant. h_i refers to a temperature T_i that is fixed, but at a pressure that may be chosen at will. Therefore, the fraction liquefied y may be varied only by varying h_i. Since

$$y = \frac{h_f - h_i}{h_f - h_L},$$

y will be a maximum when h_i is a minimum; and since h_i may be varied only by varying the pressure, the condition that it be a minimum is that

$$\left(\frac{\partial h_i}{\partial P}\right)_{T=T_i} = 0.$$

But

$$\left(\frac{\partial h}{\partial P}\right)_T = -\left(\frac{\partial h}{\partial T}\right)_P \left(\frac{\partial T}{\partial P}\right)_h = -c_P \mu;$$

hence, for y to be a maximum,

$$\mu = 0 \qquad \text{at } T = T_i,$$

or the *point* (T_i, P_i) *must lie on the inversion curve.*

In the design of a gas-liquefaction unit, a *TS* diagram showing isobars and isenthalps is particularly useful. For example, to calculate the fraction liquefied in the steady state y, the three enthalpies h_i, h_f, and h_L may be obtained directly from such a diagram. *TS* diagrams for hydrogen and for helium are shown in Figs. 13-7 and 13-8.

The use of the Joule-Kelvin effect to produce liquefaction of gases has two advantages: (1) There are no moving parts at low temperature that would be difficult to lubricate. (2) The lower the temperature, the larger the drop in temperature for a given pressure drop, as shown by the isenthalps in Figs 13-2 and 13-3. For the purpose of liquefying hydrogen and helium, however, it has a serious disadvantage: namely, the large amount of precooling that is necessary. The hydrogen must be precooled with liquid nitrogen, and the helium must be

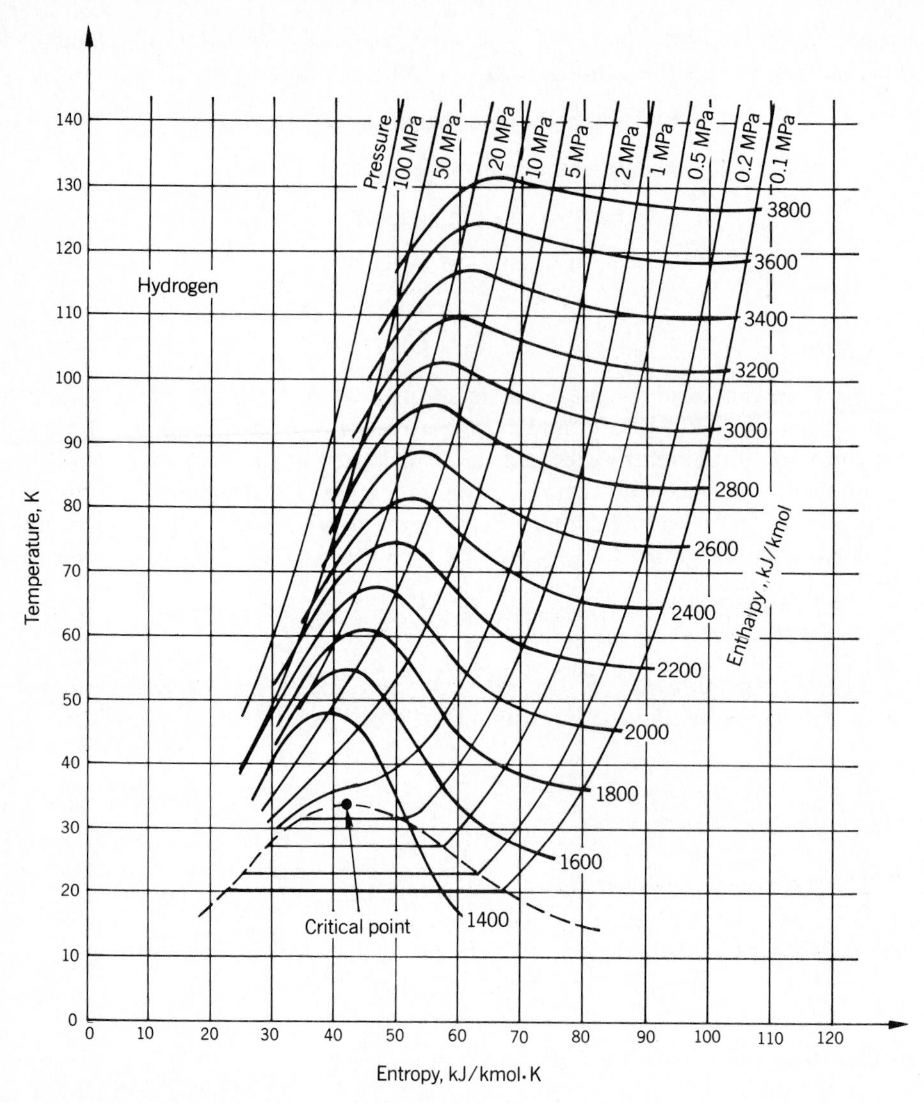

Figure 13-7 Temperature-entropy diagram for hydrogen.

precooled with liquid hydrogen, which makes the liquefaction of these gases expensive.

An approximately reversible adiabatic expansion against a piston or a turbine blade always produces a decrease in temperature, no matter what the original temperature. If, therefore, a gas like helium could be made to do external work adiabatically through the medium of an engine or a turbine, then, with the aid of a heat exchanger, the helium could be liquefied without precooling. But this method has the disadvantage that the temperature drop on adiabatic expansion decreases as the temperature decreases.

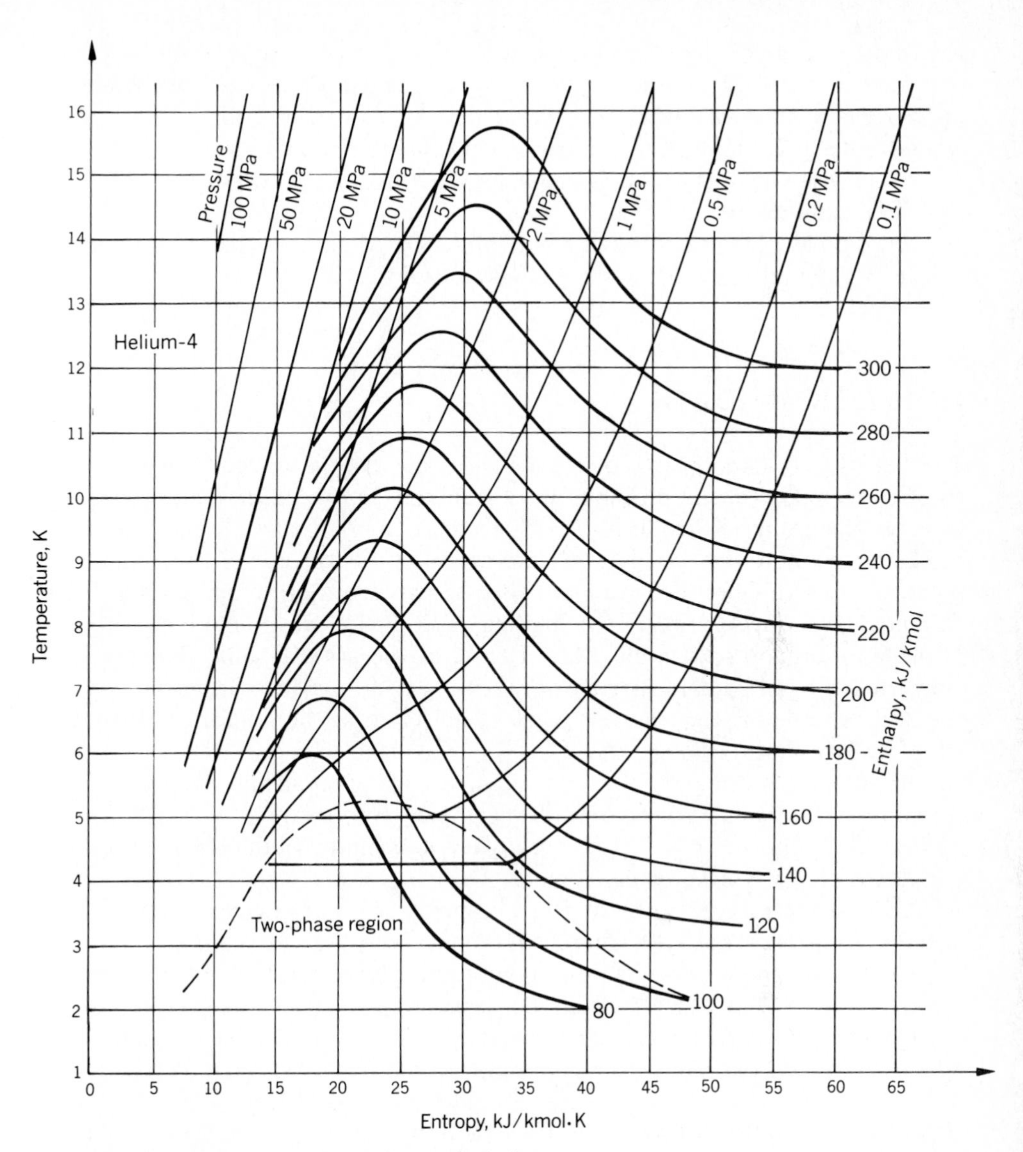

Figure 13-8 Temperature-entropy diagram for helium.

A combination of both methods has been used successfully. Thus, adiabatic reversible expansion is used to achieve a temperature within the inversion curve, and then the Joule-Kelvin effect completes the liquefaction. Kapitza was the first to liquefy helium in this way, with the aid of a small expansion engine that was lubricated by the helium itself. Later he liquefied air with the aid of a centrifugal turbine only a trifle larger than a watch.

The most significant development in the field of gas liquefaction is the Collins helium liquefier, in which helium undergoes adiabatic expansion in a reciprocating engine. The expanded gas is then used to cool the incoming gas in the usual countercurrent heat exchanger. When the temperature is low enough, the gas passes through a throttling valve, and Joule-Kelvin cooling is used to complete the liquefaction. The unit consists of a four-stage compressor, a gasholder, a purifier, and a cryostat containing the engines, heat exchangers, Dewar flasks, vacuum pumps, and gauges.

The simplest laboratory devices for producing small amounts of liquid air are the Stirling refrigerators, which were described in Art. 6-6.

13-3 CRITICAL STATE

The liquid and vapor phases of a substance may coexist in equilibrium at a constant temperature and pressure over a wide range of volumes from v'', where there is practically all liquid, to v''', where there is practically all vapor. At a higher temperature and pressure, v'' increases and v''' decreases. At the critical point, the two volumes (and therefore also the two densities) coincide. On a Pv diagram, the critical point is the limiting position as two points lying on a horizontal line approach each other. Hence, at the critical point, the critical isotherm has a horizontal tangent, or $(\partial P/\partial v)_{T_C} = 0$. As is evident from Fig. 13-9, the critical isotherm must have a point of inflection at the critical point, and therefore $(\partial^2 P/\partial v^2)_{T_C} = 0$.

At all points within the region of coexistence of the liquid and vapor phases, including the critical point, the three physical properties $c_P = T(\partial S/\partial T)_P$, $\beta = (1/v)(\partial v/\partial T)_P$, and $\kappa = -(1/v)(\partial v/\partial P)_T$ *are all infinite*. On the other hand, c_V is finite, so that $\gamma = c_P/c_V$ is also infinite.

To measure precise values of P_C, v_C, and T_C, it is necessary to determine exactly *when* the critical state is reached. For this purpose, the material is raised in temperature slowly and uniformly in a sealed tube of constant volume. Consider the tube filled so that the meniscus separating the phases is initially near the top of the tube. When the temperature is raised, the liquid will expand and the meniscus will rise to the top of the tube as the last of the vapor condenses, according to the dashed line $1 \rightarrow 2$ in Fig. 13-9. If the tube is filled so that the initial position of the meniscus is near the bottom, then the meniscus will fall to the bottom as the last of the liquid vaporizes, according to the dashed line $3 \rightarrow 4$ of Fig. 13-9. To observe the critical point, the tube must be filled so that the meniscus will remain in about the middle of the tube, as shown by the dashed line passing through the critical point. As the critical point is approached, observation becomes difficult because of the following factors:

1. Since the compressibility is infinite, the gravitational field of the earth causes large density gradients from top to bottom. These may be offset somewhat by taking observations of the tube in both vertical and horizontal positions.
2. Since the heat capacity is infinite, thermal equilibrium is difficult to achieve. It

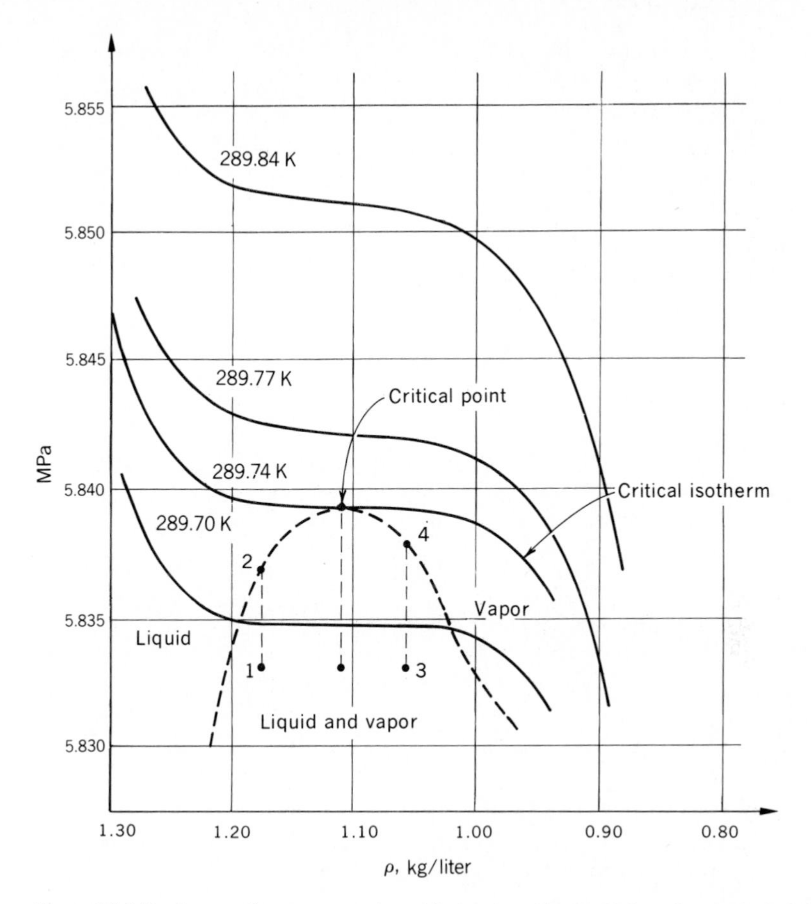

Figure 13-9 Isotherms of xenon near the critical point. *(H. W. Habgood and W. G. Schneider.)*

is necessary to keep the system at a constant temperature (within 10^{-2} or even 10^{-3} K) for a long time and to stir constantly.

3. Since the thermal expansivity is infinite, small temperature changes of a local mass element within the system produce large volume changes, and therefore there are violent density fluctuations, which give rise to a large amount of light scattering, so that the material becomes almost opaque. This phenomenon is called *critical opalescence*.

The critical pressure may be measured at the same time as the critical temperature by noting the pressure at which the meniscus disappears. The critical volume, however, is much more difficult to determine. This measurement is carried out most commonly by measuring the densities of both saturated liquid and saturated vapor as a function of temperature to as close to the critical temperature as possible. The vapor and liquid densities are then plotted against temperature, and a line representing the arithmetic average of these densities is

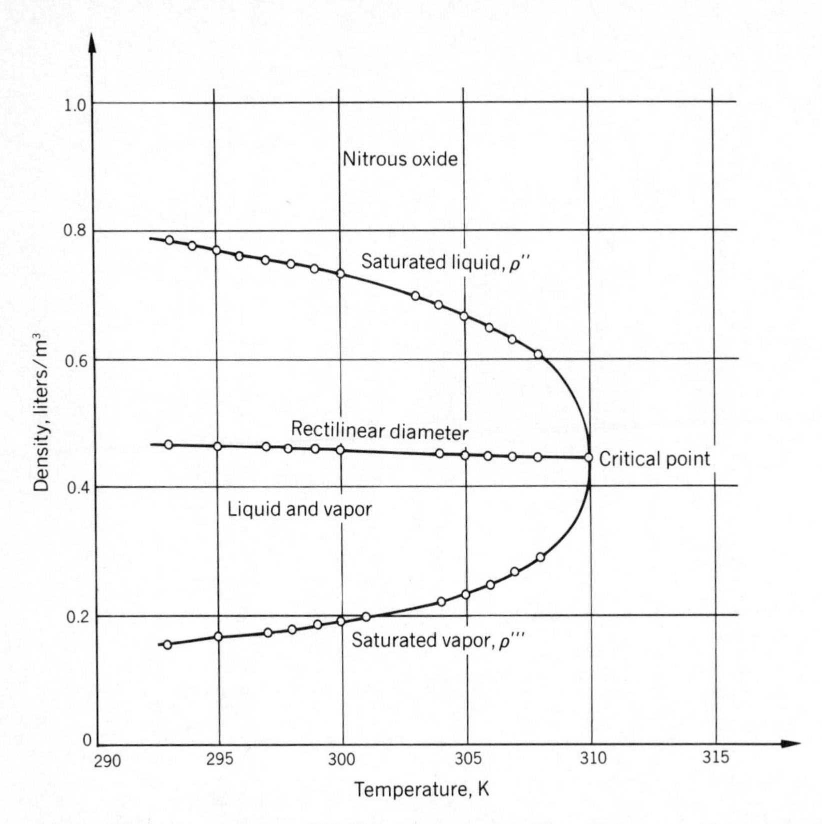

Figure 13-10 Measurements of liquid and vapor densities of nitrous oxide, and extrapolation of mean density through the critical point. *(D. Cook.)*

constructed as shown in Fig. 13-10. This last line is usually called the *rectilinear diameter*, and experiment shows it to be virtually linear. Extrapolation to the critical temperature yields the critical density.

The difference $\rho'' - \rho'''$ is called the *order parameter* and is a measure of the microscopic interactions of the particles in a hydrostatic system below the critical point. At temperatures above the critical temperature T_C the order parameter $\rho'' - \rho'''$ is zero, because the system is in the gaseous phase and presumably little interaction or ordering occurs between the particles. However, as the temperature is lowered toward the critical temperature, particles coalesce to form small "droplets" that are smaller than the wavelength of visible light. At temperatures just above T_C the droplets grow to a size comparable to the wavelength of light and light is strongly scattered by the droplets, producing the phenomenon of critical opalescence. At temperatures below T_C the droplets of higher density precipitate out of the vapor phase to form a liquid. In the region of the coexisting phases some ordering occurs among the particles, so the order parameter is nonzero.

According to the law of corresponding states, the reduced pressure

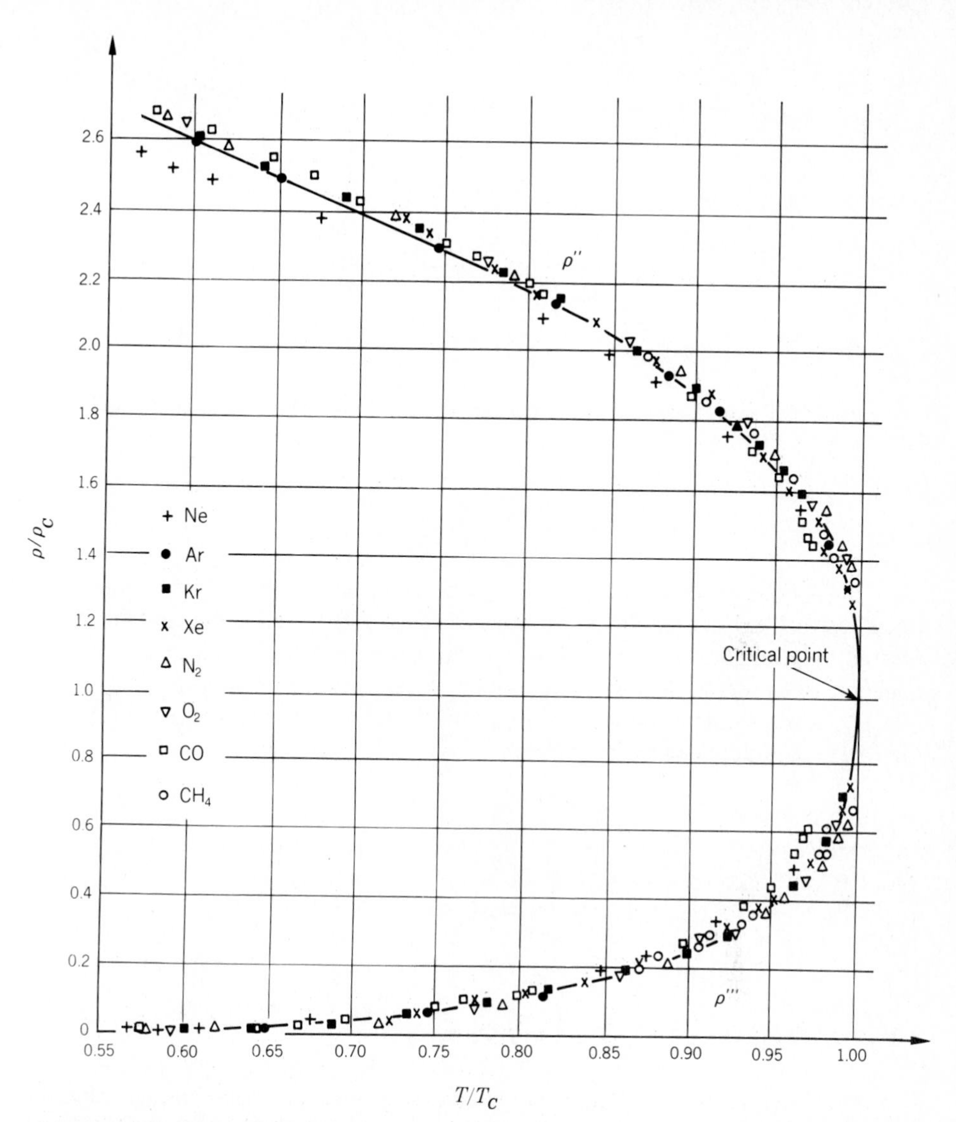

Figure 13-11 Reduced densities of coexisting liquid and vapor phases vs. reduced temperature. *(E. A. Guggenheim, Thermodynamics, Interscience, 1967.)*

$\tilde{P} = P/P_C$ and reduced density $\tilde{\rho} = \rho/\rho_C$ should be common functions of the reduced temperature $\tilde{T} = T/T_C$. Indeed, the former relationship was shown in Fig. 10-8 and the latter relationship is shown in Fig. 13-11. In the case of the density, it is customary to deal with the reduced order parameter $(\rho'' - \rho''')/\rho_C$, which is considered to be a function of the quantity ϵ, defined as

$$\epsilon \equiv \frac{T - T_C}{T_C}. \tag{13-4}$$

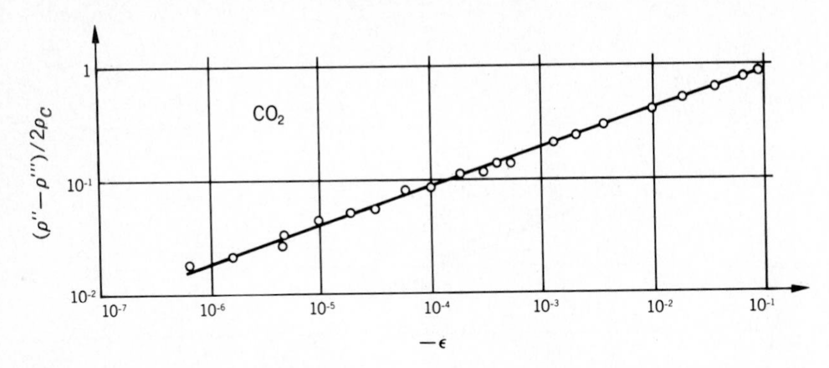

Figure 13-12 The logarithm of the reduced order parameter vs. log $(-\epsilon)$ from coexistence curve data for CO_2. *(P. Heller.)*

The dimensionless variable ϵ measures the reduced difference in temperature from the critical temperature. In Fig. 13-12 the reduced order parameter is plotted as a function of ϵ on a log-log graph for carbon dioxide and using the principle of the rectilinear diameter which states that $\rho_C \sim \frac{1}{2}(\rho'' + \rho''')$.

The straight line obtained in Fig. 13-12 means that the relationship between the order parameter and ϵ is

$$\rho'' - \rho''' \sim (-\epsilon)^{\beta}, \tag{13-5}$$

where the slope β is known as a *critical point exponent* or simply *critical exponent*. The order parameter vanishes at the critical point, but other thermodynamic variables have divergences at the critical point.

13-4 CRITICAL POINT EXPONENTS

We mentioned in Art. 13-3 that the specific heat and isothermal compressibility are infinite at the critical point. There is, however, a more detailed mathematical statement about the precise functions that govern the behavior of various systems near the critical point. We therefore define six critical point exponents, denoted by the Greek letters α, α', β, γ, γ', and δ, which are standard notation used in the asymptotic description of singular functions near the critical point.

The coefficient α characterizes the behavior of the specific heat at constant volume (density) evaluated along the critical isochore:

$$c_{V=V_C} \sim \epsilon^{-\alpha}, \qquad \epsilon \to 0+, \tag{13-6}$$

when the quantity $\epsilon \to 0+$ symbolizes the fact that the temperature T approaches the critical temperature T_C from above. One may also introduce a

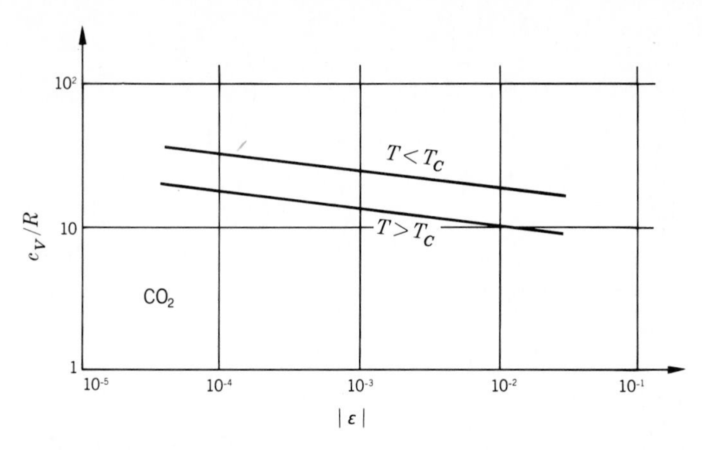

Figure 13-13 Specific heat vs. log (ϵ) for CO_2. *(Lipa, Edwards, and Buckingham.)*

similar coefficient α', which characterizes the specific heat at constant volume at temperatures below the critical point. Thus,

$$c_{V=V_C} \sim (-\epsilon)^{-\alpha'}, \qquad \epsilon \to 0-, \tag{13-7}$$

where α and α' are not necessarily equal.

The experimental values for the critical point exponents α and α' may not be the same for all simple systems, as predicted by the law of corresponding states. Kadanoff in 1967 obtained $\alpha < 0.4$ and $\alpha' < 0.25$ for argon. Later results of Lipa, Edwards, and Buckingham, who obtained very precise values of c_V for CO_2, are shown in Fig. 13-13. An analysis of Fig. 13-13 yields the values: $\alpha = 0.124 \pm 0.014$ and $\alpha' = 0.124 \pm 0.012$.

The critical point exponent, β, mentioned in Art. 13-3 and shown to characterize the behavior of the order parameter, is the exponent in the following asymptotic function:

$$\rho'' - \rho''' \sim (-\epsilon)^{\beta}, \qquad \varepsilon \to 0-, \tag{13-5}$$

where, of course, the function is meaningful only below the critical point in the region where $\rho'' - \rho'''$ is not zero. The data from the coexistence curve for CO_2 are shown in Fig. 13-12, where the results from five separate experiments are shown. Heller calculates a value of $\beta = 0.34 \pm 0.015$ for CO_2.

The pair of coefficients γ and γ' characterize the behavior of the isothermal compressibility. Just as in the case of the isochoric specific heat, there is a difference in the functions describing κ as it approaches the critical temperature from above or below. In the former case,

$$\kappa \sim \epsilon^{-\gamma}, \qquad \epsilon \to 0+, \tag{13-8}$$

and in the latter case,

$$\kappa \sim (-\epsilon)^{-\gamma'}, \qquad \epsilon \to 0-. \tag{13-9}$$

As before, these two critical exponents are not necessarily equal.

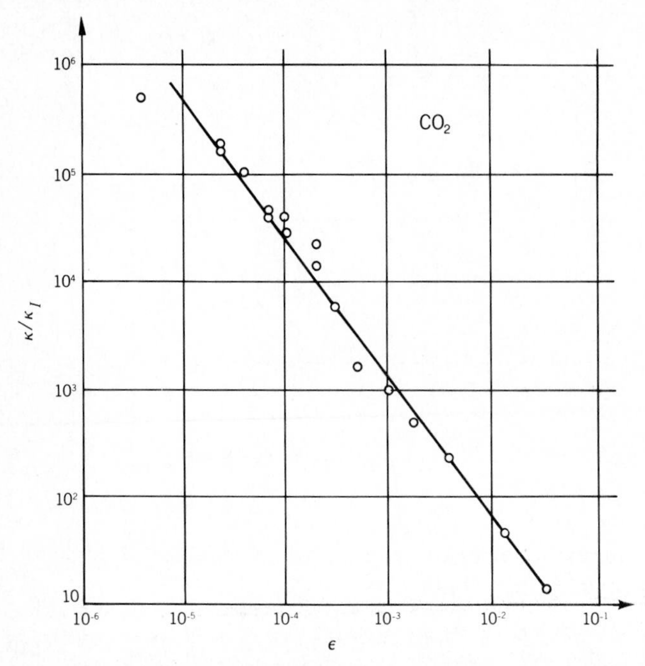

Figure 13-14 The logarithm of the reduced isothermal compressibility above T_C vs. log (ϵ) for CO_2. The normalizing factor κ_I is the compressibility of an ideal gas whose density and temperature are those of CO_2 at the critical point. *(P. Heller.)*

The isothermal compressibility data for carbon dioxide, normalized to κ of an ideal gas whose density and temperature are those of CO_2 at the critical point, are shown in Fig. 13-14. The slope of the straight line yields a value of the critical exponent $\gamma = 1.35 \pm 0.15$ for CO_2 above T_C. Below T_C the isothermal compressibility can be determined for either the liquid or the vapor at densities on the coexistence curve. The analysis for data on CO_2 results in a value of γ' equal to 1.1 ± 0.4. If subsequent experiments should show that the two values of gamma are indeed unequal, then the dependence of the isothermal compressibility as a function of reduced temperature would not be symmetric with respect to the critical temperature.

The last exponent δ describes the critical isotherm itself:

$$P - P_C \sim (\rho - \rho_C)^{\delta}, \qquad \rho \to \rho_C, \tag{13-10}$$

where the parameter $\epsilon = 0$ always. The critical exponent δ is a measure of the "flatness" of the critical isotherm at the critical point, as shown in Fig. 13-9. Because of the difficulty in obtaining accurate pressure and density measurements very close to the critical point along the coexistence curve, the shape of the coexistence curve is somewhat uncertain. Figure 13-15 shows the difference in pressures for carbon dioxide, normalized to the pressure of an ideal gas at the density and temperature of CO_2 at the critical point, as a function of the reduced

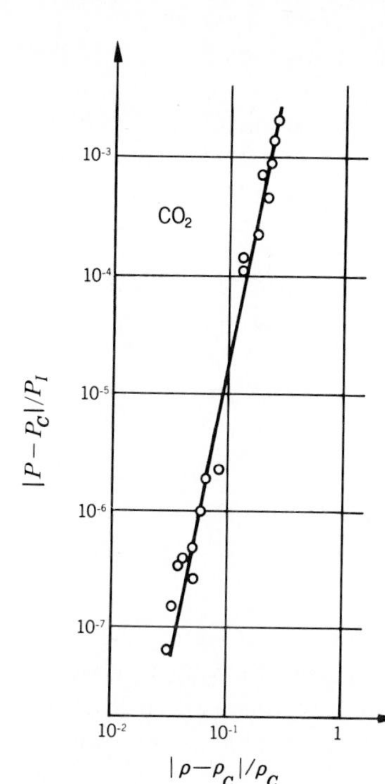

Figure 13-15 The logarithm of the reduced pressure difference vs. the logarithm of the reduced density for CO_2. The normalizing factor P_I is the pressure of an ideal gas at the critical density and temperature of CO_2. *(P. Heller.)*

density. Heller calculates the value of $\delta = 5.0 \pm 1$. For the xenon data shown in Fig. 13-9, the value of δ is 4.4 ± 0.4. Table 13-2 summarizes the definitions and conditions for the six critical point exponents for a hydrostatic system and lists experimental values for the exponents in several systems.

The only rigorous relations thus far proposed among the critical point exponents are a set of inequalities. The *Rushbrooke inequality* states that

$$\alpha' + 2\beta' + \gamma' \geq 2.$$

It will be noticed from Table 13-2 that most of the substances either fulfill the Rushbrooke inequality, or come close to fulfilling it. Large uncertainties in some

Table 13-2 Summary of critical point exponents for hydrostatic systems

Exponent	Function	Conditions			Ar	Xe	CO_2	^{3}He	^{4}He
		ϵ	$P - P_C$	$\rho - \rho_C$					
α'	$c_{V=V_C} \sim (-\epsilon)^{-\alpha'}$	< 0	0	0	< 0.25	< 0.2	0.124	0.105	0.017
α	$c_{V=V_C} \sim \epsilon^{-\alpha}$	> 0	0	0	< 0.4	—	0.124	0.105	0.017
β	$\rho'' - \rho''' \sim (-\epsilon)^{\beta}$	< 0	0	$\neq 0$	0.362	0.35	0.34	0.361	0.354
γ'	$\kappa \sim (-\epsilon)^{-\gamma'}$	< 0	0	$\neq 0$	1.20	—	1.1	1.17	1.24
γ	$\kappa \sim \epsilon^{-\gamma}$	> 0	0	0	1.20	1.3	1.35	1.17	1.24
δ	$P - P_C \sim \lvert\rho'' - \rho'''\rvert^{\delta}$	0	$\neq 0$	$\neq 0$	—	4.4	5.0	4.21	4.0

of the experimental data account for those substances that do not satisfy the inequality.

Another relationship among different critical point exponents is the *Griffiths inequality*,

$$\alpha' + \beta(1 + \delta) \geq 2,$$

which is satisfied by some of the data from Table 13-2. There are additional inequalities, many of which involve critical point exponents of singular functions not introduced here.

13-5 CRITICAL POINT EXPONENTS OF A MAGNETIC SYSTEM

We have seen that the critical point exponents describe the asymptotic behavior of singular functions near the critical point of a hydrostatic system. Another system that can be analyzed in terms of critical point exponents is a ferromagnetic material. A ferromagnet is characterized by a permanent magnetization M, which does not vanish when the magnetic intensity $\mathcal{H}$ is zero. When the temperature is raised to the Curie point T_C, the magnetization is directly proportional to the magnetic intensity; that is, the substance ceases to be ferromagnetic and becomes paramagnetic. In the region above the Curie point, the temperature dependence of the ratio $M/\mathcal{H}$ is described by the Curie-Weiss law, which is a modification of the Curie law described in Art. 2-12, namely

$$\frac{M}{\mathcal{H}} = \frac{C_C}{T - T_C}, \qquad T > T_C, \tag{13-11}$$

where the Curie constant C_C depends on the atomic properties of the substance and thus differs for each material. The Curie points of various ferromagnetic elements and compounds are listed in Table 13-3.

The form of Eq. (13-11) shows that at the Curie point one expects a singularity in the magnetic properties, namely, an infinite value of the ratio $M/\mathcal{H}$. Since the maximum value of M is finite (being limited to the saturation magnetization obtained when all the microscopic magnetic dipoles of the substance are aligned parallel to one another), we must conclude that $\mathcal{H} = 0$. In other words, the substance is magnetized even in the absence of an external magnetic field. This "spontaneous magnetization," due to an internal field, is a characteristic of ferromagnets. The Curie point T_C is the boundary between paramagnetic behavior at $T_C > T$ and ferromagnetic behavior at $T < T_C$.

It is customary to compare the similarities between the hydrostatic and magnetic systems. The application of a pressure P to a fluid increases the density

Table 13-3 Curie points of various ferromagnets

Substance	T_C, K
Co	1388
Fe	1042.5
$YFeO_3$	643
Ni	631.6
CrO_2	386.5
Gd	292.5
$CrBr_3$	32.56
EuS	16.50

ρ, and the application of a magnetic intensity $\mathscr{H}$ to a magnet increases the magnetization M. Thus $\mathscr{H}$ is analogous to P and M is analogous to ρ.

Figure 13-16 shows an $\mathscr{H}MT$ surface analogous to the surfaces of hydrostatic systems shown in Figs. 2-3 and 2-4. At temperatures very large compared to T_C the Curie law is obeyed and the isotherm is therefore given by Eq. (2-13):

$$\frac{M}{\mathscr{H}} = \frac{C_C}{T}. \tag{2-13}$$

As the temperature is decreased to near T_C, the Curie-Weiss law given by Eq. (13-11) describes the behavior of the isotherms. Below T_C there is no simple analytic expression to describe temperature dependence of $M/\mathscr{H}$. In fact, there are no stable states in the region shown as a flat plateau in Fig. 13-16. The projections of the $\mathscr{H}MT$ surface on the $\mathscr{H}M$, $\mathscr{H}T$, and MT planes are shown in Fig. 13-17.

For the hydrostatic system the order parameter was defined as the difference in the densities of the liquid and vapor. For the magnetic system the order parameter is the magnetization M at zero magnetic intensity $\mathscr{H}$. Similarly, the isothermal compressibility κ of the hydrostatic system finds its analog in the isothermal differential susceptibility χ', which is defined

$$\chi' = \left(\frac{\partial M}{\partial \mathscr{H}}\right)_T. \tag{13-12}$$

Finally, the isochoric specific heat of the fluid system is replaced by the specific heat at constant magnetic intensity.

All three magnetic functions—specific heat, order parameter, and differential susceptibility—have singular behavior at the Curie temperature T_C and, correspondingly, six critical point exponents that are, respectively, α', α, β, γ', γ, and δ. Table 13-4 provides a summary of the critical point exponents for a magnetic system and lists experimental values for some ferromagnetic substances.

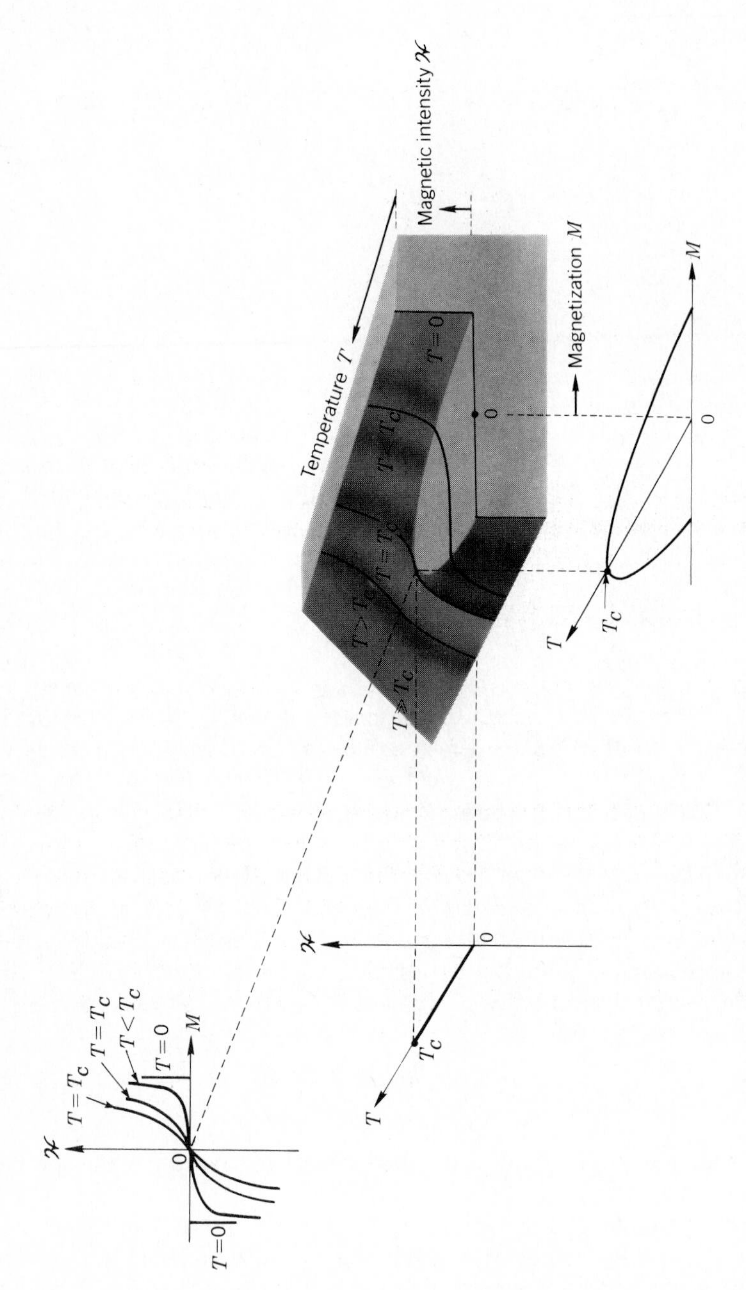

Figure 13-16 The $\mathcal{H}MT$ surface of a hypothetical ferromagnetic substance.

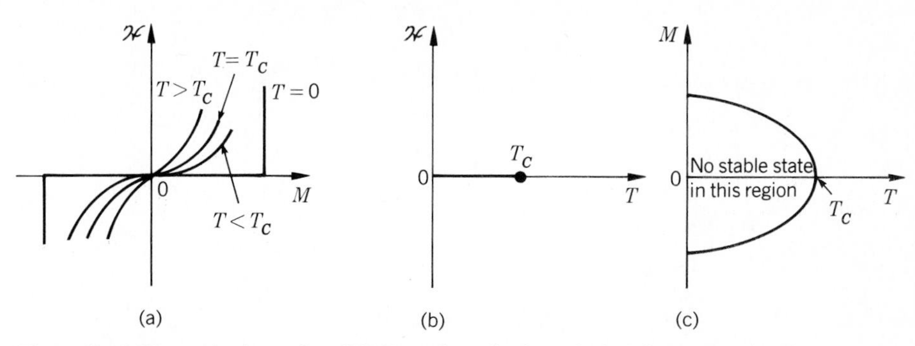

Figure 13-17 The projections of an $\mathcal{H}MT$ surface of a hypothetical ferromagnetic substance onto (*a*) the $\mathcal{H}M$ plane, (*b*) the $\mathcal{H}T$ plane, and (*c*) the MT plane.

Table 13-4 Summary of critical point exponents for magnetic systems

Exponent	Function	Conditions ϵ	Conditions $\mathcal{H}$	Conditions M	Fe	Ni	Gd	$CrBr_3$
α'	$c_{\mathcal{H}=0} \sim (-\epsilon)^{-\alpha'}$	<0	0	0	-0.12	-0.10	—	—
α	$c_{\mathcal{H}=0} \sim \epsilon^{-\alpha}$	>0	0	0	-0.12	-0.10	—	—
β	$M \sim (-\epsilon)^{\beta}$	<0	0	$\neq 0$	0.34	0.33	—	0.365
γ'	$\chi' \sim (-\epsilon)^{-\gamma'}$	<0	0	$\neq 0$	1.33	1.32	1.33	—
γ	$\chi' \sim \epsilon^{-\gamma}$	>0	0	0	1.33	1.32	1.33	1.215
δ	$\mathcal{H} \sim \lvert M \rvert^{\delta}$	0	$\neq 0$	$\neq 0$	—	4.2	4.0	4.3

13-6 HIGHER-ORDER TRANSITIONS

The processes of sublimation, vaporization, and fusion were called phase transitions of the firs order because the first-order derivatives of the Gibbs function, $S = -(\partial G/\partial T)_P$ and $V = (\partial G/\partial P)_T$, underwent finite changes during the transition. In contrast, there are many phase transitions in which the entropy and the volume are the *same* at the end of the transition as they were in the beginning. In such phase changes, T, P, G, S, and V remain unchanged, and therefore H, U, and F also remain unchanged. If C_P, κ; and β undergo *finite* changes in such a transition, then, since

$$\frac{C_P}{T} = \left(\frac{\partial S}{\partial T}\right)_P = \frac{\partial}{\partial T}\left(-\frac{\partial G}{\partial T}\right) = -\frac{\partial^2 G}{\partial T^2},$$

$$\kappa V = -\left(\frac{\partial V}{\partial P}\right)_T = -\frac{\partial}{\partial P}\left(\frac{\partial G}{\partial P}\right) = -\frac{\partial^2 G}{\partial P^2},$$

and

$$\beta V = \left(\frac{\partial V}{\partial T}\right)_P = \frac{\partial}{\partial T}\left(\frac{\partial G}{\partial P}\right) = \frac{\partial^2 G}{\partial T\,\partial P},$$

there would be finite changes of the second-order derivatives of the Gibbs function. Such a transition is called a *second-order phase transition.* This transition

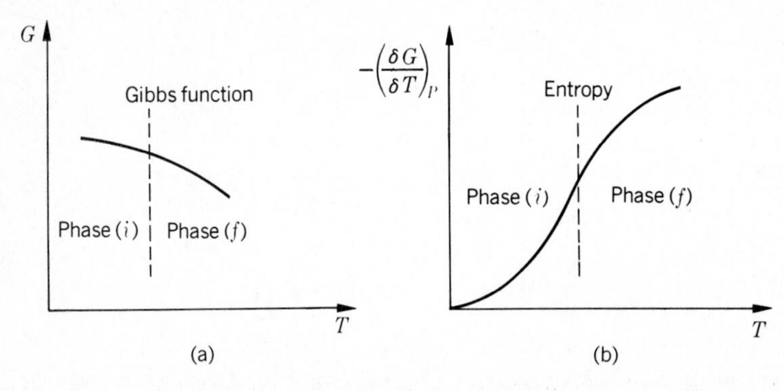

Figure 13-18 Characteristics of a second-order phase transition. (*a*) Gibbs function; (*b*) entropy.

was first suggested by Ehrenfest, who derived two simple equations that expressed the constancy of S and V. The variations of G and S with temperature for a second-order phase transition are shown in Fig. 13-18.

It was first thought that there were many examples of second-order transitions; but as experimental measurements were made closer and closer to the transition temperature (sometimes to within a millionth of a degree!), neither C_P nor β was found to achieve a finite value at the beginning or at the end of the phase transition. It may be that there is only one example of a second-order phase transition, namely, the change from superconductivity to normal conductivity in zero magnetic field.

By far the most interesting higher-order phase transition is that called the *lambda transition*, characterized by the following:

1. T, P, and G remain constant.
2. S and V (also U, H, and F) remain constant.
3. C_P, β, and κ are infinite.

A graph of C_P against T serves to distinguish among the three types of transitions, as shown in Fig. 13-19. The name "lambda transition" is accounted for by the fact that the shape of the $C_P - T$ curve in the third graph resembles the Greek letter lambda. Among the many examples of λ transitions are: (1) "order-disorder" transformations in alloys; (2) the onset of ferroelectricity in certain crystals such as Rochelle salt; (3) the transition from ferromagnetism to paramagnetism at the Curie point; (4) a change of orientation of an ion in a crystal lattice, such as the ammonium chloride transition; and—most interesting of all—(5) the transition from ordinary liquid helium (liq. He I) to superfluid helium (liq. He II) at a temperature and corresponding pressure known as a *lambda point*.

It may be seen in Fig. 13-19*a* that, as a substance in any one phase approaches the temperature at which a first-order phase transition is to occur, its C_P remains finite up to the transition temperature. It becomes infinite only when a small amount of the other phase is present, and its behavior *before* this takes

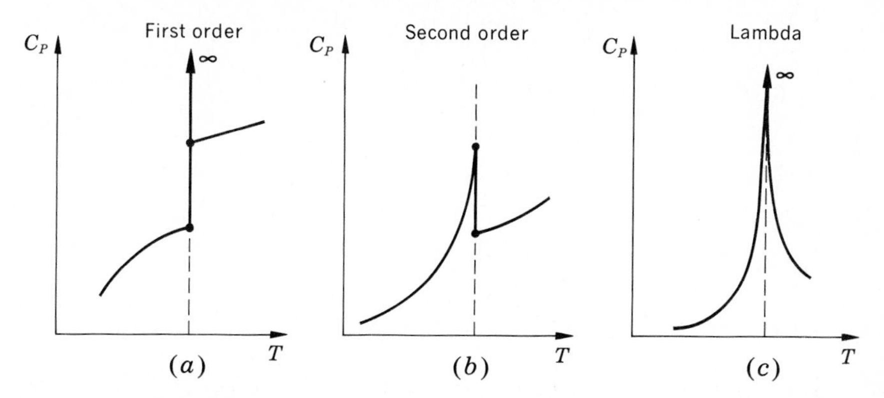

Figure 13-19 Distinguishing characteristics among the three types of phase transitions. (*a*) First-order; (*b*) second-order; (*c*) lambda.

place shows no evidence of any premonition of the coming event. In the case of a λ transition, however, as is evident in Fig. 13-19*c*, C_P starts to rise before the transition point is reached, as though the substance, in the form of only one phase, "anticipated" the coming phase transition. Because the molecules of substances undergoing λ transitions interact strongly with one another, even at distances beyond those of their nearest neighbors, a rigorous statistical treatment is very difficult. One physical process that has yielded to mathematical solution, namely, the transition from ferromagnetism to paramagnetism according to a simple model known as the *two-dimensional Ising model*, shows the typical anticipatory rise of C_P before the λ transition occurs.

The temperatures and pressures at which the λ transition takes place in ^{4}He constitute the λ line shown on the phase diagram of Fig. 13-20*a*. If we define a new variable t, where

$$t = T - T_\lambda,$$

the slope of the λ line at any point, $dP/dT(=P'_\lambda)$, will be the same at a corresponding point on the parallel curve where the temperature $T = T_\lambda + t$, since the dashed curve marked $t = t$ is merely the λ line shifted a distance t to the right. We shall limit ourselves to very small values of t.

On the entropy curve of Fig. 13-20*b*, the slope of the dashed curve marked $t = t$ is the same as that of the curve $t = 0(S'_\lambda)$ at a point at the same pressure, since t is very small. The same is true on the volume diagram of Fig. 13-20*c*. With these facts in mind, let us use the second $T\,dS$ equation to represent a small entropy change *on the dashed entropy curve* of Fig. 13-20*b*, where $T = T_\lambda + t$. Using small letters to indicate molar quantities,

$$ds = \frac{c_P}{T}\,dT - v\beta\;dP,$$

or

$$\frac{c_P}{T} = \frac{ds}{dT} + v\beta\frac{dP}{dT}.$$

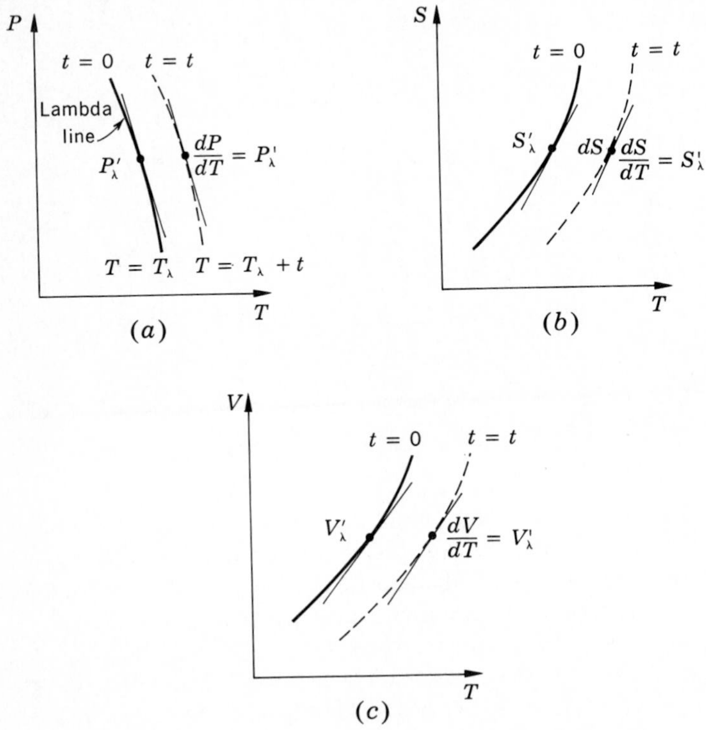

Figure 13-20 (*a*) The λ line of ^{4}He on a phase diagram. (*b*), (*c*) Entropy and volume at points on the λ line. The slope of a curve at a small temperature difference t above (or below) the λ line is the same as that at a corresponding point on the λ line.

Since ds/dT and dP/dT are the same, respectively, as the slopes at the corresponding points on the curves where $T = T_\lambda$, we have

$$\frac{c_P}{T} = s'_\lambda + v\beta P'_\lambda. \tag{13-13}$$

This equation is true whether $T > T_\lambda$ or $T < T_\lambda$, and it shows that c_P/T should vary linearly with $v\beta$ as the transition is approached (either from above or below) with a slope equal to P'_λ. Using t as a convenient parameter, experimental measurements of c_P and β as a function of t, when combined, should yield a linear relation between c_P/T and $v\beta$. Because it is not convenient to measure the c_P of liquid helium, the molar heat capacity at constant saturation c_S is usually measured, and c_P is calculated from a simple relation that can be found in the problems at the end of this chapter. In the case of liquid helium, the correction is quite small.

The variation of c_S (which is equivalent to c_P) of liquid helium II with t is shown in Fig. 13-21, and the variation of β of liquid helium II with t is seen in

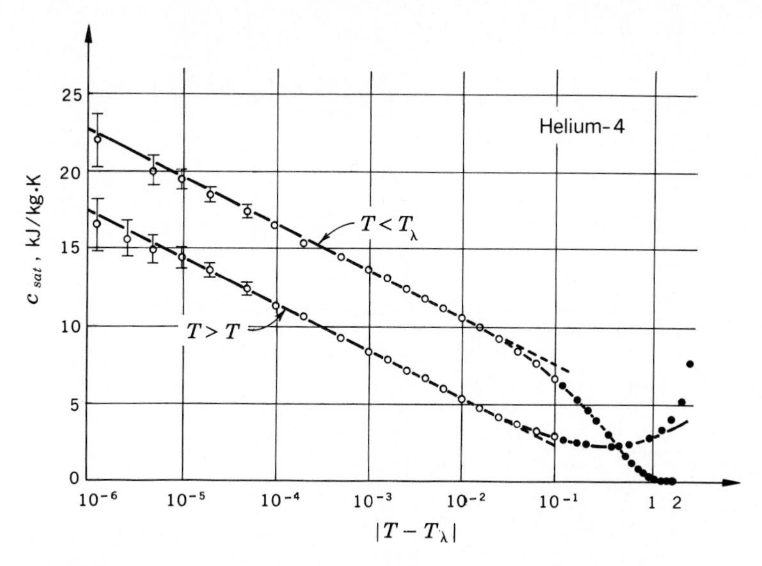

Figure 13-21 Specific heat of liquid helium II vs. log $|T - T_\lambda|$. *(Kellers, Fairbank, and Buckingham; Hill, Lounasmaa, and Kojo; Kramers, Wasscher, and Gorter.)*

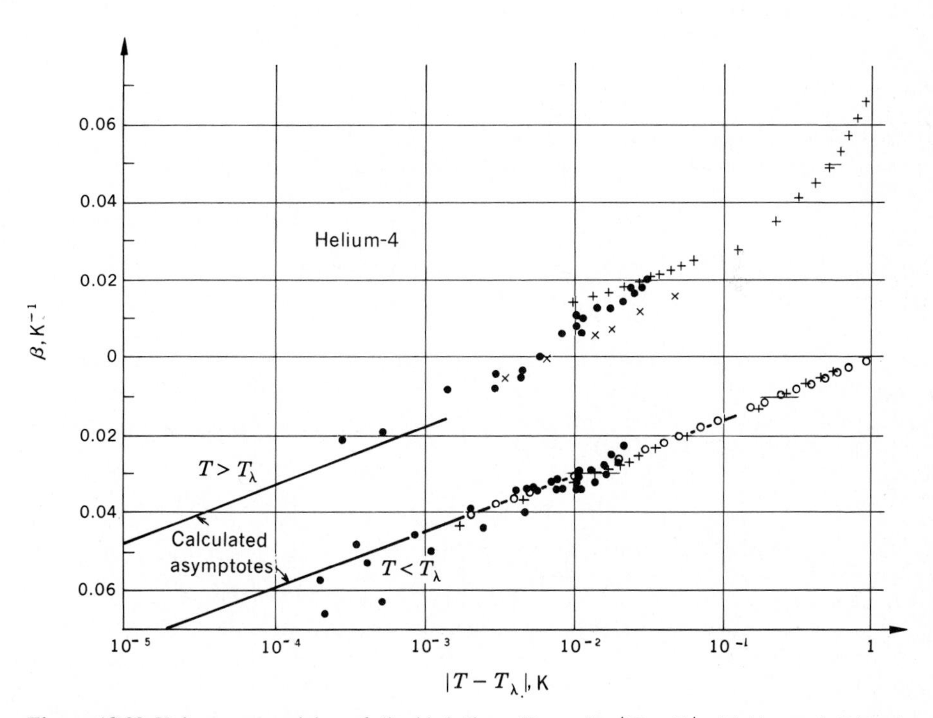

Figure 13-22 Volume expansivity of liquid helium II vs. log $|T - T_\lambda|$. *(Atkins and Edwards; Chase and Maxwell; Kerr and Taylor.)*

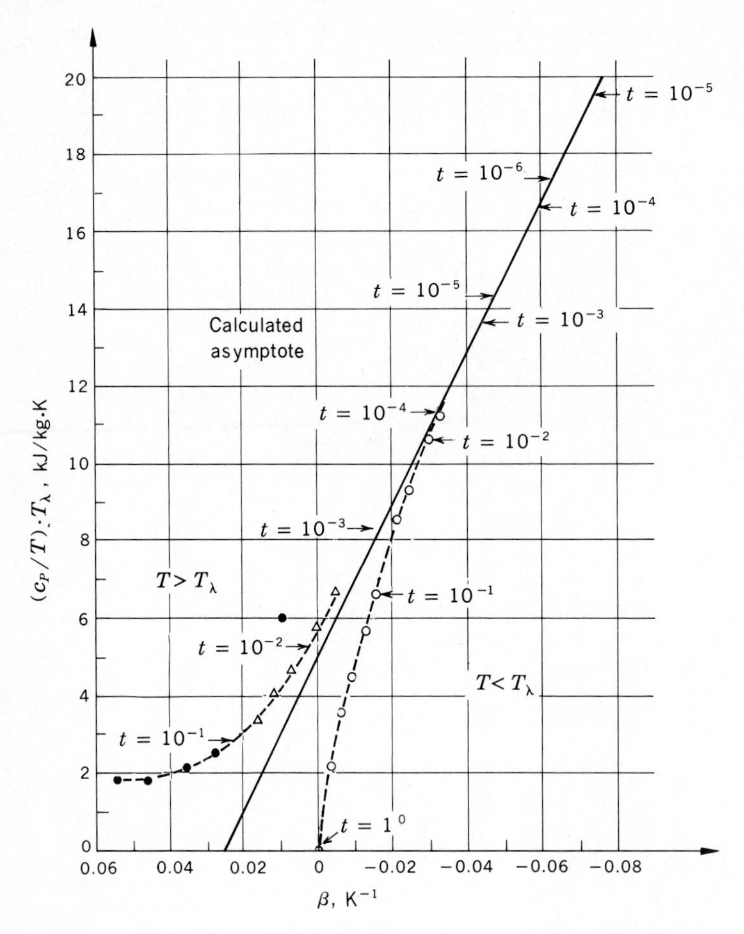

Figure 13-23 Straight-line relation between c_P/T and β of liquid helium II.

Fig. 13-22, where the dependence on the logarithm of t, characteristic of λ transitions, is displayed even down to $t = 10^{-6}$ K. The expected linear relation between c_P/T and $v\beta$ is shown in Fig. 13-23. Some numerical values at three points on the λ line are given in Table 13-5.

Table 13-5 Quantities at three points on the λ line of ^{4}He

P_λ, kPa	T_λ, K	ρ_λ, kg/m^3	$(dP/dT)_\lambda$, MPa/K	$(\partial v/\partial T)_\lambda$, l/kmol · K	$(\partial s/\partial T)_\lambda$, kJ/mol · K
5.035	2.177	146.2	−11.25	51.1	10.9
1487	2.00	167	−7.1	11.0	3.8
3013	1.763	180.4	−5.58	5.3	2.7

13-7 LIQUID AND SOLID HELIUM

Any temperature and pressure at which three phases of the same substance may coexist in equilibrium are a triple point. Water and many other substances have several triple points, but only one triple point refers to the equilibrium of solid, liquid, and vapor phases. Every material that has been studied has such a triple point, with the exception of helium. As the vapor of liquid ^{4}He is pumped away, the remaining liquid gets colder; and at the λ point, liquid helium II forms, so that the λ point may be regarded as a triple point for liquid I, liquid II, and vapor, as shown in Fig. 13-24.

The zero-point energy of liquid helium is about 210 J/mol, or about three times as large as the heat of vaporization. If a crystal were to form, it would be unstable under its own vapor pressure; therefore, as the vapor above liquid helium is pumped away, helium remains a liquid down to the lowest temperatures yet achieved ($\sim 10^{-6}$ K). To produce solid helium, it is necessary to bring the helium atoms closer together, to the point where attractive forces may pro-

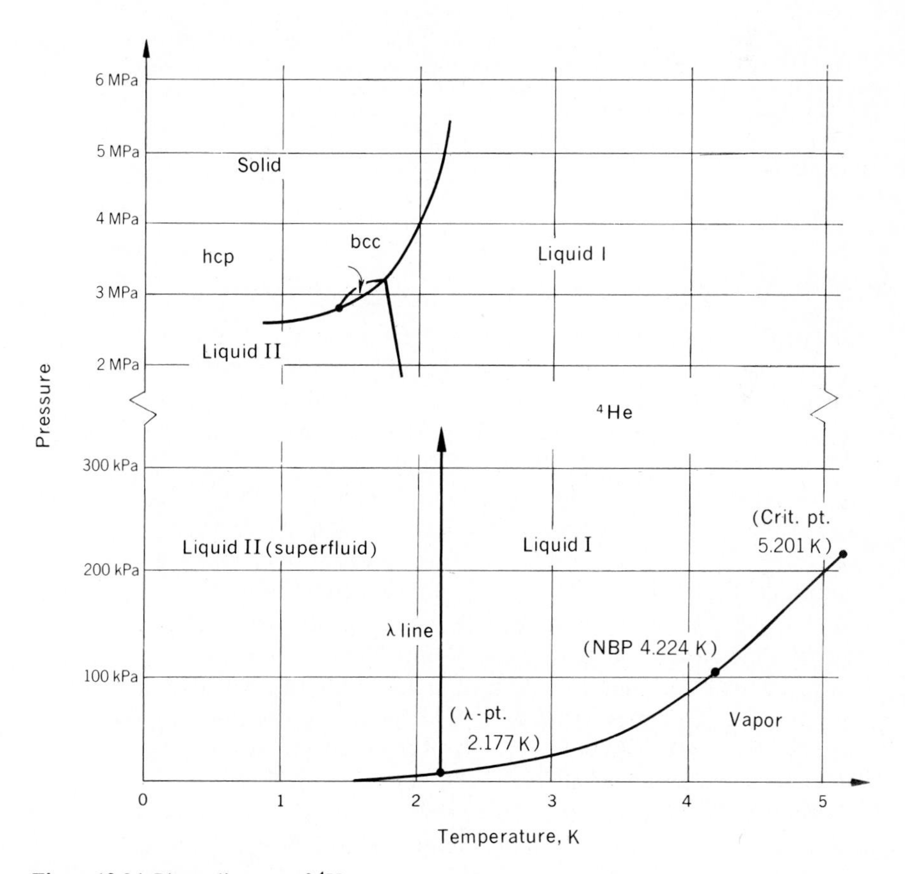

Figure 13-24 Phase diagram of ^{4}He.

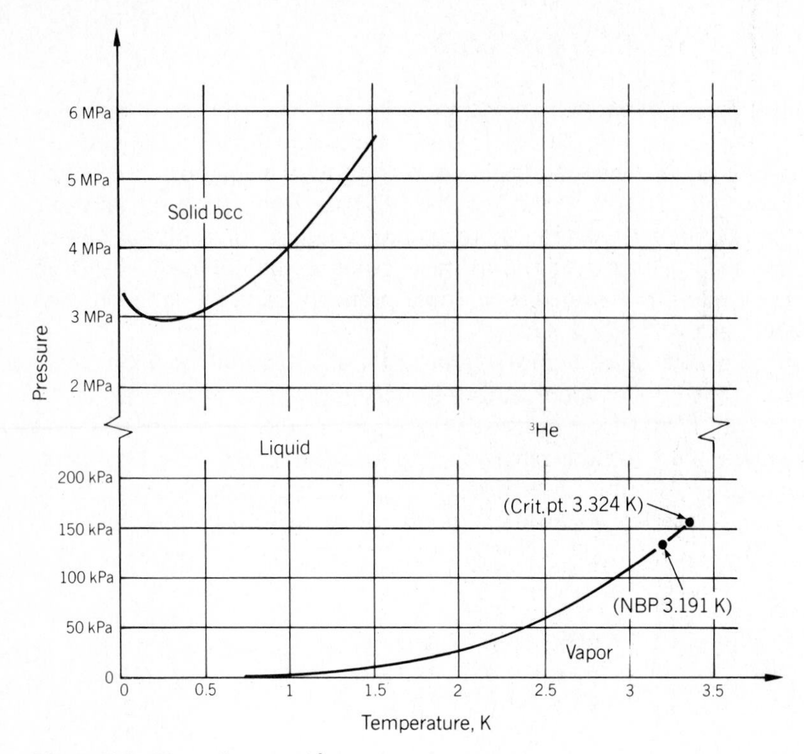

Figure 13-25 Phase diagram of ^{3}He.

duce cohesion. This requires an increase of pressure above 25 atm, so that there is an *upper triple point* where two liquids and a solid may coexist in equilibrium. As a matter of fact, the phase diagram of Fig. 13-24 shows the presence of two different crystal structures for solid ^{4}He, hexagonal close-packed (abbreviated hcp) and body-centered cubic (abbreviated bcc).

About one part in a million of ^{4}He is the light isotope ^{3}He, shown in Fig. 13-25, along with that of ^{4}He. Three features are quite striking:

1. The normal boiling point ^{3}He (at which the vapor pressure is 101 kPa) is 3.2 K, so that by pumping the ^{3}He vapor away temperatures may be produced which are lower than those obtainable with ^{4}He. Thus, by lowering the vapor pressure of ^{3}He to 3.2 Pa, a temperature of 0.4 K can be reached. To reach the same temperature with ^{4}He would require lowering the vapor pressure to 3.6×10^{-5} Pa, which is out of the question.
2. ^{3}He has no triple points at all.
3. The melting curve, which is the locus of points representing temperatures and pressures at which solid and liquid ^{3}He are in equilibrium, *shows a minimum* at about 0.3 K.

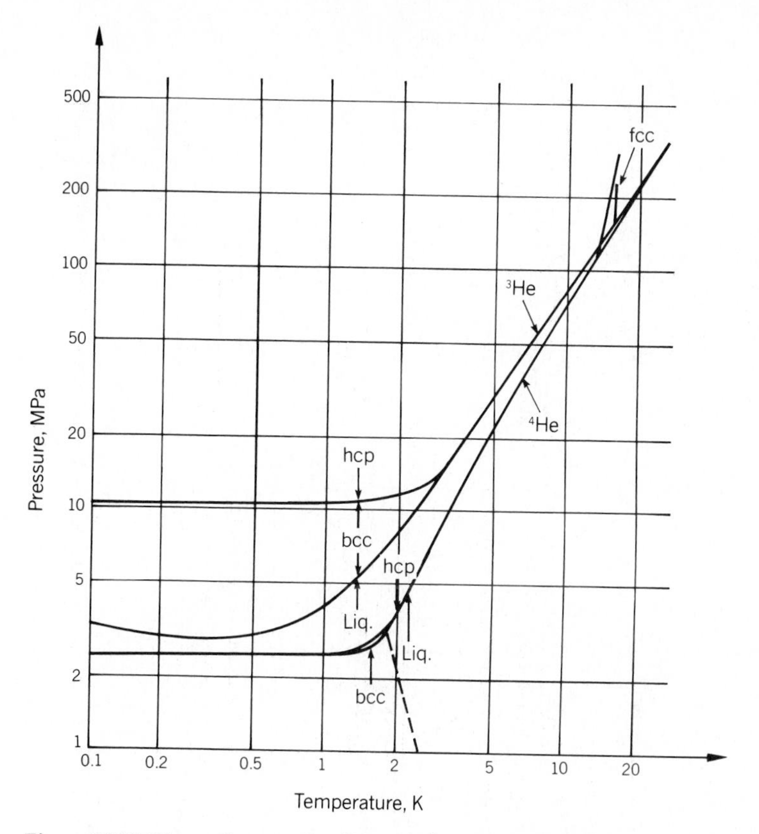

Figure 13-26 Phase diagrams for ^{4}He and ^{3}He at high pressures. *(K. H. Bennemann and J. B. Ketterson, The Physics of Liquid and Solid Helium, Interscience, 1976.)*

Both liquid ^{4}He and liquid ^{3}He show a negative thermal expansivity at temperatures above about 1 K.

In the region of high pressures, from 10 to 500 MPa, other crystal structures form in both ^{4}He and ^{3}He, as shown in Fig. 13-26. The heat capacity of these solids has been measured and found to obey a Debye T^3 law, with a Debye Θ depending both on T and on the molar volume v. The temperature dependence is such as to enable an extrapolation to be made to zero temperature, so that the volume dependence of Θ_0 can be studied. This is shown in Fig. 13-27, and the curves have a twofold purpose:

1. By using the result of Prob. 12-2, namely, that the zero-point energy of a crystal is $\frac{9}{8}R\Theta_0$, one can calculate the zero-point energy of solid helium and compare it with that of liquid helium. At $\Theta_0 = 30$ K, $\frac{9}{8}R\Theta \sim 270$ J/mol; and at $\Theta_0 = 110$ K, $\frac{9}{8}R\Theta_0 \sim 1000$ J/mol. Both of these extreme values *exceed* the zero-point energy of the liquid.

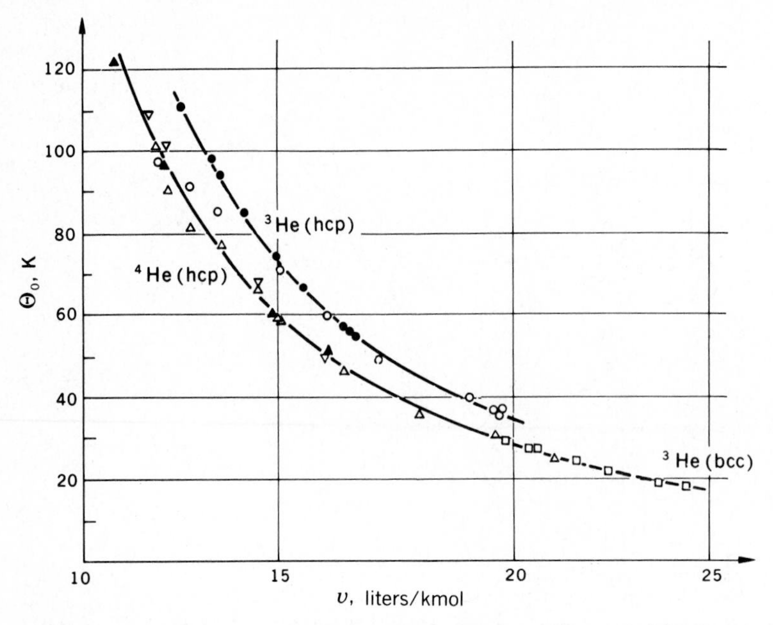

Figure 13-27 Debye Θ_0 as a function of volume for solid ^{4}He and ^{3}He. *(Heltemes and Swenson; Dugdale and Franck.)*

2. By using the result of Prob. 12-5, namely, that the Grüneisen Γ is given by

$$\Gamma = -\frac{d \ln \Theta_0}{d \ln v},$$

in conjunction with the curves in Fig. 13-27, the result is obtained that $\Gamma \sim 2.5$.

The transition between the solid and liquid phases of ^{4}He and of ^{3}He is particularly interesting. The latent heat of fusion may be calculated from experimental values of $v'' - v'$ and of dP/dT, using Clapeyron's equation. The results of such calculations by Simon and Swenson on ^{4}He and by Dugdale on ^{3}He are shown in Fig. 13-28. Notice that, in both cases, the latent heat becomes practically zero at low temperatures, whereas $P(v'' - v')$ does not. As a result, the energy difference $u'' - u' = l_F - P(v'' - v')$ becomes negative, indicating that *the energy of solid helium is greater than that of liquid helium* at the same temperature. The melting of solid helium at such temperatures is a purely mechanical process, since there is practically no latent heat in this temperature region. An isothermal reduction in pressure produces melting; conversely, an isothermal increase of pressure produces solidification.

Another interesting characteristic of solid helium is that its compressibility is greater than that of liquid helium.

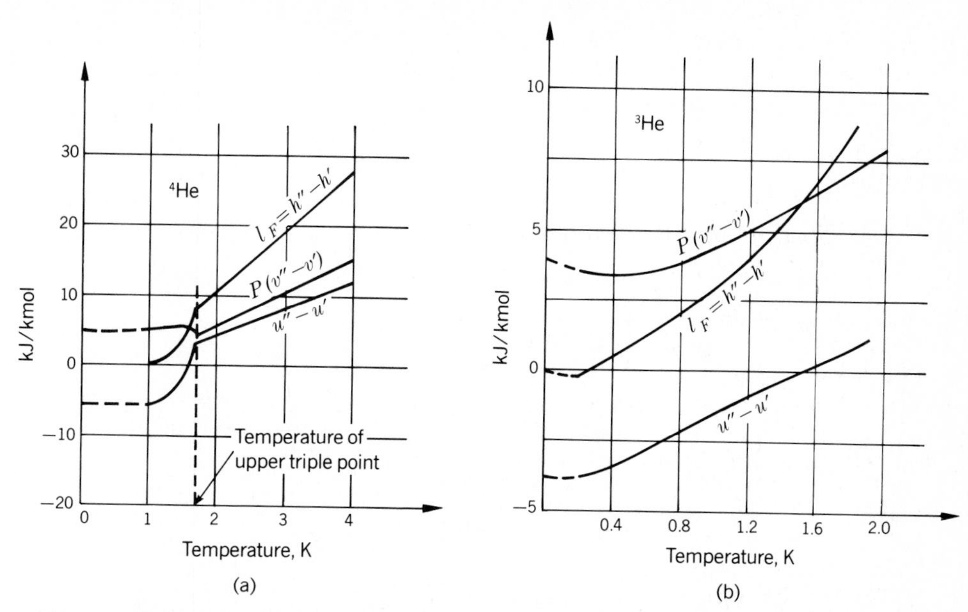

Figure 13-28 Energy relations for liquid and solid helium calculated with the aid of Clapeyron's equation and the first law of thermodynamics. (*a*) ^{4}He. *(F. E. Simon and C. A. Swenson.)* (*b*) ^{3}He. *(J. S. Dugdale.)*

PROBLEMS

13-1 (*a*) Show that in a Joule-Kelvin expansion, no temperature change occurs if $(\partial v/\partial T)_P = v/T$.
(*b*) Show that

$$\mu c_P = T^2 \left[\frac{\partial(v/T)}{\partial T}\right]_P.$$

In the region of moderate pressures, the equation of state of 1 mol of a gas may be written

$$Pv = RT + B'P + C'P^2,$$

where the second and third virial coefficients B' and C' are functions of T only.
(*c*) Show that, as the pressure approaches zero,

$$\mu c_P \to T\frac{dB'}{dT} - B'.$$

(*d*) Show that the equation of the inversion curve is

$$P = -\frac{B' - T(dB'/dT)}{C' - T(dC'/dT)}.$$

13-2 The Joule-Kelvin coefficient μ is a measure of the temperature change during a throttling process. A similar measure of the temperature change produced by an isentropic change of pressure is provided by the coefficient μ_S, where

$$\mu_S = \left(\frac{\partial T}{\partial P}\right)_S.$$

Prove that

$$\mu_S - \mu = \frac{V}{C_P}.$$

13-3 According to Hill and Lounasmaa, the equation of the helium inversion curve is

$$P = -21.0 + 5.44T - 0.132T^2,$$

where P is given in atmospheres.

(*a*) What is the maximum inversion temperature?

(*b*) What point on the inversion curve has the maximum pressure?

13-4 A substance undergoes a phase transition at the temperatures and pressures indicated by the points on the curves in Fig. 13-20*a*, *b*, and *c*. Choose any point where the slope is $(dP/dT)_\lambda$ in (*a*); $(ds/dT)_\lambda$ in (*b*); and $(dv/dT)_\lambda$ in (*c*). *Near* this point, where $T = T_\lambda + t$, show that:

(*a*)
$$\frac{c_V}{T} = \left(\frac{ds}{dT}\right)_\lambda - \frac{\beta}{\kappa}\left(\frac{dv}{dT}\right)_\lambda.$$

(*b*)
$$\beta = \frac{1}{v}\left(\frac{dv}{dT}\right)_\lambda - \kappa\left(\frac{dP}{dT}\right)_\lambda.$$

13-5 In a second-order phase transition, $s^{(i)} = s^{(f)}$ at (T, P), and $s^{(i)} + ds^{(i)} = s^{(f)} + ds^{(f)}$ at $(T + dT, P + dP)$.

(*a*) Prove that

$$\frac{dP}{dT} = \frac{1}{Tv}\frac{c_P^{(f)} - c_P^{(i)}}{\beta^{(f)} - \beta^{(i)}}.$$

In a second-order phase transition, $v^{(i)} = v^{(f)}$ at (T, P), and $v^{(i)} + dv^{(i)} = v^{(f)} + dv^{(f)}$ at $(T + dT, P + dP)$.

(*b*) Prove that

$$\frac{dP}{dT} = \frac{\beta^{(f)} - \beta^{(i)}}{\kappa^{(f)} - \kappa^{(i)}}.$$

(These are *Ehrenfest's equations.*)

13-6 For a two-phase system in equilibrium, P is a function of T only; therefore,

$$\left(\frac{\partial P}{\partial T}\right)_V = \left(\frac{\partial P}{\partial T}\right)_S = \frac{dP}{dT}.$$

For such a system show that

$$\frac{C_V}{\kappa_S} = TV\left(\frac{dP}{dT}\right)^2,$$

regardless of the type of transition between the phases.

CHAPTER

FOURTEEN

CHEMICAL EQUILIBRIUM

14-1 DALTON'S LAW

Imagine a homogeneous mixture of inert ideal gases at a temperature T, a pressure P, and a volume V. Suppose that there are n_1 mol of gas A_1, n_2 mol of gas A_2, etc., up to n_c mol of gas A_c. Since there is no chemical reaction, the mixture is in a state of equilibrium with the equation of state

$$PV = (n_1 + n_2 + \cdots + n_c)RT,$$

or

$$P = \frac{n_1}{V}RT + \frac{n_2}{V}RT + \cdots + \frac{n_c}{V}RT.$$

It is clear that the expression

$$\frac{n_k}{V}RT$$

represents the pressure that the kth gas would exert if it occupied the volume V alone. This is called the *partial pressure* of the kth gas and is denoted by p_k. Thus,

$$p_1 = \frac{n_1}{V}RT, \qquad p_2 = \frac{n_2}{V}RT, \qquad \ldots, \qquad p_c = \frac{n_c}{V}RT,$$

and

$$P = p_1 + p_2 + \cdots + p_c. \tag{14-1}$$

Equation (14-1) expresses the fact that the total pressure of a mixture of inert ideal gases is equal to the sum of the partial pressures. This is *Dalton's law*.

Now

$$V = (n_1 + n_2 + \cdots + n_c)\frac{RT}{P}$$

$$= \sum n_k \frac{RT}{P},$$

and the partial pressure of the kth gas is

$$p_k = \frac{n_k}{V} RT.$$

Substituting the value for V, we get

$$p_k = \frac{n_k}{\sum n_k} P.$$

The ratio $n_k/\sum n_k$ is called the *mole fraction* of the kth gas and is denoted by x_k. Thus,

$$x_1 = \frac{n_1}{\sum n_k}, \qquad x_2 = \frac{n_2}{\sum n_k}, \qquad \ldots, \qquad x_c = \frac{n_c}{\sum n_k},$$

and

$$p_1 = x_1 P, \qquad p_2 = x_2 P, \qquad \ldots, \qquad p_c = x_c P.$$

The mole fractions are convenient dimensionless quantities with which to express the composition of a mixture. It is clear that

$$x_1 + x_2 + \cdots + x_c = \frac{n_1}{\sum n_k} + \frac{n_2}{\sum n_k} + \cdots + \frac{n_c}{\sum n_k}$$

$$= 1.$$

Hence, if all but one mole fraction is determined, the last can be calculated from the preceding equation.

14-2 SEMIPERMEABLE MEMBRANE

If a narrow tube of palladium is closed at one end and the open end is sealed into a glass tube, as shown in Fig. 14-1, the system may be pumped to a very high vacuum. If the palladium remains at room temperature, the vacuum can be maintained indefinitely. If however, an ordinary bunsen burner is placed so that the blue cone surrounds part of the tube, with the rest of the flame causing the palladium tube to become red-hot, hydrogen present in the blue cone will pass through the tube, but other gases will not. Red-hot palladium is said to be a *semipermeable membrane*, permeable to hydrogen only. This is the simplest laboratory method of obtaining pure dry hydrogen.

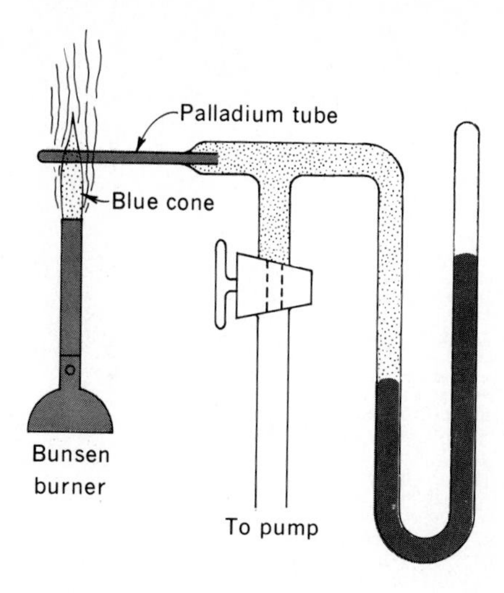

Figure 14-1 Palladium tube permeable to hydrogen.

Experiment shows that hydrogen continues to flow through the red-hot palladium until the pressure of hydrogen in the vessel reaches a value equal to the partial pressure of the hydrogen in the flame. When the flow stops, *membrane equilibrium* is said to exist. Membrane equilibrium is achieved when the partial pressure of the gas to which the membrane is permeable is the same on both sides of the membrane. We shall suppose that there exists a special membrane permeable to each gas with which we have to deal. Whether this is actually so is not important. We shall make use of the principle of the semipermeable membrane as an ideal device for theoretical purposes.

14-3 GIBBS' THEOREM

With the aid of a device equipped with two semipermeable membranes, it is possible to conceive of separating in a reversible manner a mixture of two inert ideal gases. The vessel depicted in Fig. 14-2 is divided into two equal compartments by a rigid wall, which is a membrane permeable only to the gas A_1. Two pistons coupled so that they move together at a constant distance apart are constructed of materials such that one is impermeable to all gases, and the other is permeable only to the gas A_2. The initial state is depicted in Fig. 14-2*i*. A mixture of A_1 and A_2 is in the left-hand chamber, and the right-hand chamber is evacuated.

Now imagine pushing the coupled pistons to the right in such a manner that the following conditions are satisfied:

1. The motion is infinitely slow, so that membrane equilibrium exists at all times.
2. There is no friction.
3. The whole system is kept at constant temperature.

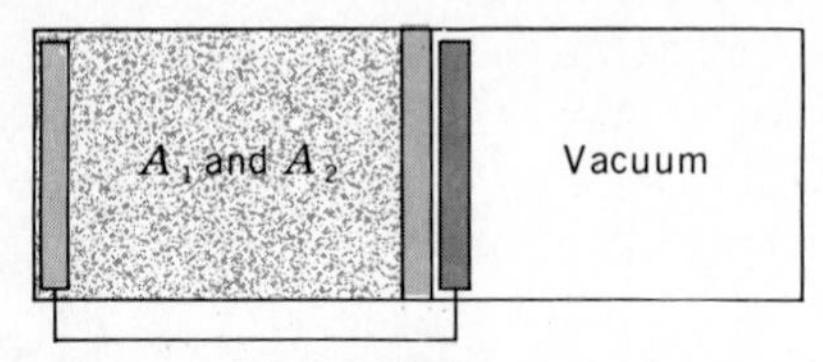

(*i*) Initial equilibrium state

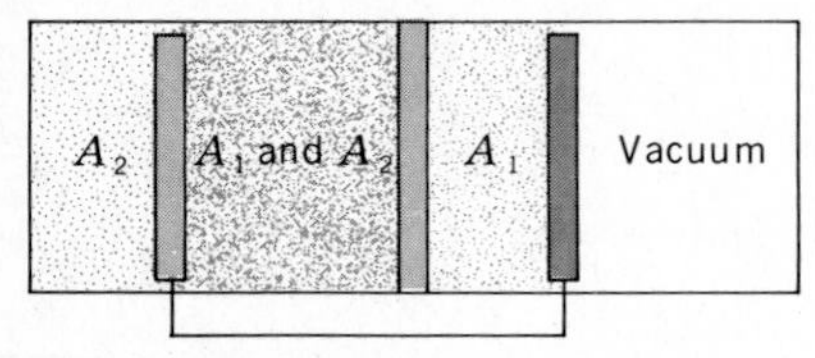

(*k*) Intermediate equilibrium state

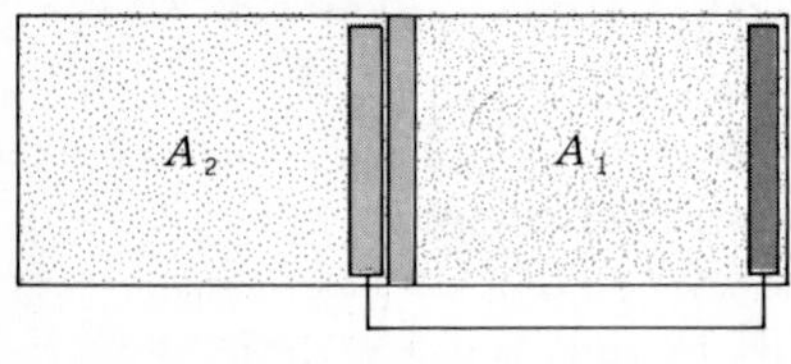

(*f*) Final equilibrium state

Figure 14-2 Reversible isothermal separation of two inert ideal gases.

These conditions define a *reversible isothermal process.* Consider the system at any intermediate state such as that depicted in Fig. 14-2*k*. If p_1 and p_2 are the partial pressures, respectively, of A_1 and A_2 in the mixture, P_1 is the pressure of A_1 alone, and P_2 the pressure of A_2 alone, then the forces acting on the coupled pistons are:

$$\text{Force to the left} = (p_1 + p_2) \times \text{area},$$

and

$$\text{Sum of the forces to the right} = (P_1 + P_2) \times \text{area}.$$

Since membrane equilibrium exists, $p_1 = P_1$ and $p_2 = P_2$; whence the resultant force acting on the coupled pistons is zero. After the pistons have moved all the way to the right, the gases are completely separated as shown in Fig. 14-2*f*.

Since the resultant force was infinitesimal in the beginning and zero throughout the remainder of the process, $W = 0$. Also, since the process was isothermal and the internal energy of an ideal gas is a function of T only, $U_f = U_i$. Finally, since the process was both reversible and isothermal, the heat transferred Q is equal to $T(S_f - S_i)$. We have, therefore, the result that

$$T(S_f - S_i) = 0,$$

and since T is not zero,

$$S_i = S_f.$$

Now S_i is the entropy of the mixture at the temperature T and the volume V, while S_f is the sum of the entropies of the two gases, each at the same temperature and each occupying the volume V alone. If we define the *partial entropy* of one of the gases of a mixture as the entropy that the gas would have if it occupied the whole volume alone at the same temperature, we obtain the result that the *entropy of a mixture of ideal gases is the sum of the partial entropies.* This is known as *Gibbs' theorem.* The generalization for any number of gases is obvious.

14-4 ENTROPY OF A MIXTURE OF INERT IDEAL GASES

Imagine a number of inert ideal gases separated from one another by suitable partitions, all the gases being at the same temperature T and pressure P. Suppose that there are n_1 mol of gas A_1, n_2 mol of A_2, etc., up to n_c mol of A_c. Before the partitions are removed, the entropy of the whole system S_i is the sum of the separate entropies. The entropy of 1 mol of the kth gas at temperature T and pressure P is

$$s_k = \int c_{Pk} \frac{dT}{T} + s_{0k} - R \ln P;$$

therefore,

$$S_i = \sum n_k \left(\int c_{Pk} \frac{dT}{T} + s_{0k} - R \ln P \right)$$

$$= R \sum n_k \left(\frac{1}{R} \int c_{Pk} \frac{dT}{T} + \frac{s_{0k}}{R} - \ln P \right).$$

It is convenient to represent the first two terms within the parentheses by σ, thus:

$$\sigma_k = \frac{1}{R} \int c_{Pk} \frac{dT}{T} + \frac{s_{0k}}{R}. \tag{14-2}$$

Then

$$S_i = R \sum n_k(\sigma_k - \ln P). \tag{14-3}$$

After the partitions are removed, the temperature and pressure remain the same because there is no chemical reaction, but the gases diffuse and, by Gibbs' theorem, the entropy of the mixture is the sum of the partial entropies. The partial entropy of the kth gas is the entropy that the kth gas would have if it occupied the whole volume alone at the same temperature, in which case it would exert a pressure equal to the partial pressure p_k. Therefore, the total entropy of the mixture is

$$S_f = R \sum n_k(\sigma_k - \ln p_k).$$

Since $p_k = x_k P$,

$$\boxed{S_f = R \sum n_k(\sigma_k - \ln P - \ln x_k).} \tag{14-4}$$

The change of entropy due to the diffusion of any number of inert gases is therefore equal to

$$\boxed{S_f - S_i = -R \sum n_k \ln x_k.} \tag{14-5}$$

Each of the mole fractions is a number less than unity with a negative logarithm. The whole expression is therefore positive, as it should be. Since

$$x_k = \frac{n_k}{\sum n} = \frac{n_k RT}{\sum nRT} = \frac{n_k Pv}{PV} = \frac{n_k v}{V},$$

we may write

$$S_f - S_i = n_1 R \ln \frac{V}{n_1 v} + n_2 R \ln \frac{V}{n_2 v} + \cdots.$$

The foregoing result shows that the entropy change due to the diffusion of any number of ideal gases is the same as that which would take place if each gas were caused to undergo a free expansion from the volume that it occupies alone at T and P to the volume of the mixture at the same T and P. The validity of this result was assumed in Chap. 8 in order to calculate the entropy change of the universe when two ideal gases diffuse. The assumption is therefore seen to be justified.

As an example, consider the diffusion of 1 mol of helium and 1 mol of neon. Then,

$$\begin{aligned} S_f - S_i &= -R(1 \ln \tfrac{1}{2} + 1 \ln \tfrac{1}{2}) \\ &= 2R \ln 2. \end{aligned}$$

In this expression there are no quantities such as heat capacities or entropy constants that distinguish one gas from another. The result is the same for the diffusion of any two inert ideal gases, no matter how similar or dissimilar they are. If the two gases are identical, however, the concept of diffusion has no meaning, and there is no entropy change. From the microscopic point of view, this means that the diffusion of any two dissimilar gases brings about the same degree of disorder, whereas the diffusion of two identical gases introduces no element of disorder.

The application of mathematics to the macroscopic processes of nature usually gives rise to continuous results. Our experience suggests that, as the two diffusing gases become more and more alike, the entropy change due to diffusion should get smaller and smaller, approaching zero as the gases become identical. The fact that this is not the case is known as *Gibbs' paradox*. The paradox has been resolved by Bridgman in the following way: To recognize that two gases are dissimilar requires a set of experimental operations. These operations become more and more difficult as the gases become more and more alike; but, at least in principle, the operations are possible. In the limit, when the gases become identical, there is a discontinuity in the instrumental operations inasmuch as no instrumental operation exists by which the gases may be distinguished. Hence, a discontinuity in a function, such as that of an entropy change, is to be expected.

14-5 GIBBS FUNCTION OF A MIXTURE OF INERT IDEAL GASES

The enthalpy and the entropy of 1 mol of an ideal gas at temperature T and pressure P are, respectively,

$$h = h_0 + \int c_P \, dT,$$

and

$$s = \int c_P \frac{dT}{T} + s_0 - R \ln P;$$

therefore, the molar Gibbs function $g = h - Ts$ is equal to

$$g = h_0 + \int c_P \, dT - T \int c_P \frac{dT}{T} - Ts_0 + RT \ln P.$$

Applying the formula for integration by parts,

$$\int c_P \, dT - T \int c_P \frac{dT}{T} = -T \int \frac{\int c_P \, dT}{T^2} \, dT,$$

we get

$$g = h_0 - T \int \frac{\int c_P \, dT}{T^2} \, dT - Ts_0 + RT \ln P$$

$$= RT\left(\frac{h_0}{RT} - \frac{1}{R} \int \frac{\int c_P \, dT}{T^2} \, dT - \frac{s_0}{R} + \ln P\right).$$

It is convenient to denote the first three terms within the parentheses by ϕ, thus:

$$\phi = \frac{h_0}{RT} - \frac{1}{R} \int \frac{\int c_P \, dT}{T^2} \, dT - \frac{s_0}{R}. \tag{14-6}$$

The molar Gibbs function of an ideal gas may therefore be written as

$$\boxed{g = RT(\phi + \ln P),} \tag{14-7}$$

where ϕ is a function of T only.

Consider a number of inert ideal gases separated from one another, all at the same T and P. Suppose that there are n_1 mol of gas A_1, n_2 mol of A_2, etc., up to n_c mol of A_c. Before the gases are mixed, the Gibbs function of the system G is the sum of the separate Gibbs functions, or

$$G_i = \sum n_k g_k$$

$$= RT \sum n_k(\phi_k + \ln P),$$

where the summation extends from $k = 1$ to $k = c$. To express the Gibbs function after mixing G_f, it is necessary merely to replace the total pressure P by the partial pressure p_k. (Why?) Thus,

$$G_f = RT \sum n_k(\phi_k + \ln p_k)$$
$$= RT \sum n_k(\phi_k + \ln P + \ln x_k).$$

Therefore,
$$G_f - G_i = RT \sum n_k \ln x_k,$$

where the expression on the right is a negative quantity. It is seen therefore that the Gibbs function after diffusion is less than the Gibbs function before diffusion. This will be shown later to be an expression of a general law that holds for all irreversible processes which take place at constant T and P.

We have shown that the Gibbs function of a mixture of inert ideal gases at temperature T and pressure P is

$$\boxed{G = RT \sum n_k(\phi_k + \ln P + \ln x_k),} \tag{14-8}$$

where ϕ is given by Eq. (14-6).

14-6 CHEMICAL EQUILIBRIUM

Consider a homogeneous mixture of 1 mol of hydrogen and 1 mol of oxygen at room temperature and at atmospheric pressure. It is a well-known fact that this mixture will remain indefinitely at the same temperature, pressure, and composition. The most careful measurements over a long period of time will disclose no appreciable spontaneous change of state. One might be inclined to deduce from this that such a mixture represents a system in a state of thermodynamic equilibrium. However, this is not the case, If a small piece of platinized asbestos is introduced or if an electric spark is created across two electrodes, an explosion takes place involving a sudden change in the temperature, the pressure, and the composition. If at the end of the explosion the system is brought back to the same temperature and pressure, it will be found that the composition is now $\frac{1}{2}$ mol of oxygen, no measurable amount of hydrogen, and 1 mol of water vapor.

The piece of material such as platinized asbestos by whose agency a chemical reaction is started is known as a *catalyst*. If chemical combination is started in a mixture of 1 mol of hydrogen and 1 mol of oxygen with different amounts and different kinds of catalysts and if the final composition of the mixture is measured in each case, it is found that (1) the final composition does not depend on the amount of catalyst used; (2) the final composition does not depend on the kind of catalyst used; and (3) the catalyst itself is the same at the end of the reaction as at the beginning. These results lead us to the following conclusions:

1. The initial state of the mixture is a state of mechanical and thermal equilibrium but not of chemical equilibrium.

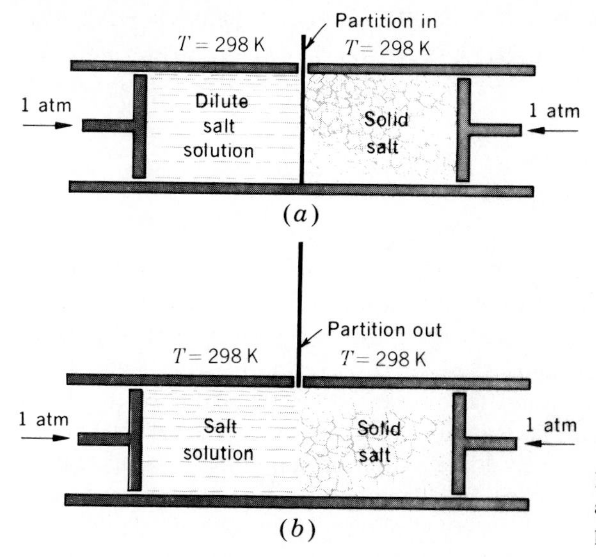

Figure 14-3 Transport of matter across the boundary between two phases.

2. The final state is a state of thermodynamic equilibrium.
3. The transition from the initial nonequilibrium state to the final equilibrium state is accompanied by a chemical reaction that is too slow to be measured when it takes place spontaneously. Through the agency of the catalyst, the reaction is caused to take place more rapidly.

Imagine a vessel divided into two compartments by a removable partition as shown in Fig. 14-3*a*. Suppose that one compartment contains a dilute solution of sodium chloride and water maintained at standard atmospheric pressure and at a temperature of 298 K—the mole fraction of the salt being, say, 0.01. Under these conditions the solution is in thermodynamic equilibrium. Suppose that the other compartment contains solid salt in equilibrium also at standard atmospheric pressure and a temperature of 298 K. Now imagine that the partition is removed (Fig. 14-3*b*) and that the pressure and temperature of the whole system are kept constant at the original values. Experiment shows that some solid salt dissolves; i.e., the mole fraction of the salt in the solution increases spontaneously at constant pressure and temperature. After a while, the change ceases and the mole fraction is found to be about 0.1.

Focusing our attention on the solution from the moment it was put in contact with the solid salt, we are led to the following conclusions:

1. The initial state of the solution (at the moment it was put in contact with the solid salt) is one of mechanical and thermal equilibrium but not of chemical equilibrium.
2. The final state of the solution is a state of thermodynamic equilibrium.

3. The transition from the initial nonequilibrium state to the final equilibrium state is accompanied by a transport of a chemical constituent into the solution.

14-7 THERMODYNAMIC DESCRIPTION OF NONEQUILIBRIUM STATES

A *phase* is defined as a system or a portion of a system composed of any number of chemical constituents satisfying the requirements (1) that it is homogeneous and (2) that it has a definite boundary. The hydrogen-oxygen mixture described in Art. 14-6 is a gaseous phase of two chemical constituents and of constant mass. The salt solution is a liquid phase of two chemical constituents whose mass, when in contact with the solid-salt phase, is variable. Although the initial states of both these phases are nonequilibrium states, it is possible to describe them in terms of thermodynamic coordinates. Since each phase is in mechanical and thermal equilibrium, a definite P and T may be ascribed to each; since each has a definite boundary, each has a definite volume; and since each is homogeneous, the composition of each phase may be described by specifying the number of moles of each constituent. In general, a phase consisting of c chemical constituents in mechanical and thermal equilibrium may be described with the aid of the coordinates P, V, T, $n_1, n_2, \ldots, n_c$.

Under a given set of conditions, a phase may undergo a change of state in which some or all of these coordinates change. While this is going on, the phase passes through states not of thermodynamic equilibrium but of mechanical and thermal equilibrium only. These states are connected by an equation of state that is a relation among P, V, T, and the n's. Whether a phase is in chemical equilibrium or not, it has a definite internal energy and enthalpy. Both U and H may be regarded as functions of P, V, T, and the n's; and upon eliminating one of the coordinates by means of the equation of state, U and H may be expressed as a function of any two of P, V, and T and all the n's. Since entropy is a measure of the molecular disorder of the system, the entropy of a phase that is not in chemical equilibrium must have a meaning. We shall assume that the entropy of a phase, and therefore the Helmholtz and Gibbs functions also, can be expressed as functions of any two of P, V, and T and all the n's.

During a change of state the n's, which determine the composition of the phase, change either by virtue of a chemical reaction or by virtue of a transport of matter across the boundaries between phases, or both. In general, under given conditions, there is a set of values of the n's for which the phase is in chemical and therefore in thermodynamic equilibrium. The functions that express the properties of a phase when it is not in chemical equilibrium must obviously reduce to those for thermodynamic equilibrium when the equilibrium values of the n's are substituted. We are therefore led to assume that *any property of a phase in mechanical and thermal equilibrium can be represented by a function of any two of P, V, T, and the n's of the same form as that used to denote the same property when the phase is in thermodynamic equilibrium.*

Consider, for example, a phase consisting of a mixture of ideal gases. When the gases are inert, the equation of state is

$$PV = \sum n_k RT,$$

the entropy is

$$S = R \sum n_k(\sigma_k - \ln P - \ln x_k),$$

and the Gibbs function is

$$G = RT \sum n_k(\phi_k + \ln P + \ln x_k).$$

According to the assumption just made, these same equations may be used in connection with an ideal-gas phase in mechanical and thermal equilibrium when the gases are chemically active, when the phase is in contact with other phases, or under both conditions, whether chemical equilibrium exists or not. Under these conditions the n's and x's are variables. Whether they are all independent variables or not is a question that cannot be answered until the conditions under which a change of state takes place are specified. It is clear that, if the mass of the phase remains constant and the gases are inert, the n's and x's are constants. If the mass of the phase remains constant and the gases are chemically active, then it will be shown that each n (and therefore each x) is a function of only one independent variable, the degree of reaction. If the mass of the phase is variable, the number of n's that are independent depends on the number of other phases in contact with the original phase and on the chemical constituents of these other phases.

A system composed of two or more phases is called a *heterogeneous system.* Any extensive property such as V, U, S, H, F, or G of any one of the phases may be expressed as a function of, say, T, P, and the n's of that phase. Thus, for the Gibbs function of the first phase,

$$G^{(1)} = \text{function of } (T, P, n_1^{(1)}, n_2^{(1)}, \ldots);$$

for the second phase,

$$G^{(2)} = \text{function of } (T, P, n_1^{(2)}, n_2^{(2)}, \ldots);$$

and so on. The Gibbs function of the whole heterogeneous system is therefore

$$G = G^{(1)} + G^{(2)} + \cdots.$$

This result holds for any extensive property of a heterogeneous system.

14-8 CONDITIONS FOR CHEMICAL EQUILIBRIUM

Consider any hydrostatic system of constant mass, either homogeneous or heterogeneous, in mechanical and thermal equilibrium but not in chemical equilibrium. Suppose that the system is in contact with a reservoir at temperature T and undergoes an infinitesimal irreversible process involving an exchange of heat đQ with the reservoir. The process may involve a chemical reaction or a transport of matter between phases or both. Let dS denote the entropy change of the system and dS_0 the entropy change of the reservoir. The total entropy change of the universe is therefore $dS_0 + dS$; and since the performance of an irreversible process is attended by an increase in the entropy of the universe, we may write

$$dS_0 + dS > 0.$$

Since

$$dS_0 = -\frac{đQ}{T},$$

we have

$$-\frac{đQ}{T} + dS > 0,$$

or

$$đQ - T\,dS < 0.$$

During the infinitesimal irreversible process, the internal energy of the system changes by an amount dU, and an amount of work $P\,dV$ is performed. The first law can therefore be written in its usual form,

$$đQ = dU + P\,dV,$$

and the inequality becomes

$$\boxed{dU + P\,dV - T\,dS < 0.} \tag{14-9}$$

This inequality holds during any infinitesimal portion and, therefore, during all infinitesimal portions of the irreversible process. According to the assumption made in the preceding article, U, V, and S may all be regarded as functions of thermodynamic coordinates.

During the irreversible process for which inequality (14-9) holds, some or all of the coordinates may change. If we restrict the irreversible process by imposing the condition that two of the thermodynamic coordinates remain constant, then the inequality can be reduced to a simpler form. Suppose, for example, that the internal energy and the volume remain constant. Then the inequality reduces to $dS > 0$, which means that the entropy of a system at constant U and V increases during an irreversible process, approaching a maximum at the final state of equilibrium. This result, however, is obvious from the entropy principle, since a system at constant U and V is isolated and is therefore, so to speak, its own universe. The two most important sets of conditions are the following:

1. *If T and V are constant*, the inequality reduces to

$$d(U - TS) < 0,$$

or

$$dF < 0, \tag{14-10}$$

expressing the result that *the Helmholtz function of a system at constant T and V decreases during an irreversible process* and becomes a minimum at the final equilibrium state.

2. *If T and P are constant*, the inequality reduces to

$$d(U + PV - TS) < 0,$$

or

$$dG < 0, \tag{14-11}$$

expressing the result that *the Gibbs function of a system at constant T and P decreases during an irreversible process* and becomes a minimum at the final equilibrium state.

The student will recall that the condition for equilibrium of a conservative mechanical system is that the potential energy shall be a minimum. The Helmholtz and Gibbs functions are therefore seen to play a similar role in thermodynamics. For this reason, Gibbs called the function F the *thermodynamic potential at constant volume* and the function G the *thermodynamic potential at constant pressure.*

14-9 CONDITION FOR MECHANICAL STABILITY

In discussing the important equation

$$C_P - C_V = -T\left(\frac{\partial V}{\partial T}\right)_P^2 \left(\frac{\partial P}{\partial V}\right)_T$$

in Art. 9-7, the remark was made that $(\partial P/\partial V)_T$ is always negative. This will now be proved for any system of constant mass. Consider the system depicted symbolically in Fig. 14-4. At first, each half of the system is in equilibrium as well as the entire system. Suppose, now, that the left half is compressed by an amount δv and the right half is expanded by the same amount, with each half remaining at constant temperature and the *total* volume remaining constant.

The Helmholtz function f_L of the left half may now be expanded in a Taylor series about its equilibrium value $F_{\min}/2$ as follows:

$$f_L = \frac{F_{\min}}{2} - \left(\frac{\partial f}{\partial v}\right)_T \delta v + \frac{1}{2}\left(\frac{\partial^2 f}{\partial v^2}\right)_T (\delta v)^2 - \cdots,$$

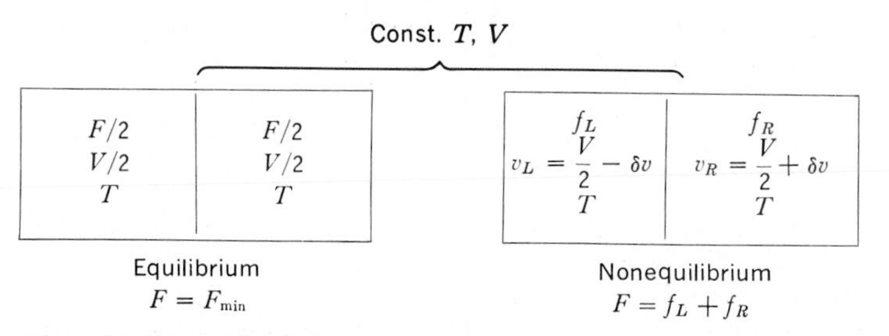

Figure 14-4 Transition from a state of thermodynamic equilibrium to a state characterized by a lack of mechanical equilibrium.

which may be terminated after the squared term if δv is small enough. But there is no difference between the behavior of half the system and that of the whole system, or

$$\left(\frac{\partial f}{\partial v}\right)_T = \left(\frac{\partial F}{\partial V}\right)_T.$$

Hence
$$f_L = \frac{F_{\min}}{2} - \left(\frac{\partial F}{\partial V}\right)_T \delta v + \frac{1}{2}\left(\frac{\partial^2 F}{\partial V^2}\right)_T (\delta v)^2.$$

Similarly, for the right half,

$$f_R = \frac{F_{\min}}{2} + \left(\frac{\partial F}{\partial V}\right)_T \delta v + \frac{1}{2}\left(\frac{\partial^2 F}{\partial V^2}\right)_T (\delta v)^2.$$

Adding these two equations, the total Helmholtz function of the system *in the nonequilibrium state* is

$$f_L + f_R = F_{\min} + \left(\frac{\partial^2 F}{\partial V^2}\right)_T (\delta v)^2,$$

or
$$f_L + f_R - F_{\min} = \delta F_{T,V} = \left(\frac{\partial^2 F}{\partial V^2}\right)_T (\delta v)^2.$$

The relation between F and the volume v of either half of the system at constant T and V is shown in Fig. 14-5. Since $\delta F_{T,V}$ is positive, it follows that

$$\left(\frac{\partial^2 F}{\partial V^2}\right)_T > 0.$$

But
$$\left(\frac{\partial F}{\partial V}\right)_T = -P,$$

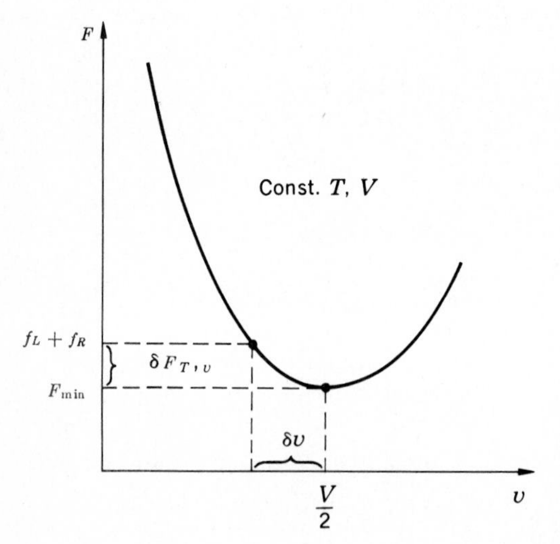

Figure 14-5 Dependence of Helmholtz function of the entire system on the volume of half of it, when the total volume and the temperature are constant.

and hence, finally,

$$\left(\frac{\partial P}{\partial V}\right)_T < 0.$$

This is *the condition for mechanical stability.*

14-10 THERMODYNAMIC EQUATIONS FOR A PHASE

Consider a phase composed of c chemical constituents, of which there are n_1 mol of substance A_1, n_2 mol of A_2, etc., n_c mol of A_c; the phase is in thermal equilibrium at temperature T and in mechanical equilibrium at pressure P. The Gibbs function of the phase can be written as follows:

$$G = \text{function of } (T, P, n_1, n_2, \ldots, n_c).$$

If the constituents are inert, the phase is in chemical and therefore in thermodynamic equilibrium. Imagine the performance of an infinitesimal *reversible* process in which the temperature and pressure are changed by dT and dP, respectively, and the numbers of moles of the various constituents are altered by the amounts dn_1, dn_2, ..., dn_c. Since we have assumed that the constituents are inert, the changes in the n's are to be regarded as accomplished by the reversible addition or withdrawal of the constituents with the aid of suitable semipermeable membranes. The resulting change in the Gibbs function of the phase is given by

$$dG = \frac{\partial G}{\partial T}\,dT + \frac{\partial G}{\partial P}\,dP + \frac{\partial G}{\partial n_1}\,dn_1 + \frac{\partial G}{\partial n_2}\,dn_2 + \cdots + \frac{\partial G}{\partial n_c}\,dn_c,$$

where it is understood that G is a function of T, P, and the n's and that each partial derivative implies that all variables other than the one indicated are to be kept constant.

As a special case, consider an infinitesimal reversible process in which all the dn's are zero. Under these conditions, the composition and the mass of the phase remain constant, and the equation becomes

$$dG = \frac{\partial G}{\partial T}\,dT + \frac{\partial G}{\partial P}\,dP \qquad \text{(for constant composition and mass).}$$

But, for this case, it has already been shown that

$$dG = -S\,dT + V\,dP.$$

It follows, therefore, that

$$\frac{\partial G}{\partial T} = \left(\frac{\partial G}{\partial T}\right)_{P, n_1, n_2, \ldots, n_c} = -S, \tag{14-12}$$

$$\frac{\partial G}{\partial P} = \left(\frac{\partial G}{\partial P}\right)_{T, n_1, n_2, \ldots, n_c} = V, \tag{14-13}$$

and $$dG = -S\,dT + V\,dP + \frac{\partial G}{\partial n_1}\,dn_1 + \frac{\partial G}{\partial n_2}\,dn_2 + \cdots + \frac{\partial G}{\partial n_c}\,dn_c.$$

Now consider the effect upon the Gibbs function when a small amount of one of the constituents (say, the kth constituent, A_k) is introduced into the phase, T, P, and the other n's remaining constant. If dn_k moles of A_k are introduced, the effect on the Gibbs function is expressed by the partial derivative

$$\mu_k = \frac{\partial G}{\partial n_k}, \tag{14-14}$$

where μ_k is called the *chemical potential* of the kth constituent of the phase in question. A chemical potential of one constituent is a function of T, P, and *all* the n's. If a substance is not present in a phase, it does not follow that its chemical potential is zero. The chemical potential is a measure of the effect on the Gibbs function when a substance *is* introduced. Even though the substance is not present in the phase, there is always the possibility of introducing it, in which case the Gibbs function would be altered and the value of μ would be finite. We may now write

$$dG = -S\,dT + V\,dP + \mu_1\,dn_1 + \mu_2\,dn_2 + \cdots + \mu_c\,dn_c$$

for an infinitesimal change in the Gibbs function of any phase consisting of inert constituents.

Suppose, now, that the constituents are chemically active. Changes in the n's may then take place because of a chemical reaction. Although the phase is always considered to be in thermal and mechanical equilibrium, an infinitesimal process involving a change in T, P, and the n's will, in general, be irreversible, since chemical equilibrium may not exist. In accordance with our previous assumption as to the form of the expressions denoting properties of a phase in thermal and mechanical equilibrium but not in chemical equilibrium, *we shall assume that the equation*

$$\boxed{\begin{aligned} dG &= -S\,dT + V\,dP + \mu_1\,dn_1 + \mu_2\,dn_2 + \cdots + \mu_c\,dn_c \\ &= -S\,dT + V\,dP + \sum \mu_k\,dn_k \end{aligned}} \tag{14-15}$$

correctly expresses the change in the Gibbs function for any *infinitesimal process in which the n's may be caused to change either by the transfer of constituents to or from the phase or by the agency of a chemical reaction, or both.*

Imagine a phase at constant T and P in which all constituents are increased in the same proportion. Since the Gibbs function is an extensive quantity, it also will be increased in the same proportion. Infinitesimal changes in the mole numbers in the same proportion are represented by

$$dn_1 = n_1\,d\lambda, \qquad dn_2 = n_2\,d\lambda, \qquad \ldots, \qquad dn_c = n_c\,d\lambda,$$

and the corresponding change in G is

$$dG = G\,d\lambda,$$

where $d\lambda$ is the proportionality factor. Since

$$dG_{T,P} = \mu_1\,dn_1 + \mu_2\,dn_2 + \cdots + \mu_c\,dn_c,$$

we have

$$G\,d\lambda = \mu_1 n_1\,d\lambda + \mu_2 n_2\,d\lambda + \cdots + \mu_c n_c\,d\lambda,$$

or

$$\boxed{\begin{aligned} G &= \mu_1 n_1 + \mu_2 n_2 + \cdots + \mu_c n_c,\\ &= \sum \mu_k n_k. \end{aligned}} \tag{14-16}$$

Equation (14-16) shows that *the chemical potentials are intensive quantities*, for if all the n's are increased in the same proportion at constant T and P, the μ's must remain constant in order that G increase in the same proportion.

14-11 CHEMICAL POTENTIALS

The chemical potentials play a fundamental role in chemical thermodynamics. The chemical potential of the kth constituent of a phase is defined as

$$\mu_k = \frac{\partial G}{\partial n_k}$$

and is a function of T, P, and all the n's. In order that μ_k may be an intensive quantity, it is clear that the n's must be combined in such a way that, when all of them are multiplied by the same factor, the value of μ_k remains the same. The mole fraction of the kth constituent,

$$x_k = \frac{n_k}{\sum n},$$

satisfies that requirement; hence it is to be expected that μ_k is a function of T, P, and x_k. The actual form of the function depends, of course, on the nature of the phase. Consider the following phases:

1. *Phase consisting of only one constituent.* In this simple but not trivial case,

$$G = \mu n$$

and

$$\mu = \frac{G}{n} = g; \tag{14-17}$$

i.e., the chemical potential is the molar Gibbs function and is a function of T and P only.

2. *Phase consisting of a mixture of ideal gases.* In this case we have, from Art. 14-5,

$$G = RT \sum n_k(\phi_k + \ln P + \ln x_k),$$

which, when compared with the general equation

$$G = \sum n_k \mu_k,$$

shows that the chemical potential of one ideal gas in a mixture of ideal gases is

$$\mu = RT(\phi + \ln P + \ln x), \tag{14-18}$$

which may be written in the two alternative forms

$$\mu = RT(\phi + \ln p),$$

and

$$\mu = g + RT \ln x. \tag{14-19}$$

3. *Phase consisting of an ideal solution.* An ideal solution is defined as one in which the chemical potential of each constituent is of the form

$$\mu_k = g_k + RT \ln x_k, \tag{14-20}$$

where g_k is the Gibbs function of 1 mol of the kth constituent in the pure state, expressed as a function of T and P.

4. *Phase consisting of a dilute solution.* In the case of a dilute solution in which the mole fraction of the solvent x_0 is very much larger than each of the mole fractions of the solutes $x_1, x_2, \ldots$, it can be shown that, for the solvent,

$$\mu_0 = g_0 + RT \ln x_0, \tag{14-21}$$

where g_0 is the molar Gibbs function of the solvent in the pure state, expressed as a function of T and P. For any one of the solutes,

$$\mu_k = g_{0k} + RT \ln x_k, \tag{14-22}$$

where g_{0k} is a function of T and P only but depends upon the nature of the solvent as well as upon the solute.

5. *Other phases.* By defining functions known as *fugacity* and *activity coefficients*, it is possible to express the chemical potentials of constituents of a mixture of real gases and also of concentrated solutions. This is beyond the scope of this book.

We shall assume that it is always possible to express the chemical potential of any constituent of any phase as a function of T, P, and the x of that constituent. It will be a surprise to the student to discover how much valuable information can be obtained with the aid of this assumption, without knowing the exact expressions for the chemical potentials of the constituents of a phase.

14-12 DEGREE OF REACTION

If we introduce into a vessel a mixture of any arbitrary number of moles of water vapor, hydrogen, and oxygen, the chemical reaction capable of taking place is indicated by the notation

$$H_2O \rightleftharpoons H_2 + \tfrac{1}{2}O_2,$$

where the quantity on the left is called an *initial constituent* and those on the right *final constituents*. The numbers which precede the chemical symbols and which "balance" the equation (it is understood that both H_2O and H_2 are preceded by unity) are called the *stoichiometric coefficients* and are proportional to the number of moles of the constituents that *change* during the reaction. Thus, if 1 mol of water vapor dissociates, then 1 mol of hydrogen and $\frac{1}{2}$ mol of oxygen are formed; or if 0.1 mol of water vapor dissociates, then 0.1 mol of hydrogen and 0.05 mol of oxygen are formed; or if n_0 mol of water vapor dissociate (n_0 being any number whatever), then n_0 mol of hydrogen and $n_0/2$ mol of oxygen are formed. Similarly, if the reaction proceeds to the left to the extent that n_0' mol of hydrogen combine with $n_0'/2$ mol of oxygen, then n_0' mol of water vapor are formed.

In general, suppose that we have a mixture of four substances whose chemical symbols are A_1, A_2, A_3, and A_4. Let A_1 and A_2 be the initial constituents and A_3 and A_4 the final constituents, with the reaction being represented by

$$\nu_1 A_1 + \nu_2 A_2 \rightleftharpoons \nu_3 A_3 + \nu_4 A_4.$$

We have chosen four substances only for convenience. The equations to be developed are of such a character that they can be applied to reactions in which any number of substances participate. The ν's are the stoichiometric coefficients, which are always positive integers or fractions.

Let us start arbitrary amounts of *both* initial and final constituents. If we imagine the reaction to proceed completely to the right, at least one of the initial constituents (say, A_1) will completely disappear. Then it is possible to find a positive number n_0 such that the original number of moles of each of the initial constituents is expressed in the form

$$n_1(\text{original}) = n_0 \nu_1$$

and

$$n_2(\text{original}) = n_0 \nu_2 + N_2,$$

where N_2 is a constant representing the number of moles of A_2 that cannot combine. If we imagine the reaction to proceed completely to the left, at least one of the final constituents (say, A_3) will completely disappear. In this event, another positive number n_0' may be found such that the original number of moles of each final constituent is expressed in the form

$$n_3(\text{original}) = n_0' \nu_3$$

and

$$n_4(\text{original}) = n_0' \nu_4 + N_4.$$

If the reaction is imagined to proceed completely to the left, there is the maximum amount possible of each initial constituent and the minimum amount of each final constituent:

$$n_1(\max) = (n_0 + n_0')\nu_1, \qquad n_3(\min) = 0,$$

$$n_2(\max) = (n_0 + n_0')\nu_2 + N_2, \qquad n_4(\min) = N_4.$$

If the reaction is imagined to proceed completely to the right, there is the minimum amount possible of each initial constituent and the maximum amount of each final constituent:

$$n_1(\text{min}) = 0, \qquad n_3(\text{max}) = (n_0 + n'_0)\nu_3,$$

$$n_2(\text{min}) = N_2, \qquad n_4(\text{max}) = (n_0 + n'_0)\nu_4 + N_4.$$

Suppose that the reaction proceeds partially either to the right or to the left to such an extent that there are n_1 mol of A_1, n_2 mol of A_2, n_3 mol of A_3, and n_4 mol of A_4 present at a given moment. We define the *degree of reaction* ϵ in terms of any one of the initial constituents (say, A_1) as the fraction

$$\boxed{\epsilon = \frac{n_1(\text{max}) - n_1}{n_1(\text{max}) - n_1(\text{min})}.} \tag{14-23}$$

It follows from this definition that $\epsilon = 0$ when the reaction is completely to the left and $\epsilon = 1$ when the reaction is completely to the right. When the reaction consists in the dissociation of one initial constituent, ϵ is called the *degree of dissociation;* when it consists in the ionization of one initial constituent, ϵ is called the *degree of ionization.* Expressing $n_1(\text{max})$ and $n_1(\text{min})$ in terms of the constants that express the original amounts of the constituents, we get

$$\epsilon = \frac{(n_0 + n'_0)\nu_1 - n_1}{(n_0 + n'_0)\nu_1};$$

and, solving for n_1,

$$n_1 = (n_0 + n'_0)\nu_1(1 - \epsilon).$$

The number of moles of each of the constituents is therefore given by the following expressions:

$$\begin{aligned} n_1 &= (n_0 + n'_0)\nu_1(1 - \epsilon), & n_3 &= (n_0 + n'_0)\nu_3\,\epsilon, \\ n_2 &= (n_0 + n'_0)\nu_2(1 - \epsilon) + N_2, & n_4 &= (n_0 + n'_0)\nu_4\,\epsilon + N_4. \end{aligned} \tag{14-24}$$

When a chemical reaction takes place, all the n's change, but not independently. The restrictions imposed on the n's are given by the relations (14-24). These equations therefore are examples of *equations of constraint.*

The equations of constraint are equally valid whether the system is heterogeneous or homogeneous. If each constituent is present in, say, φ different phases, with $n_1^{(1)}$ mol of constituent A_1 in phase 1 and and $n_1^{(2)}$ mol of the same constituent in phase 2, etc., then the total number of moles of constituent A_1 is

$$n_1 = n_1^{(1)} + n_1^{(2)} + \cdots + n_1^{(\varphi)} = (n_0 + n'_0)\nu_1(1 - \epsilon),$$

etc., for the other constituents. For the present, however, we shall limit ourselves to homogeneous systems, reserving heterogeneous systems for Chap. 16.

Since all the n's are functions of ϵ only, it follows that in a homogeneous

Table 14-1 $H_2O \rightleftharpoons H_2 + \frac{1}{2}O_2$

A	ν	n	x
$A_1 = H_2O$	$\nu_1 = 1$	$n_1 = n_0(1 - \epsilon)$	$x_1 = \dfrac{1 - \epsilon}{1 + \epsilon/2}$
$A_3 = H_2$	$\nu_3 = 1$	$n_3 = n_0 \epsilon$	$x_3 = \dfrac{\epsilon}{1 + \epsilon/2}$
$A_4 = O_2$	$\nu_4 = \frac{1}{2}$	$n_4 = \dfrac{n_0 \epsilon}{2}$	$x_4 = \dfrac{\epsilon/2}{1 + \epsilon/2}$
		$\sum n = n_0(1 + \epsilon/2)$	

system all the mole fractions are functions of ϵ only. An example will show how simple these expressions are when the starting conditions are simple. Consider a vessel containing n_0 mol of water vapor only, with no hydrogen or oxygen present. If dissociation occurs until the degree of dissociation is ϵ, then the n's and x's are shown as functions of ϵ in Table 14-1. Since the chemical potential of each gas in the mixture is a function of T, P, and x, it follows that every chemical potential is a function of T, P, and ϵ.

If the reaction is imagined to proceed to an infinitesimal extent, the degree of reaction changing from ϵ to $\epsilon + d\epsilon$, the various n's will change by the following amounts:

$$dn_1 = -(n_0 + n_0')\nu_1 \, d\epsilon, \qquad dn_3 = (n_0 + n_0')\nu_3 \, d\epsilon,$$

$$dn_2 = -(n_0 + n_0')\nu_2 \, d\epsilon, \qquad dn_4 = (n_0 + n_0')\nu_4 \, d\epsilon$$

These equations show that the changes in the n's are proportional to the ν's, with the factor of proportionality being, for the initial constituents, $-(n_0 + n_0')\, d\epsilon$ and, for the final constituents, $+(n_0 + n_0')\, d\epsilon$. Another way of writing these is as follows:

$$\frac{dn_1}{-\nu_1} = \frac{dn_2}{-\nu_2} = \frac{dn_3}{\nu_3} = \frac{dn_4}{\nu_4} = (n_0 + n_0')\, d\epsilon, \tag{14-25}$$

which shows perhaps more clearly that the dn's are proportional to the ν's.

14-13 EQUATION OF REACTION EQUILIBRIUM

Consider a homogeneous phase consisting of arbitrary amounts of the four constituents A_1, A_2, A_3, and A_4, capable of undergoing the reaction

$$\nu_1 A_1 + \nu_2 A_2 \rightleftharpoons \nu_3 A_3 + \nu_4 A_4 .$$

Suppose that the phase is at a uniform temperature T and pressure P. If n_1, n_2, n_3, and n_4 denote the numbers of moles of each constituent that are present at any moment and μ_1, μ_2, μ_3, and μ_4 are the respective chemical potentials, then the Gibbs function of the mixture is

$$G = \mu_1 n_1 + \mu_2 n_2 + \mu_3 n_3 + \mu_4 n_4.$$

The n's are given by the equations of constraint:

$$n_1 = (n_0 + n_0')\nu_1(1 - \epsilon), \qquad n_3 = (n_0 + n_0')\nu_3 \epsilon,$$

$$n_2 = (n_0 + n_0')\nu_2(1 - \epsilon) + N_2, \qquad n_4 = (n_0 + n_0')\nu_4 \epsilon + N_4.$$

The μ's are functions of T, P, and ϵ. It therefore follows that G is a function of T, P, and ϵ.

Let us imagine that the reaction is allowed to take place at constant T and P. Under these conditions, the Gibbs function decreases; i.e., during an infinitesimal change in ϵ from ϵ to $\epsilon + d\epsilon$,

$$dG_{T,P} < 0.$$

We have shown that, for any infinitesimal change to which a phase in thermal and mechanical equilibrium is subjected,

$$dG = -S\,dT + V\,dP + \mu_1\,dn_1 + \mu_2\,dn_2 + \cdots.$$

Therefore, for this mixture of four constituents,

$$dG_{T,P} = \mu_1\,dn_1 + \mu_2\,dn_2 + \mu_3\,dn_3 + \mu_4\,dn_4,$$

with the equations of constraint in differential form:

$$dn_1 = -(n_0 + n_0')\nu_1\,d\epsilon, \qquad dn_3 = (n_0 + n_0')\nu_3\,d\epsilon,$$

$$dn_2 = -(n_0 + n_0')\nu_2\,d\epsilon, \qquad dn_4 = (n_0 + n_0')\nu_4\,d\epsilon.$$

Substituting, we obtain a general expression for an infinitesimal change of the Gibbs function at constant T and P. Thus,

$$dG_{T,P} = (n_0 + n_0')(-\nu_1\mu_1 - \nu_2\mu_2 + \nu_3\mu_3 + \nu_4\mu_4)\,d\epsilon. \tag{14-26}$$

It follows from this equation that when the reaction proceeds spontaneously to the right, so that $d\epsilon$ is positive, then, in order that $dG_{T,P} < 0$,

$$\nu_1\mu_1 + \nu_2\mu_2 > \nu_3\mu_3 + \nu_4\mu_4 \qquad \text{(reaction to right)}.$$

Conversely, if the reaction proceeds spontaneously to the left,

$$\nu_1\mu_1 + \nu_2\mu_2 < \nu_3\mu_3 + \nu_4\mu_4 \qquad \text{(reaction to left)}.$$

The mixture will be in equilibrium at the given T and P when the Gibbs function is a minimum at which an infinitesimal change in ϵ will produce no change in the Gibbs function. Therefore, for $dG_{T,P} = 0$ at equilibrium, we have

$$\nu_1\mu_1 + \nu_2\mu_2 = \nu_3\mu_3 + \nu_4\mu_4 \qquad \text{(at equilibrium)} \tag{14-27}$$

which is called the *equation of reaction equilibrium*. It should be noted that this equation contains only intensive variables. Evidently, to determine the composition of a homogeneous mixture after the reaction has come to equilibrium, it is necessary merely to substitute the appropriate expressions for the chemical potentials into the equation of reaction equilibrium. This will be done in the case of ideal gases in the next chapter.

PROBLEMS

14-1 Prove Gibbs' theorem by using the apparatus depicted in Fig. 14-2 in such a way that the gases are separated reversibly and adiabatically.

14-2 What is the minimum amount of work required to separate 1 mole of air at 27°C and 1 atm pressure (assumed to be composed of $\frac{1}{5}O_2$ and $\frac{4}{5}N_2$) into O_2 and N_2, each at 27°C and 1 atm pressure?

14-3 Calculate the entropy change of the universe due to the diffusion of two ideal gases (1 mol of each) at the same temperature and pressure, by calculating $\int đQ/T$ over a series of reversible processes involving the use of the apparatus depicted in Fig. 14-2.

14-4 n_1 moles of an ideal monatomic gas at temperature T_1 and pressure P are in one compartment of an insulated container. In an adjoining compartment separated by an insulating partition are n_2 moles of another ideal monatomic gas at temperature T_2 and pressure P. When the partition is removed:

(*a*) Show that the final pressure of the mixture is P.

(*b*) Calculate the entropy change when the gases are identical.

(*c*) Calculate the entropy change when the gases are different.

14-5 n_1 moles of an ideal gas at pressure P_1 and temperature T are in one compartment of an insulated container. In an adjoining compartment separated by a partition, are n_2 moles of an ideal gas at pressure P_2 and temperature T. When the partition is removed:

(*a*) Calculate the final pressure of the mixture.

(*b*) Calculate the entropy change when the gases are identical.

(*c*) Calculate the entropy change when the gases are different.

(*d*) Prove that the entropy change in part (*c*) is the same as that which would be produced by two independent free expansions.

14-6 Suppose that we have 1 mol of a monatomic gas A whose nuclei are in their lowest energy state and 1 mol of a monatomic gas B consisting of exactly the same atoms as A, except that the nuclei are in an excited state whose energy ϵ is much larger that kT and whose lifetime is much larger than the time for diffusion to take place. Both gases are at the same pressure and are maintained at the same temperature T by a heat reservoir.

(*a*) Immediately after the nuclei of gas B have been excited, diffusion takes place. Calculate the entropy change of the universe.

(*b*) After the nuclei of gas B have been excited, a time much larger than the lifetime of the excited state is allowed to elapse, and then diffusion takes place. Calculate the entropy change of the universe.

(*c*) Show that the answer to part (*b*) is larger than that to (*a*). (This problem is due to M. J. Klein.)

14-7 Consider the system depicted in Fig. P14-1, where the whole system and also each half are in equilibrium. Consider a process to take place in which each half of the system remains at the constant volume $V/2$, but a small amount of heat δu (at constant volume $đQ = dU$) is extracted from the left half and transferred to the right half, as shown in Fig. P14-1. Realizing that

$$\left(\frac{\partial s}{\partial u}\right)_v = \left(\frac{\partial S}{\partial U}\right)_V,$$

(*a*) Expand the entropy s_L of the left half by means of a Taylor series about the equilibrium value $S_{max}/2$, terminating the series after the squared term. Do the same for s_R.

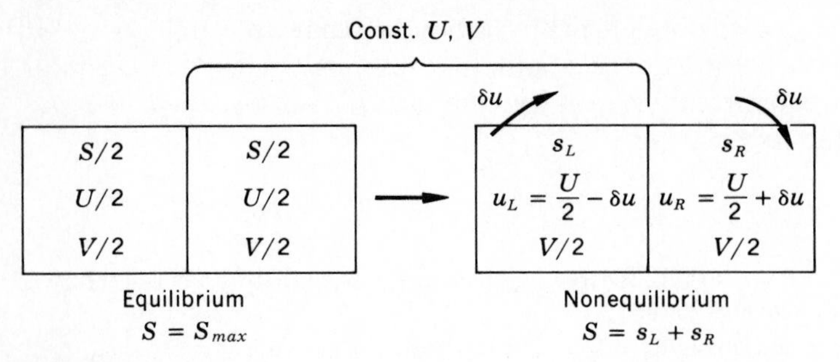

Figure P14-1

(*b*) Show that

$$\delta S_{U,V} = s_L + s_R - S_{\max} = \left(\frac{\partial^2 S}{\partial U^2}\right)_V (\delta u)^2.$$

(*c*) Show that $C_V > 0$, which is the *condition for thermal stability*.

14-8 (*a*) Show that the molar Helmholtz function of an ideal gas is

$$f = u_0 - T\int \frac{\int c_V\, dT}{T^2}\, dT - Ts_0 - RT \ln v.$$

(*b*) Show that the Helmholtz function of a mixture of inert ideal gases is

$$F = \sum n_k(f_k + RT \ln x_k).$$

(*c*) Show that the change in the Helmholtz function due to diffusion is

$$F_f - F_i = RT \sum n_k \ln x_k.$$

14-9 Consider a hydrostatic system of constant mass maintained in thermal and mechanical equilibrium. It is not in chemical equilibrium, however, because of chemical reactions and transport of matter between phases. Under these circumstances, the system undergoes an irreversible cycle. Prove that

$$\oint_I \frac{đQ}{T} < 0.$$

14-10 By means of the equations $dG = -S\,dT + V\,dP + \sum \mu_k\, dn_k$ and $G = \sum \mu_k n_k$, prove that:

(*a*) $$-S\,dT + V\,dP = \sum n_k\, d\mu_k.$$

(*b*) $$-s\,dT + v\,dP = \sum x_k\, d\mu_k.$$

14-11 Show that, if the internal energy of a phase is expressed as a function of $S, V, n_1, n_2, \ldots, n_c$, then:

(*a*) $$dU = T\,dS - P\,dV + \sum \mu_k\, dn_k \qquad \text{where } \mu_k = (\partial U/\partial n_k)_{S,V,\text{other } n\text{'s}}.$$

(*b*) $$U = TS - PV + \sum \mu_k n_k.$$

(*c*) $$-S\,dT + V\,dP = \sum n_k\, d\mu_k.$$

14-12 Show that, if the Helmholtz function of a phase is expressed as a function of $T, V, n_1, n_2, \ldots, n_c$, then:

(*a*) $$dF = -S\,dT - P\,dV + \sum \mu_k\, dn_k \qquad \text{where } \mu_k = (\partial F/\partial n_k)_{T,V,\text{other } n\text{'s}}.$$

(*b*) $$F = -PV + \sum \mu_k n_k.$$

(*c*) $$-S\,dT + V\,dP = \sum n_k\, d\mu_k.$$

14-13 Prove that:

(a)
$$Td\left(\frac{F}{T}\right) = -\frac{U}{T}\,dT - P\,dV + \sum \mu_k\,dn_k.$$

(b)
$$d(PV) = S\,dT + P\,dV + \sum n_k\,d\mu_k.$$

14-14 Show that, for an ideal gas in a mixture of ideal gases,

$$d\mu_k = \frac{\mu_k - h_k}{T}\,dT + v_k\,dP + RT\,d\ln x_k.$$

14-15 Consider a uniform substance of variable mass. Suppose that at any moment there are n moles of substance and that the system undergoes an infinitesimal reversible process in which n changes by dn. For any *1 mol* of substance,

$$đq = T\,ds = du + P\,dv,$$

and the total heat transfer $đQ = n\,đq = nT\,ds$.

(a) Prove that

$$đQ = dU + P\,dV - h\,dn,$$

where U and V refer to the entire system, but h is the molar enthalpy.

(b) From Prob. 14-11*a*, show that

$$T\,dS = dU + P\,dV - g\,dn.$$

(c) How does $T\,dS$ differ from $đQ$?

(d) Show that, if T and P remain constant, $đQ = 0$ but $T\,dS$ does not.

14-16 Starting with n_0 moles of CO and n_0 moles of H_2O capable of undergoing the reaction

$$CO + H_2O \rightleftharpoons CO_2 + H_2$$

In the gaseous phase, set up a table of values of A, v, n, and x, similar to that of Table 14-1.

14-17 Starting with n_0 moles of H_2S and $2n_0$ moles of H_2O, capable of undergoing the reaction

$$H_2S + 2H_2O \rightleftharpoons 3H_2 + SO_2$$

in the gaseous phase, set up a table of values of A, v, n, and x, similar to that of Table 14-1.

CHAPTER

FIFTEEN

IDEAL-GAS REACTIONS

15-1 LAW OF MASS ACTION

It has been shown that when four substances which are capable of undergoing the reaction

$$\nu_1 A_1 + \nu_2 A_2 \rightleftharpoons \nu_3 A_3 + \nu_4 A_4,$$

and which constitute a homogeneous phase at constant temperature and pressure, come to equilibrium, the equation of reaction equilibrium

$$\nu_1 \mu_1 + \nu_2 \mu_2 = \nu_3 \mu_3 + \nu_4 \mu_4$$

must be satisfied. If the constituents are ideal gases, then the chemical potentials are given by expressions of the type

$$\mu_k = RT(\phi_k + \ln P + \ln x_k),$$

where the ϕ's are functions of the temperature only. Substituting in the equation of reaction equilibrium, we get

$$\nu_1(\phi_1 + \ln P + \ln x_1) + \nu_2(\phi_2 + \ln P + \ln x_2)$$
$$= \nu_3(\phi_3 + \ln P + \ln x_3) + \nu_4(\phi_4 + \ln P + \ln x_4).$$

Rearranging terms yields

$$\nu_3 \ln x_3 + \nu_4 \ln x_4 - \nu_1 \ln x_1 - \nu_2 \ln x_2$$
$$+ (\nu_3 + \nu_4 - \nu_1 - \nu_2) \ln P = -(\nu_3 \phi_3 + \nu_4 \phi_4 - \nu_1 \phi_1 - \nu_2 \phi_2),$$

or
$$\ln \frac{x_3^{\nu_3} x_4^{\nu_4}}{x_1^{\nu_1} x_2^{\nu_2}} P^{\nu_3 + \nu_4 - \nu_1 - \nu_2} = -(\nu_3 \phi_3 + \nu_4 \phi_4 - \nu_1 \phi_1 - \nu_2 \phi_2).$$

The right-hand member is a quantity whose value depends only on the temperature. Denoting it by ln K, where K is known as the *equilibrium constant*,

$$\ln K = -(\nu_3\phi_3 + \nu_4\phi_4 - \nu_1\phi_1 - \nu_2\phi_2), \tag{15-1}$$

we get finally

$$\boxed{\left(\frac{x_3^{\nu_3}x_4^{\nu_4}}{x_1^{\nu_1}x_2^{\nu_2}}\right)_{\epsilon=\epsilon_e} P^{\nu_3+\nu_4-\nu_1-\nu_2} = K} \tag{15-2}$$

This form of the equation of reaction equilibrium was first presented by two Norwegian chemists, Guldberg and Waage, in 1863. For some unknown reason, it is called the *law of mass action*, a misnomer that is harmless. The fraction involving the equilibrium values of the x's is a function of the equilibrium value of ϵ—that is, of ϵ_e—and hence the law of mass action is seen to be a relation among ϵ_e, P, and T.

It is obvious that, if there are more than two initial constituents and two final constituents, the law of mass action becomes

$$\left(\frac{x_3^{\nu_3}x_4^{\nu_4}\cdots}{x_1^{\nu_1}x_2^{\nu_2}\cdots}\right)_{\epsilon=\epsilon_e} P^{\nu_3+\nu_4+\cdots-\nu_1-\nu_2-\cdots} = K,$$

where K is given by

$$\ln K = -(\nu_3\phi_3 + \nu_4\phi_4 + \cdots - \nu_1\phi_1 - \nu_2\phi_2 - \cdots).$$

15-2 EXPERIMENTAL DETERMINATION OF EQUILIBRIUM CONSTANTS

It has been pointed out that a mixture of hydrogen and oxygen will remain indefinitely without reacting at atmospheric pressure and at room temperature. If the temperature is raised considerably, however, water vapor forms and equilibrium takes place quickly. If the mixture is then cooled very suddenly so as not to disturb the equilibrium, an analysis of the composition of the mixture yields the values of the mole fractions corresponding to equilibrium at the high temperature. The equilibrium has, so to speak, been "frozen." Sometimes a flow method is used. The reacting gases are mixed in known proportions at a low temperature, and the mixture is caused to flow slowly through a long reacting tube at a desired temperature. The gases remain at this temperature a sufficient time for equilibrium to take place. The mixture is then allowed to flow through a capillary, where it is suddenly cooled. The equilibrium values of the mole fractions are then measured by the methods of chemical analysis.

It is a consequence of the law of mass action that the equilibrium constant corresponding to a given temperature is independent of the amounts of products

Table 15-1 Equilibrium data for the water-gas reaction
($CO_2 + H_2 \rightleftharpoons CO + H_2O$ at 1259 K)

Original mixture		Equilibrium mixture			$K = \frac{x_{CO}x_{H_2O}}{x_{CO_2}x_{H_2}}$
x_{CO_2}	x_{H_2}	x_{CO_2}	$x_{CO} = x_{H_2O}$	x_{H_2}	
0.101	0.899	0.0069	0.094	0.805	1.60
0.310	0.699	0.0715	0.2296	0.4693	1.58
0.491	0.509	0.2122	0.2790	0.2295	1.60
0.609	0.391	0.3443	0.2645	0.1267	1.60
0.703	0.297	0.4750	0.2282	0.0685	1.60

which are originally mixed. For example, in the case of the "water-gas" reaction,

$$CO_2 + H_2 \rightleftharpoons CO + H_2O,$$

the law of mass action requires that, at equilibrium,

$$\frac{x_{CO}x_{H_2O}}{x_{CO_2}x_{H_2}} P^{1+1-1-1} = K,$$

or

$$\frac{x_{CO}x_{H_2O}}{x_{CO_2}x_{H_2}} = \text{const.}$$

at a constant temperature, independent of the starting conditions. For conditions that involve starting with arbitrary amounts of CO_2 and H_2 and maintaining the temperature constant at 1259 K, the equilibrium values of the mole fractions are given in Table 15-1, where it may be seen that K is quite constant.

The water-gas reaction is an example of a reaction that does not involve a change in the total number of moles. If such a reaction were to take place at constant temperature and pressure, there would be no change in volume. There are, however, many reactions in which the total number of moles varies. In such cases, it is possible to measure the value of the degree of reaction at equilibrium by merely measuring the volume (or the density) of the mixture at equilibrium. If ϵ_e is known, the equilibrium constant can then be calculated. As an example of this procedure, consider the dissociation of nitrogen tetroxide according to the equation

$$N_2O_4 \rightleftharpoons 2NO_2.$$

If we start with n_0 mol of N_2O_4 at temperature T and pressure P, the initial volume is expressed as

$$V_0 = n_0 \frac{RT}{P}.$$

If V_e denotes the volume at equilibrium, with the temperature and pressure remaining the same, then

$$V_e = [n_0(1 - \epsilon_e) + 2n_0 \epsilon_e]\frac{RT}{P},$$

where ϵ_e is the value of the degree of dissociation at equilibrium. This can be written as

$$V_e = (1 + \epsilon_e)V_0,$$

or

$$\epsilon_e = \frac{V_e}{V_0} - 1.$$

Since the density ρ is inversely proportional to the volume, we get finally

$$\epsilon_e = \frac{\rho_0}{\rho_e} - 1.$$

Now, at equilibrium,

$$x_{N_2O_4} = \frac{n_0(1 - \epsilon_e)}{n_0(1 + \epsilon_e)} \quad \text{and} \quad x_{NO_2} = \frac{2n_0 \epsilon_e}{n_0(1 + \epsilon_e)};$$

therefore, the law of mass action becomes

$$\frac{[2\epsilon_e/(1 + \epsilon_e)]^2}{(1 - \epsilon_e)/(1 + \epsilon_e)} P = K,$$

or

$$\frac{4\epsilon_e^2}{1 - \epsilon_e^2} P = K.$$

The pressure is by custom measured in atmospheres, even though the "atmosphere" is not an SI unit. Numerical data for this reaction are given in Table 15-2, where it is seen that at the constant temperature of 323 K the equilibrium constant remains fairly constant for three different values of the pressure.

There are many other methods of measuring equilibrium constants. For a complete account of this important branch of physical chemistry, the student is referred to an advanced treatise.

Table 15-2 $N_2O_4 \rightleftharpoons 2NO_2$

Temp., K	P, atm	ρ_0, kg/m³	ρ_e, kg/m³	$\epsilon_e = \dfrac{\rho_0}{\rho_e} - 1$	$K = \dfrac{4\epsilon_e^2}{1 - \epsilon_e^2} P$, atm
	0.124	1.093	0.615	0.777	0.756
323	0.241	1.093	0.651	0.679	0.825
	0.655	1.093	0.737	0.483	0.797

15-3 HEAT OF REACTION

The equilibrium constant is defined by the equation

$$\ln K = -(\nu_3\phi_3 + \nu_4\phi_4 - \nu_1\phi_1 - \nu_2\phi_2).$$

Differentiating $\ln K$ with respect to T, we get

$$\frac{d}{dT}\ln K = -\left(\nu_3\frac{d\phi_3}{dT} + \nu_4\frac{d\phi_4}{dT} - \nu_1\frac{d\phi_1}{dT} - \nu_2\frac{d\phi_2}{dT}\right).$$

Since
$$\phi = \frac{h_0}{RT} - \frac{1}{R}\int\frac{\int c_P\,dT}{T^2}\,dT - \frac{s_0}{R},$$

we have
$$\frac{d\phi}{dT} = -\frac{h_0}{RT^2} - \frac{\int c_P\,dT}{RT^2}$$

$$= -\frac{1}{RT^2}\left(h_0 + \int c_P\,dT\right),$$

and
$$\frac{d\phi}{dT} = -\frac{h}{RT^2}. \tag{15-3}$$

Therefore,
$$\frac{d}{dT}\ln K = \frac{1}{RT^2}(\nu_3 h_3 + \nu_4 h_4 - \nu_1 h_1 - \nu_2 h_2),$$

where all the h's refer to the same temperature T and the same pressure P. The right-hand term has a simple interpretation. If ν_1 moles of A_1 and ν_2 moles of A_2 are converted at constant temperature and pressure into ν_3 moles of A_3 and ν_4 moles of A_4, the heat transferred will be equal to the final enthalpy $\nu_3 h_3 + \nu_4 h_4$ minus the initial enthalpy $\nu_1 h_1 + \nu_2 h_2$. Calling this heat the *heat of reaction* and denoting it by ΔH, we have

$$\Delta H = \nu_3 h_3 + \nu_4 h_4 - \nu_1 h_1 - \nu_2 h_2, \tag{15-4}$$

and
$$\boxed{\frac{d}{dT}\ln K = \frac{\Delta H}{RT^2}.} \tag{15-5}$$

The preceding equation, called the *van't Hoff isobar*, is one of the most important equations in chemical thermodynamics.

Rewriting the equation as

$$\frac{d\ln K}{dT/T^2} = \frac{\Delta H}{R},$$

or
$$\frac{d\ln K}{d(1/T)} = -\frac{\Delta H}{R},$$

we get
$$\Delta H = -2.30R\frac{d\log K}{d(1/T)}. \tag{15-6}$$

The van't Hoff isobar enables one to calculate the heat of reaction at any desired temperature or within any desired temperature range, once the temperature variation of the equilibrium constant is known. The slope of the curve obtained by plotting log K against $1/T$, multiplied by $2.30R$, is the heat of reaction at the temperature corresponding to the point chosen. As a rule, log K can be measured only within a small temperature range, in which case the curve is usually a straight line.

As an example of this procedure, consider the dissociation of water vapor according to the reaction

$$H_2O \rightleftharpoons H_2 + \tfrac{1}{2}O_2.$$

Starting with n_0 mol of water vapor and no hydrogen or oxygen, the mole fractions corresponding to any value ϵ of the degree of dissociation are shown in Table 14-1. At equilibrium,

$$\left(\frac{x_3^{\nu_3}x_4^{\nu_4}}{x_1^{\nu_1}}\right)_{\epsilon=\epsilon_e} P^{\nu_3+\nu_4-\nu_1} = K,$$

or

$$\frac{\dfrac{\epsilon_e}{1+\epsilon_e/2}\left(\dfrac{\epsilon_e/2}{1+\epsilon_e/2}\right)^{1/2}}{\dfrac{1-\epsilon_e}{1+\epsilon_e/2}} P^{1/2} = K,$$

or

$$\frac{\epsilon_e^{3/2}}{(2+\epsilon_e)^{1/2}(1-\epsilon_e)} P^{1/2} = K.$$

When ϵ_e is very much smaller than unity, this equation reduces to

$$K = \sqrt{\frac{\epsilon_e^3 P}{2}}.$$

In Table 15-3 experimental values of ϵ_e are given at a number of temperatures and at constant atmospheric pressure, along with the corresponding values of K,

Table 15-3 $H_2O \rightleftharpoons H_2 + \tfrac{1}{2}O_2 (P = 1\text{ atm})$

Temp., K	ϵ_e (measured)	$K = \dfrac{\epsilon_e^{3/2}P^{1/2}}{(2+\epsilon_e)^{1/2}(1-\epsilon_e)}$, (atm)$^{1/2}$	Log K	$\dfrac{1}{T}$
1500	1.97×10^{-4}	1.95×10^{-6}	-5.71	6.67×10^{-4}
1561	3.4×10^{-4}	4.48×10^{-5}	-5.36	6.41×10^{-4}
1705	1.2×10^{-3}	2.95×10^{-5}	-4.53	5.87×10^{-4}
2155	1.2×10^{-2}	9.0×10^{-4}	-3.05	4.64×10^{-4}
2257	1.77×10^{-2}	1.67×10^{-3}	-2.78	4.43×10^{-4}
2300	2.6×10^{-2}	2.95×10^{-3}	-2.53	4.35×10^{-4}

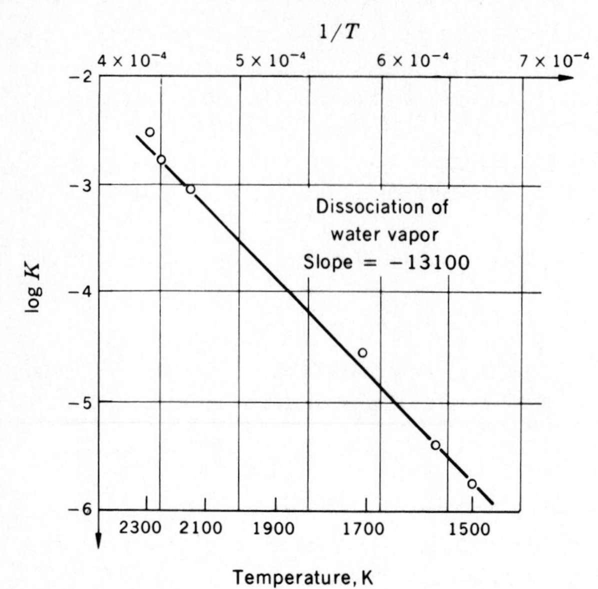

Figure 15-1 Graph of log K against $1/T$ for dissociation of water vapor.

log K, and $1/T$. The graph of log K against $1/T$ is shown in Fig. 15-1, where it is seen that the points lie on a straight line with a slope equal to $-13{,}100$. It follows, therefore, that the heat of dissociation at the average temperature of about 1900 K is equal to

$$\Delta H = -2.30R \times -13{,}100$$

$$= 250 \text{ kJ/mol}.$$

It is rather difficult to measure the heat of reaction accurately by a direct calorimetric method. Most heats of reaction are obtained either with the aid of the van't Hoff isobar or by means of a reversible cell. In a few cases in which it is possible to measure the equilibrium constant over a very wide temperature range, the graph of log K against $1/T$ is found to have a variable slope, indicating that the heat of reaction depends on the temperature. Such a graph is impossible to obtain in most cases, however, since at low temperatures a reaction either does not proceed or, if it does, the equilibrium value of the degree of reaction is too small to measure. Thus, in the case of the dissociation of water vapor, the degree of dissociation at room temperature and atmospheric pressure is about 10^{-27}, which means that under these conditions not even one molecule of H_2O dissociates.

To obtain the heat of reaction at any desired temperature, it is necessary to know the temperature dependence of the heat capacities of all the reacting gases. Since

$$\Delta H = \nu_3 h_3 + \nu_4 h_4 - \nu_1 h_1 - \nu_2 h_2$$

and

$$h = h_0 + \int c_P \, dT,$$

then

$$\Delta H = v_3 h_{03} + v_4 h_{04} - v_1 h_{01} - v_2 h_{02} + \int (v_3 c_{P3} + v_4 c_{P4} - v_1 c_{P1} - v_2 c_{P2})\, dT.$$

Denoting the constant party by ΔH_0, thus:

$$\Delta H_0 = v_3 h_{03} + v_4 h_{04} - v_1 h_{01} - v_2 h_{02}, \tag{15-7}$$

and defining

$$\Delta C_P = v_3 c_{P3} + v_4 c_{P4} - v_1 c_{P1} - v_2 c_{P2}, \tag{15-8}$$

we get

$$\boxed{\Delta H = \Delta H_0 + \int \Delta C_P\, dT.} \tag{15-9}$$

The integral can be determined by substituting for the c_P's the empirical equations expressing their temperature dependence. If ΔH is known at one temperature, therefore, ΔH_0 can be calculated and the equation may be used to provide ΔH at any temperature.

15-4 NERNST'S EQUATION

The equilibrium constant is defined by the equation

$$\ln K = -(v_3 \phi_3 + v_4 \phi_4 - v_1 \phi_1 - v_2 \phi_2),$$

where

$$\phi = \frac{h_0}{RT} - \frac{1}{R} \int \frac{\int c_P\, dT}{T^2}\, dT - \frac{s_0}{R}.$$

We have, therefore,

$$\ln K = -\frac{1}{RT}(v_3 h_{03} + v_4 h_{04} - v_1 h_{01} - v_2 h_{02}) + \frac{1}{R} \int \frac{\int (v_3 c_{P3} + v_4 cP_4 - v_1 c_{P1} - v_2 c_{P2})\, dT}{T^2}\, dT + \frac{1}{R}(v_3 s_{03} + v_4 s_{04} - v_1 s_{01} - v_2 s_{02}).$$

Defining

$$\Delta S_0 = v_3 s_{03} + v_4 s_{04} - v_1 s_{01} - v_2 s_{02},$$

the equation becomes

$$\boxed{\ln K = -\frac{\Delta H_0}{RT} + \frac{1}{R} \int \frac{\int \Delta C_P\, dT}{T^2}\, dT + \frac{\Delta S_0}{R},} \tag{15-10}$$

which is known as *Nernst's equation*.

Table 15-4 $A \rightleftharpoons A^+ + e$

A	ν	n	x
$A_1 = A$	$\nu_1 = 1$	$n_1 = n_0(1 - \epsilon_e)$	$x_1 = \dfrac{1 - \epsilon_e}{1 + \epsilon_e}$
$A_3 = A^+$	$\nu_3 = 1$	$n_3 = n_0 \epsilon_e$	$x_3 = \dfrac{\epsilon_e}{1 + \epsilon_e}$
$A_4 = e$	$\nu_4 = 1$	$n_4 = n_0 \epsilon_0$	$x_4 = \dfrac{\epsilon_e}{1 + \epsilon_e}$
	$\nu_3 + \nu_4 - \nu_1 = 1$	$\sum n = n_0(1 + \epsilon_e)$	

An interesting application of Nernst's equation was made by Megh Nad Saha to the thermal ionization of a monatomic gas. If a monatomic gas is heated to a high enough temperature, some ionization occurs and the atoms, ions, and electrons may be regarded as a mixture of three ideal monatomic gases undergoing the reaction

$$A \rightleftharpoons A^+ + e.$$

Starting with n_0 moles of atoms alone, the state of affairs at equilibrium is shown in Table 15-4. For this reaction,

$$\ln K = \ln \frac{x_3^{\nu_3} x_4^{\nu_4}}{x_1^{\nu_1}} P^{\nu_3 + \nu_4 - \nu_1}$$

$$= \ln \frac{[\epsilon_e/(1 + \epsilon_e)][\epsilon_e/(1 + \epsilon_e)]}{(1 - \epsilon_e)/(1 + \epsilon_e)} P,$$

or

$$\ln K = \ln \frac{\epsilon_e^2}{1 - \epsilon_e^2} P.$$

ΔH_0 is the amount of energy necessary to ionize 1 mol of atoms. If we denote the ionization potential of the atom in volts by E, then $\Delta H_0 = N_F E$, where N_F is Faraday's constant. Since the three gases are monatomic, each c_P is equal to $\frac{5}{2}R$. Therefore, $\Delta C_P = \frac{5}{2}R$, and

$$\frac{1}{R} \int \frac{\int \Delta C_P \, dT}{T^2} dT = \tfrac{5}{2} \ln T.$$

Let us put

$$\frac{\Delta S_0}{R} = \ln B.$$

Introducing these results into Nernst's equation, we get

$$\ln \frac{\epsilon_e^2}{1 - \epsilon_e^2} P = -\frac{N_F E}{RT} + \tfrac{5}{2} \ln T + \ln B.$$

Table 15-5 Values of E and ω

Element	E, V	ω_a	ω_i
Na	5.12	2	1
Cs	3.87	2	1
Ca	6.09	1	2
Cd	8.96	1	2
Zn	9.36	1	2
Tl	6.07	2	1

Expressing P in atmospheres, changing to common logarithms, and introducing the value of B from statistical mechanics, Saha finally obtained the formula

$$\log \frac{\epsilon_e^2}{1 - \epsilon_e^2} P(\text{atm}) = -(5050 \text{ K/V}) \frac{E}{T} + \tfrac{5}{2} \log T + \log \frac{\omega_i \omega_e}{\omega_a} - 6.491, \tag{15-11}$$

where ω_i, ω_e, and ω_a are constants that refer, respectively, to the ion, the electron, and the atom.

In order to apply Saha's equation to a specific problem, it is necessary to know the ionization potential and the ω's. A complete discussion of these quantities is beyond the scope of this book. Values of these constants for a few elements are listed in Table 15-5. The constant ω for an electron is 2.

Saha applied his equation to the determination of the temperature of a stellar atmosphere. The spectrum of a star contains lines which originate from atoms (*arc lines*) and also those which originate from ions (*spark lines*). A comparison of the intensity of a spark line with that of an arc line, both referring to the same element, gives rise to a value of the degree of ionization ϵ_e. Treating a star as a sphere of ideal gas, it is possible to obtain an estimate of the pressure of a stellar atmosphere. Since all the other quantities are known, the temperature can be calculated.

15-5 AFFINITY

It was shown in Art. 14-5 that the molar Gibbs function of an ideal gas at temperature T and pressure P is equal to

$$g = RT(\phi + \ln P).$$

If we have four gases that can engage in the reaction

$$\nu_1 A_1 - \nu_2 A_2 \rightleftharpoons \nu_3 A_3 + \nu_4 A_4,$$

we define the quantity $\sum \nu g$ by the expression

$$\sum \nu g = \nu_3 g_3 + \nu_4 g_4 - \nu_1 g_1 - \nu_2 g_2, \tag{15-12}$$

where the g's refer to the gases completely separated at T, P. It should be emphasized that $\sum \nu g$ is defined in terms of the separate Gibbs functions of the gases, not in terms of the mixture. The connection between $\sum \nu g$ and the behavior of the gases when mixed is shown by introducing the values of the g's. Thus,

$$\sum \nu g = RT(\nu_3 \phi_3 + \nu_4 \phi_4 - \nu_1 \phi_1 - \nu_2 \phi_2) + RT \ln P^{\nu_3 + \nu_4 - \nu_1 - \nu_2}.$$

But
$$\ln K = -(\nu_3 \phi_3 + \nu_4 \phi_4 - \nu_1 \phi_1 - \nu_2 \phi_2);$$

therefore,
$$\sum \nu g = -RT \ln K + RT \ln P^{\nu_3 + \nu_4 - \nu_1 - \nu_2}. \tag{15-13}$$

The student will recall that K also contains the factor P raised to the $(\nu_3 + \nu_4 - \nu_1 - \nu_2)$th power. It follows, therefore, that the above equation will be satisfied when both P's are measured in the same units, whatever the units are. If we express P in atmospheres as usual and calculate $\sum \nu g$ when each gas is at a pressure of 1 atm, the second term on the right drops out. Under these conditions, $\sum \nu g$ is denoted by $\sum (\nu g)°$. Therefore,

$$\boxed{\sum (\nu g)° = -RT \ln K.} \tag{15-14}$$

Let us imagine that ν_1 mol of A_1 and ν_2 mol of A_2 are mixed at uniform temperature T and pressure P and that chemical reaction takes place, thereby forming constituents A_3 and A_4. At any moment when the degree of reaction is ϵ, the Gibbs function of mixture is

$$G = \mu_1 n_1 + \mu_2 n_2 + \mu_3 n_3 + \mu_4 n_4,$$

where
$$n_1 = \nu_1(1 - \epsilon), \qquad n_3 = \nu_3 \epsilon,$$
$$n_2 - \nu_2(1 - \epsilon), \qquad n_4 = \nu_4 \epsilon,$$

and each chemical potential is a function of T, P, and ϵ. It follows that G is a function of T, P, and ϵ; therefore, at constant T and P, G is a function of ϵ only. The graph of G against ϵ has somewhat the form shown in Fig. 15-2.

At the equilibrium point where $\epsilon = \epsilon_e$, the curve has a minimum at which

$$\left(\frac{\partial G}{\partial \epsilon}\right)_{T,P} = 0 \qquad (\text{at } \epsilon = \epsilon_e).$$

The slopes of the curve at the points $\epsilon = 0$ and $\epsilon = 1$ may be calculated from the equation derived in Art. 14-13, namely,

$$dG_{T,P} = (n_0 + n_0')(\nu_3 \mu_3 + \nu_4 \mu_4 - \nu_1 \mu_1 - \nu_2 \mu_2)\, d\epsilon,$$

which in this case becomes

$$\left(\frac{\partial G}{\partial \epsilon}\right)_{T,P} = \nu_3 \mu_3 + \nu_4 \mu_4 - \nu_1 \mu_1 - \nu_2 \mu_2. \tag{15-15}$$

Since
$$\mu_k = RT(\phi_k + \ln P + \ln x_k)$$

and
$$g_k = RT(\phi_k + \ln P),$$

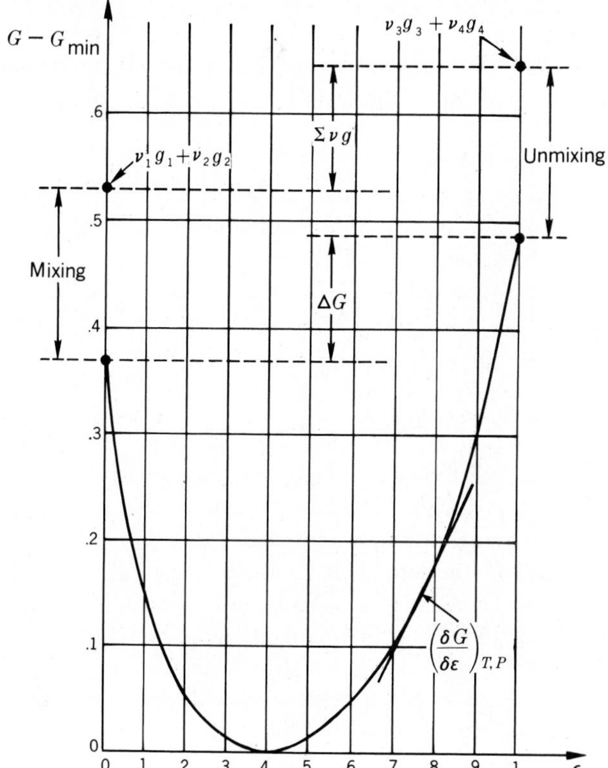

Figure 15-2 Graph of G against ϵ at constant T and P.

the chemical potential may be written in convenient form

$$\mu_k = g_k + RT \ln x_k .$$

Therefore,

$$\left(\frac{\partial G}{\partial \epsilon}\right)_{T,P} = \nu_3 g_3 + \nu_4 g_4 - \nu_1 g_1 - \nu_2 g_2$$

$$+ RT(\nu_3 \ln x_3 + \nu_4 \ln x_4 - \nu_1 \ln x_1 - \nu_2 \ln x_2),$$

or

$$\left(\frac{\partial G}{\partial \epsilon}\right)_{T,P} = \sum \nu g + RT \ln \frac{x_3^{\nu_3} x_4^{\nu_4}}{x_1^{\nu_1} x_2^{\nu_2}}. \tag{15-16}$$

It should be borne in mind that the x's in this equation are not equilibrium values but correspond to *any* value of ϵ. Now, when $\epsilon = 0$, there are no final constituents, and therefore both x_3 and x_4 are zero. Hence,

$$\left(\frac{\partial G}{\partial \epsilon}\right)_{T,P} = -\infty \qquad (\text{at } \epsilon = 0).$$

On the other hand, when $\epsilon = 1$, there are no initial constituents and therefore x_1 and x_2 are zero. Hence,

$$\left(\frac{\partial G}{\partial \epsilon}\right)_{T,P} = +\infty \qquad (\text{at } \epsilon = 1).$$

The graph in Fig. 15-2 has these properties.

Consider the point $\epsilon = \frac{1}{2}$ at which there are $\nu_1/2$ mol of A_1, $\nu_2/2$ mol of A_2, $\nu_3/2$ mol of A_3, and $\nu_4/2$ mol of A_4. At this point the constituents are present in proportion to their stoichiometric coefficients, and the mole fractions are

$$x_1 = \frac{\nu_1}{\sum \nu}, \qquad x_3 = \frac{\nu_3}{\sum \nu},$$

$$x_2 = \frac{\nu_2}{\sum \nu}, \qquad x_4 = \frac{\nu_4}{\sum \nu}.$$

The slope of the curve at this point indicates whether such a mixture is in equilibrium or not. If this slope is positive, the equilibrium point is to the left; i.e., when both the initial and final constituents are mixed, there will be a tendency for the reaction to proceed to the left, causing initial constituents to be formed. Conversely, if the slope of the curve at $\epsilon = \frac{1}{2}$ is negative, the equilibrium point is to the right; or when both initial and final constituents are mixed, there will be a tendency for the reaction to proceed to the right, causing final constituents to be formed. Finally, if this slope is zero, a mixture of both initial and final constituents is in equilibrium, and there is no tendency for the reaction to proceed at all.

It is seen, therefore, that the *sign* of the slope of the $G\epsilon$ curve at $\epsilon = \frac{1}{2}$ is an indication of the behavior of a system composed of initial and final constituents mixed in proportion to their stoichiometric coefficients. Furthermore, it is obvious from the curve that the *magnitude* of this slope is a measure of the departure from or nearness to equilibrium of such a mixture. We shall call the slope of the $G\epsilon$ curve at the point $\epsilon = \frac{1}{2}$ the *affinity* of the reaction, which is equal to

$$\left(\frac{\partial G}{\partial \epsilon}\right)_{T,P}(\epsilon = \tfrac{1}{2}) = \sum \nu g + RT \ln \frac{(\nu_3/\sum \nu)^{\nu_3}(\nu_4/\sum \nu)^{\nu_4}}{(\nu_1/\sum \nu)^{\nu_1}(\nu_2/\sum \nu)^{\nu_2}}. \qquad (15\text{-}17)$$

This equation is true at all temperatures and pressures. Let us choose $T = 298$ K and $P = 1$ atm. The last term on the right is a constant for each reaction. When all the ν's are unity, this term is zero. In general, its value is much less than $\sum (\nu g)^\circ_{298}$. Consequently, $\sum (\nu g)^\circ_{298}$ has more effect on the quantity $\partial G/\partial \epsilon$ than the other term. We may therefore write

$$\boxed{\frac{\partial G}{\partial \epsilon}\begin{bmatrix} \epsilon = \frac{1}{2} \\ T = 298 \text{ K} \\ P = 1 \text{ atm} \end{bmatrix} \sim \sum (\nu g)^\circ_{298}.} \qquad (15\text{-}18)$$

Therefore, *the quantity* $\sum (\nu g)^\circ_{298}$ *is an indication of the direction and amount to which a reaction will proceed at this temperature.* For example, $\sum (\nu g)^\circ_{298}$ for the water-vapor reaction is a large positive number, meaning that the equilibrium point is far to the left of $\epsilon = \frac{1}{2}$, and therefore ϵ_e is very small. Again, $\sum (\nu g)^\circ_{298}$ is a large negative number for the reaction $NO \rightleftharpoons \frac{1}{2}N_2 + \frac{1}{2}O_2$; therefore the equilibrium point is far to the right of $\epsilon = \frac{1}{2}$, and ϵ_e is almost unity.

15-6 DISPLACEMENT OF EQUILIBRIUM

The equilibrium value of the degree of reaction was obtained by setting $(\partial G/\partial \epsilon)_{T,P}$ equal to zero. This, however, is the condition that G be a maximum as well as a minimum. In order to verify that G is a minimum, it is necessary to show that $(\partial^2 G/\partial \epsilon^2)_{T,P}$ is positive at the equilibrium point. For a mixture of arbitrary amounts of four chemically active ideal gases, we have

$$\left(\frac{\partial G}{\partial \epsilon}\right)_{T,P} = (n_0 + n_0')\left(\sum \nu g + RT \ln \frac{x_3^{\nu_3} x_4^{\nu_4}}{x_1^{\nu_1} x_2^{\nu_2}}\right)$$

for all values of ϵ. Therefore, since $\sum \nu g$ is a function of T and P only,

$$\left(\frac{\partial^2 G}{\partial \epsilon^2}\right)_{T,P} = (n_0 + n_0')RT \frac{d}{d\epsilon} \ln \frac{x_3^{\nu_3} x_4^{\nu_4}}{x_1^{\nu_1} x_2^{\nu_2}}. \tag{15-19}$$

The right-hand member of this equation can be easily evaluated for any value of ϵ (the details of this calculation constitute Prob. 15-14*a*), and is found to be

$$\frac{d}{d\epsilon} \ln \frac{x_3^{\nu_3} x_4^{\nu_4}}{x_1^{\nu_1} x_2^{\nu_2}} = \frac{n_0 + n_0'}{\sum n_k} \left[\frac{\nu_1^2}{x_1} + \frac{\nu_2^2}{x_2} + \frac{\nu_3^2}{x_3} + \frac{\nu_4^2}{x_4} - (\Delta \nu)^2\right], \tag{15-20}$$

where $\Delta \nu = \nu_3 + \nu_4 - \nu_1 - \nu_2$. It may be proved rigorously that the expression in brackets is positive for all values of the ν's and x's. The proof, however, is rather lengthy. If we do not insist on complete generality but choose as starting conditions $n_0 \nu_1$ moles of A_1 and $n_0 \nu_2$ moles of A_2, with no amount of A_3 and A_4, then the preceding equation reduces to the following simple form (this calculation constitutes Prob. 15-14*b*):

$$\frac{d}{d\epsilon} \ln \frac{x_3^{\nu_3} x_4^{\nu_4}}{x_1^{\nu_1} x_2^{\nu_2}} = \frac{n_0}{\sum n_k} \frac{(\nu_1 + \nu_2)(\nu_3 + \nu_4)}{\epsilon(1 - \epsilon)}. \tag{15-21}$$

Since the right-hand member of this equation is always positive for all ν's and all values of ϵ, it follows that $(\partial G^2/\partial \epsilon^2)_{T,P}$ is always positive; hence, when $\epsilon = \epsilon_e$, G is a minimum and not a maximum. It will be seen that this plays an important role in determining the displacement of equilibrium when the temperature or the pressure is changed.

Let us consider first the effect on the equilibrium value of the degree of reaction of a change of temperature at constant pressure. We know that at equilibrium the law of mass action provides us with a relation among T, ϵ_e, and P, which may be written in the form

$$\ln K = \ln \left(\frac{x_3^{\nu_3} x_4^{\nu_4}}{x_1^{\nu_1} x_2^{\nu_2}} \right)_{\epsilon=\epsilon_0} + (\nu_3 + \nu_4 - \nu_1 - \nu_2) \ln P,$$

where $\ln K$ is a function of T only, and the first term on the right is a function of ϵ_e only. Now,

$$\left(\frac{\partial \epsilon_e}{\partial T} \right)_P = \left(\frac{\partial \epsilon_e}{\partial \ln K} \right)_P \left(\frac{\partial \ln K}{\partial T} \right)_P$$

$$= \frac{(d \ln K)/dT}{(\partial \ln K/\partial \epsilon_e)_P}.$$

Using the van't Hoff isobar to evaluate the numerator and the law of mass action to evaluate the denominator, we get

$$\boxed{\left(\frac{\partial \epsilon_e}{\partial T} \right)_P = \frac{\Delta H}{RT^2\, d[\ln (x_3^{\nu_3} x_4^{\nu_4}/x_1^{\nu_1} x_2^{\nu_2})]/d\epsilon_e}.} \tag{15-22}$$

Since we have already mentioned that the denominator on the right is positive, it follows that the sign of $(\partial \epsilon_e/\partial T)_P$ is determined by the sign of ΔH. Therefore, *an increase of temperature at constant pressure causes a shift in the equilibrium value of the degree of reaction in the direction in which the heat of reaction is absorbed.*

To determine the effect of a change of pressure at constant temperature, we write

$$\left(\frac{\partial \epsilon_e}{\partial P} \right)_T = - \left(\frac{\partial \epsilon_e}{\partial \ln K} \right)_P \left(\frac{\partial \ln K}{\partial P} \right)_{\epsilon_e}$$

$$= - \frac{(\partial \ln K/\partial P)_{\epsilon_e}}{(\partial \ln K/\partial \epsilon_e)_P}.$$

Using the law of mass action to evaluate both numerator and denominator, we get

$$\boxed{\left(\frac{\partial \epsilon_e}{\partial P} \right)_T = - \frac{\nu_3 + \nu_4 - \nu_1 - \nu_2}{P\, d[\ln (x_3^{\nu_3} x_4^{\nu_4}/x_1^{\nu_1} x_2^{\nu_2})]/d\epsilon_e}.} \tag{15-23}$$

The numerator on the right is proportional to the change in the number of moles of the constituents as the reaction proceeds to the right. If it is positive, this means that the volume increases at constant T and P. Therefore, *an increase of pressure at constant temperature causes a shift in the equilibrium value of the degree of reaction in the direction in which a decrease of volume takes place.*

15-7 HEAT CAPACITY OF REACTING GASES IN EQUILIBRIUM

Let us consider, as usual, a mixture of arbitrary amounts of four ideal gases capable of undergoing the reaction

$$\nu_1 A_1 + \nu_2 A_2 \rightleftharpoons \nu_3 A_3 + \nu_4 A_4 .$$

At equilibrium, the enthalpy of the mixture is

$$H = \sum n_k h_k ,$$

where

$$n_1 = (n_0 + n_0')\nu_1(1 - \epsilon_e), \qquad n_3 = (n_0 + n_0')\nu_3 \epsilon_e ,$$

$$n_2 = (n_0 + n_0')\nu_2(1 - \epsilon_e) + N_2 , \qquad n_4 = (n_0 + n_0')\nu_4 \epsilon_e + N_4 ,$$

and ϵ_e is the equilibrium value of the degree of reaction. Suppose that an infinitesimal change of temperature takes place *at constant pressure* in such a way that equilibrium is maintained. Then ϵ_e will change to the value $\epsilon_e + d\epsilon_e$, and the enthalpy will change by the amount

$$dH_P = \sum n_k \, dh_k + \sum h_k \, dn_k .$$

Since $dh_k = c_{Pk} \, dT$ and $dn_k = \pm(n_0 + n_0')\nu_k \, d\epsilon_e$,

$$dH_P = \sum n_k c_{Pk} \, dT + (n_0 + n_0')(\nu_3 h_3 + \nu_4 h_4 - \nu_1 h_1 - \nu_2 h_2) \, d\epsilon_e ,$$

and the heat capacity of the reacting gas mixture is

$$C_P = \left(\frac{\partial H}{\partial T}\right)_P = \sum n_k c_{Pk} + (n_0 + n_0') \, \Delta H \left(\frac{\partial \epsilon_e}{\partial T}\right)_P .$$

From the preceding article,

$$\left(\frac{\partial \epsilon_e}{\partial T}\right)_P = \frac{\Delta H}{RT^2 \, d[\ln (x_3^{\nu_3} x_4^{\nu_4} / x_1^{\nu_1} x_2^{\nu_2})]/d\epsilon_e} ;$$

consequently,

$$C_P = \sum n_k c_{Pk} + (n_0 + n_0') \frac{(\Delta H)^2}{RT^2 \, d[\ln (x_3^{\nu_3} x_4^{\nu_4} / x_1^{\nu_1} x_2^{\nu_2})]/d\epsilon_e} . \tag{15-24}$$

As an example, consider the equilibrium mixture of H_2O vapor, H_2, and O_2 caused by the dissociation of 1 mol of H_2O at 1 atm and 1900 K. We have

$n_0 = 1,\quad n_0' = 0,\quad \Delta H = 250\text{ kJ/mol},\quad R = 8.31\text{ J/mol}\cdot\text{K},\quad \epsilon_e = 3.2\times 10^{-3}$, $\sum n_k = n_0(1+\epsilon_e/2)$, $\nu_1 = 1$, $\nu_2 = 0$, $\nu_3 = 1$, $\nu_4 = \frac{1}{2}$, and

$$\frac{d}{d\epsilon_e}\ln\frac{x_3^{\nu_3}x_4^{\nu_4}}{x_1^{\nu_1}x_2^{\nu_2}} = \frac{n_0}{\sum n_k}\frac{(\nu_1+\nu_2)(\nu_3+\nu_4)}{\epsilon_e(1-\epsilon_e)}.$$

Hence,

$$C_P - \sum n_k c_{Pk} = \frac{(\Delta H)^2(1+\epsilon_e/2)\epsilon_e(1-\epsilon_e)}{RT^2(\nu_1+\nu_2)(\nu_3+\nu_4)}$$

$$= \frac{(250{,}000)^2\times 3.2\times 10^{-3}}{8.31\times(1900)^2\times\frac{3}{2}}$$

$$= 4.32\text{ J/K}.$$

PROBLEMS

15-1 Show that the law of mass action may be written

$$\frac{p_3^{\nu_3}p_4^{\nu_4}}{p_1^{\nu_1}p_2^{\nu_2}} = K,$$

where the p's are the equilibrium values of the partial pressures.

15-2 If we start with n_0 moles of NH_3, which dissociates according to the equation $NH_3 \rightleftharpoons \frac{1}{2}N_2 + \frac{3}{2}H_2$, show that at equilibrium

$$K = \frac{\sqrt{27}}{4}\frac{\epsilon_e^2}{1-\epsilon_e^2}P.$$

15-3 Starting with n_0 moles of CO and $3n_0$ moles of H_2, which react according to the equation $CO + 3H_2 \rightleftharpoons CH_4 + H_2O$, show that at equilibrium

$$K = \frac{4\epsilon_e^2(2-\epsilon_e)^2}{27(1-\epsilon_e)^4P^2}.$$

15-4 A mixture of $n_0\nu_1$ moles of A_1 and $n_0\nu_2$ moles of A_2 at temperature T and pressure P occupies a volume V_0. When the reaction

$$\nu_1A_1 + \nu_2A_2 \rightleftharpoons \nu_3A_3 + \nu_4A_4$$

has come to equilibrium at the same T and P, the volume is V_e. Show that

$$\epsilon_e = \frac{V_e - V_0}{V_0}\frac{\nu_1+\nu_2}{\nu_3+\nu_4-\nu_1-\nu_2}.$$

15-5 At 35°C and 1 atm, the degree of dissociation of N_2O_4 at equilibrium is 0.27.

(*a*) Calculate K.

(*b*) Calculate ϵ_e at the same temperature when the pressure is 100 mm Hg.

(*c*) The equilibrium constant for the dissociation of N_2O_4 has the values 0.664 and 0.141 at the temperatures 318 K and 298 K, respectively. Calculate the average heat of reaction within this temperature range.

15-6 The equilibrium constant of reaction $SO_3 \rightleftharpoons SO_2 + \frac{1}{2}O_2$ has the following values:

Kelvin temperature	800	900	1000	1105
Equilibrium constant	0.0319	0.153	0.540	1.59

Determine the average heat of dissociation graphically.

15-7 Calculate the degree of ionization of cesium vapor at 10^{-6} atm at the two temperatures 2260 K and 2520 K.

15-8 Calculate the degree of ionization of calcium vapor in the sun's chromosphere. The temperature and pressure of the sun's chromosphere are approximately 6000 K and 10^{-10} atm, respectively.

15-9 (*a*) Show that

$$\Delta G = \Delta H + T\left(\frac{\partial\, \Delta G}{\partial T}\right)_P.$$

(*b*) Show that

$$\Delta G = -RT \ln \frac{x_3^{\nu_3} x_4^{\nu_4}}{x_1^{\nu_1} x_2^{\nu_2}},$$

where the x's are equilibrium values.

15-10 Calculate the heat capacity of the equilibrium mixture of Prob. 15-7 at the temperature of 2260 K.

15-11 When 1 mol of HI dissociates according to the reaction

$$HI \rightleftharpoons \tfrac{1}{2}H_2 + \tfrac{1}{2}I_2$$

at $T = 675$ K, $K = 0.132$ and $\Delta H = 2950$ J/mol. Calculate $(\partial \epsilon_e/\partial T)_P$ at this temperature.

15-12 Starting with ν_1 moles of A_1 and ν_2 moles of A_2, show that:

(*a*) At any value of ϵ

$$G = \epsilon(\nu_3 \mu_3 + \nu_4 \mu_4 - \nu_1 \mu_1 - \nu_2 \mu_2) + \nu_1 \mu_1 + \nu_2 \mu_2 .$$

(*b*) At equilibrium,

$$G(\min) = \nu_1 \mu_{1_e} + \nu_2 \mu_{2_e},$$

where the subscript e denotes an equilibrium value.

(*c*) $$\frac{G - G(\min)}{RT} = \epsilon\left(\ln \frac{x_3^{\nu_3} x_4^{\nu_4}}{x_1^{\nu_1} x_2^{\nu_2}} - \ln \frac{x_{3_e}^{\nu_3} x_{4_e}^{\nu_4}}{x_{1_e}^{\nu_1} x_{2_e}^{\nu_2}}\right) + \ln x_1^{\nu_1} x_2^{\nu_2} - \ln x_{1_e}^{\nu_1} x_{2_e}^{\nu_2}.$$

(*d*) At $\epsilon = 0$,

$$\frac{G_0 - G(\min)}{RT} = \ln \left(\frac{\nu_1}{\nu_1 + \nu_2}\right)^{\nu_1} \left(\frac{\nu_2}{\nu_1 + \nu_2}\right)^{\nu_2} - \ln x_{1_e}^{\nu_1} x_{2_e}^{\nu_2}.$$

(*e*) At $\epsilon = 1$,

$$\frac{G_1 - G(\min)}{RT} = \ln \left(\frac{\nu_3}{\nu_3 + \nu_4}\right)^{\nu_3} \left(\frac{\nu_4}{\nu_3 + \nu_4}\right)^{\nu_4} - \ln x_{3_e}^{\nu_3} x_{4_e}^{\nu_4}.$$

15-13 In the case of the ionization of a monatomic gas, show that:

(*a*) $$\frac{G - G(\min)}{RT} = \epsilon\left(\ln\frac{\epsilon^2}{1-\epsilon^2} - \ln\frac{\epsilon_e^2}{1-\epsilon_e^2}\right) + \ln\frac{1-\epsilon}{1+\epsilon} - \ln\frac{1-\epsilon_e}{1+\epsilon_e}.$$

(*b*) At $\epsilon = 0$,

$$\frac{G_0 - G(\min)}{RT} = -\ln\frac{1-\epsilon_e}{1+\epsilon_e}.$$

(*c*) At $\epsilon = 1$,

$$\frac{G_1 - G(\min)}{RT} = \ln\frac{1}{4} - \ln\frac{\epsilon_e^2}{(1+\epsilon_e)^2}.$$

(*d*) Plot $[G - G(\min)]/2.30RT$ against ϵ for the ionization of cesium vapor at 2260 K and 10^{-6} atm, using the result of Prob. 15-7.

15-14 (*a*) Prove that, for a mixture of reacting ideal gases,

$$\frac{d}{d\epsilon}\ln\frac{x_3^{\nu_3}x_4^{\nu_4}}{x_1^{\nu_1}x_2^{\nu_2}} = \frac{n_0 + n_0'}{\sum n_k}\frac{1}{\psi},$$

where $$\frac{1}{\psi} = \frac{\nu_1^2}{x_1} + \frac{\nu_2^2}{x_2} + \frac{\nu_3^2}{x_3} + \frac{\nu_4^2}{x_4} - (\Delta\nu)^2$$

and $$\Delta\nu = \nu_3 + \nu_4 - \nu_1 - \nu_2.$$

(*b*) If we start with $n_0\nu_1$ moles of A_1, $n_0\nu_2$ moles of A_2, and no A_3 or A_4, show that

$$\psi = \frac{\epsilon(1-\epsilon)}{(\nu_1+\nu_2)(\nu_3+\nu_4)}.$$

15-15 Prove that, for a mixture of reacting ideal gases *in equilibrium*,

(*a*) $$\left(\frac{\partial V}{\partial P}\right)_T = -\frac{V}{P} - \frac{(n_0 + n_0')RT(\Delta\nu)^2}{P^2(d/d\epsilon_e)\ln(x_3^{\nu_3}x_4^{\nu_4}/x_1^{\nu_1}x_2^{\nu_2})}.$$

(*b*) $$\left(\frac{\partial V}{\partial T}\right)_P = \frac{V}{T} + \frac{(n_0 + n_0')\,\Delta\nu\,\Delta H}{PT(d/d\epsilon_e)\ln(x_3^{\nu_3}x_4^{\nu_4}/x_1^{\nu_1}x_2^{\nu_2})}.$$

(*c*) $$\left(\frac{\partial P}{\partial T}\right)_{\epsilon_e} = -\frac{P\,\Delta H}{RT^2\,\Delta\nu}.$$

15-16 Prove that, for a mixture of reacting ideal gases *in equilibrium*,

$$dS = \sum n_k\left[\sum x_k c_{Pk} + \frac{\psi(\Delta H)^2}{RT^2}\right]\frac{dT}{T} - R\sum n_k\left[1 + \frac{\psi\,\Delta H\,\Delta\nu}{RT}\right]\frac{dP}{P}.$$

CHAPTER

SIXTEEN

HETEROGENEOUS SYSTEMS

16-1 THERMODYNAMIC EQUATIONS FOR A HETEROGENEOUS SYSTEM

It was shown in Art. 14-10 that the Gibbs function of any homogeneous phase, consisting of c constituents and in thermal and mechanical equilibrium at temperature T and pressure P, is equal to

$$G = \sum \mu_k n_k,$$

where each chemical potential is a function of T, P, and the mole fraction of the respective constituent, and the summation is taken over all the constituents. Furthermore, if the phase undergoes an infinitesimal process involving a change of temperature dT, a change of pressure dP, and changes in each of the n's, the accompanying change in the Gibbs function is equal to

$$dG = -S\,dT + V\,dP + \sum \mu_k\,dn_k.$$

Suppose that we have a heterogeneous system of φ phases, all homogeneous and all at the uniform temperature T and pressure P. Let us denote constituents, as usual, by subscripts and phases by superscripts. The total Gibbs function of the heterogeneous system $\boldsymbol{G}$ is the sum of the Gibbs functions of all the phases; i.e.,

$$\begin{aligned} \boldsymbol{G} = &\sum \mu_k^{(1)} n_k^{(1)} \text{ over all the constituents of the 1st phase} \\ &+ \sum \mu_k^{(2)} n_k^{(2)} \text{ over all the constituents of the 2d phase} \\ &\cdots\cdots\cdots\cdots\cdots\cdots\cdots\cdots\cdots\cdots \qquad (16\text{-}1) \\ &+ \sum \mu_k^{(\varphi)} n_k^{(\varphi)} \text{ over all the constituents of the } \varphi\text{th phase.} \end{aligned}$$

If an infinitesimal process takes place in which *all* the phases undergo a change in temperature dT and a change in pressure dP, then the change in the Gibbs function is

$$\begin{aligned} dG = & -S^{(1)}\,dT + V^{(1)}\,dP + \sum \mu_k^{(1)}\,dn_k^{(1)} && \text{(for 1st phase)} \\ & -S^{(2)}\,dT + V^{(2)}\,dP + \sum \mu_k^{(2)}\,dn_k^{(2)} && \text{(for 2d phase)} \\ & \cdots\cdots\cdots\cdots\cdots\cdots\cdots\cdots\cdots\cdots\cdots\cdots \\ & -S^{(\varphi)}\,dT + V^{(\varphi)}\,dP + \sum \mu_k^{(\varphi)}\,dn_k^{(\varphi)} && \text{(for } \varphi\text{th phase).} \end{aligned}$$

This equation evidently reduces to

$$dG = -S\,dT + V\,dP + \sum \mu_k^{(1)}\,dn_k^{(1)} + \sum \mu_k^{(2)}\,dn_k^{(2)} + \cdots + \sum \mu_k^{(\varphi)}\,dn_k^{(\varphi)}, \qquad (16\text{-}2)$$

where S and V are the entropy and volume, respectively, of the whole heterogeneous system.

The problem of the equilibrium of a heterogeneous system is to obtain an equation or a set of equations among the μ's that hold when all the phases are in chemical equilibrium. If the system is assumed to approach equilibrium at constant T and P, then G is a minimum at equilibrium, and the problem can be stated thus: *To render* G *a minimum at constant* T *and* P, *subject to whatever conditions are imposed on the* n's *by virtue of the constraints of the system.* The mathematical condition that G be a minimum at constant T and P is that

$$dG_{T,\,P} = 0.$$

Hence, the equation that must be satisfied at equilibrium is

$$dG_{T,\,P} = \sum \mu_k^{(1)}\,dn_k^{(1)} + \sum \mu_k^{(2)}\,dn_k^{(2)} + \cdots + \sum \mu_k^{(\varphi)}\,dn_k^{(\varphi)} = 0, \qquad (16\text{-}3)$$

where the dn's are not all independent but are connected by equations of constraint.

To return for a moment to the system treated in Chap. 14—namely, one phase consisting of a mixture of chemically active substances—we found that the equations of constraint were of such a simple form that, by direct substitution, G could be expressed as a function of T, P, and only one other independent variable ϵ and that $dG_{T,\,P}$ could be expressed in terms of only one differential $d\epsilon$. thus:

$$dG_{T,\,P} = (n_0 + n_0')(\nu_3\mu_3 + \nu_4\mu_4 - \nu_1\mu_1 - \nu_2\mu_2)\,d\epsilon.$$

At equilibrium, when $dG_{T,\,P} = 0$, only one equation, the equation of reaction equilibrium, was obtained.

In the case of a heterogeneous system, however, the situation is more complicated. In the first place there is, as a rule, more than one independent variable besides T and P. In the second place, the equations of constraint are usually of such a nature that it is either impossible or exceedingly cumbersome to attempt by direct substitution to express G in terms of the independent variables only and dG in terms of differentials of these independent variables. Finally, instead of only one equation of equilibrium, there may be several, depending on the type of heterogeneous system.

We are therefore confronted with a type of problem that requires the use of Lagrange's method of undetermined multipliers.

16-2 PHASE RULE WITHOUT CHEMICAL REACTION

Consider a heterogeneous system of c chemical constituents that do not combine chemically with one another. Suppose that there are φ phases, each of which is in contact with every other phase in such a way that there are no impediments to the transport of any constituent from one phase to another. Let us assume temporarily that every constituent is present in every phase. As usual, constituents will be denoted by subscripts and phases by superscripts. As we have shown previously, the Gibbs function of the whole heterogeneous system is

$$G = \sum_1^c n_k^{(1)}\mu_k^{(1)} + \sum_1^c n_k^{(2)}\mu_k^{(2)} + \cdots + \sum_1^c n_k^{(\varphi)}\mu_k^{(\varphi)},$$

where all the summations extend from $k = 1$ to $k = c$, since all the constituents are present in all the phases. G is a function of T, P, and the n's, of which there are $c\varphi$ in number. Not all these n's, however, are independent. Since there are no chemical reactions, the only way in which the n's may change is by the transport of constituents from one phase to another, in which case the total number of moles of each constituent remains constant. We have as our equations of constraint, therefore,

$$\begin{aligned} n_1^{(1)} + n_1^{(2)} + \cdots + n_1^{(\varphi)} &= \text{const.} \\ n_2^{(1)} + n_2^{(2)} + \cdots + n_2^{(\varphi)} &= \text{const.} \\ \cdots\cdots\cdots\cdots&\cdots\cdots \\ n_c^{(1)} + n_c^{(2)} + \cdots + n_c^{(\varphi)} &= \text{const.} \end{aligned}$$

In order to find the equations of chemical equilibrium, it is necessary to render G a minimum at constant T and P, subject to these equations of constraint. Applying Lagrange's method, we have

$$\begin{aligned} dG = \mu_1^{(1)}\,dn_1^{(1)} + \cdots + \mu_c^{(1)}\,dn_c^{(1)} + \cdots + \mu_1^{(\varphi)}\,dn_1^{(\varphi)} + \cdots + \mu_c^{(\varphi)}\,dn_c^{(\varphi)} &= 0 \\ \lambda_1\,dn_1^{(1)} \qquad\qquad + \cdots + \lambda_1\,dn_1^{(\varphi)} \qquad\qquad &= 0 \\ \cdot \qquad\qquad\qquad \cdot \qquad\qquad &\;\cdot \\ \cdot \qquad\qquad\qquad \cdot \qquad\qquad &\;\cdot \\ \cdot \qquad\qquad\qquad \cdot \qquad\qquad &\;\cdot \\ \lambda_c\,dn_c^{(1)} + \cdots \qquad\qquad + \lambda_c\,dn_c^{(\varphi)} &= 0 \end{aligned}$$

where there are c Lagrangian multipliers, one for each equation of constraint. Adding and equating each coefficient of each dn to zero, we get

$$\begin{array}{llll}
\mu_1^{(1)} = -\lambda_1 & \mu_1^{(2)} = -\lambda_1 & \cdots & \mu_1^{(\varphi)} = -\lambda_1 \\
\mu_2^{(1)} = -\lambda_2 & \mu_2^{(2)} = -\lambda_2 & \cdots & \mu_2^{(\varphi)} = -\lambda_2 \\
\multicolumn{4}{c}{\cdots\cdots\cdots\cdots\cdots\cdots\cdots\cdots\cdots\cdots} \\
\mu_c^{(1)} = -\lambda_c & \mu_c^{(2)} = -\lambda_c & \cdots & \mu_c^{(\varphi)} = -\lambda_c
\end{array}$$

or

$$\boxed{\begin{array}{l}
\mu_1^{(1)} = \mu_1^{(2)} = \cdots = \mu_1^{(\varphi)} \\
\mu_2^{(1)} = \mu_2^{(2)} = \cdots = \mu_2^{(\varphi)} \\
\cdots\cdots\cdots\cdots\cdots\cdots \\
\mu_c^{(1)} = \mu_c^{(2)} = \cdots = \mu_c^{(\varphi)}.
\end{array}} \tag{16-4}$$

These are the *equations of phase equilibrium*, which express the important fact that at equilibrium the chemical potential of a constituent in one phase must be equal to the chemical potential of the same constituent in every other phase.

As a simple example, suppose that we have only one coefficient present in two phases. Then

$$dG_{T,P} = \mu_1^{(1)}\, dn_1^{(1)} + \mu_1^{(2)}\, dn_1^{(2)};$$

and since $dn_1^{(2)} = -dn_1^{(1)}$,

$$dG_{T,P} = (\mu_1^{(1)} - \mu_1^{(2)})\, dn_1^{(1)}.$$

Now, before equilibrium is reached, suppose there is a flow of matter from phase 1 to phase 2. Then $dn_1^{(1)}$ is negative; and since this flow is irreversible, $dG_{T,P}$ must be negative. Therefore, *while the flow is taking place*,

$$\mu_1^{(1)} > \mu_1^{(2)} \qquad \text{(flow of matter from phase 1 to phase 2).}$$

Obviously, the transfer of matter ceases when the two chemical potentials become equal. The chemical potentials of a constituent in two neighboring phases may be compared with the temperatures and pressures of these phases, thus:

1. If the temperature of phase 1 is greater than that of phase 2, there is a flow of heat that ceases when the temperatures are equal, i.e., when thermal equilibrium is established.
2. If the pressure of phase 1 is greater than that of phase 2, there is a "flow" of work that ceases when the pressures are equal, i.e., when mechanical equilibrium is established.
3. If the chemical potential of a constituent of phase 1 is greater than that of phase 2, there is a flow of that constituent which ceases when the chemical potentials are equal, i.e., when chemical equilibrium is established.

The equations of phase equilibrium expressing the equality of the chemical potentials of any one constituent in all the φ phases are obviously $\varphi - 1$ in number. Therefore, for c constituents, there are altogether $c(\varphi - 1)$ equations among the μ's.

Following out the procedure of Lagrange's method, we should complete our solution of the problem by solving the $c(\varphi - 1)$ equations of phase equilibrium and the c equations of constraint for the $c\varphi$ values of the n's that make G a minimum. These values should, of course, be functions of the parameters T and P. We find, however, that the equations do not contain the n's in such fashion as to determine their values, because of the fact that the equations of equilibrium are equations among the chemical potentials, which are intensive quantities and depend on the x's, which contain the n's in the special form

$$x_k = \frac{n_k}{\sum n}.$$

This is another way of saying that the chemical potential for a constituent in a phase depends on the composition of that phase but not on its total mass.

There are many *different* sets of n's that satisfy the equations of phase equilibrium and give rise to the same minimum value of the Gibbs function. This may be seen from the fact that

$$G = \mu_1^{(1)} n_1^{(1)} + \cdots + \mu_c^{(1)} n_c^{(1)} + \cdots + \mu_1^{(\varphi)} n_1^{(\varphi)} + \cdots + \mu_c^{(\varphi)} n_c^{(\varphi)},$$

but at equilibrium the chemical potentials of the same constituent are the same in all phases and hence may be written without any superscripts. Factoring out the μ's, we get

$$G(\text{min} = \mu_1 (n_1^{(1)} + \cdots + n_1^{(\varphi)}) + \cdots + \mu_c (n_c^{(1)} + \cdots + n_c^{(\varphi)}).$$

Therefore, *the minimum value of the Gibbs function is the same for many different distributions of the total mass among the phases.* Since we cannot find the values of the n's at equilibrium, therefore, we may inquire as to whether we can obtain *any* precise information about a heterogeneous system in equilibrium.

As we have seen, the state of the system at equilibrium is determined by the temperature, the pressure, and $c\varphi$ mole fractions. Hence.

$$\text{Total number of variables} = c\varphi + 2.$$

Among these variables there are two types of equations: (1) equations of phase equilibrium, which are $c(\varphi - 1)$ in number; and (2) equations of the type $\sum x = 1$, for each phase, and therefore φ such equations altogether. Hence,

$$\text{Total number of equations} = c(\varphi - 1) + \varphi.$$

If there are as many equations as there are variables, then the temperature, pressure, and composition of the whole system at equilibrium are determined. Such a system is called *nonvariant* and is said to have zero variance. If the number of variables exceeds the number of equations by one, then the equilibrium of the system is not determined until one of the variables is arbitrarily chosen. Such a system is called *monovariant* and is said to have a variance of 1. In general, *the excess of variables over equations is called the variance f*. Thus,

$$\text{Variance} = (\text{number of variables}) - (\text{number of equations}),$$

or

$$f = (c\varphi + 2) - [c(\varphi - 1) + \varphi];$$

whence

$$\boxed{f = c - \varphi + 2.} \tag{16-5}$$

This is known as the *phase rule*, which was first derived in 1875 by Josiah Willard Gibbs, who was then professor of mathematical physics at Yale University. The phase rule arose from a general theory of the equilibrium of heterogeneous systems that Gibbs developed during the years 1875 to 1878 and published in an obscure journal, *The Transactions of the Connecticut Academy*. The original paper, entitled "On the Equilibrium of Heterogeneous Substances," was almost 300 pages long. In it, Gibbs considered not only chemical effects but also those produced by gravity, capillarity, and non-homogeneous strains. It stands today as one of the most profound contributions to the world of human thought and, along with Gibbs' researches in vector analysis and statistical mechanics, places him with the greatest of the world's geniuses.

It is a simple matter to remove the restriction that every constituent must be present in every phase. Suppose that constituent A_1 is absent from phase 1. Then the equation of equilibrium that exists when the constituent is present, namely,

$$\mu_1^{(1)} = -\lambda_1,$$

is now lacking. However, to describe the composition of the first phase, we need one mole fraction fewer than before. Therefore, since both the number of equations and the number of variables have been reduced by one, the difference is the same and the phase rule remains unchanged.

To remove the second restriction—that no chemical reaction take place—is more difficult and requires solving the problem *de novo*. Before this is done, however, it is worth while to consider a few simple applications of the phase rule in its present form.

16-3 SIMPLE APPLICATIONS OF THE PHASE RULE

As simple examples of the use of the phase rule, we shall consider a pure substance, a simple eutectic, and a freezing mixture.

1. Pure substance. In the case of a pure substance such as water, the phase rule merely confirms what is already known. If there are two phases in equilibrium (say, solid and vapor), the variance is 1. There is one equation of equilibrium, namely,

$$\mu'(T, P) = \mu'''(T, P),$$

where one prime stands for solid and three primes for vapor. We have already shown that, when a phase consists of only one constituent, the chemical potential is equal to the molar Gibbs function. Hence,

$$g' = g'''$$

is the equation of equilibrium among the two coordinates T and P, which will be recognized as the equation of the sublimation curve. If three phases are in equilibrium, the system is nonvariant, and the two equations of equilibrium

$$g' = g''' \qquad \text{and} \qquad g'' = g'''$$

serve to determine both T and P. The phase rule shows that the maximum number of phases of a one-constituent system which can exist in equilibrium is three. The various triple points of water confirm this result.

2. Simple eutectic. Let us consider a system of two constituents which neither combine to produce a compound nor form a solid solution but which, in the liquid phase, are miscible in all proportions. A mixture of gold and thallium has these properties. Suppose that we have a liquid alloy consisting of 40 percent thallium and 60 percent gold in an evacuated chamber originally at about 1000°C. A mixture of thallium and gold vapors will constitute the vapor phase, and we shall have $c = 2$ and $\varphi = 2$. It follows that the variance is 2; hence, the composition and temperature having been chosen, the vapor pressure is determined. If the temperature is now progressively lowered, a solid phase of pure gold will separate from the liquid at a temperature of about 600°C, and the percentage of thallium in the solution is thus increased. At any given concentration, there will be one and only one temperature at which the three phases—vapor mixture, liquid solution, and solid gold—will be in equilibrium, because now $c = 2$ and $\varphi = 3$; therefore, $f = 1$.

By covering the metals with a piston on which any desired pressure may be exerted, we may exclude the vapor phase and study the variance of the system when only solid and liquid phases are present. In this way, the temperatures and compositions at which equilibrium exists among various phases may be measured and the results plotted on a phase diagram such as that shown in Fig. 16-1.

Point A is the melting point (strictly speaking, the triple point) of pure gold, and B that of pure thallium. When two phases, solution and vapor, are present, the system is divariant; and equilibrium may exist at any temperature and composition represented by a point in the *region* above AEB. When the three phases (solution, vapor, and solid gold) are present, the system is monovariant; and equilibrium may exist only at those temperatures and compositions represented by points on the *curve* AE. Similarly, curve BE represents temperatures and compositions at which the monovariant system consisting of three phases—solution, vapor, and solid thallium—is in equilibrium. The complete curve AEB is known as the *liquidus*.

At E, there are four phases present: solution, vapor, solid gold, and solid

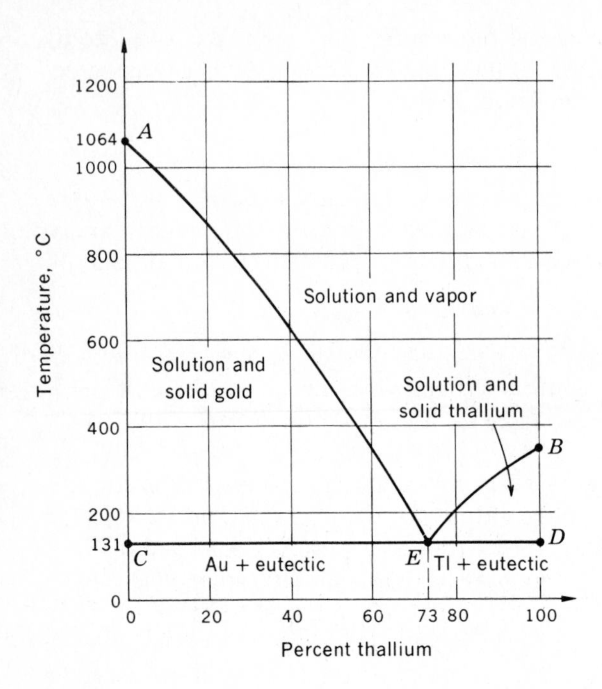

Figure 16-1 Phase diagram for eutectic system of gold and thallium.

thallium. Hence, $c = 2$, $\varphi = 4$, and $f = 0$, or the system is nonvariant. This is known as the *eutectic point*, and the composition at this point as the *eutectic composition*.

Solution and solid gold may coexist at all temperatures and compositions represented by points in region *ACE*. Below line *CE*, however, no liquid can exist, and the system consists of a solid with the eutectic mixture plus free gold. Solution and solid thallium coexist in region *BED*, and below *ED* we have eutectic plus free thallium. Line *CED* is known as the *solidus*.

There are many different types of eutectic system, each with phase diagrams of different character. All of them, however, may be understood completely in terms of the phase rule.

3. Freezing mixture. Many years ago, before the commercial use of solid carbon dioxide ("dry ice") as a cooling agent, foods such as ice cream were packed in a container surrounded by a mixture of ice and common salt. If the mixture was thermally insulated and covered, it maintained a constant temperature of about −21°C. Another practice that is still current is to melt the ice which forms on the sidewalk by sprinkling salt on it. These phenomena may be clearly understood on the basis of the phase rule.

Consider the phase diagram of NaCl and water shown in Fig. 16-2. *A* is the triple point of pure water, and *B* is the transition point where the dihydrate $NaCl \cdot 2H_2O$ changes into NaCl. Except for the upper right-hand portion of the

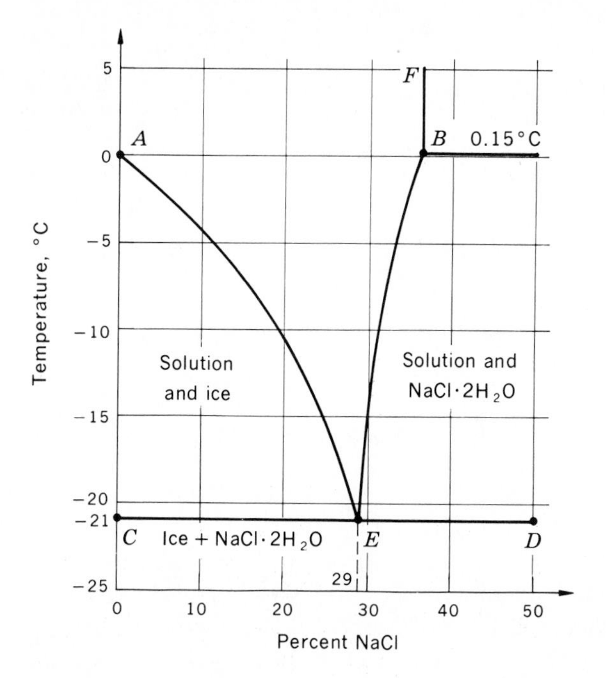

Figure 16-2 Phase diagram for mixture of NaCl and H_2O.

figure, the diagram is in all respects similar to the simple eutectic diagram of Fig. 16-1. At all points on *AE*, the system is monovariant and consists of three phases: solution, vapor, and ice. Similarly, on *EB* we have solution, vapor, and dihydrate. Point *E* is the eutectic point, which, since in this case the system contains water, is called the *cryohydric point*. The mixture of dihydrate and water that forms at the cryohydric point is called a *cryohydrate*.

Only at points below the solidus *CED* can ice and $NaCl \cdot 2H_2O$ exist as solids together. Consequently, when they are mixed at a temperature above −21°C (as on the sidewalk), they are not in equilibrium, and as a result the ice melts and the salt dissolves. It should be mentioned at this point that, if the system is open to the air at atmospheric pressure, then there is one more constituent (air), which would ordinarily increase the variance by 1. Since the pressure is constant, however, this extra variance is used up, and the system behaves as before.

If ice, salt, water, and vapor at 0°C are together in a thermally insulated container, they are not in equilibrium, and some ice will melt and dissolve some salt. But this saturated solution will be too concentrated to be in equilibrium with ice. Ice will therefore melt, lowering the concentration of the solution, which will then dissolve more salt. While this is going on, the temperature of the whole system automatically and spontaneously decreases until the temperature of −21°C is reached. Such a system is known as a *freezing mixture*.

At the transition point *B*, where NaCl forms, there are three constituents;

Table 16-1 Freezing mixtures

First constituent	Second constituent	Lowest temperature, °C
NH_4Cl	Ice	−15.4
NaCl	Ice	−21
Alcohol	Ice	−30
$CaCl_2 \cdot 6H_2O$	Ice	−55
Alcohol	Solid CO_2	−72
Ether	Solid CO_2	−77

hence one might expect a maximum of five phases to coexist. This is not the case, however, because a chemical reaction

$$NaCl \cdot 2H_2O \rightleftharpoons NaCl + \text{solution}$$

takes place. We shall see in the next article that the presence of this reaction causes the system to behave as if there were only two constituents, so that only four phases coexist at the point *B*: solid NaCl, solution, and vapor.

There are a number of freezing mixtures that are often used for preserving materials at low temperatures. These are listed in Table 16-1.

16-4 PHASE RULE WITH CHEMICAL REACTION

Let us consider a heterogeneous system composed of arbitrary amounts of *c* constituents, assuming for the sake of simplicity that four of the constituents are

$$\begin{array}{lr}
\mu_1^{(1)}\, dn_1^{(1)} + \cdots + \mu_4^{(1)}\, dn_4^{(1)} + \cdots + \mu_c^{(1)}\, dn_c^{(1)} & + \cdots + \\
\lambda_1\, dn_1^{(1)} & + \cdots + \\
\quad \lambda_2\, dn_2^{(1)} & + \cdots + \\
\qquad \lambda_3\, dn_3^{(1)} & + \cdots + \\
\qquad\quad \lambda_4\, dn_4^{(1)} & + \cdots + \\
\qquad\qquad \lambda_5\, dn_5^{(1)} & + \cdots + \\
\qquad\qquad\quad \cdot & \cdot \\
\qquad\qquad\qquad \cdot & \cdot \\
\qquad\qquad\qquad\quad \cdot & \cdot \\
\qquad\qquad\qquad\qquad \lambda_c\, dn_c^{(1)} & + \cdots +
\end{array}$$

chemically active, capable of undergoing the reaction

$$\nu_1 A_1 + \nu_2 A_2 \rightleftharpoons \nu_3 A_3 + \nu_4 A_4 .$$

Suppose that there are φ phases and, as only a temporary assumption, that all the constituents are present in all the phases. As before, the Gibbs function of the system is

$$G = \sum_1^c \mu_k^{(1)} n_k^{(1)} + \sum_1^c \mu_k^{(2)} n_k^{(2)} + \cdots + \sum_1^c \mu_k^{(3)} n_k^{(\varphi)}.$$

The equations of constraint for those constituents which do not react are of the same type as before; i.e., they express the fact that the total number of moles of each inert constituent is constant. In the case of the chemically active constituents, however, the total number of moles of any one is not constant but is a function of the degree of reaction. Hence, the equations of constraint are

$$\begin{aligned}
n_1^{(1)} + n_1^{(2)} + \cdots + n_1^{(\varphi)} &= (n_0 + n_0')\nu_1(1 - \epsilon) \\
n_2^{(1)} + n_2^{(2)} + \cdots + n_2^{(\varphi)} &= (n_0 + n_0')\nu_2(1 - \epsilon) + N_2 \\
n_3^{(1)} + n_3^{(2)} + \cdots + n_3^{(\varphi)} &= (n_0 + n_0')\nu_3 \epsilon \\
n_4^{(1)} + n_4^{(2)} + \cdots + n_4^{(\varphi)} &= (n_0 + n_0')\nu_4 \epsilon + N_4 \\
n_5^{(1)} + n_5^{(2)} + \cdots + n_5^{(\varphi)} &= \text{const.} \\
&\cdots\cdots\cdots\cdots \\
n_c^{(1)} + n_c^{(2)} + \cdots + n_c^{(\varphi)} &= \text{const.,}
\end{aligned}$$

where n_0, n_0', N_2, and N_4 have their usual meaning. Applying Lagrange's method, we get the equations below (reading across pages 410 and 411).

$$\begin{aligned}
&\mu_1^{(\varphi)}\, dn_1^{(\varphi)} + \cdots + \mu_4^{(\varphi)}\, dn_4^{(\varphi)} + \cdots + \mu_c^{(\varphi)}\, dn_c^{(\varphi)} && = 0 \\
&\lambda_1\, dn_1^{(\varphi)} && + \lambda_1(n_0 + n_0')\nu_1\, d\epsilon = 0 \\
&\quad \lambda_2\, dn_2^{(\varphi)} && + \lambda_2(n_0 + n_0')\nu_2\, d\epsilon = 0 \\
&\quad\quad \lambda_3\, dn_3^{(\varphi)} && - \lambda_3(n_0 + n_0')\nu_3\, d\epsilon = 0 \\
&\quad\quad\quad \lambda_4\, dn_4^{(\varphi)} && - \lambda_4(n_0 + n_0')\nu_4\, d\epsilon = 0 \\
&\quad\quad\quad\quad \lambda_c\, dn_5^{(\varphi)} && = 0 \\
&\quad\quad\quad\quad\quad \cdot && \cdot \\
&\quad\quad\quad\quad\quad\quad \cdot && \cdot \\
&\quad\quad\quad\quad\quad\quad\quad \cdot && \cdot \\
&\quad\quad\quad\quad\quad\quad\quad\quad \lambda_c\, dn_c^{(\varphi)} && = 0
\end{aligned}$$

Adding, and equating coefficients of the dn's to zero, we get the usual $c(\varphi - 1)$ equations of phase equilibrium,

$$\mu_1^{(1)} = \mu_1^{(2)} = \cdots = \mu_1^{(\varphi)}$$
$$\mu_2^{(1)} = \mu_2^{(2)} = \cdots = \mu_2^{(\varphi)}$$
$$\cdots\cdots\cdots\cdots\cdots\cdots$$
$$\mu_c^{(1)} = \mu_c^{(2)} = \cdots = \mu_c^{(\varphi)}.$$

Equating the coefficient of $d\epsilon$ to zero, we get an extra equation of equilibrium, namely,

$$\lambda_1 \nu_1 + \lambda_2 \nu_2 - \lambda_3 \nu_3 - \lambda_4 \nu_4 = 0,$$

which, since $\lambda_1 = -\mu_1$ of any phase, $\lambda_2 = -\mu_2$ of any phase, etc., becomes

$$\nu_1 \mu_1 + \nu_2 \mu_2 = \nu_3 \mu_3 + \nu_4 \mu_4 .$$

This equation will be recognized as the equation of reaction equilibrium, which in the case of ideal gases was found to lead to the law of mass action.

The rest of the argument follows the same lines as before. There are $c(\varphi - 1)$ equations of phase equilibrium, 1 equation of reaction equilibrium, and φ equations of the type $\sum x = 1$. Hence, the total number of equations is

$$c(\varphi - 1) + 1 + \varphi.$$

Since the variables are the same as before, namely, T, P, and the x's, of which there are $c\varphi + 2$ in number, the variance is

$$f = c\varphi + 2 - [c(\varphi - 1) + 1 + \varphi],$$

or

$$f = (c - 1) - \varphi + 2. \tag{16-6}$$

The phase rule in this case is seen to be different, in that $c - 1$ now stands where c formerly stood. For the reason given before, this form of the phase rule remains unchanged when every constituent is not present in every phase. We see therefore that when there are c constituents, present in arbitrary amounts, and only one chemical reaction, there is one extra equation of equilibrium and the variance is reduced by 1. It is obvious that, if there were two independent chemical reactions, there would be two extra equations of equilibrium; whence the phase rule would become $f = (c - 2) - \varphi + 2$. For r independent reactions, we would have

$$f = (c - r) - \varphi + 2. \tag{16-7}$$

The argument up to now has been based on the fact that only three kinds of equations have existed among the variables T, P, and the x's: equations of phase equilibrium, equations of reaction equilibrium, and equations of the type $\sum x = 1$. It often happens, however, that a chemical reaction takes place in such a manner that additional equations expressing further restrictions upon the x's are at hand. Suppose, for example, that we put an arbitrary amount of solid NH_4HS into an evacuated chamber and two new constituents form according to the reaction

$$NH_4HS \rightleftharpoons NH_3 + H_2S.$$

Since gaseous NH_3 and H_2S are in the same phase, the restriction always exists that

$$x_{NH_3} = x_{H_2S}.$$

This constitutes a fourth type of equation, to be added to the three listed at the beginning of this paragraph.

Another example of additional restricting equations among the x's is provided by the phenomenon of dissociation in solution. Suppose that we have a heterogeneous system one of whose phases is a solution of salt 2 in solvent 1. Suppose that the salt dissociates according to the scheme shown in Fig. 16-3*a*. All the ions, of course, remain in the liquid phase, and no precipitate is formed. Consequently, we have the equation

$$x_3 = x_4,$$

which expresses the fact that the solution is electrically neutral. If multiple dissociation takes place according to the scheme shown in Fig. 16-3*b*, then there are three independent restricting equations among the x's, namely,

$$x_3 = x_8 + x_6 + x_4,$$

$$x_5 = x_8 + x_6,$$

and

$$x_7 = x_8.$$

Adding these equations, we get the *dependent* equation

$$x_3 + x_5 + x_7 = 3x_8 + 2x_6 + x_4,$$

expressing the fact of electrical neutrality.

Let us call equations of the preceding type *restricting equations*, and let us

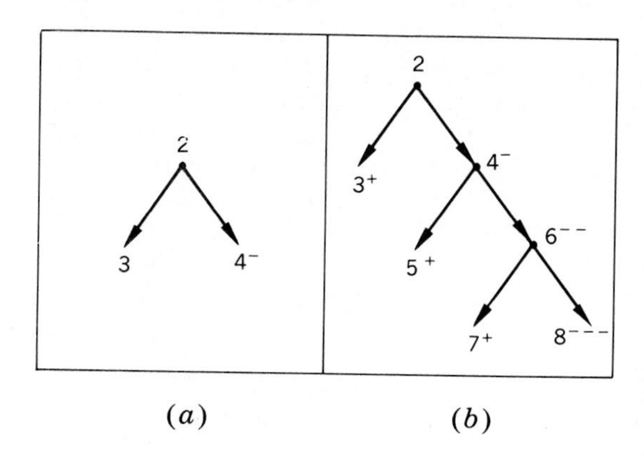

Figure 16-3 Dissociations of a salt that give rise to additional restricting equations among the mole fractions. (*a*) Single dissociation; (*b*) multiple dissociation.

suppose that z of them are independent. Then we may list four types of equations among T, P, and the x's, thus:

1. Equations of phase equilibrium [$c(\varphi - 1)$ in number].
2. Equations of reaction equilibrium (r in number).
3. Equations of the type $\sum x = 1$ (φ in number).
4. Restricting equations (z in number).

Hence, the total number of equations is

$$c(\varphi - 1) + r + \varphi + z;$$

and, as usual, the total number of variables is $c\varphi + 2$. Therefore,

$$f = c\varphi + 2 - [c(\varphi - 1) + r + \varphi + z],$$

or

$$f = (c - r - z) - \varphi + 2.$$

If we define the *number of components* c' *as the total number of constituents minus the number of independent reactions minus the number of independent restricting equations, i.e.,*

$$\boxed{c' = c - r - z,} \tag{16-8}$$

we may always write the phase rule in the same form, thus:

$$\boxed{f = c' - \varphi + 2.} \tag{16-9}$$

16-5 DETERMINATION OF THE NUMBER OF COMPONENTS

The problem of determining the number of components in a heterogeneous system may be somewhat difficult for the beginner. As a result of experience with the behavior of typical heterogeneous systems, the physical chemist is able to determine the number of components by counting the smallest number of constituents whose specification is sufficient to determine the composition of every phase. The validity of this working rule rests upon a few fundamental facts whose truth we shall demonstrate rigorously in this article.

Example 1 First let us consider a heterogeneous system consisting of a liquid phase composed of a solution of the salt NaH_2PO_4 in water and a vapor phase composed of water vapor. It is important to show that, *so long as no precipitate is formed by virtue of a reaction between the salt and the water, no matter what else we assume to take place in the solution, the number of components is two.*

1. *Neglecting all dissociation.* There are two constituents, no chemical reactions, and no restricting equations. Hence,

$$c' = 2 - 0 - 0 = 2.$$

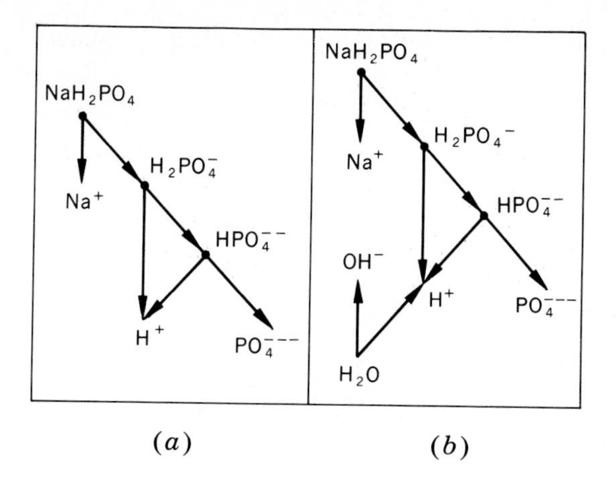

Figure 16-4 Multiple dissociation of NaH_2PO_4. (*a*) Without dissociation of H_2O; (*b*) with dissociation of H_2O.

2. *Assuming single dissociation of the salt.* There are four constituents, NaH_2PO_4, H_2O, Na^+, $H_2PO_4^-$; one chemical reaction,

$$NaH_2PO_4 \rightleftharpoons Na^+ + H_2PO_4^-;$$

and one restricting equation,

$$x_{Na^+} = x_{H_2PO_4^-}$$

Hence, $c = 4 - 1 - 1 = 2$.

3. *Assuming multiple dissociation of the salt.* There are seven constituents and three independent chemical reactions, as shown in Fig. 16-4*a*, and two independent restricting equations,

$$x_{Na^+} = x_{PO_4{}^{3-}} + x_{HPO_4{}^{2-}} + x_{H_2PO_4{}^-}$$

and

$$x_{H^+} = 2x_{PO_4{}^{3-}} + x_{HPO_4{}^{2-}}.$$

(By adding the two equations, the *dependent* equation expressing electrical neutrality of the solution is obtained.) Hence, $c' = 7 - 3 - 2 = 2$.

4. *Assuming dissociation of the water also.* There are eight constituents and four independent chemical reactions, as shown in Fig. 16-4*b*, and two independent restricting equations,

$$x_{Na^+} = x_{PO_4{}^{3-}} + x_{HPO_4{}^{2-}} + x_{H_2PO_4{}^-}$$

and

$$x_{H^+} = 2x_{PO_4{}^{3-}} + x_{HPO_4{}^{2-}} + x_{OH^-}.$$

Hence, $c' = 8 - 4 - 2 = 2$.

5. *Assuming association of the water.* A ninth constituent, $(H_2O)_2$, is formed as a result of a fifth independent reaction,

$$H_2O \rightleftharpoons \tfrac{1}{2}(H_2O)_2.$$

There are still the same two independent restricting equations, and hence $c' = 9 - 5 - 2 = 2$.

It is clear, therefore, that it is a matter of indifference as to what chemical changes take place in the solution. The number of components is always two, provided that no precipitate forms.

Example 2 To investigate the effect of precipitate, let us consider a mixture of $AlCl_3$ and water. In this case, the $AlCl_3$ combines with the water to form $Al(OH)_3$, some of which precipitates out of the solution, according to the reaction shown in Fig. 16-5. There are eight constituents and only four independent reactions. At first thought, one might imagine that there are five

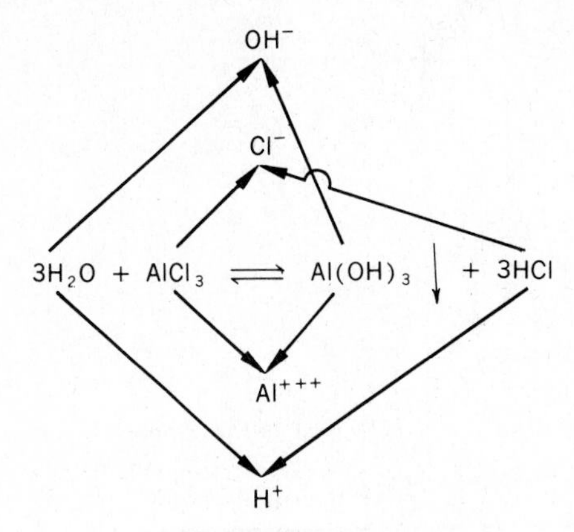

Figure 16-5 Dissociation, reaction, and precipitation that occur when $AlCl_3$ is dissolved in water.

independent reactions, but if we write the equations of reaction equilibrium corresponding to the four dissociations:

$$3\mu_{H_2O} = 3\mu_{H^+} + 3\mu_{OH^-},$$

$$\mu_{AlCl_3} = \mu_{Al^{3+}} + 3\mu_{Cl^-},$$

$$\mu_{Al(HO)_3} = \mu_{Al^{3+}} + 3\mu_{OH^-},$$

and

$$3\mu_{HCl} = 3\mu_{H^+} + 3\mu_{Cl^-},$$

and add the first two and subtract the sum of the last two, we get

$$3\mu_{H_2O} + \mu_{AlCl_3} = \mu_{Al(OH)_3} + 3\mu_{HCl},$$

which is the equation of reaction equilibrium corresponding to the reaction

$$3H_2O + AlCl_3 \rightleftharpoons Al(OH)_3 + 3HCl,$$

which is therefore seen to be a *dependent* reaction.

Since some of the $Al(OH)_3$ has precipitated out, there is only one restricting equation, namely, that expressing the electrical neutrality of the solution,

$$3x_{Al^{3+}} + x_{H^+} = x_{OH^-} + x_{Cl^-}.$$

Consequently, $c' = 8 - 4 - 1 = 3$, and we have an interesting situation in which a heterogeneous system, formed originally by mixing two substances, has three components.

Example 3 As a last example, let us consider a system consisting of water vapor and a solution containing arbitrary amounts of NaCl and KNO_3 in water.

1. *Neglecting all dissociations.* There are five constituents, H_2O, NaCl, KNO_3, $NaNO_3$, KCl; one reaction,

$$NaCl + KNO_3 \rightleftharpoons NaNO_3 + KCl;$$

and one restricting equation,

$$x_{NaNO_3} = x_{KCl}.$$

Hence, $c' = 5 - 1 - 1 = 3$.

2. *Considering all reactions.* There are eleven constituents, which undergo these reactions:

$$H_2O \rightleftharpoons H^+ + OH^-$$

$$NaCl \rightleftharpoons Na^+ + Cl^-$$

$$KNO_3 \rightleftharpoons K^+ + NO_3^-$$

$$NaNO_3 \rightleftharpoons Na^+ + NO_3^-$$

$$KCl \rightleftharpoons K^+ + Cl^-.$$

It should be noticed that the reaction

$$NaCl + KNO_3 \rightleftharpoons NaNO_3 + KCl$$

is not independent of the preceding five but that its equation of reaction equilibrium may be obtained by adding the second and third and subtracting the sum of the fourth and fifth.

There are three restricting equations. The first,

$$x_{Na^+} + x_{NaNO_3} = x_{Cl^-} + x_{KCl},$$

expresses the fact that the amount of sodium lost by the NaCl (to form Na^+ and $NaNO_3$) is equal to the amount of chlorine lost by the NaCl (to form Cl^- and KCl). The second,

$$x_{K^+} + x_{KCl} = x_{NO_3} + x_{NaNO_3},$$

expresses the corresponding fact concerning the loss of potassium and nitrate from KNO_3. The third is

$$x_{H^+} = x_{OH^-}.$$

(The dependent equation of electrical neutrality is obtained by adding these three equations.) Hence, $c' = 11 - 5 - 3 = 3$.

In the event that we start with arbitrary amounts of all five substances (H_2O, NaCl, KNO_3, $NaNO_3$, and KCl), there are still eleven constituents and five independent reactions but only two restricting equations, namely,

$$x_{Na^+} + x_{K^+} + x_{H^+} = x_{Cl^-} + x_{NO_3^-} + x_{OH^-},$$

which expresses electrical neutrality, and

$$x_{H^+} = x_{OH^-}$$

Hence, $c' = 11 - 5 - 2 = 4$.

16-6 DISPLACEMENT OF EQUILIBRIUM

Consider a heterogeneous system of φ phases and c constituents, four of which undergo the reaction

$$\nu_1 A_1 + \nu_2 A_2 \rightleftharpoons \nu_3 A_3 + \nu_4 A_4.$$

Any infinitesimal process involving a change in temperature, pressure, and composition of the phases is accompanied by a change in the Gibbs function equal to

$$dG = -S\,dT + V\,dP + \mu_1^{(1)}\,dn_1^1 + \cdots + \mu_c^{(1)}\,dn_c^{(1)} + \cdots + \mu_1^{(\varphi)}\,dn_1^{(\varphi)} + \cdots + \mu_c^{(\varphi)}\,dn_c^{(\varphi)}.$$

In general, during such an infinitesimal change, there is neither equilibrium among the phases nor equilibrium with regard to the chemical reaction. Complete chemical equilibrium would require both phase equilibrium and reaction equilibrium. *Suppose we assume phase equilibrium only.* Then

$$\mu_1^{(1)} = \mu_1^{(2)} = \cdots = \mu_1^{(\varphi)}$$
$$\mu_2^{(2)} = \mu_2^{(2)} = \cdots = \mu_2^{(\varphi)}$$
$$\cdots\cdots\cdots\cdots\cdots\cdots\cdots$$
$$\mu_c^{(1)} = \mu_c^{(2)} = \cdots = \mu_c^{(\varphi)},$$

and the change in the Gibbs function becomes

$$dG = -\varphi S\,dT + V\,dP + \mu_1(dn_1^{(1)} + \cdots + dn_1^{(\varphi)}) + \cdots + \mu_4(dn_4^{(1)} + \cdots + dn_4^{(\varphi)}) + \mu_5(dn_5^{(1)} + \cdots + dn_5^{(\varphi)}) + \cdots + \mu_c(dn_c^{(1)} + \cdots + dn_c^{(\varphi)}).$$

But

$$dn_1^{(1)} + \cdots + dn_1^{(\varphi)} = -(n_0 + n_0')\nu_1\,d\epsilon$$
$$dn_2^{(1)} + \cdots + dn_2^{(\varphi)} = -(n_0 + n_0')\nu_2\,d\epsilon$$
$$dn_3^{(1)} + \cdots + dn_3^{(\varphi)} = +(n_0 + n_0')\nu_3\,d\epsilon$$
$$dn_4^{(1)} + \cdots + dn_4^{(\varphi)} = +(n_0 + n_0')\nu_4\,d\epsilon$$
$$dn_5^{(1)} + \cdots + dn_5^{(\varphi)} = 0$$

$$dn_c^{(1)} + \cdots + dn_c^{(\varphi)} = 0.$$

Therefore, the change in the Gibbs function during an infinitesimal process in which there is phase equilibrium but not reaction equilibrium is given by

$$dG = -S\,dT + V\,dP + (n_0 + n_0')(\nu_3\mu_3 + \nu_4\mu_4 - \nu_1\mu_1 - \nu_2\mu_2)\,d\epsilon.$$

Since, under these circumstances G is a function of T, P, and ϵ, it follows that

$$\frac{\partial G}{\partial T} = -S,$$

$$\frac{\partial G}{\partial P} = V,$$

and

$$\frac{\partial G}{\partial \epsilon} = (n_0 + n_0')(\nu_3\mu_3 + \nu_4\mu_4 - \nu_1\mu_1 - \nu_2\mu_2).$$

When reaction equilibrium exists at temperature T and pressure P, we must have $\partial G/\partial\epsilon = 0$ at $\epsilon = \epsilon_e$. If we go to a slightly different equilibrium state at temperature $T + dT$ and pressure $P + dP$, then the new degree of reaction will be $\epsilon_e + d\epsilon_e$ and the change in $\partial G/\partial\epsilon$ during this process is zero. Therefore,

$$d\left(\frac{\partial G}{\partial \epsilon}\right) = 0.$$

But
$$d\left(\frac{\partial G}{\partial \epsilon}\right) = \frac{\partial^2 G}{\partial T\,\partial \epsilon}\,dT + \frac{\partial^2 G}{\partial P\,\partial \epsilon}\,dP + \frac{\partial^2 G}{\partial \epsilon^2}\,d\epsilon = 0$$
$$= \frac{\partial}{\partial \epsilon}\left(\frac{\partial G}{\partial T}\right) dT + \frac{\partial}{\partial \epsilon}\left(\frac{\partial G}{\partial P}\right) dP + \frac{\partial^2 G}{\partial \epsilon^2}\,d\epsilon = 0$$
$$= -\frac{\partial S}{\partial \epsilon}\,dT + \frac{\partial V}{\partial \epsilon}\,dP + \frac{\partial^2 G}{\partial \epsilon^2}\,d\epsilon = 0.$$

Solving for $d\epsilon = d\epsilon_e$, we get

$$d\epsilon_e = \frac{\partial S/\partial \epsilon}{\partial^2 G/\partial \epsilon^2}\,dT - \frac{\partial V/\partial \epsilon}{\partial^2 G/\partial \epsilon^2}\,dP. \tag{16-10}$$

Recognizing that, at thermodynamic equilibrium, $đQ = T\,dS$ or

$$\left(\frac{đQ}{d\epsilon}\right)_{T,P} = T\left(\frac{\partial S}{\partial \epsilon}\right)_{T,P}, \tag{16-11}$$

we get
$$\left(\frac{\partial \epsilon_e}{\partial T}\right)_P = \frac{(đQ/d\epsilon)_{T,P}}{T(\partial^2 G/\partial \epsilon^2)_{T,P}} \tag{16-12}$$

$$\left(\frac{\partial \epsilon_e}{\partial P}\right)_T = -\frac{(\partial V/\partial \epsilon)_{T,P}}{(\partial^2 G/\partial \epsilon^2)_{T,P}}. \tag{16-13}$$

Since G is a minimum at thermodynamic equilibrium, $\partial^2 G/\partial \epsilon^2$ is positive. Equation (16-12) therefore states that an increase of temperature at constant pressure always causes a reaction to proceed in the direction in which heat is absorbed at constant T and P; whereas from Eq. (16-13), we see that an increase of pressure at constant temperature causes a reaction to proceed in the direction in which the volume decreases at constant T and P.

PROBLEMS

16-1 All the lettered points in Fig. P16-1 lie in one plane. The line CD separates the plane into two regions: on the left a wave has the speed v and on the right the speed v'. Show by Lagrange's method that the time for the wave to travel the path APB is a minimum when $v/v' = \sin\phi/\sin\phi'$.

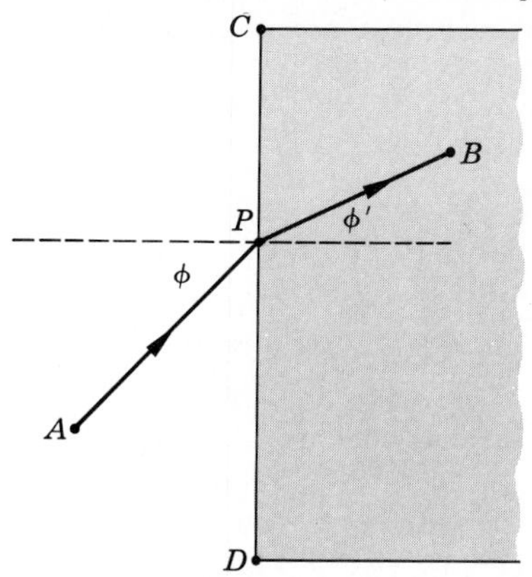

Figure P16-1

16-2 A hot metal of mass m, specific heat c_P, and temperature T_i is immersed in a cooler liquid of mass m', specific heat c'_P, and temperature T'_i. The entire system is thermally insulated. If the final temperature of the metal is T_f and that of the liquid is T'_f, show by Lagrange's method that the condition for the entropy change of the universe to be a maximum is that $T_f = T'_f$.

16-3 Consider a homogeneous mixture of four ideal gases capable of undergoing the reaction

$$\nu_1 A_1 + \nu_2 A_2 \rightleftharpoons \nu_3 A_3 + \nu_4 A_4.$$

How many components are there if one starts with:

(*a*) Arbitrary amounts of A_1 and A_2 only.
(*b*) Arbitrary amounts of all four gases.
(*c*) ν_1 moles of A_1 and ν_2 moles of A_2 only.

16-4 Consider a system composed of a solid phase of calcium carbonate ($CaCO_3$), a solid phase of calcium oxide (CaO), and a gaseous phase consisting of a mixture of CO_2, $CaCO_3$ vapor, and CaO vapor, all three constituents being present initially in arbitrary amounts. These are the substances which are present in a limekiln, where the reaction

$$CaCO_3 \rightleftharpoons CaO + CO_2$$

takes place.

(*a*) How many components are there, and what is the variance?
(*b*) Assuming the gaseous phase to be a mixture of ideal gases, show that

$$\frac{p_{NH_3} p_{H_2S}}{p_{NH_4HS}} = K.$$

(*c*) If solid $CaCO_3$ is introduced into an evacuated chamber, how many components are there, and what is the variance?

16-5 Solid ammonium hydrosulphide (NH_4HS) is mixed with arbitrary amounts of gaseous NH_3 and H_2S, forming a three-constituent system of two phases, undergoing the reaction

$$NH_4HS \rightleftharpoons NH_3 + H_2S.$$

(*a*) How many components are there, and what is the variance?
(*b*) Assuming that the gaseous phase is a mixture of ideal gases, show that

$$\frac{p_{NH_3} p_{H_2S}}{p_{NH_4HS}} = K.$$

(*c*) If solid NH_4HS is placed in an evacuated chamber, how many components are there, and what is the variance?

16-6 How many components are there in a system composed of arbitrary amounts of water, sodium chloride, and barium chloride?

16-7 At high temperature the following reactions take place:

$$C + CO_2 \rightleftharpoons 2CO$$

and $CO_2 + H_2 \rightleftharpoons CO + H_2O$. How many components are there if we start with

(*a*) Arbitrary amounts of C, CO_2, and H_2?
(*b*) Arbitrary amounts of C, CO_2, H_2, CO, and H_2O?

16-8 Consider a system consisting of a pure liquid phase in equilibrium with a gaseous phase composed of a mixture of the vapor of the liquid and an inert gas that is insoluble in the liquid. Suppose that the inert gas (sometimes called the *foreign gas*) can flow into or out of the gaseous phase, so that the total pressure can be varied at will.

(*a*) How many components are there, and what is the variance?

(*b*) Assuming the gaseous phase to be a mixture of ideal gases, show that

$$g'' = RT(\phi - \ln p),$$

where g'' is the molar Gibbs function of the liquid, and ϕ and p refer to the vapor.

(*c*) Suppose that a little more foreign gas is added, thus increasing the pressure from P to $P + dP$, at constant temperature. Show that

$$v''\,dP = RT\frac{dp}{p},$$

where v'' is the molar volume of the liquid, which is practically constant.

(*d*) Integrating at constant temperature from an initial state where there is no foreign gas to a final state where the total pressure is P and the partial vapor pressure is p, show that

$$\ln\frac{p}{P_0} = \frac{v''}{RT}(P - P_0) \qquad \text{(Gibbs equation)},$$

where P_0 is the vapor pressure when no foreign gas is present.

(*e*) In the case of water at 0°C, at which $P_0 = 4.57$ mm Hg, show that, when there is sufficient air above the water to make the total pressure equal to 10 atm, $p = 4.61$ mm Hg.

16-9 The Gibbs function G of a liquid phase consisting of a solvent and very small amounts of several solutes is

$$G = \mu_0 n_0 + \mu_1 n_1 + \mu_2 n_2 + \cdots,$$

where the subscript zero refers to the solvent, and

$$\mu_k = g_k + RT\ln x_k.$$

(*a*) Using the relation $V = (\partial G/\partial P)_T$, show that

$$V = \sum n_k v_k.$$

(*b*) Using the relation $H = G - T(\partial G/\partial T)_P$, show that

$$H = \sum n_k h_k,$$

which means that there is no heat of dilution.

16-10 A very small amount of sugar is dissolved in water, and the solution is in equilibrium with pure water vapor.

(*a*) Show that the equation of phase equilibrium is

$$g''' = g'' + RT\ln(1 - x),$$

where g''' is the molar Gibbs function of water vapor, g'' the molar Gibbs function of pure liquid water, and x the mole fraction of the sugar in solution.

(*b*) For an infinitesimal change in x at constant temperature, show that

$$(v''' - v'')\,dP = RT\,d\ln(1 - x).$$

(*c*) Assuming the vapor to behave like an ideal gas and regarding v'' as constant, integrate the preceding equation at constant temperature from an initial state $x = 0$, $P = P_0$, to a final state $x = x$, $P = P_x$, and derive

$$\ln\frac{P_x}{P_0} = \ln(1 - x) + \frac{v''}{RT}(P_x - P_0).$$

P_0 is the vapor pressure of the pure liquid, and P_x the vapor pressure of the dilute solution.

(*d*) Justify neglecting the last term on the right, and show that

$$P_x = P_0(1 - x) \qquad \text{(Raoult's law)},$$

or

$$\frac{P_0 - P_x}{P_0} = x.$$

16-11 Consider the system of Prob. 16-10, and let x stand for the mole fraction of the sugar.

(*a*) For an infinitesimal change in x at constant pressure, show that

$$-s''' \, dT = -s'' \, dT + R \ln (1 - x) \, dT + RT \, d \ln (1 - x).$$

(*b*) Substituting for $R \ln (1 - x)$ the value obtained from the equation of phase equilibrium, show that (*a*) reduces to

$$0 = \frac{h''' - h''}{T} \, dT + RT \, d \ln (1 - x).$$

(*c*) Taking into account that $x \ll 1$ and calling $h''' - h''$ the latent heat of vaporization l_V, show that the elevation of the boiling point is

$$\Delta T = \frac{RT^2}{l_V} x.$$

16-12 A very small amount of sugar is dissolved in water, and the solution is in equilibrium with pure ice. The equation of phase equilibrium is

$$g' = g'' + RT \ln (1 - x),$$

where g' = molar Gibbs function of pure ice,
g'' = molar Gibbs function of pure water,
and x = mole fraction of sugar in solution.

(*a*) For an infinitesimal change in x at constant pressure, show that

$$-s' \, dT = -s'' \, dT + R \ln (1 - x) \, dT + RT \, d \ln (1 - x).$$

(*b*) Substituting for $R \ln (1 - x)$ the value obtained from the equation of phase equilibrium, show that (*a*) reduces to

$$\frac{h'' - h'}{T} \, dT = RT \, d \ln (1 - x).$$

(*c*) Taking into account that $x \ll 1$ and calling $h'' - h'$ the heat of fusion l_F, show that the depression of the freezing point is

$$\Delta T = -\frac{RT^2}{l_F} x.$$

16-13 In the osmotic pressure apparatus depicted in Fig. P16-2, let the pressure of the pure solvent be P_0 and that of the dilute solution be P, the temperature being T throughout. The molar Gibbs function of the pure solvent is g''.

(*a*) Show that, at equilibrium,

$$g''(T, P_0) = g''(T, P) + RT \ln(1 - x),$$

where x is the mole-fraction of the solute.

(*b*) For an infinitesimal change of x at constant T, show that

$$0 = v'' \, dP + RTd \ln(1 - x).$$

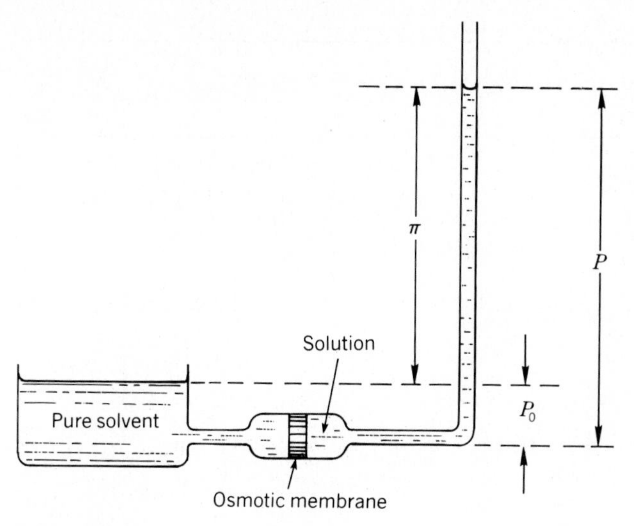

Figure P16-2

(*c*) Integrating P from P_0 to P, and x from 0 to x, show that

$$\Pi v'' = xRT.$$

Compare this equation with the ideal gas law.

CHAPTER

SEVENTEEN

SPECIAL TOPICS

17-1 STRETCHED WIRE

In the case of an infinitesimal reversible process of a stretched wire, the two laws of thermodynamics yield the equation

$$T\,dS = dU - \mathcal{J}\,dL.$$

Therefore, to obtain any desired equation for a stretched wire, it is necessary merely to choose the corresponding equation for a hydrostatic system and replace P by $-\mathcal{J}$ and V by L. Thus, the second $T\,dS$ equation becomes

$$T\,dS = C_{\mathcal{J}}\,dT + T\left(\frac{\partial L}{\partial T}\right)_{\mathcal{J}} d\mathcal{J}. \tag{17-1}$$

The application of this equation to isothermal and adiabatic processes is exemplified by some of the problems at the end of the chapter.

17-2 SURFACE FILM

To obtain any desired equation for a surface film, it is necessary merely to replace V by A and P by $-\mathcal{S}$ in the corresponding equation for a hydrostatic system. Thus, the first $T\,dS$ equation becomes

$$T\,dS = C_A\,dT - T\left(\frac{\partial \mathcal{S}}{\partial T}\right)_A dA. \tag{17-2}$$

Imagine a small amount of liquid in the form of a droplet with a very small surface film. Suppose that, with the aid of a suitable wire framework, the surface film is expanded isothermally until the area A is very much larger than the original value. Then, *if the surface tension is a function of the temperature only*, the heat transferred is

$$Q = -T\frac{d\mathscr{S}}{dT}(A - 0),$$

and the work done is

$$W = \mathscr{S}(A - 0).$$

From the first law,

$$U - U_0 = \left(\mathscr{S} - T\frac{d\mathscr{S}}{dT}\right)A,$$

where U_0 is the energy of the liquid with practically no surface and U is the energy of the liquid with the surface of area A. Hence,

$$\boxed{\frac{U - U_0}{A} = \mathscr{S} - T\frac{d\mathscr{S}}{dT}.} \tag{17-3}$$

The left-hand member is interpreted as the energy per unit area associated with the surface only, i.e., the *surface energy per unit area*. It is seen that $(U - U_0)/A$ has the same dimensions as $\mathscr{S}$; that is, joules per square meter equals newtons per meter. With the aid of Eq. (17-3), the surface energy per unit area of a film may be calculated once the surface tension has been measured as a function of temperature.

Table 17-1 Surface tension, surface energy, and heat of vaporization of water†

T, K	$\mathscr{S}$, dyn/cm	$\frac{U - U_0}{A}$, ergs/cm^2	l_V, kJ/kg
273	75.50	143	2501
373	58.91	138	2257
473	37.77	129	1939
523	26.13	122	1714
573	14.29	111	1403
623	3.64	80	893
647	0	0	0

† One dyn/cm equals 10^{-3} N/m; 1 erg/cm^2 equals 10^{-3} J/m^2.

At the critical temperature, the surface tension of all liquids is zero. The surface tension of a pure liquid in equilibrium with its vapor can usually be represented by a formula of the type

$$\mathscr{S} = \mathscr{S}_0\left(1 - \frac{t}{t_C}\right)^n,$$

where $\mathscr{S}_0$ is the surface tension at 0°C, t_C is the critical temperature, and n is a constant between 1 and 2. For example, in the case of water, $\mathscr{S}_0 = 75.5$ dyn/cm, $t_C = 374$°C, and $n = 1.2$. Since $d\mathscr{S}/dT$ is the same as $d\mathscr{S}/dt$, all the quantities necessary to calculate $(U - U_0)/A$ are at hand. The data for water are given in Table 17-1.

17-3 REVERSIBLE CELL

Equations for a reversible cell composed of solids and liquids only may be obtained from corresponding equations for a chemical system by replacing V by Z and P by $-\mathcal{E}$. Thus, the first $T\,dS$ equation becomes

$$T\,dS = C_Z\,dT - T\left(\frac{\partial \mathcal{E}}{\partial T}\right)_Z dZ; \tag{17-4}$$

and for a saturated reversible cell whose emf depends on the temperature only, the equation becomes

$$T\,dS = C_Z\,dT - T\frac{d\mathcal{E}}{dT}\,dZ.$$

In the case of a *reversible isothermal transfer of a quantity of electricity* $Z_f - Z_i$,

$$Q = -T\frac{d\mathcal{E}}{dT}(Z_f - Z_i),$$

where $Z_f - Z_i$ is negative when positive electricity is transferred externally from the positive to the negative electrode. During this process, the cell delivers an amount of work

$$W = \mathcal{E}(Z_f - Z_i).$$

If jN_F coulombs of positive electricity are transferred from positive to negative externally, where j is the valence and N_F is Faraday's constant, then

$$Z_f - Z_i = -jN_F;$$

whence

$$Q = jN_F T\frac{d\mathcal{E}}{dT},$$

and

$$W = -jN_F\ .$$

From the first law, the change in internal energy is

$$U_f - U_i = -jN_F\left(\mathcal{E} - T\frac{d\mathcal{E}}{dT}\right).$$

When a process takes place at constant pressure with a negligible volume change, the change of internal energy is equal to the change of enthalpy. For

$$H = U + PV$$

and

$$dH = dU + P\,dV + V\,dP,$$

whence, under the conditions mentioned,

$$dH = dU.$$

For the reversible transfer of jN_F coulombs of electricity through a reversible cell whose volume does not change appreciably at constant atmospheric pressure, we may therefore write

$$H_f - H_i = -jN_F\left(\mathcal{E} - T\frac{d\mathcal{E}}{dT}\right).$$

In order to interpret this change of enthalpy, let us take the particular case of the Daniell cell. The transfer of positive electricity externally from the copper to the zinc electrode is accompanied by the reaction

$$Zn + CuSO_4 \rightarrow Cu + ZnSO_4.$$

When jN_F coulombs are transferred, 1 mol of each of the initial constituents disappears, and 1 mol of each of the final constituents is formed. The change of enthalpy in this case is equal to the enthalpy of 1 mol of each of the final constituents minus the enthalpy of 1 mol of each of the initial constituents at the same temperature and pressure. This is called the *heat of reaction* and is denoted by ΔH. Therefore,

$$\Delta H = -jN_F\left(\mathcal{E} - T\frac{d\mathcal{E}}{dT}\right). \tag{17-5}$$

In the case of a saturated reversible cell in which gases are liberated, it can be shown rigorously (see Prob. 17-20) that

$$\boxed{\Delta H = -jN_F\left[\mathcal{E} - T\left(\frac{\partial\mathcal{E}}{\partial T}\right)_P\right].} \tag{17-6}$$

The important feature of this equation is that it provides a method of measuring the heat of reaction of a chemical reaction without resorting to calorimetry. If the reaction can be made to proceed in an electric cell, all that is necessary is to measure the emf of the cell as a function of the temperature at constant atmospheric pressure. The heat of reaction is therefore measured with a potentiometer and a thermometer. Both measurements can be made with great

Table 17-2 Reversible cells (N_F = 96,500 C/mol)

Reaction	T, K	Valence, j	Emf $\mathscr{E}$, V	$\frac{d\mathscr{E}}{dT}$ mV/K	ΔH (electric method) kJ/mol	ΔH (calorimetric method), kJ/mol
$Zn + CuSO_4 = Cu + ZnSO_4$	273	2	1.0934	−0.453	−235	−232
$Zn + 2AgCl = 2Ag + ZnCl_2$	273	2	1.0171	−0.210	−207	−206
$Cd + 2AgCl = 2Ag + CdCl_2$	298	2	0.6753	−0.650	−168	−165
$Pb + AgI = 2Ag + PbI_2$	298	2	0.2135	−0.173	− 51.1	− 51.1
$Ag + \frac{1}{2}Hg_2Cl_2 = Hg + AgCl$	298	1	0.0455	+0.338	+ 5.45	+ 3.77
$Pb + Hg_2Cl_2 = 2Hg + PbCl_2$	298	2	0.5356	+0.145	− 96.0	− 98.0
$Pb + 2AgCl = 2Ag + PbCl_2$	298	2	0.4900	−0.186	−105	−104

accuracy; hence this method yields by far the most accurate values of the heat of reaction. It is interesting to compare values of ΔH obtained electrically with those measured calorimetrically. This is shown for a number of cells in Table 17-2. Most of the values of ΔH are negative. A negative ΔH indicates a rejection of heat, i.e., an exothermic reaction.

17-4 FUEL CELL

In the reversible cells listed in Table 17-2 and in the well-known automobile battery, the reactants and the products of reaction are stored within the cell itself. In the modern development called a *fuel cell*, the reactants are fed to the cell in a steady stream and the products are continually withdrawn, while the cell delivers a steady current to an external load. A typical fuel cell is depicted schematically in Fig. 17-1. The electrolyte is an aqueous solution of potassium hydroxide chosen to provide an abundant supply of hydroxyl ions. The electrodes are of metal or carbon specially baked, impregnated, and treated so that multitudinous pores are produced, each of which is a few microns in diameter. The pores are small enough to allow the electrolyte to be drawn in by capillary action and to prevent the hydrogen present at one electrode and the air at the other electrode from blowing through.

Within the pores of the negative electrode, hydrogen gas is absorbed in the form of hydrogen atoms, which combine with OH ions to form water and electrons. The electrons move through the electrode to the external circuit while the water mixes with the hydrogen stream and is removed. Within the pores of the positive electrode, adsorbed oxygen molecules combine with water and electrons to form hydroxyl ions and perhydroxyl ions O_2H^-. By action of a suitable catalyst, the perhydroxyl ions decompose into an additional hydroxyl ion and an oxygen atom. The oxygen atoms thus formed unite to form oxygen molecules.

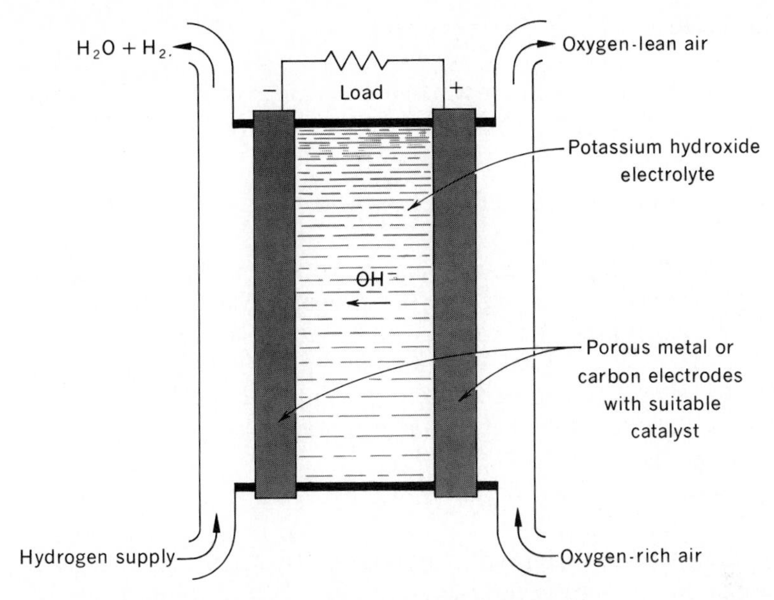

Figure 17-1 Schematic diagram of a hydrogen-oxygen fuel cell.

The net effect of the reactions within the pores of both electrodes is the reaction

$$H_2 + \tfrac{1}{2}O_2 = H_2O,$$

for which the following quantities apply: The heat of reaction ΔH at 1 atm and 298 K is -286 kJ/mol. To maintain isothermal conditions at 298 K, heat Q must be transferred between the cell and its surroundings equal to -48 kJ/mol. The work output $W(=jN_F\mathcal{E})$ of the fuel cell is therefore

$$jN_F\mathcal{E} = Q - \Delta H$$

$$= (-48 + 286)\ \text{kJ/mol};$$

whence

$$\mathcal{E} = \frac{238{,}000\ \text{J/mol}}{2 \times 96{,}500\ \text{C/mol}}$$

$$= 1.23\ \text{V}.$$

The actual terminal voltage of this cell when delivering current is lower than 1.23 V, not only because of the Ir drop within the cell but also because of the failure to achieve isothermal conditions. The operation of the cell requires that the electrodes have pores of just the right size, that they be impregnated with just the right catalyst to initiate the decomposition of perhydroxyl ions, and that no impurities poison the surfaces and stop the reaction. To solve all these problems is a very difficult task, and much work remains to be done toward their solution.

17-5 DIELECTRIC

The thermodynamic equations appropriate to a dielectric may be easily obtained from those of a hydrostatic system by replacing P by $-E$, and V by Π. Thus, the second $T\,dS$ equation becomes

$$T\,dS = C_E\,dT + T\left(\frac{\partial \Pi}{\partial T}\right)_E dE. \tag{17-7}$$

This equation may be applied to (1) a reversible isothermal change of electric field or (2) a reversible adiabatic change of electric field. The temperature change accompanying the second process is known as the *electrocaloric effect*.

The thermodynamic coordinates of a dielectric are E, Π, and T, and an equation of state is a relation among them. If the temperature is not too low, a typical equation of state for a dilute gas of polar molecules is

$$\frac{\Pi}{V} = \left(a + \frac{b}{T}\right)E, \tag{17-8}$$

where a and b are constants. Changes of polarization that accompany changes of temperature are known as *pyroelectric effects*.

Some materials possess a permanent polarization Π in the absence of an electric field E, a phenomenon known as *ferroelectricity* in analogy to ferromagnetism. The simplest ferroelectric substance is barium titanate, $BaTiO_3$, which has a spontaneous polarization at room temperature. As the temperature of the ferroelectric is raised, the substance passes through a transition temperature T_C, called the *ferroelectric Curie temperature*, which separates the low-temperature polarized state from the high-temperature unpolarized state. The equation of state for a ferroelectric material at a temperature above T_C is

$$\frac{\Pi}{V} = \frac{\alpha'}{T - T_C}, \tag{17-9}$$

where α' is a constant. Equation (17-9) is sometimes called the Curie-Weiss law for ferroelectrics because of its similarity to Eq. (13-11). Ferroelectric behavior near the transition temperature T_C may be analyzed by the methods discussed in Art. 13-5. Curie temperatures for ferroelectrics are listed in Table 17-3.

An elastic substance describable with the aid of the coordinates $\mathcal{J}$, L, T undergoes adiabatic temperature changes or isothermal entropy changes when the tension or the length is changed. Such effects may be called *thermoelastic*. An isotropic dielectric whose coordinates are E, Π, T undergoes adiabatic temperature changes or isothermal entropy changes when the electric intensity or the polarization is changed. Such effects may be called *pyroelectric*. If a system in an electric field undergoes isothermal or adiabatic changes of polarization when the tension is varied, or isothermal or adiabatic changes of tension when the electric intensity is varied, the system is said to be *piezoelectric*. These phenomena are called *piezoelectric effects*. (The first two syllables are pronounced like "pie" and "ease.")

Table 17-3 Curie temperatures for ferroelectrics

Ferroelectric substance	Ferroelectric Curie temperature, T_C, K
KH_2AsO_3	96
PbH_2AsO_4	111
KH_2PO_4	123
RbH_2PO_4	147
KD_2PO_4	213
$BaTiO_3$	393
$KNbO_3$	712
$PbTiO_3$	763
$LiTaO_3$	890
$LiNbO_3$	1470

It may be seen from Fig. 17-2 that piezoelectric effects are really a combination of thermoelastic and pyroelectric effects. The simplest type of piezoelectric material under the simplest type of stress—say, a pure tension in only one direction and in a uniform electric field in that direction—is described with the aid of the five coordinates $\mathcal{J}$, L, E, Π, T. In actual cases of crystals such as rochelle salt, quartz, and ammonium dihydrogen phosphate, one is concerned with several components of stress and of strain, with each component related to the field as well as to the several components of the other, so that the equations are quite involved.

17-6 THERMOELECTRIC PHENOMENA

When two dissimilar metals or semiconductors are connected and the junctions held at different temperatures, there are five phenomena that take place simultaneously: the Seebeck effect, the Joule effect, the Fourier effect, the Peltier effect, and the Thomson effect. Let us consider each of these effects briefly.

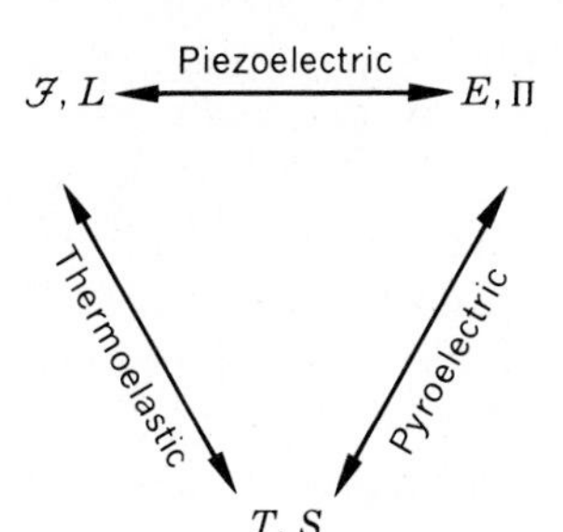

Figure 17-2 Relation between piezoelectric, thermoelastic, and pyroelectric effects.

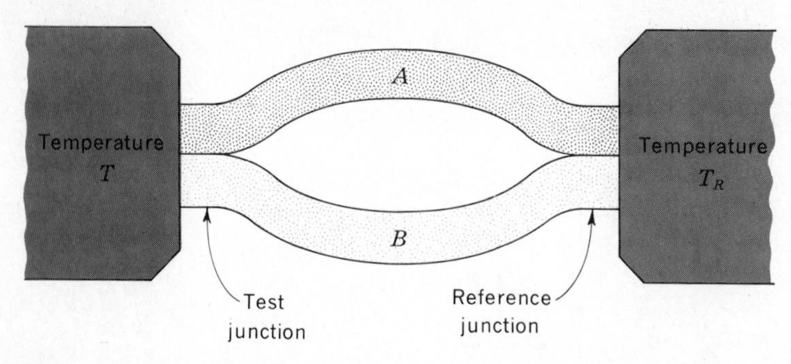

Figure 17-3 Thermocouple of conductors A and B with junctions at T and T_R.

1. Seebeck effect. In Fig. 17-3 a thermocouple consisting of two different conductors A and B has its junctions in contact with different heat reservoirs. We call the warmer junction at T the *test junction*, and the other at temperature T_R the *reference* junction. The existence of a thermal emf $\mathcal{E}_{AB}$ in the circuit is known as the *Seebeck effect*. When the temperature of the reference junction is kept constant, the thermal emf is found to be a function of the temperature T of the test junction. This fact enables the thermocouple to be used as a thermometer, as described in Chap. 1.

The Seebeck effect arises from the fact that the density of charge carriers (electrons in a metal) differs from one conductor to another and depends on the temperature. When two different conductors are connected to form two junctions and the two junctions are maintained at different temperatures, diffusion of the charge carriers takes place at the junctions at different rates. There is a net motion of the charge carriers as though they were driven by a nonelectrostatic field. The line integral of this field around the thermocouple is the Seebeck emf.

For a given T_R, $\mathcal{E}_{AB}$ is a function of T. If T_R is changed to another constant value, the relation between $\mathcal{E}_{AB}$ and T is the same except for an additive constant. It follows therefore that the value of $d\mathcal{E}_{AB}/dT$ is independent of T_R and depends only on the nature of A and B and upon T. The derivative $d\mathcal{E}_{AB}/dT$, at any value of T_R, is known as the *thermoelectric power* of the thermocouple.

2. Joule effect. If the thermal emf $\mathcal{E}_{AB}$ is not balanced by an external emf, a current I exists whose value may be adjusted by varying the external emf. As long as a current exists, electrical energy of amount I^2R is dissipated within the circuit. This is the well-known *Joule effect*.

3. Fourier effect. Imagine a thermocouple whose junctions are at temperatures T and T_R, respectively ($T > T_R$), and which has been broken at one point, its two ends being maintained at some intermediate temperature T_i by means of an insulating reservoir. There is no thermoelectric current, and therefore no Joule effect; but heat is lost by the reservoir at T, conducted along both wires, and gained by the reservoir at T_R, with no net gain or loss to the reservoir at T_i. The wires can be imagined to be suitably lagged so that there is no appreciable lateral transfer of heat across the surfaces of the wires. The phenomenon of heat conduction is called the *Fourier effect*.

4. Peltier effect. Imagine a thermocouple with its junctions at the same temperature. If, by means of an outside battery, a current is produced in the thermocouple, the temperatures of the junctions are changed by an amount that is not entirely due to the Joule effect. This additional temperature change is the *Peltier effect*. Allowing for the Joule effect, the heat that must be either supplied or extracted to restore a junction to its initial temperature is called the *Peltier heat*. The Peltier effect takes place whether the current is provided by an outside source or is generated by the thermocouple itself.

The Peltier heat is measured by creating a known current in a junction initially at a known temperature and measuring the rate at which the temperature of the junction changes. The junction itself is used as a sort of calorimeter. From the rate of change of temperature and the heat capacity of the junction, the rate at which heat is transferred is calculated. After subtracting the I^2R loss and correcting for the conducted heat, which was determined from previous experiments, the Peltier heat is finally obtained. Extensive measurements have yielded the following results:

a. The rate at which Peltier heat is transferred is proportional to the first power of the current or equal to πI. The quantity π is called the *Peltier coefficient* and is equal to the heat transferred when unit quantity of electricity traverses the junction.
b. The Peltier heat is reversible. When the direction of the current is reversed, with the magnitude remaining the same, the Peltier heat is the same but in the opposite direction.
c. The Peltier coefficient depends on the temperature and the materials of a junction, being independent of the temperature of the other junction.
d. It is the accepted convention to regard π_{AB} as positive when an electric current from A to B causes an *absorption* of heat by the junction.

5. Thomson effect. The conduction of heat along the wires of a thermocouple carrying no current gives rise to a uniform temperature distribution in each wire. If a current exists, the temperature distribution in each wire is altered by an amount that is not entirely due to the Joule effect. This additional change in the temperature distribution is called the *Thomson effect*. Allowing for the Joule effect, the heat that must be either supplied or extracted laterally *at all places along the wires* to restore the initial temperature distribution is called the *Thomson heat*.

To measure the Thomson heat at a small region of any one wire, it is necessary to produce a known temperature gradient in the region and to pass a known current either up or down the gradient. The rate at which Thomson heat is transferred is equal to the rate at which electrical energy is dissipated minus the rate at which heat is conducted. Since the Joule effect can be calculated and the conducted heat is known from previous experiments, the Thomson heat can be obtained. The following conclusions may be drawn from such measurements:

a. The rate at which Thomson heat is transferred into a small region of a wire carrying a current I and supporting a temperature difference dT is equal to $\sigma I\, dT$, where σ is called the *Thomson coefficient*.
b. The Thomson heat is reversible.
c. The Thomson coefficient depends on the material of the wire and on the mean temperature of the small region under consideration.
d. It is the accepted convention to regard σ as positive when a current opposite to the direction of the temperature gradient (low to high temperature) causes an absorption of heat by the conductor.

17-7 SIMULTANEOUS ELECTRIC AND HEAT CURRENTS IN A CONDUCTOR

The application of thermodynamics to the thermocouple has had a long and interesting history. Lord Kelvin was the first to realize that the two irreversible phenomena, the Joule effect and the conduction of heat, could not be eliminated by merely choosing wires of proper dimensions. For, if the wires are made very thin in order to cut down heat conduction, the electric resistance increases; whereas if the wires are made thick to cut down the electric resistance, the heat conduction increases. In spite of this, Kelvin assumed that the irreversible effects could be ignored on the ground that they seemed to be independent of the reversible Peltier and Thomson effects. By considering the purely reversible transfer of unit quantity of electricity through a thermocouple circuit, Kelvin set the sum of all the entropy changes equal to zero and derived relations which have been amply checked and which are undoubtedly correct. The stubborn fact remains, however, that the Seebeck, Peltier, and Thomson effects are inextricably linked with the irreversible effects.

Attempts to resolve these difficulties were made by Bridgman, by Tolman and Fine, and by Meixner, but the results were not entirely free of objection. The solution is to be found in the macroscopic treatment of irreversible coupled flows developed by Onsager, which was introduced briefly in Chap. 8. The following is a simplified version of Onsager's method, based on the work of H. B. Callen.

A small temperature difference ΔT established across a wire disturbs the thermal equilibrium and gives rise to a heat current I_Q. Since a cool reservoir at one end of the wire is gaining entropy *from the wire* at a greater rate than that at which a warmer reservoir at the other end is losing it *to the wire*, we say that entropy is being produced *in the wire* at a rate

$$\frac{dS}{d\tau} = I_Q \frac{\Delta T}{T^2} = I_S \frac{\Delta T}{T},$$

where I_S is the entropy current equal to I_Q/T.

A small potential difference $\Delta\mathcal{E}$ established across a wire disturbs the electrical equilibrium and gives rise to an electric current I. Since a reservoir at temperature T which maintains the wire at a uniform temperature is gaining entropy and there is no entropy input to the wire, we say that entropy is being produced in the wire at a rate

$$\frac{dS}{d\tau} = I\frac{\Delta\mathcal{E}}{T}.$$

When both a temperature difference ΔT and a potential difference $\Delta\mathcal{E}$ exist across the wire, the rate of entropy production is the sum, or

$$\frac{dS}{d\tau} = I_S\frac{\Delta T}{T} + I\frac{\Delta\mathcal{E}}{T}.$$

If departure from equilibrium is not too great, the entropy and electricity flow are coupled in a simple manner, both flows depending *linearly* on both $\Delta T/T$ and $\Delta\mathcal{E}/T$, thus:

$$\boxed{I_S = L_{11}\frac{\Delta T}{T} + L_{12}\frac{\Delta\mathcal{E}}{T},} \tag{17-10}$$

and

$$\boxed{I = L_{21}\frac{\Delta T}{T} + L_{22}\frac{\Delta\mathcal{E}}{T}.} \tag{17-11}$$

The coefficients L_{11} and L_{22} have simple interpretations in terms of thermal conductivity and electrical conductivity, respectively. The quantities L_{12} and L_{21} are coupling coefficients. They represent the effect of a potential difference on an entropy current and the effect of a temperature difference on an electric current, respectively. Onsager proved by means of the *microscopic* point of view that

$$\boxed{L_{12} = L_{21},} \tag{17-12}$$

which is known as *Onsager's reciprocal relation.*

If ΔT is set equal to zero in both Eqs. (17-10) and (17-11), and then the equations are divided, we get

$$\left(\frac{I_S}{I}\right)_{\Delta T=0} = \frac{L_{12}}{L_{22}}.$$

Also, in the absence of an electric current, Eq. (17-11) provides the relation

$$-\left(\frac{\Delta\mathcal{E}}{\Delta T}\right)_{I=0} = \frac{L_{21}}{L_{22}}.$$

Since Onsager's reciprocal relation provides that $L_{12} = L_{21}$, let us write

$$\epsilon = \frac{L_{12}}{L_{22}} = \frac{L_{21}}{L_{22}}.$$

We therefore have two different physical interpretations of the quantity ϵ:

$$\epsilon = \begin{cases} \left(\dfrac{I_S}{I}\right)_{\Delta T=0} = \left(\dfrac{I_Q/T}{I}\right)_{\Delta T=0} \\ -\left(\dfrac{\Delta \mathcal{E}}{\Delta T}\right)_{I=0} = -\left(\dfrac{d\mathcal{E}}{dT}\right)_{I=0}. \end{cases} \tag{17-13}$$

Thus, ϵ may be regarded as the *entropy current per unit electric current* at a given temperature, or as the *change of potential difference per unit change of temperature* at zero electric current. Because of this latter physical interpretation, ϵ is called the *Seebeck coefficient* of a substance and is a function of the temperature and the nature of the substance.

17-8 SEEBECK AND PELTIER EFFECTS

Consider the thermocouple depicted in Fig. 17-4. The test junction e of the conductors A and B is maintained at the temperature T while the two junctions c and d, each with copper, are maintained at temperature T_R (usually that of an ice bath). The two copper wires marked C are connected to the brass binding posts of a potentiometer, forming two more junctions, each of which is at room temperature T_0. The potentiometer is supposed to be balanced, so that $I = 0$ and $\mathcal{E}_a - \mathcal{E}_b$ is the Seebeck emf $\mathcal{E}_{AB}$.

The second relation in Eq. (17-13), namely,

$$\left(\frac{d\mathcal{E}}{dT}\right)_{I=0} = -\epsilon,$$

may be applied to each of the conductors A, B, and C—the Seebeck coefficients being ϵ_A, ϵ_B, and ϵ_C. Integrating from one end of each wire to the other, we have

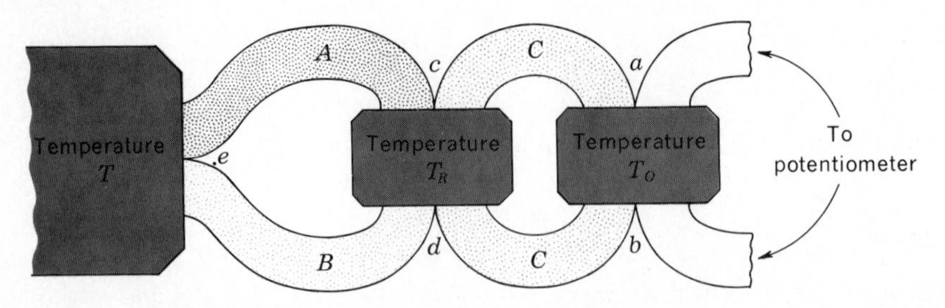

Figure 17-4 Thermocouple consisting of wires A and B connected to copper wires C and C, and thence to the binding posts of a potentiometer. (See Fig. 1.10.)

$$\mathcal{E}_a - \mathcal{E}_c = \int_{T_0}^{T_R} \epsilon_C \, dT,$$

$$\mathcal{E}_c - \mathcal{E}_e = \int_{T_R}^{T} \epsilon_A \, dT,$$

$$\mathcal{E}_e - \mathcal{E}_d = \int_{T}^{T_R} \epsilon_B \, dT,$$

and

$$\mathcal{E}_d - \mathcal{E}_b = \int_{T_R}^{T_0} \epsilon_C \, dT.$$

When these equations are added, the left side becomes $\mathcal{E}_a - \mathcal{E}_b = \mathcal{E}_{AB}$. On the right side, the first and last terms cancel, so that

$$\boxed{\mathcal{E}_{AB} = \int_{T_R}^{T} (\epsilon_A - \epsilon_B) \, dT.} \tag{17-14}$$

If the Seebeck coefficients of A and B are known as functions of T, the emf of the thermocouple made by joining A and B and holding the junctions at T_R and T could be calculated by performing the integration indicated in Eq. (17-14).

A thermocouple consisting of two metals is well-suited as a thermometer. At low temperatures a highly sensitive potentiometer must be used, and at high temperatures metals with high melting points are essential. Junctions may be made with very low heat capacity, so that they react quickly to changes of temperature. As a result, thermometry has been the main practical application of thermoelectricity since its discovery by Seebeck in 1821. Since

$$\mathcal{E}_{AB} = \int_{T_R}^{T} (\epsilon_A - \epsilon_B) \, dT,$$

it follows that

$$\mathcal{E}_{AB} = \int_{T_R}^{T} (\epsilon_A - \epsilon_C) \, dT - \int_{T_R}^{T} (\epsilon_B - \epsilon_C) \, dT,$$

or

$$\mathcal{E}_{AB} = \mathcal{E}_{AC} - \mathcal{E}_{BC}. \tag{17-15}$$

Therefore, if C is chosen to be platinum, two tables of values of thermal emf—one for metal A and platinum and the other for metal B and platinum—enable one to find $\mathcal{E}_{AB}$. Such tables are given in the *AIP Handbook* for many different metals.

When an electric current traverses a junction of two different conductors at a uniform temperature, the heat that must be supplied or withdrawn, over and above the Joule heat, to keep the junction at a constant temperature is what we have already described above as the Peltier heat. This may be calculated in terms of Seebeck coefficients as follows. Consider the thermojunction e of Fig. 17-4, depicted in greater detail in Fig. 17-5. Even though the junction is at a uniform

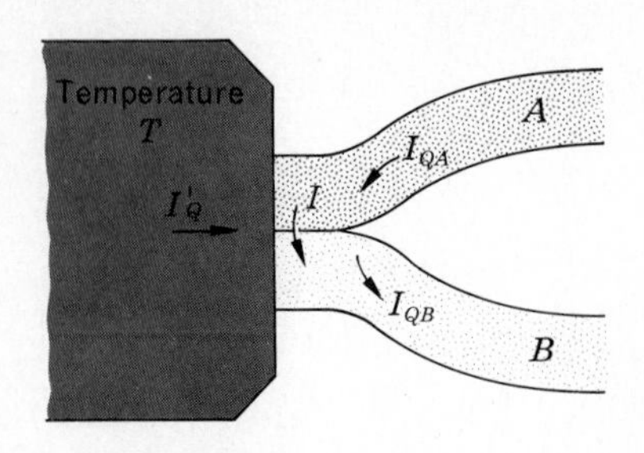

Figure 17-5 Thermojunction of conductors A and B at a uniform constant temperature T. The Peltier heat is the difference between I'_Q and I^2R_j.

temperature, there is a heat current I_{QA} into the junction and a different heat current I_{QB} out of the junction, both carried along with the electric current I, which is taken arbitrarily in the direction from A to B. The difference between I_{QA} and I_{QB} must be transferred, along with I^2R_j (where R_j is the resistance of the junction), in order to keep the temperature constant. Calling this heat current I'_Q, we have

$$I'_Q = I^2R_j + (I_{QA})_{\Delta T=0} - (I_{QB})_{\Delta T=0}.$$

By definition, the Peltier heat is

$$\pi_{AB}I = I'_Q - I^2R_j = (I_{QA})_{\Delta T=0} - (I_{QB})_{\Delta T=0}.$$

Since the first part of Eq. (17-13) defines ϵ to be $(I_S/I)_{\Delta T=0}$ and since $I_S = I_Q/T$, it follows that

$$I_{QA} = IT\epsilon_A \qquad \text{and} \qquad I_{QB} = IT\epsilon_B.$$

The Peltier coefficient therefore becomes

$$\boxed{\pi_{AB} = T(\epsilon_A - \epsilon_B)} \qquad (17\text{-}16)$$

17-9 THOMSON EFFECT AND KELVIN EQUATIONS

Consider the thermocouple of Fig. 17-6 consisting of wires A and B with the test junction at temperature T' and the reference junction at T_R. If the thermocouple is on open circuit or is connected to a potentiometer that is balanced, there will be no electric current, but there will be a heat current and a temperature distribution throughout the wires. Suppose that wire A is now placed in contact *at each point with a heat reservoir of the same temperature as that point*, so that there is no heat exchange between A and the reservoirs. This situation is depicted in Fig. 17-6a, but only one of the reservoirs is shown—the one at temperature T in contact with a small portion of wire A, across which there is a temperature difference ΔT and a potential difference $\Delta\mathcal{E}_A = \epsilon_A \Delta T$. The heat current I_Q entering this portion is equal to the heat current leaving the portion.

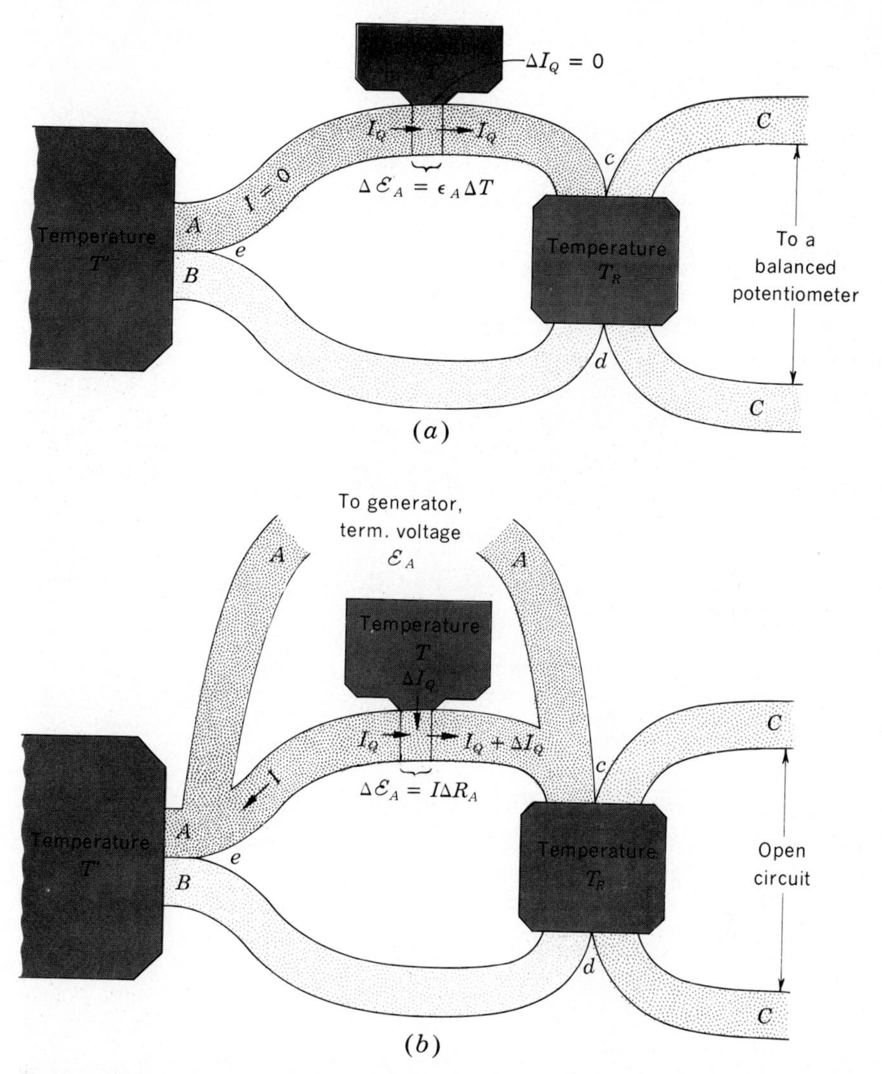

Figure 17-6 (*a*) In the *absence* of an electric current, there is no heat flow between a portion of wire A at temperature T and its heat reservoir at T. (*b*) In the *presence* of an electric current, the heat transfer between a portion of wire A at temperature T and its heat reservoir at T is equal to $\Delta I_Q = I^2 \, \Delta R - I\sigma_A \, \Delta T$.

Now suppose that the potentiometer circuit is opened and an outside generator is connected to wire A (with the aid of connecting wires of the same material as A) and that the terminal voltage of the generator is adjusted until it has exactly the value $\mathcal{E}_A = \mathcal{E}_c - \mathcal{E}_e = \int_{T_0}^{T'} \epsilon_A \, dT$. A current will now exist in wire A of magnitude $I = \mathcal{E}_A / R_A$, and the temperature distribution in wire A will be altered both by Joulean dissipation and by the Thomson effect. Referring to

Fig. 17-6*b*, we see that the small portion of wire A in contact with the reservoir at T has the following characteristics:

1. It supports a temperature difference ΔT.
2. It supports a potential difference $\Delta \mathcal{E}_A = \epsilon_A \, \Delta T$.
3. It carries an electric current $I = \Delta \epsilon_A / \Delta R_A$.
4. A heat current I_Q enters it.
5. A different heat current $I_Q + \Delta I_Q$ leaves it.
6. It exchanges heat $I^2 \, \Delta R_A$ (Joule effect) and $I \sigma_A \, \Delta T$ (Thomson effect) with the heat reservoir in contact with it. Therefore,

$$\Delta I_Q = I^2 \, \Delta R_A - I\sigma_A \, \Delta T$$
$$= I \, \Delta \mathcal{E}_A - I\sigma_A \, \Delta T,$$

or

$$\Delta I_Q = I\epsilon_A \, \Delta T - I\sigma_A \, \Delta T. \tag{17-17}$$

As before, $I_Q = IT\epsilon_A$, and therefore a small change is given by

$$\Delta I_Q = I\epsilon_A \, \Delta T + IT \, \Delta \epsilon_A. \tag{17-18}$$

It follows from Eqs. (17-17) and (17-18) that

$$-I\sigma_A \, \Delta T = IT \, \Delta \epsilon_A,$$

or

$$\sigma_A = -T \frac{d\epsilon_A}{dT}.$$

Similarly, at a point on wire B that is at temperature T,

$$\sigma_B = -T \frac{d\epsilon_B}{dT};$$

and finally,

$$\boxed{\sigma_A - \sigma_B = -T \frac{d}{dT}(\epsilon_A - \epsilon_B).} \tag{17-19}$$

Let us collect the three equations for the three reversible thermoelectric effects:

Seebeck effect

$$\mathcal{E}_{AB} = \int_{T_R}^{T} (\epsilon_A - \epsilon_B) \, dT. \tag{17-14}$$

Peltier effect

$$\pi_{AB} = T(\epsilon_A - \epsilon_B). \tag{17-16}$$

Thomson effect

$$\sigma_A - \sigma_B = -T \frac{d}{dT}(\epsilon_A - \epsilon_B). \tag{17-19}$$

Notice that they are all expressed in terms of the difference of the two Seebeck coefficients, so that if ϵ_A and ϵ_B are known as functions of T, then $\mathcal{E}_{AB}$ may be obtained by integration, π_{AB}/T without further calculation, and $(\sigma_A - \sigma_B)/T$ by differentiation.

If we differentiate Eq. (17-14) with respect to T, while holding T_R constant, we get $d\,\mathcal{E}_{AB}/dT = \epsilon_A - \epsilon_B$. Comparing this with Eq. (17-16), we see that

$$\boxed{\frac{\pi_{AB}}{T} = \frac{d\mathcal{E}_{AB}}{dT},} \tag{17-20}$$

which is *Kelvin's first equation.*

If we substitute the value $d\mathcal{E}_{AB}/dT$ for $\epsilon_A - \epsilon_B$ in Eq. (17-19), we get

$$\boxed{\frac{\sigma_A - \sigma_B}{T} = -\frac{d^2\mathcal{E}_{AB}}{dT^2},} \tag{17-21}$$

which is *Kelvin's second equation.* The Kelvin equations have been often verified by experiment.

17-10 THERMOELECTRIC REFRIGERATION

Metal thermocouples are not suited to extract heat by the Peltier effect because the difference in the Seebeck coefficients is so small. To produce ice cubes, T would have to be about 270 K. Using a current of about 20 A and about 10 thermocouples in series, we could obtain a Peltier heat current equal to

$$\dot{Q} = 270\ \text{K}(\epsilon_A - \epsilon_B) \times 20\ \text{A} \times 10.$$

If the thermojunctions were of Cu and Fe, $\epsilon_{\text{Cu}} - \epsilon_{\text{Fe}} = 13.7\ \ \mu\text{V/K}$, and $\dot{Q}$ would be only 0.74 W.

In 1838, Lenz tried thermocouples of Sb and Bi, with

$$\epsilon_{\text{Sb}} - \epsilon_{\text{Bi}} = 109\ \mu\text{V/K},$$

and succeeded in converting a drop of water into ice. The large currents needed, the large amount of Fourier conduction, and the large Joulean heat dissipation resulted in a very low coefficient of performance, so that no commercial Peltier refrigerator appeared on the market for over a century after Lenz's experiment. The major breakthrough occurred when semiconducting compounds were found to have large Seebeck coefficients, good electrical conductivity, and poor thermal conductivity. Thus, a thermocouple of *p*-type $\text{Si}_{0.78}\text{Ge}_{0.22}$ joined to *n*-type $\text{Si}_{0.78}\text{Ge}_{0.22}$ has a value of $\epsilon_A - \epsilon_B$ equal to 646 μV/K, so that 10 couples at 270 K with a current of 20 A give rise to a Peltier heat current of 35 W. These values make a Peltier refrigerator economically feasible. Such a refrigerator is depicted schematically in Fig. 17-7. Progress in Peltier refrigeration since 1821 is shown in Table 17-4.

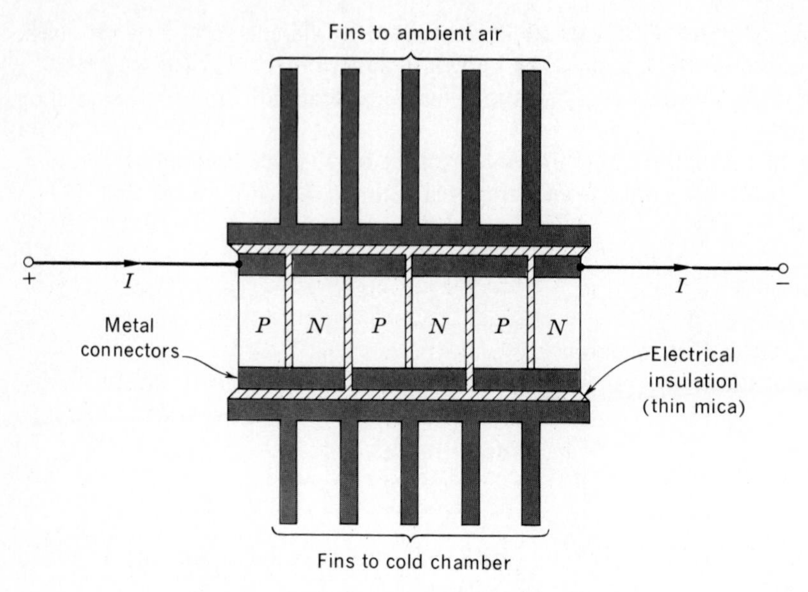

Figure 17-7 Schematic diagram of a thermoelectric refrigerator.

Table 17-4 Peltier refrigeration

Thermojunction	T, K	I, A	No. of couples	$\epsilon_A - \epsilon_B$, μV/K	Peltier heat rate, W
Fe–Cu (1821)	270	20	10	13.7	0.74
Sb–Bi (1838)	270	20	10	109	5.9
$Bi_2Te_3(p)$–$Bi_2Te_3(n)$ (1963)	270	20	10	423	23
$Si_{0.78}Ge_{0.22}(p) - Si_{0.78}Ge_{0.22}(n)$ (1978)	270	20	10	646	35

17-11 PROPERTIES OF A SYSTEM OF PHOTONS

The electromagnetic radiation in equilibrium with the interior walls of an evacuated cavity is called *blackbody radiation.* If the walls are at a uniform temperature, the distribution of frequencies and the energy of each frequency band are independent of the nature of the walls and depend only on the temperature and the volume. The quantum picture of the radiation in the cavity is that of a system of photons with many different frequencies, but all moving with a constant speed c and completely independent of one another. All photons of the same frequency constitute a set of particles that satisfies the conditions of *indistinguishability* and *independence* (absence of interaction) better than any system of molecules or electrons.

The most striking peculiarity of a system of photons is that *the total number of photons of all frequencies is not constant.* As equilibrium is approached, photons are absorbed by atoms, and other photons, perhaps of different frequencies, are reemitted. Therefore, during the approach to equilibrium, although the total *energy* of the photons may remain constant, the total *number* does not.

The number of quantum states associated with the translational kinetic energy of the molecules of a gas is enormously larger than the number of molecules that can occupy these states. Very few of the available states can be occupied at one time, and when a state is occupied, it is most unlikely that it contains more than one molecule. With electrons, however, the number of quantum states and the number of electrons are comparable. The operation of the Pauli exclusion principle, which limits each state to at most two electrons, provides therefore the complete filling of all low-lying states up to a level known as the *Fermi level.* With photons the situation is still different. Since the total number of photons is not constant, there is no clear-cut relation between the number of photons of frequency ν and the number of quantum states g_ν available to these photons. The Pauli principle does not hold for photons, so that *there may be any number of photons in the same energy state.*

The translational kinetic energy of a molecule or an electron moving in a cubical box of side L is given by the familiar expression in Eq. (11-1),

$$\epsilon = \frac{h^2}{8mL^2}(n_x^2 + n_y^2 + n_z^2),$$

where the n's are large integers. If the n's are taken to be rectangular coordinates, the locus of all points in this space corresponding to one value ϵ_i of the energy is the surface of the sphere

$$n_x^2 + n_y^2 + n_z^2 = \frac{8mL^2\epsilon_i}{h^2},$$

with a radius $r = 2L(2m\epsilon_i)^{1/2}/h$. Since the n's are integers, a unit volume in this space contains one quantum state. The number of quantum states $g_i\,d\epsilon_i$ corresponding to an energy lying between ϵ_i and $\epsilon_i + d\epsilon_i$ is $\frac{1}{8}$ the volume of a spherical shell of radius r. Thus,

$$\begin{aligned} g_i\,d\epsilon_i &= \tfrac{1}{8}\cdot 4\pi r^2\,dr \\ &= \tfrac{1}{8}\cdot 4\pi\frac{4L^2}{h^2}2m\epsilon_i\frac{2L}{h}(2m)^{1/2}\tfrac{1}{2}\epsilon_i^{-1/2}\,d\epsilon_i \\ &= 2\pi L^3\left(\frac{2m}{h^2}\right)^{3/2}\epsilon_i^{1/2}\,d\epsilon_i. \end{aligned}$$

This result appeared in the discussion of free electrons in metals given in Eq. (12-22).

This equation must also hold for photons, but it is not in an appropriate form, since photons have zero rest mass. To apply it to photons, let us rewrite it in terms of momentum p, where $\epsilon = p^2/2m$. Let $g_p\, dp$ represent the number of quantum states corresponding to momenta lying between p and $p + dp$. Then, since $L^3 = V$,

$$g_p\, dp = 2\pi V \left(\frac{2m}{h^2}\right)^{3/2} \frac{p}{(2m)^{1/2}} \frac{1}{2m} 2p\, dp$$

$$= \frac{4\pi V}{h^3} p^2\, dp.$$

Having gotten rid of m, let us now express our results in terms of frequency ν by using the de Broglie equation,

$$p = \frac{h}{\lambda} = \frac{h\nu}{c}.$$

Let $g_\nu\, d\nu$ be the number of quantum states between ν and $\nu + d\nu$. Then

$$g_\nu\, d\nu = \frac{4\pi V}{h^3} \frac{h^2 \nu^2}{c^2} \frac{h}{c} d\nu.$$

Finally, doubling this result so as to include photons with both kinds of polarization,

$$\boxed{g_\nu\, d\nu = \frac{8\pi V}{c^3} \nu^3\, d\nu.} \tag{17-22}$$

17-12 BOSE-EINSTEIN STATISTICS APPLIED TO PHOTONS

Consider an evacuated cavity with perfectly reflecting walls containing a total of N photons, of which N_ν have a frequency ν. Suppose that within the cavity there is a speck of coal with negligible heat capacity. As the coal particle absorbs and reemits photons, any resulting change of energy will be so small that the total energy of the radiation may be regarded as constant. We require an expression for the number of ways in which N_ν photons may be distributed among g_ν quantum states with no restriction on the number of photons in any one state. Let us represent identical indistinguishable photons by crosses, $\times \times \times$; and quantum states by the spacing between vertical lines, $\times \times \times | \times | \times \times$, etc. A typical distribution of photons among states would thus be

$$\times \times \times | \times | \times \times | \times \times \times \times \times | \times \times | \times \times \times .$$

This symbolism represents three photons in the first state, one in the second, two in the third, etc. Notice that only five vertical lines are needed to represent six states; therefore the total number of symbols needed to represent N_ν photons and g_ν states is $N_\nu + g_\nu - 1$. The total number of permutations of $N_\nu + g_\nu - 1$ symbols is $(N_\nu + g_\nu - 1)!$, but this expression is much too large. To take care of the fact that the photons are indistinguishable, we must divide by $N_\nu!$; and to show that the lines representing partitions are indistinguishable, we must divide by $(g_\nu - 1)!$. Very little error is introduced by replacing $(g_\nu - 1)$ by g_ν; therefore, summing over all the frequencies, the thermodynamic probability becomes

$$\Omega_{\text{BE}} = \prod^{\nu} \frac{(N_\nu + g_\nu)!}{N_\nu! g_\nu!}. \tag{17-23}$$

This is the probability appropriate to the *Bose-Einstein statistics*.

To find the distribution of photons among the frequencies at equilibrium, we render $\ln \Omega_{\text{BE}}$ a maximum subject to *only one* restriction, namely, that $\sum N_\nu h\nu = U = \text{const}$. We have

$$\ln \Omega_{\text{BE}} = \sum \ln (N_\nu + g_\nu)! - \sum \ln N_\nu! - \sum \ln g_\nu!;$$

and using Stirling's approximation, $\ln x! = x \ln x - x$,

$$\ln \Omega_{\text{BE}} = \sum (N_\nu + g_\nu) \ln (N_\nu + g_\nu) - \sum (N_\nu + g_\nu)$$
$$- \sum N_\nu \ln N_\nu + \sum N_\nu - \sum g_\nu \ln g_\nu + \sum g_\nu.$$

After some cancellation,

$$\ln \Omega_{\text{BE}} = \sum (N_\nu + g_\nu) \ln (N_\nu + g_\nu) - \sum N_\nu \ln N_\nu - \sum g_\nu \ln g_\nu,$$

and

$$U = \sum N_\nu h\nu.$$

Applying the method of Lagrange multipliers, we set

$$d \ln \Omega_{\text{BE}} = \sum [1 + \ln (N_\nu + g_\nu)] \, dN_\nu - \sum (1 + \ln N_\nu) \, dN_\nu = 0,$$

or

$$d \ln \Omega_{\text{BE}} = \sum \ln \frac{N_\nu + g_\nu}{N_\nu} \, dN_\nu = 0, \tag{17-24}$$

with one restriction, namely,

$$dU = \sum h\nu \, dN_\nu = 0.$$

Multiplying this last equation by $-\beta$ and adding, we get

$$\ln \left(1 + \frac{g_\nu}{N_\nu}\right) - \beta h\nu = 0, \tag{17-25}$$

$$1 + \frac{g_\nu}{N_\nu} = e^{\beta h\nu},$$

and

$$\boxed{N_\nu = \frac{g_\nu}{e^{\beta h\nu} - 1}.} \tag{17-26}$$

To discover the physical significance of β, we consider the system of photons as a thermodynamic system with an entropy $S = k \ln \Omega_{\text{BE}}$. Then, from Eq. (17-24),

$$dS = k\, d \ln \Omega_{\text{BE}} = k \sum \ln \left(1 + \frac{g_\nu}{N_\nu}\right) dN_\nu;$$

and from Eq. (17-25), at equilibrium,

$$\ln \left(1 + \frac{g_\nu}{N_\nu}\right) = \beta h\nu.$$

Substituting, we get

$$dS = k \sum \beta h\nu \, dN_\nu .$$

But since $U = \sum N_\nu h\nu$ and $dU = \sum h\nu \, dN_\nu$,

$$dS = k\beta \, dU.$$

Inasmuch as the volume is constant, $dS/dU = 1/T$ and

$$\beta = \frac{1}{kT},$$

as always. Substituting into Eq. (17-26) the values of g_ν and β, we get

$$N_\nu = \frac{8\pi V}{c^3} \frac{\nu^2}{e^{h\nu/kT} - 1}.$$

If the energy lying in the frequency range between ν and $\nu + d\nu$ is denoted by $U_\nu \, d\nu$, then

$$U_\nu \, d\nu = N_\nu h\nu \, d\nu$$

$$= \frac{8\pi V h}{c^3} \frac{\nu^3 \, d\nu}{e^{h\nu/kT} - 1}.$$

The *spectral energy density* u_ν is defined to be U_ν/V, so that finally

$$\boxed{u_\nu = \frac{8\pi h}{c^3} \frac{\nu^3}{e^{h\nu/kT} - 1},} \tag{17-27}$$

which is *Planck's radiation equation.*

17-13 OPTICAL PYROMETER

To measure temperatures above the range of thermocouples and resistance thermometers, an *optical pyrometer* is used. As shown in Fig. 17-8, it consists essentially of a telescope T in the tube of which is mounted a filter of red glass and a

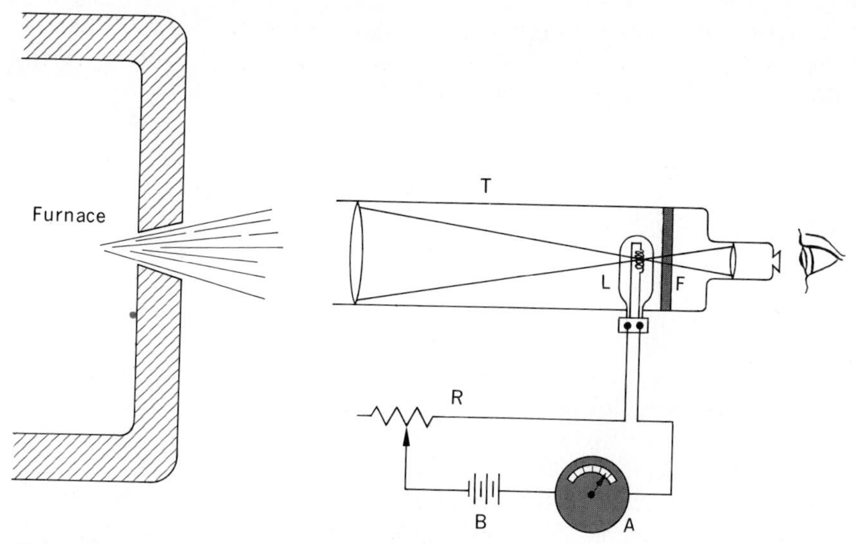

Figure 17-8 Main features of optical pyrometer.

small electric lamp bulb L. When the pyrometer is directed toward a furnace, an observer looking through the telescope sees the dark lamp filament against the bright background of the furnace. The lamp filament is connected to a battery B and a rheostat R. By turning the rheostat knob, the current in the filament—and hence its brightness—may be gradually increased until the brightness of the filament just matches the brightness of the background. From previous calibration of the instrument at known temperatures, the scale of the ammeter A in the circuit may be marked to read the unknown temperature directly. Since no part of the instrument needs to come into contact with the hot body, the optical pyrometer may be used at temperatures above the melting points of many metals.

To understand how an optical pyrometer is calibrated at temperatures above the gold point, let us rewrite the Planck equation, Eq. (17-27), in terms of wavelength instead of frequency. Let $u_\lambda \, d\lambda$ be the energy density of the radiation within the wavelength range λ and $\lambda + d\lambda$. Then, since $\nu = c/\lambda$ and $|d\nu| = (c/\lambda^2)|d\lambda|$,

$$u_\lambda \, d\lambda = 8\pi hc \frac{\lambda^{-5} \, d\lambda}{e^{hc/\lambda kT} - 1}. \tag{17-28}$$

Suppose that an optical pyrometer is sighted on a blackbody at the normal melting point of gold, known as the *gold point* (1337.58 K), and that the current necessary to make the lamp filament disappear is noted. Now suppose that another blackbody at a higher temperature T is sighted *with the same pyrometer at the same current setting*, but through a disk equipped with a sector-shaped opening whose angle θ can be varied at will. As the sectored disk rotates rapidly, it transmits the same fraction $\theta/2\pi$ of the radiation of all wavelengths. As θ is made smaller and smaller, the radiation gets dimmer and dimmer until it is the same as that from the gold-point blackbody. When this is the case,

$$\frac{u_\lambda(T_{\mathrm{Au}})}{u_\lambda(T)} = \frac{\theta}{2\pi};$$

and using the Planck equation,

$$\frac{e^{c_2/\lambda T} - 1}{e^{c_2/\lambda T_{\mathrm{Au}}} - 1} = \frac{\theta}{2\pi}. \tag{17-29}$$

where $c_2 = hc/k = 0.014388$ m/K. The wavelength transmitted by the filter is often chosen in the red region, where $\lambda = 6.5 \times 10^{-7}$ m and, as mentioned before, $T_{\mathrm{Au}} = 1337.58$ K. Solving Eq. (17-29) for T gives the temperature of the unknown blackbody.

If the red filter transmitted only a very narrow band of wavelengths and if all unknown bodies whose temperatures must be measured were blackbodies, everything would be fine and no further physical ideas or instruments would be needed. This is not the case, however. Elaborate and difficult corrections must be made in order to take into account the finite bandwidth needed for a visual match, as well as the departure from blackbody conditions. These details, however, are for the expert.

17-14 THE LAWS OF WIEN AND OF STEFAN-BOLTZMANN

The Planck equation, expressed in terms of wavelength, takes the form

$$u_\lambda \, d\lambda = 8\pi hc \frac{\lambda^{-5}\, d\lambda}{e^{hc/\lambda kT} - 1}.$$

This is plotted in Fig. 17-9 for seven different temperatures. It can be seen that, the higher the temperature, the smaller the wavelength λ_m at which the energy density is a maximum. To see how λ_m depends on T, we note that u_λ is a maximum when $\lambda^5(e^{hc/\lambda kT} - 1)$ is a minimum. Thus,

$$\frac{d}{d\lambda}[\lambda^5(e^{hc/\lambda kT} - 1)] = 0,$$

or

$$5\lambda^4(e^{hc/\lambda kT} - 1) + \lambda^5 e^{hc/\lambda kT}\left(-\frac{hc}{kT}\frac{1}{\lambda^2}\right) = 0.$$

The maximum λ must therefore satisfy the equation

$$1 - e^{-hc/\lambda kT} = \frac{1}{5}\frac{hc}{\lambda kT},$$

or

$$1 - e^{-x} = \frac{x}{5}.$$

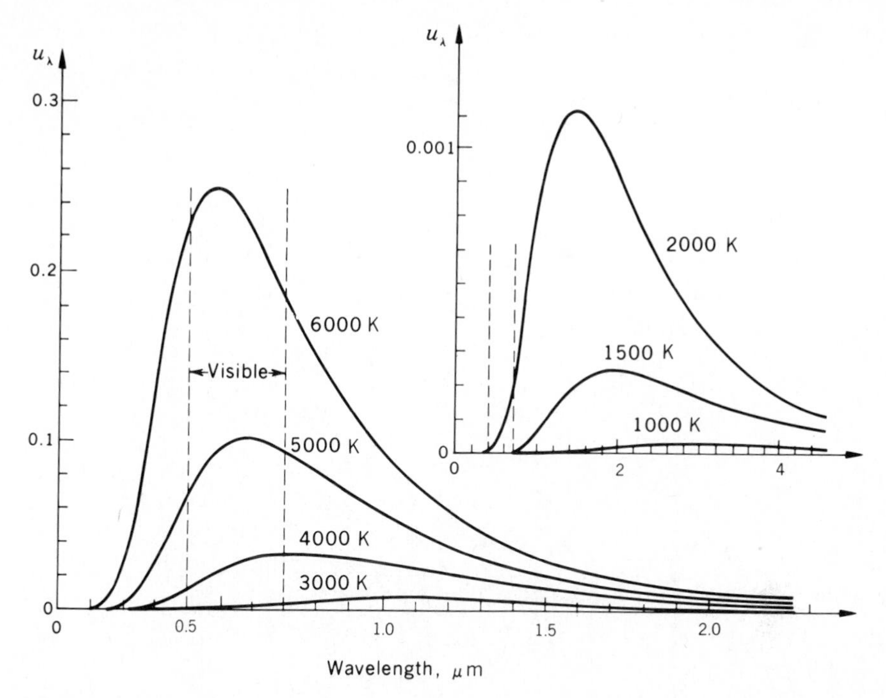

Figure 17-9 Variation of spectral density of blackbody radiation at different temperatures, according to Planck equation.

This is a transcendental equation whose solution is obtained most simply by plotting two curves, $y = 1 - e^{-x}$ and $y = x/5$, and noting the value x_m at the point of intersection. This is done in Fig. 17-10, where the point of intersection is found to 4.96. Hence,

$$x_m = \frac{hc}{\lambda_m kT} = 4.96,$$

or
$$\lambda_m T = \frac{hc}{4.96k} = \frac{6.63 \times 10^{-34}\ \mathrm{J \cdot s} \times 3 \times 10^{8}\ \mathrm{m/s}}{4.96 \times 1.38 \times 10^{-23}\ \mathrm{J/K}};$$

whence
$$\boxed{\lambda_m T = 2.90\ \mathrm{mm \cdot K},} \tag{17-30}$$

which is known as *Wien's law* (name is pronounced "Veen"). At $T = 2000$ K, $\lambda_m = 1.45\ \mu$m, whereas at $T = 6000$ K, $\lambda_m \approx 0.5\ \mu$m, as shown in Fig. 17-9.

Let us go back to Planck's equation expressed in terms of frequency,

$$u_\nu = \frac{8\pi h}{c^3} \frac{\nu^3}{e^{h\nu/kT} - 1}. \tag{17-27}$$

The *total energy density* u is

$$u = \int_0^\infty u_\nu\, d\nu = \frac{8\pi h}{c^3} \int_0^\infty \frac{\nu^3\, d\nu}{e^{h\nu/kT} - 1}.$$

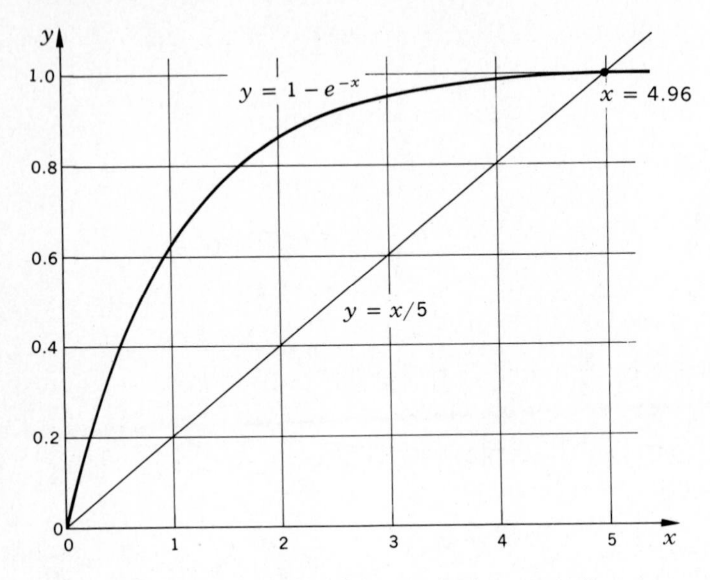

Figure 17-10 Graphical solution of the equation $1 - e^{-x} = x/5$.

If we let $x = h\nu/kT$, then

$$u = \frac{8\pi k^4 T^4}{h^3 c^3} \int_0^\infty \frac{x^3\, dx}{e^x - 1}.$$

It was mentioned in Art. 12-2, in evaluating the heat capacity of a crystal in the limit as T approaches zero, that

$$\int_0^\infty \frac{x^3\, dx}{e^x - 1} = 3!\ \zeta(4)$$

$$= 6\frac{\pi^4}{90} = \frac{\pi^4}{15}.$$

Hence,

$$u = \frac{8\pi^5 k^4}{15 h^3 c^3} T^4. \tag{17-31}$$

Let us accept without proof that the radiant emittance $\mathscr{R}_B$ of the walls of a cavity in which blackbody radiation is in equilibrium is connected with the total energy density u by the following relation:

$$\mathscr{R}_B = \frac{c}{4} u = \frac{2\pi^5 k^4}{15 h^3 c^2} T^4.$$

This is the famous *Stefan-Boltzmann law* mentioned in Chap. 4 and written $\mathcal{R}_B = \sigma T^4$. We have

$$\sigma = \frac{2\pi^5 k^4}{15h^3c^2}$$

$$= \frac{2 \times (3.14)^5 \times (1.38)^4 \times 10^{-92}(\text{J/K})^4}{15 \times (6.63)^3 \times 10^{-102}(\text{J} \cdot \text{s})^3 \times 9 \times 10^{16}(\text{m/s})^2}$$

$$= 56.3 \frac{\text{nW}}{\text{m}^2 \cdot \text{K}^4},$$

which is in good agreement with experimental values.

17-15 RADIATION PRESSURE; BLACKBODY RADIATION AS A THERMODYNAMIC SYSTEM

The quantum picture of the radiation in an enclosure is very similar to that of a gas. If the volume is V, we imagine photons of many different frequencies and pick out as a sample N_ν photons of frequency ν. Each photon has an energy $h\nu$ and a constant speed c. A photon has no rest mass. Its mass is due entirely to its speed, so that $mc^2 = h\nu$, or

$$m = \frac{h\nu}{c^2}.$$

In the kinetic theory of an ideal gas that was treated in Chap. 5, the pressure exerted by N molecules on a wall of an enclosure of volume V was shown to be [see Eq. 5-21)]

$$P = \frac{1}{3}\left(\frac{N}{V}\right)m\langle w^2 \rangle,$$

where $\langle w^2 \rangle$ was the average of the square of the speed. Since all photons have speed c, we get for the contribution to the pressure P_ν of those photons with frequency ν the expression

$$P_\nu = \frac{1}{3}\frac{N_\nu}{V}\frac{h\nu}{c^2}c^2$$

$$= \frac{1}{3}\frac{N_\nu h\nu}{V} = \frac{1}{3}u_\nu$$

and, for the total pressure,

$$P = \tfrac{1}{3}\sum u_\nu,$$

or

$$\boxed{P = \frac{u}{3}.} \qquad (17\text{-}32)$$

Blackbody radiation is therefore completely specified by the pressure of the radiation, the volume of the radiation, and the temperature of the walls with which the radiation is in equilibrium. This temperature is sometimes called the *temperature of the radiation,* for convenience. Strictly speaking, it is not the radiation to which the temperature applies, but the matter in equilibrium with the radiation.

Since blackbody radiation is described by the coordinates P, V, and T, it may be treated as a hydrostatic system, and any one of the equations derived in Chap. 9 may be applied to it. A particularly important result can be obtained from the energy equation

$$\left(\frac{\partial U}{\partial V}\right)_T = T\left(\frac{\partial P}{\partial T}\right)_V - P. \tag{9-18}$$

Since $U = Vu$ and $P = u/3$, where u is a function of T only, the energy equation becomes

$$u = \frac{T}{3}\frac{du}{dT} - \frac{u}{3}$$

and reduces to

$$\frac{du}{u} = 4\frac{dT}{T}.$$

Integrating the above, we get

$$\ln u = \ln T^4 + \ln b,$$

or

$$u = bT^4,$$

which is in agreement with Eq. (17-31).

Since $P = u/3$ and $u = bT^4$, the equation of state of blackbody radiation takes the interesting form

$$P = \frac{b}{3}T^4, \tag{17-33}$$

and

$$\left(\frac{\partial P}{\partial T}\right)_V = \frac{dP}{dT} = \frac{4}{3}bT^3.$$

Also, since $U = VbT^4$,

$$C_V = \left(\frac{\partial U}{\partial T}\right)_V = 4VbT^3. \tag{17-34}$$

The first $T\,dS$ equation,

$$T\,dS = C_V\,dT + T\left(\frac{\partial P}{\partial T}\right)_V dV,$$

therefore becomes

$$T\,dS = 4VbT^3\,dT + \tfrac{4}{3}bT^4\,dV, \tag{17-35}$$

which may be used in two ways.

1. *Reversible isothermal change of volume.* If blackbody radiation, in equilibrium with the walls of a cavity at temperature T, is caused to expand isothermally, heat will have to be supplied to the walls to keep the temperature constant. Thus,

$$T\,ds = \tfrac{4}{3}bT^4\,dV,$$

and

$$Q = \tfrac{4}{3}bT^4(V_f - V_i).$$

2. *Reversible adiabatic change of volume.* Suppose that blackbody radiation, in equilibrium with an extremely small piece of matter such as a grain of coal, is contained in a cylinder with perfectly reflecting walls. If the radiation is caused to expand against a piston, the expansion will be adiabatic, since there is no exchange of energy between the walls and the radiation. The work done on the surroundings is accomplished at the expense of the internal energy of both the piece of coal and the radiation. If the grain of coal has an extremely minute mass, its heat capacity may be regarded as negligible in comparison with that of the radiation. During the expansion, the radiation is always in equilibrium with the coal; but since the energy density of the radiation is decreasing, the temperature of the coal also decreases. The final temperature of the coal can be found by setting $dS = 0$ in the first $T\,dS$ equation. Thus,

$$\tfrac{4}{3}bT^4\,dV = -4VbT^3\,dT,$$

or

$$\frac{dV}{V} = -3\frac{dT}{T};$$

and after integration,

$$VT^3 = \text{const.} \tag{17-36}$$

Equation (17-36) shows that, if the volume of blackbody radiation were increased adiabatically by a factor of 8, the radiation would then be capable of existing in equilibrium with matter at a temperature which is one-half the original temperature.

PROBLEMS

17-1 The tension in a steel wire of length 1 m, diameter 1 mm, and temperature 300 K is increased reversibly and isothermally from zero to 10^3 N. Assume the following quantities to remain constant: $\rho = 7.86 \times 10^3$ kg/m^3; $\alpha = 12.0 \times 10^{-6}$ K^{-1}; $Y = 2.00 \times 10^{11}$ N/m^2; $C_{\mathcal{J}} = 0.482$ J/g · K.

(*a*) How much heat in joules is transferred?

(*b*) How much work in joules is done?

(*c*) What is the change in internal energy?

(*d*) What would be the temperature change if the process were performed isentropically?

17-2 The equation of state of an ideal elastic cylinder is

$$\mathcal{J} = KT\left(\frac{L}{L_0} - \frac{L_0^2}{L^2}\right),$$

where K is a constant and L_0, the length at zero tension, is a function of T only. If the cylinder is stretched reversibly and isothermally from $L = L_0$ to $L = 2L_0$, show that:

(*a*) The heat transferred is

$$Q = -KTL_0(1 - \tfrac{5}{2}\alpha_0 T),$$

where α_0, the linear expansivity at zero tension, is expressed as

$$\alpha_0 = \frac{1}{L_0}\frac{dL_0}{dT}.$$

(*b*) The change of internal energy is

$$\Delta U = +\tfrac{5}{2}KT^2 L_0 \alpha_0.$$

17-3 In the case of the ideal elastic substance whose equation of state is given in Prob. 17-2, prove that:

(*a*) $$\left(\frac{\partial U}{\partial L}\right)_T = AY\alpha_0 T.$$

(*b*) $$\left(\frac{\partial U}{\partial \mathcal{J}}\right)_T = L\alpha_0 T.$$

17-4 An ideal elastic cylinder has the equation of state given in Prob. 17-2. When the length is changed isentropically, a temperature change occurs. This is called the *elastocaloric effect*. The magnitude of this effect is expressed by the quantity $(\partial T/\partial L)_S$.

(*a*) Derive the expression

$$\left(\frac{\partial T}{\partial L}\right)_S = \frac{KT}{C_L}\left[\left(\frac{L}{L_0} - \frac{L_0^2}{L^2}\right) - \alpha_0 T\left(\frac{L}{L_0} + \frac{2L_0^2}{L^2}\right)\right],$$

where C_L is the heat capacity at constant length of the entire cylinder.

(*b*) Assume the following values for a certain sample of rubber: $T = 300$ K (practically constant), $K = 1.33 \times 10^{-2}$ N/K, $\alpha_0 = 5 \times 10^{-4}$ K^{-1}, $C_L = 2$ J/K. Calculate $(\partial T/\partial L)_S$ for values of L/L_0 equal to 1, 1.1, 1.5, and 2.

17-5 When rubber is unstretched, X-ray diffraction experiments indicate an amorphous structure. When it is stretched isothermally, a crystalline structure is found, indicating that the large chainlike molecules are oriented.

(*a*) Is $(\partial S/\partial \mathcal{J})_T$ positive or negative?

(*b*) Prove that the linear expansivity of *stretched* rubber is negative.

17-6 A reversible cell whose emf depends only on T undergoes an isothermal transfer of jN_F coulombs of electricity.

(*a*) Show that $$\Delta S = jN_F \frac{d\mathcal{E}}{dT}.$$

(*b*) Show that $$\Delta G = -jN_F\mathcal{E}.$$

(*c*) If ΔG and ΔH are found to approach the same value as T approaches zero, what conclusion may be drawn concerning $\lim_{T\to 0} \Delta S$?

17-7 When zinc sulfate (valence = 2) reacts chemically with copper, at a temperature of 273 K and at atmospheric pressure, the heat evolved in the reaction is 2.31×10^8 J/kmol. The emf of a Daniell cell at 273 K is 1.0934 V, and the emf decreases with temperature at the rate at 4.533×10^{-4} V/K. How does the calculated heat of reaction compare with the measured value?

17-8 The emf of the cell Zn, $ZnCl_2$, Hg_2Cl_2, Hg is given by the equation

$$\mathcal{E} = 1.0000 + 0.000094(t - 15).$$

Write the reaction, and calculate the heat of reaction at 100°C.

17-9 (*a*) Prove that, for a dielectric,

$$\left(\frac{\partial U}{\partial E}\right)_T = T\left(\frac{\partial \Pi}{\partial T}\right)_E + E\left(\frac{\partial \Pi}{\partial E}\right)_T.$$

(*b*) Assuming the equation of state to be

$$\Pi = \epsilon_0 \chi V E,$$

where the susceptibility χ is a function of T only, and the volume V is constant, show that the energy per unit volume of the dielectric is equal to

$$\frac{U}{V} = f(T) + \frac{\epsilon_0 E^2}{2}\left(\chi + T\frac{d\chi}{dT}\right),$$

where $f(T)$ is an undetermined function of temperature.

(*c*) Prove that the energy per unit volume of an electric field in a vacuum is

$$\frac{\epsilon_0 E^2}{2}.$$

(*d*) Using the relation $\epsilon = \epsilon_0(1 + \chi)$, show that the *total* energy per unit volume

$$\frac{U}{V}\text{(dielectric plus field)} = f(T) + \frac{\epsilon E^2}{2}\left(1 + \frac{T}{\epsilon}\frac{d\epsilon}{dT}\right).$$

17-10 (*a*) Show that for a dielectric the difference in the heat capacities is given by

$$C_E - C_\Pi = T\frac{(\partial \Pi/\partial T)_E^2}{(\partial \Pi/\partial E)_T}.$$

(*b*) If $\Pi = \epsilon_0 \chi V E$, show that

$$\frac{C_E - C_\Pi}{V} = \frac{\epsilon_0 T}{\chi}\left(\frac{d\chi}{dT}\right)^2 E^2.$$

(*c*) If $\chi = C/T$, show that

$$\frac{C_E - C_\Pi}{V} = \frac{\epsilon_0 \chi}{T} E^2.$$

17-11 In the case of a dielectric whose equation of state is $\Pi = CVE/T$, show that:

(*a*) The heat transferred in a reversible isothermal change of field is

$$Q = -\frac{CV}{2T}(E_f^2 - E_i^2).$$

(*b*) The small temperature change accompanying a reversible adiabatic change of field is

$$\Delta T = \frac{CV}{2C_E T}(E_f^2 - E_i^2).$$

17-12 A KCl crystal doped with Li ions was found to have a heat capacity equal to AT^3 and a polarization $\Pi = (N\mu^2/3kT)E$, where N is the concentration of Li ions, μ the electric dipole moment of each Li ion, and k is Boltzmann's constant. If the initial electric field E_i is adiabatically reduced to zero, show that the resulting temperature change ΔT (with $\Delta T \ll T_i$) is given by

$$\Delta T = \frac{N\mu^2 E_i^2}{6kAT^4}.$$

This result, called the *electrocaloric effect*, was observed by G. Lombardo and R. O. Pohl in 1965. If $T = 2$ K and $N\mu^2/6kA = 1.6 \times 10^{-12}$ m$^2 \cdot$ K^5/V^2, calculate ΔT at the following values of E_i: 5×10^4 V/m, 1×10^5 V/m, 5×10^5 V/m, and 1×10^6 V/m.

17-13 A piezoelectric crystal has a direction in which a change of tension $\mathcal{F}$ causes a change of length L. In this same direction a change of electric field intensity E is accompanied by a change of polarization Π. The first and second laws yield the equation

$$T\,dS = dU - \mathcal{F}\,dL - E\,d\Pi.$$

Taking the entropy to be a function of T, $\mathcal{F}$, and E, we may write

$$T\,dS = T\left(\frac{\partial S}{\partial T}\right)_{\mathcal{F}, E} dT + T\left(\frac{\partial S}{\partial \mathcal{F}}\right)_{E, T} d\mathcal{F} + T\left(\frac{\partial S}{\partial E}\right)_{T, \mathcal{F}} dE.$$

Defining a *piezoelectric Gibbs function* G' as

$$G' = U - TS - \mathcal{F}L - E\Pi,$$

show that

$$dG' = -S\,dT - L\,d\mathcal{F} - \Pi\,dE.$$

Prove that:

(*a*) At constant $\mathcal{F}$, $$\left(\frac{\partial S}{\partial E}\right)_{T, \mathcal{F}} = \left(\frac{\partial \Pi}{\partial T}\right)_{E, \mathcal{F}}$$

(*b*) At constant E, $$\left(\frac{\partial S}{\partial \mathcal{F}}\right)_{T, E} = \left(\frac{\partial L}{\partial T}\right)_{\mathcal{F}, E}.$$

(*c*) At constant T, $$\left(\frac{\partial L}{\partial E}\right)_{T, \mathcal{F}} = \left(\frac{\partial \Pi}{\partial \mathcal{F}}\right)_{T, E}.$$

(*d*) $$T\,dS = C_{\mathcal{F}, E}\,dT + T\left(\frac{\partial L}{\partial T}\right)_{\mathcal{F}, E} d\mathcal{F} + T\left(\frac{\partial \Pi}{\partial T}\right)_{\mathcal{F}, E} dE.$$

17-14 The difference between the Seebeck coefficients for Bi and Pb is given by -43.7 μV/K $-$ (0.47 μV/K^2)t, where t is the Celsius temperature.

(*a*) Calculate the thermal emf of a Bi-Pb thermocouple with the reference junction at 0°C and the test junction at 100°C.

(*b*) What is the Peltier coefficient of the test junction at 100°C? What Peltier heat would be transferred at this junction by an electric current of 10 A in 5 min? Would this heat go into or out of the junction?

(*c*) What is the difference between the Thomson coefficients at points on the Bi and Pb wires which are both at 50°C?

17-15 The thermal emf of a Ni-Pb thermocouple is given by (19.1 μV/K)t $-$ (0.030 μV/K^2)t^2 when the reference junction is at 0°C.

(*a*) What is π at 127°C?

(*b*) What is $\sigma_{\text{Ni}} - \sigma_{\text{Pb}}$ at 57°C?

17-16 (*a*) What would be the characteristics of a thermocouple where $\sigma_A - \sigma_B = 0$?

(*b*) Would this thermocouple make a good thermometer?

17-17 Integrate Eq. (17-19) from a reference temperature T_R to any temperature T, and derive the equation

$$\mathcal{E}_{AB} = (\pi_{AB})_T - (\pi_{AB})_{T_R} + \int_{T_R}^{T} (\sigma_A - \sigma_B)\, dT.$$

(*a*) Interpret this equation in terms of a Seebeck emf, two Peltier emf's, and two Thomson emf's.

(*b*) Interpret this equation in terms of the first law of thermodynamics, neglecting irreversible effects.

(*c*) If three different wires A, B, and C are joined in series, with all junctions at the same temperature, prove that

$$(\pi_{AB})_T + (\pi_{BC})_T + (\pi_{CA})_T = 0.$$

(*d*) Two wires A and B form a junction at the temperature T. The other ends of A and B are joined by a wire C forming two junctions, both at T_R. Prove that

$$\mathcal{E}_{ABC} = \mathcal{E}_{AB}.$$

What is the importance of this result?

17-18 In Fig. P17-1 is depicted an idealized apparatus similar to that used to explain a throttling process. There is, however, a big difference. The throttling process is strictly adiabatic, whereas the gas flowing through the porous plug in Fig. P17-1 is maintained at a *constant temperature* and pressure on the left side and at a constant *higher* temperature on the right. If the pores in the plug are small enough, the pressure on the right will be found to be higher than that on the left. This process is called *thermal effusion* or the *Knudsen effect*.

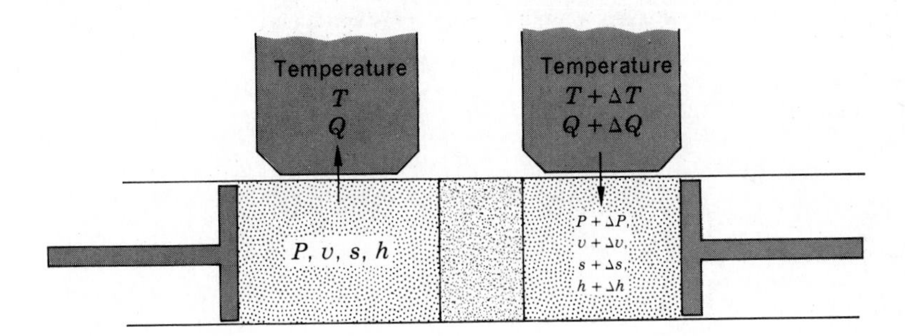

Figure P17-1

Suppose that both pistons are moved to the right simultaneously, thereby maintaining a constant pressure P on the left and a constant pressure $P + \Delta P$ on the right. Under these circumstances, let n moles of gas be transferred in time τ. In this time, suppose the gas loses heat Q at temperature T and gains heat $Q + \Delta Q$ at the temperature $T + \Delta T$.

(*a*) Show that the entropy produced in time τ is

$$\Delta S = Q \frac{\Delta T}{T^2} - \frac{\Delta Q}{T} + n\, \Delta s.$$

(*b*) Apply the first law and obtain

$$\Delta Q = n\, \Delta h.$$

(*c*) Show that

$$\frac{\Delta S}{\tau} = \frac{Q}{\tau}\frac{\Delta T}{T^2} - \frac{n}{\tau}\frac{v}{T}\Delta P$$

$$= I_S \frac{\Delta T}{T} - I_n v \frac{\Delta P}{T}.$$

(*d*) Express I_S and $-I_n$ as linear functions of $\Delta T/T$ and $\Delta P/T$.

(*e*) Show that

$$\frac{L_{12}}{L_{22}} = -\left(\frac{I}{I_n v}\right)_{\Delta T=0},$$

and

$$\frac{L_{21}}{L_{22}} = -\left(\frac{\Delta P}{\Delta T}\right)_{I_n=0}.$$

(*f*) Show that

$$\left(\frac{\Delta P}{\Delta T}\right)_{I_n=0} = \frac{(I_S/I_n)_{\Delta T=0}}{v} = \frac{S^*}{v}.$$

(*g*) The quantity S^* is the entropy accompanying the transport of 1 mol of gas through the plug. It is found from kinetic theory to be $R/2$. Prove that

$$\frac{P_1}{P_2} = \sqrt{\frac{T_1}{T_2}},$$

where the subscripts 1 and 2 refer to opposite sides of the plug.

(*h*) Derive the preceding equation by using the result of Prob. 11-10*b*.

17-19 A thermodynamic system is described with the aid of the five coordinates P, V, T, $\mathcal{Y}$, X, where

$$T\,dS = dU + P\,dV - \mathcal{Y}\,dX.$$

Prove that, if $G = U + PV - TS$, as usual:

(*a*)
$$dG = -S\,dT + V\,dP + \mathcal{Y}\,dX.$$

(*b*)
$$-\left(\frac{\partial S}{\partial P}\right)_{X,T} = \left(\frac{\partial V}{\partial T}\right)_{X,P}.$$

(*c*)
$$-\left(\frac{\partial S}{\partial X}\right)_{P,T} = \left(\frac{\partial \mathcal{Y}}{\partial T}\right)_{P,X}.$$

(*d*)
$$\left(\frac{\partial V}{\partial X}\right)_{T,P} = \left(\frac{\partial \mathcal{Y}}{\partial P}\right)_{T,X}.$$

17-20 Using the five coordinates P, V, T, $\mathcal{E}$, Z to describe a reversible cell in which gases may be liberated, show that:

(*a*)
$$dG = -S\,dT + V\,dP + \mathcal{E}\,dZ.$$

(*b*)
$$-\left(\frac{\partial S}{\partial Z}\right)_{T,P} = \left(\frac{\partial \mathcal{E}}{\partial T}\right)_{P,Z}.$$

(*c*)
$$dH = T\,dS + V\,dP + \mathcal{E}\,dZ.$$

(*d*)
$$\left(\frac{\partial H}{\partial Z}\right)_{T,P} = \mathcal{E} - T\left(\frac{\partial \mathcal{E}}{\partial T}\right)_{P,Z}.$$

(*e*) For a saturated cell $\Delta H = -\dfrac{jN_F}{J}\left[\mathcal{E} - T\left(\dfrac{\partial \mathcal{E}}{\partial T}\right)_P\right].$

17-21 Using the five coordinates P, V, T, $\mathcal{S}$, A to describe a surface film and its accompanying liquid, show that:

(a)
$$dG = -S\,dT + V\,dP + \mathcal{S}\,dA.$$

(b)
$$\left(\frac{\partial \mathcal{S}}{\partial P}\right)_{T,A} = \left(\frac{\partial V}{\partial A}\right)_{T,P}.$$

17-22 Fill in the empty boxes in this table:

	Ideal-gas molecules	Conduction electrons	Photons in an evacuated cavity
How does N_i compare with g_i?			
What is the expression for g_i?			
What fraction of these states is filled?			
How many particles are in each filled state?			
Is $\sum N_i = \text{const.}$?			
What kind of statistics is appropriate?			

17-23 (a) The spectral energy curve of sunlight has a maximum at a wavelength of 4.84×10^{-7} m. Assuming the sun to be a blackbody, what is the temperature of its emitting surface?

(b) What is the energy density of the sun's radiation?

(c) What is the radiation pressure exerted by the sun's radiation?

17-24 The volume of the blackbody radiation contained in a cavity with walls at a temperature of 2000 K is increased reversibly and isothermally from 10 to 1000 cm^3.

(a) How much heat is transferred?

(b) What is the change of energy of the radiation?

(c) How much work was done?

(d) If the expansion were performed reversibly and adiabatically, what would be the final temperature of the radiation?

17-25 Prove that, for blackbody radiation:

(a)
$$PV^{4/3} = \text{const., for an isentropic process.}$$

(b)
$$\frac{S}{V} = \tfrac{4}{3}bT^3.$$

(c)
$$C_P = \infty.$$

(d)
$$G = 0.$$

CHAPTER

EIGHTEEN

IONIC PARAMAGNETISM AND CRYOGENICS

18-1 ATOMIC MAGNETISM

An atom or molecule in its lowest energy state and uninfluenced by its neighbors or by an external magnetic field may have a magnetic moment by virtue of two kinds of electronic processes: (1) orbital motion of one or more electrons, and (2) uncompensated spins of one or more electrons. These two effects are specified by the resultant orbital quantum number L and the net-spin quantum number S. The resultant of L and S is denoted by J. It is often the case that the orbital quantum number L is zero, in which case the atomic magnetism is due entirely to electron spin, since $J = S$.

When atoms or molecules are moving freely in the *gaseous* phase, the magnetic moment of the gas is determined by the L and S values of each *neutral* atom or molecule. Thus, oxygen gas is paramagnetic, because the resultant orbital quantum number of each neutral oxygen molecule is zero, and $S = 1$ for the ground state of the molecule. At lower temperatures, when atomic particles coalesce to form crystals, the atomic structures that occupy the various lattice sites are rarely neutral. In the sodium chloride crystal, the lattice sites are occupied by positive sodium ions and negative chlorine ions. Each of these ions has an inert-gas configuration of closed electron shells, with $L = 0$, $S = 0$, and $J = 0$. In a crystal where two or more valence electrons are removed from one atom and taken up by another, the resulting structures may again have an inert-gas configuration with no resultant magnetic moment.

There are some atoms, however, such as Cr and Fe, each with three valence electrons which exist as *trivalent ions* in the crystalline phase and do *not* have an inert-gas structure because *the third electron shell is incomplete*. These trivalent ions Cr^{3+} and Fe^{3+} have magnetic moments. The effect of orbital motions is said to be *quenched* by the fields of neighboring ions, but there remains a net electron spin S, so that $J = S$. The rare-earth atoms are particularly interesting in that, when their valence electrons are removed, the resulting ion lacks an inert-gas configuration because of incomplete filling of the $4f$ electron shell. In the particular case of gadolinium, which in the crystalline state is in the form of a trivalent ion Gd^{3+}, the ground state has an L value equal to zero, so that again $J = S$.

When an atom, molecule, or ion possesses a resultant magnetic moment in its lowest energy level and is sufficiently removed from its neighbors to be only weakly influenced by neighboring magnetic fields, it behaves in a simple way when subjected to an external magnetic field. The level splits into a number of separate states, each of which is characterized by a *magnetic quantum number m*. The allowed, discrete values of m correspond to discrete orientations of the magnetic-moment vector with respect to the external magnetic field. These values of m vary in *integral* steps from $-J$ to $+J$. Thus, when Fe^{3+} in its lowest energy level, with $J = \frac{5}{2}$, is subjected to a magnetic field $\mathcal{H}$, this level is split into six different energy states with m equal to

$$-\tfrac{5}{2}, -\tfrac{3}{2}, -\tfrac{1}{2}, +\tfrac{1}{2}, +\tfrac{3}{2}, +\tfrac{5}{2}.$$

When $\mathcal{H} = 0$, these six states all have the same energy, and therefore the lowest energy level of Fe^{3+} has a degeneracy of 6, or $2J + 1$. The effect of an external magnetic field is to remove this degeneracy.

If the energy of the lowest level is arbitrarily set equal to zero, then the magnetic potential energy of the ion in any one of the magnetic states is equal to

$$\epsilon_i = -g\mu_B\mu_0\mathcal{H}m_i, \qquad (18\text{-}1)$$

where μ_B is the Bohr magneton,

$$\mu_B = \frac{eh}{4\pi m} = 9.274 \times 10^{-24} \text{ J/T},$$

μ_0 is the permeability of free space, and $\mathcal{H}$ is the external magnetic field when the sample of crystal is a long cylinder oriented parallel to $\mathcal{H}$. The quantity g is called the *Landé splitting factor*. In the case of a crystal whose internal magnetic field (created by other magnetic ions) is *isotropic*, $g = 2$ when $L = 0$. The minus sign in Eq. (18-1) merely indicates that positive values of m correspond to orientations of the magnetic-moment vector with a component *in the direction* of $\mathcal{H}$, and these orientations involve a *decrease* of energy. When the orientations of the ion are such as to have components antiparallel to the field, then m is negative, and the potential energy of the ion is increased. Think of the magnetic ion as a small compass needle. If held perpendicular to a magnetic field and then allowed to undergo a quasi-static rotation till its magnetic moment is parallel to the field, it will do work on its surroundings and undergo a decrease in energy. To point a compass needle antiparallel to a magnetic field, work would have to be done *on*

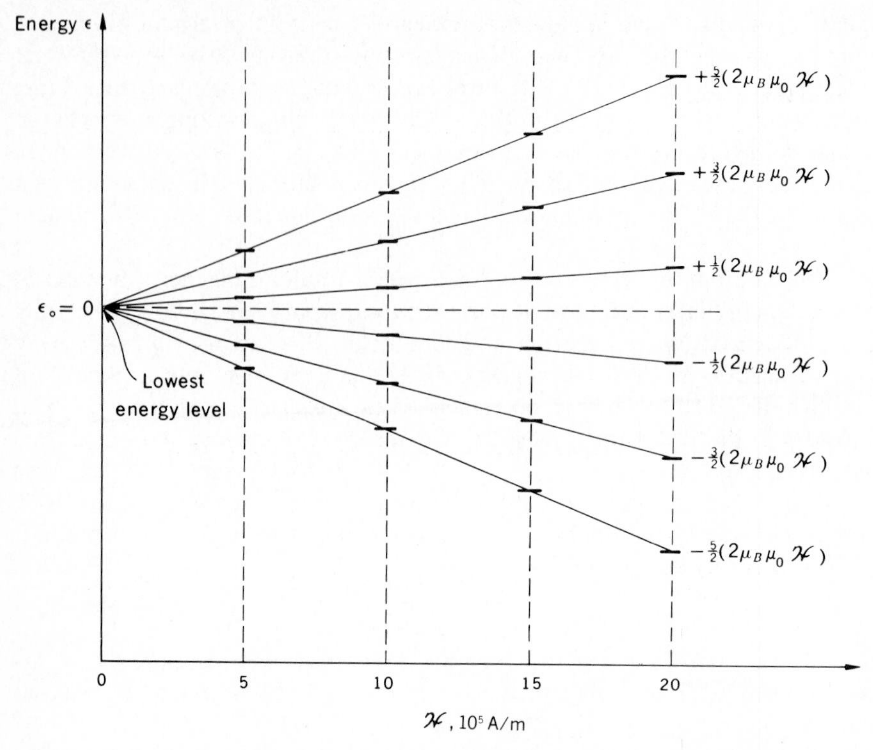

Figure 18-1 Splitting of the lowest level of Fe^{3+} ion in a magnetic field into six separate energy states.

the needle in twisting it against an opposing torque, and its potential energy would therefore increase.

The magnetic splitting, proportional to the external magnetic field $\mathcal{H}$, is shown for Fe^{3+} in Fig. 18-1, and some of the electronic data for the lowest energy level of four important trivalent magnetic ions existing inside a crystal lattice are listed in Table 18-1.

The upper energy levels of an atom or ion are also split into a number of separate magnetic energy states by an external magnetic field. In the case of a gas supporting an electric discharge in a narrow tube placed between the poles of a strong electromagnet, atoms in one of the magnetic energy states of an upper energy level may undergo a transition to a magnetic state of a lower energy level and emit a spectral line. What was, in the absence of the magnetic field, a single spectral line now becomes a number of spectral lines. This apparent "splitting" of a spectral line into a pattern of lines is called the *Zeeman effect*.

It may be seen in Fig. 18-1 that at a low value of $\mathcal{H}$ the six magnetic states are close together, so that, if $\Delta\epsilon$ is the energy difference between any two adja-

Table 18-1 Magnetic ions in paramagnetic salts

Paramagnetic salt	Magnetic ion	S	L	J	g
$Cr_2(SO_4)_3 \cdot K_2SO_4 \cdot 24H_2O$ (chromium potassium alum)	Cr^{3+}	$\frac{3}{2}$	0 (quenched)	$\frac{3}{2}$	2
$Fe_2(SO_4)_3 \cdot (NH_4)_2SO_4 \cdot 24H_2O$ (iron ammonium alum)	Fe^{3+}	$\frac{5}{2}$	0 (quenched)	$\frac{5}{2}$	2
$Gd_2(SO_4)_3 \cdot 8H_2O$ (gadolinium sulphate)	Gd^{3+}	$\frac{7}{2}$	0	$\frac{7}{2}$	2
$2Ce(NO_3)_3 \cdot 3Mg(NO_3)_2 \cdot 24H_2O$ (cerium magnesium nitrate, CMN)	Ce^{3+}	$\frac{1}{2}$	3	$\frac{5}{2}$	$(\perp)1.84$ $(\parallel)0.02$

cent states, $\Delta\epsilon$ is very small. If $\mathcal{H}$ is 10^5 A/m,

$$\Delta\epsilon = 2\mu_B\mu_0\mathcal{H} = 2 \times 9.27 \times 10^{-24}\ \text{J/T} \times 4\pi \times 10^{-7}\ \text{H/m} \times 10^5\ \text{A/m}$$
$$= 0.23 \times 10^{-23}\ \text{J}.$$

At a temperature of 1 K,

$$kT = 1.38 \times 10^{-24}\ \text{J/K} \times 1\ \text{K}$$
$$= 1.38 \times 10^{-23}\ \text{J}.$$

Hence, at $T = 1$ K and $\mathcal{H} = 10^5$ A/m,

$$\Delta\epsilon < kT \qquad (\text{for } \mathcal{H} = 10^5\ \text{A/m}).$$

In a large magnetic field, however, say, $\mathcal{H} = 2 \times 10^6$ A/m, $\Delta\epsilon = 4.66 \times 10^{-23}$ J, and kT at 1 K is still 1.38×10^{-23} J, so that

$$\Delta\epsilon > kT \qquad (\text{for } \mathcal{H} = 2 \times 10^6\ \text{A/m}).$$

In the paramagnetic crystals listed in Table 18-1, the distribution of magnetic ions among the various magnetic states is given by the Boltzmann equation,

$$N_i \propto e^{-\epsilon_i/kT}.$$

Therefore, when $\Delta\epsilon < kT$, all the six closely packed states that exist in a weak field may be populated almost equally. Since each state corresponds to a different orientation with regard to the external magnetic field, the *net magnetic polarization* or *magnetization* M of the crystal would then be very small. When the field is strong enough to make $\Delta\epsilon > kT$, however, only the lowest state, with energy $-J \cdot 2\mu_B\mu_0\mathcal{H}$, will be populated. Since this state corresponds to the ions lining up parallel to the external magnetic field, the magnetization M would have its largest, or *saturation*, value M_{sat}.

18-2 STATISTICAL MECHANICS OF A MAGNETIC-ION SUBSYSTEM

The ions occupying lattice sites in a typical crystal such as sodium chloride are localized (and therefore distinguishable) and tremendously influenced by their neighbors. When the vibration characteristics of the crystal *as a whole* are analyzed, it is found that there are three times as many normal modes as lattice points and that these $3N$ normal modes may be treated statistically as $3N$ distinguishable but *weakly interacting* harmonic oscillators. Upon making use of the appropriate expression for the thermodynamic probability (see Art. 12-1) and performing the usual operations, the Boltzmann equation was found to apply.

The paramagnetic crystals of greatest use in practical thermodynamics contain paramagnetic ions surrounded by a very large number of non-magnetic particles, as shown in Table 18-1. Each chromium ion in $Cr_2(SO_4)_3 \cdot K_2SO_4 \cdot 24H_2O$ is surrounded by 1 potassium atom, 2 sulfur atoms, 20 oxygen atoms, and 24 hydrogen atoms—a total of 47 nonmagnetic particles. In other words, the chromium ions are very dilute and far apart, so that magnetic interaction among them is very weak. The same is true of the magnetic ions in the other three salts listed in Table 18-1. The behavior of the magnetic ions in a dilute paramagnetic salt is almost gaslike in the weakness of interaction, but the ions are, of course, distinguishable by their positions. We may therefore apply the statistical method appropriate to these distinguishable, weakly interacting particles.

Consider a piece of crystal containing N magnetic ions and about fifty times as many nonmagnetic particles. Let us suppose that the temperature of the crystal is so low (in the neighborhood of 1 K) that *the vibrational energy and the heat capacity of everything but the magnetic ions may be ignored*. The N magnetic ions constitute a *subsystem* with which there is associated a temperature T of its own, which may or may not be the same as the temperature of the rest of the crystal. The *magnetization* (or *total magnetic moment*) M of the ions has nothing to do with the rest of the crystal, and the external magnetic field $\mathcal{H}$ produces no effect on the rest of the crystal. In other words, *the magnetic ions form a subsystem with its own identity, describable with the aid of its coordinates* $\mathcal{H}$, M, *and* T, *as though the rest of the crystal were a container*. The magnetic potential energy ϵ_i of any ion is equal to $-g\mu_B\mu_0\mathcal{H}m_i$, but this is not the only energy possessed by the ion.

It has been mentioned that the lowest energy level of a magnetic ion in the absence of an external magnetic field has a $(2J + 1)$-fold degeneracy which is removed by the splitting action of the magnetic field. *Even in the absence of an external magnetic field, there are internal fields which provide some splitting of the lowest energy level*. There are, of course, the very weak magnetic fields provided by the other relatively distant magnetic ions, but the effect of these fields is small and sometimes may be neglected in comparison with the much stronger effect of the *electric field* in the crystal. This field is due to the positive and negative charges of the ions in the lattice, and the interaction between an atom or ion and an electric field, known as the *Stark effect*, is similar to the Zeeman effect in that some or all of the degeneracy of an energy level may be removed by the *electric splitting* of the level into a number of electric states.

We have seen that a magnetic field $\mathcal{H}$ gives rise to magnetic states whose separation is given by $\Delta\epsilon = 2\mu_B\mu_0\mathcal{H}$. It is often convenient to express this "splitting energy" in terms of a temperature T, where $kT = 2\mu_B\mu_0\mathcal{H}$. The magnetic splitting of a field of, say, 2×10^6 A/m would correspond to a temperature

$$T = \frac{2\mu_B\mu_0\mathcal{H}}{k} = \frac{2 \times 9.27 \times 10^{-24}\ \text{J/T} \times 4\pi \times 10^{-7}\ \text{H/m} \times 2 \times 10^6\ \text{A/m}}{1.38 \times 10^{-23}\ \text{J/K}},$$

or $\quad T \approx 3$ K.

The Stark-effect splitting in most crystals is of the order of a tenth or a hundredth of a degree or less. Sometimes the entire degeneracy of the lowest energy level is not completely removed by the crystalline field, and the level is split into only a few electric states. This splitting is much too small to show in Fig. 18-1 at the point where $\mathcal{H} = 0$. Suppose, for the sake of simplicity, we assume that the effect of the crystalline field is to provide only two states whose energies are zero and δ_1, and whose degeneracies are g_0 and g_1. *In zero magnetic field* and at temperatures far below δ_1/k, most of the ions would be in the lower energy state; whereas at temperatures much higher than δ_1/k, the two states would have populations proportional to g_0/g_1.

To apply statistical mechanics rigorously to a paramagnetic crystal, taking into account (1) the magnetic-ion subsystem in a field $\mathcal{H}$, (2) its magnetic and electric interaction with the lattice when $\mathcal{H} = 0$, and (3) its mechanical and thermal interaction with the lattice, is a complicated problem. Much can be done and understood, however, by applying the simple method of dealing with localized, weakly interacting particles that was used in Chap. 12 with a crystal.

Let ϵ_i stand for the *total* energy of a paramagnetic ion, equal to the sum of the energy δ_i due to crystalline field splitting and the magnetic potential energy $-g\mu_B\mu_0\mathcal{H}m_i$ due to the presence of an external magnetic field $\mathcal{H}$. Thus

$$\epsilon_i = \delta_i - g\mu_B\mu_0\mathcal{H}m_i, \tag{18-2}$$

where δ_i has only two values, 0 and δ_1, and m_i ranges in integral steps from $-J$ to $+J$. Let $N_1, N_2, \ldots N_i, \ldots$ be the instantaneous, nonequilibrium populations of the various energy states. In the systems we have studied up to this point, namely, an ideal gas, a nonmetallic crystal, a metallic crystal, and a photon gas, it was found possible to interpret the sum $\sum N_i\epsilon_i$ unambiguously as the internal energy U. This is *not* the case, however, with a paramagnetic ion subsystem in a magnetic field, where

$$\sum N_i\epsilon_i = \sum N_i\delta_i - \mu_0\mathcal{H}\sum N_i g\mu_B m_i.$$

The sum in the second term on the right is seen to be the magnetization M, so that the entire second term, $-\mu_0\mathcal{H}M$, is the *magnetic potential energy* of a paramagnetic system of magnetization M in a field $\mathscr{H}$, a quantity which cannot be included within an expression for internal energy. We are forced to conclude, therefore, that the first sum on the right is the internal energy, or

$$U = \sum N_i \delta_i .$$

The sum $\sum N_i \epsilon_i$ is therefore

$$\sum N_i \epsilon_i = U = \mu_0 \mathscr{H} M, \tag{18-3}$$

which is the analog of the enthalpy and is therefore the *magnetic enthalpy* H^*. If our system of paramagnetic ions approaches equilibrium (1) adiabatically ($đQ = 0$) and (2) at constant field ($d\mathscr{H} = 0$), then

$$\begin{aligned} d \sum N_i \epsilon_i &= dU - \mu_0 \mathscr{H} dM - M \mu_0 \, d\mathscr{H} \\ &= đQ - M \mu_0 \, d\mathscr{H}, \\ &= 0, \end{aligned}$$

or $\sum N_i \epsilon_i$ *remains constant.*

For a system of N *localized*, weakly interacting ions, the entropy is given by

$$S = k \ln \frac{N!}{N_1! N_2! \ldots},$$

where $\sum N_i = N =$ constant. Using the Stirling approximation, and taking the differential, we get

$$dS = -k \sum \ln N_i \, dN_i .$$

If equilibrium is approached adiabatically, at constant $\mathscr{H}$ and at constant N, we have the usual three equations to satisfy:

$$\begin{aligned} -k \sum \ln N_i \, dN_i &= 0, \\ \sum \epsilon_i \, dN_i &= 0, \\ \sum dN_i &= 0. \end{aligned}$$

Using the Lagrange method, we get

$$N_i = A e^{-\beta \epsilon_i},$$

where

$$A = \frac{N}{Z},$$

and

$$Z = \sum e^{-\beta \epsilon_i}.$$

Substituting the equilibrium values of the N's into the expression for dS, it is a simple matter to show that

$$\beta = \frac{1}{kT},$$

and therefore

$$N_i = \frac{N}{Z} e^{-\epsilon_i / kT}.$$

Substituting these values of the N's into the expression for the entropy, we get

$$S = Nk \ln Z + \frac{\sum N_i \epsilon_i}{T}.$$

Since $\sum N_i \epsilon_i = U - \mu_0 \mathscr{H} M$, we get

$$S = Nk \ln Z + \frac{U - \mu_0 \mathscr{H} M}{T},$$

or

$$U - TS - \mu_0 \mathscr{H} M = -NkT \ln Z.$$

The function on the left is the analog of the Gibbs function and is therefore the *magnetic Gibbs function*,

$$G^* = U - TS - \mu_0 \mathscr{H} M, \tag{18-4}$$

and we get finally

$$\boxed{G^* = -NkT \ln Z.} \tag{18-5}$$

Now,

$$dG^* = dU - T\,dS - \mu_0 \mathscr{H}\,dM - S\,dT - M\mu_0\,d\mathscr{H},$$

and since

$$T\,dS = dU - \mu_0 \mathscr{H}\,dM,$$

$$dG^* = -S\,dT - M\mu_0\,d\mathscr{H}.$$

Therefore,

$$M = -\frac{1}{\mu_0}\left(\frac{\partial G^*}{\partial \mathscr{H}}\right)_T;$$

consequently,

$$\boxed{M = \frac{NkT}{\mu_0}\left(\frac{\partial \ln Z}{\partial \mathscr{H}}\right)_T.} \tag{18-6}$$

Also,

$$S = -\left(\frac{\partial G^*}{\partial T}\right)_{\mathscr{H}},$$

and therefore

$$\boxed{S = NkT\left(\frac{\partial \ln Z}{\partial T}\right)_{\mathscr{H}} + Nk \ln Z.} \tag{18-7}$$

Finally,

$$U = G^* + TS + \mu_0 \mathscr{H} M,$$

$$= -NkT \ln Z + NkT^2\left(\frac{\partial \ln Z}{\partial T}\right)_{\mathscr{H}} + NkT \ln Z + \mu_0 \mathscr{H} M,$$

and

$$\boxed{U = NkT^2\left(\frac{\partial \ln Z}{\partial T}\right)_{\mathscr{H}} + \mu_0 \mathscr{H} M.} \tag{18-8}$$

The four boxed equations will enable us to find all the desirable information about our magnetic-ion subsystem, once we have our partition function expressed as a function of T and $\mathcal{H}$. Since $\epsilon_i = \delta_i - g\mu_B\mu_0\mathcal{H}m_i$,

$$Z = \sum g_i e^{-(\delta_i - g\mu_B\mu_0\mathcal{H}m_i)/kT}$$

$$= \sum g_i e^{-\delta_i/kT} \sum e^{g\mu_B\mu_0\mathcal{H}m_i/kT},$$

or

$$Z = Z_{\text{int}} Z_{\mathcal{H}}. \tag{18-9}$$

We have simplified matters by assuming that the crystal field splitting gives rise to only two states: one of energy zero and degeneracy g_0, and the other of energy δ_1 and degeneracy g_1. Then,

$$Z_{\text{int}} = g_0 + g_1 e^{-\delta_1/kT}, \tag{18-10}$$

which is a function of T only.

In the next section, we shall evaluate $Z_{\mathcal{H}}$, which will be found to be a function of $\mathcal{H}/T$.

18-3 MAGNETIC MOMENT OF A MAGNETIC-ION SUBSYSTEM

The magnetic part of the partition function of a magnetic-ion subsystem is given by

$$Z_{\mathcal{H}} = \sum e^{g\mu_B\mu_0\mathcal{H}m_i/kT}. \tag{18-11}$$

If we let

$$a = \frac{g\mu_B\mu_0\mathcal{H}}{kT}, \tag{18-12}$$

then, since m_i can take on the values $-J, -J+1, \ldots, J-1, J$, we find

$$Z_{\mathcal{H}} = \sum_{m_i=-J}^{m_i=+J} e^{am_i} = e^{-aJ} + e^{-a(J-1)} + \cdots + e^{aJ}.$$

This is a finite geometric progression with a ratio e^a; hence,

$$Z_{\mathcal{H}} = \frac{e^{-aJ} - e^{a(J+1)}}{1 - e^a}.$$

Multiplying numerator and denominator by $e^{-a/2}$, we get

$$Z_{\mathcal{H}} = \frac{e^{-a(J+1/2)} - e^{a(J+1/2)}}{e^{-a/2} - e^{a/2}},$$

or

$$Z_{\mathcal{H}} = \frac{\sinh\,(J+\frac{1}{2})a}{\sinh\,\frac{1}{2}a}. \tag{18-13}$$

Therefore,

$$\boxed{\ln Z = \ln \sinh (J + \tfrac{1}{2})a - \ln \sinh \frac{a}{2} + \ln (g_0 + g_1 e^{-\delta_1/kT}).} \tag{18-14}$$

Since $a = g\mu_B\mu_0 \mathcal{H}/kT$, ln Z is seen to be a function of $\mathcal{H}$ and T.

We are now in a position to evaluate M with the aid of Eq. (18-6).

$$\begin{aligned}
M &= \frac{NkT}{\mu_0}\left(\frac{\partial \ln Z}{\partial \mathcal{H}}\right)_T = \frac{NkT}{\mu_0}\left(\frac{\partial \ln Z_{\mathcal{H}}}{\partial \mathcal{H}}\right)_T \\
&= \frac{NkT}{\mu_0}\left[\frac{d \ln Z_{\mathcal{H}}}{da}\left(\frac{\partial a}{\partial \mathcal{H}}\right)_T\right] = \frac{NkT}{\mu_0}\left(\frac{g\mu_B\mu_0}{kT}\frac{d \ln Z_{\mathcal{H}}}{da}\right) \\
&= Ng\mu_B \frac{d}{da}[\ln \sinh (J + \tfrac{1}{2})a - \ln \sinh \tfrac{1}{2}a] \\
&= Ng\mu_B \left[\frac{(J + \frac{1}{2}) \cosh (J + \frac{1}{2})a}{\sinh (J + \frac{1}{2})a} - \frac{\frac{1}{2} \cosh \frac{1}{2}a}{\sinh \frac{1}{2}a}\right].
\end{aligned}$$

Finally,

$$M = Ng\mu_B J \left\{\frac{1}{J}[(J + \tfrac{1}{2}) \coth (J + \tfrac{1}{2})a - \tfrac{1}{2} \coth \tfrac{1}{2}a]\right\}. \tag{18-15}$$

The quantity in curved brackets is called the *Brillouin function* $B_J(a)$, named after L. Brillouin, who first extended the classical theory of paramagnetism (due to Langevin) to include quantum ideas. Therefore,

$$\boxed{M = Ng\mu_B J B_J(a),} \tag{18-16}$$

where

$$B_J(a) = \frac{1}{J}[(J + \tfrac{1}{2}) \coth (J + \tfrac{1}{2})a - \tfrac{1}{2} \coth \tfrac{1}{2}a]. \tag{18-17}$$

Before we consider the consequences of Brillouin's equation for M, let us examine the mathematical features of the Brillouin function, which is plotted in Fig. 18-2 for a number of values of J. By definition,

$$\coth x = \frac{\cosh x}{\sinh x} = \frac{e^x + e^{-x}}{e^x - e^{-x}},$$

and for $x \gg 1$, $e^x \gg e^{-x}$, so that $\coth x = 1$. Therefore, for large values of a,

$$B_J(a) = \frac{1}{J}[(J + \tfrac{1}{2}) - \tfrac{1}{2}] = 1 \qquad (\text{for } a \gg 1).$$

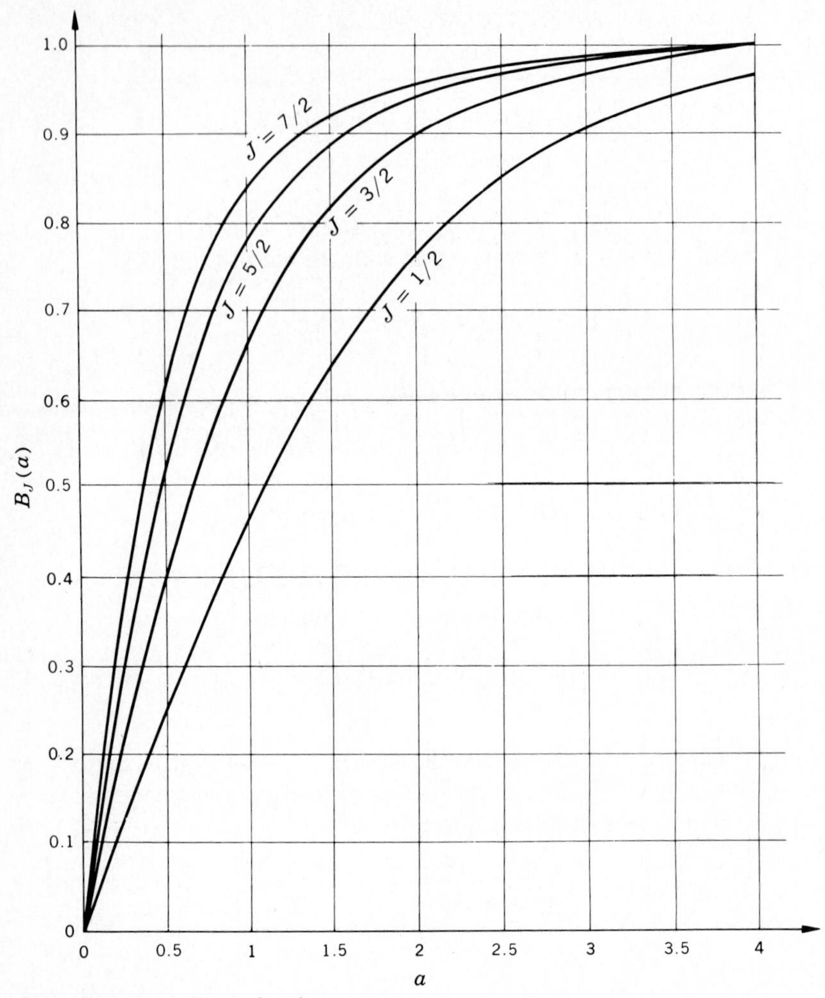

Figure 18-2 The Brillouin function.

Since $a = g\mu_B\mu_0 \quad /kT$, large values of a are achieved when

$$\frac{g\mu_B\mu_0\,\mathscr{H}}{kT} \gg 1,$$

or

$$\frac{\mathscr{H}}{T} \gg \frac{k}{g\mu_B\mu_0} \gg \frac{1.38 \times 10^{-23}\ \text{J/K}}{2 \times 9.27 \times 10^{-24}\ \text{J/T} \times 4\pi \times 10^{-7}\ \text{H/m}},$$

or

$$\frac{\mathscr{H}}{T} \gg 6 \times 10^5\ \text{A/m}.$$

Since $M = N\mu_B gJB_J(a)$, it follows that at values of $\mathscr{H}/T$ greatly in excess of 6×10^5 A/m/K, the magnetization M has its saturation value M_{sat} where

$M_{sat} = Ng\mu_B J$, and the *magnetization per magnetic ion expressed in terms of Bohr magnetons*, $M_{sat}/N\mu_B$, becomes

$$\frac{M_{sat}}{N\mu_B} = gJ \qquad (\text{for } \mathscr{H}/T \gg 6 \times 10^5 \text{ A/m/K}).$$

This conclusion was tested by W. E. Henry with the first three salts listed in Table 18-1. The experimental results, plotted in Fig. 18-3, agree very well with the Brillouin equation and with the limiting value of $M/N\mu_B$.

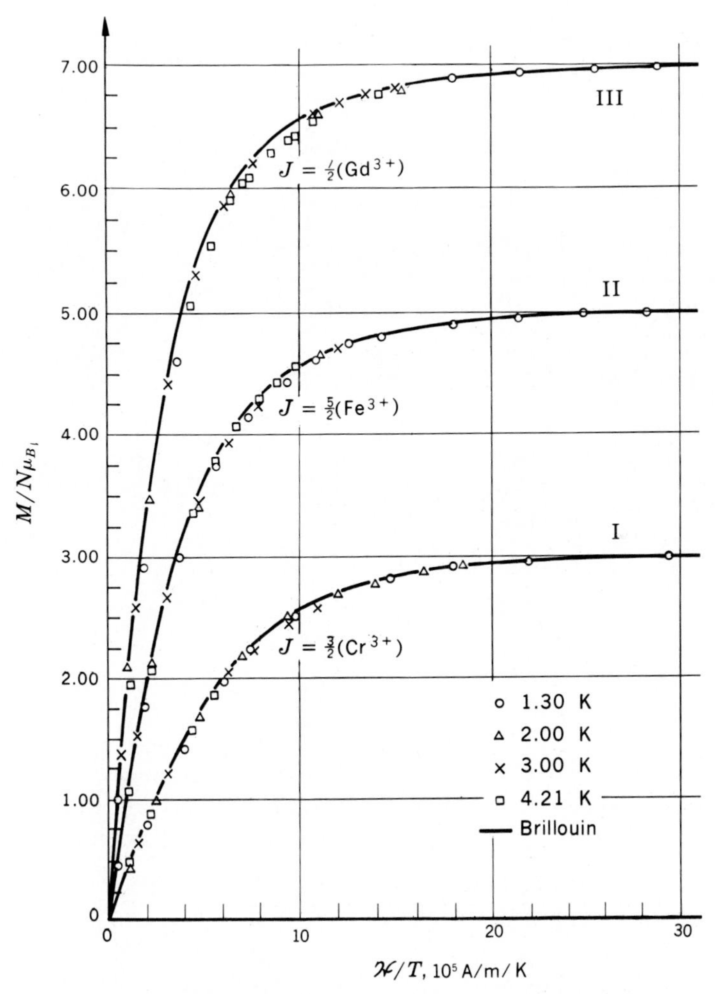

Figure 18-3 The magnetization divided by the number of Bohr magnetons plotted against $\mathscr{H}/T$ for (I) chromium potassium alum ($J = \frac{3}{2}$); (II) iron ammonium alum ($J = \frac{5}{2}$); and (III) gadolinium sulfate ($J = \frac{7}{2}$). (*The points are experimental results of W. E. Henry (1952), and the solid curves are graphs of the Brillouin equation.*)

When $x \ll 1$, it is easy to show that

$$\coth x = \frac{1}{x} + \frac{x}{3}.$$

Therefore, for small values of a,

$$B_J(a) = \frac{1}{J}\left\{(J+\tfrac{1}{2})\left[\frac{1}{(J+\frac{1}{2})a} + \tfrac{1}{3}(J+\tfrac{1}{2})a\right] - \frac{1}{2}\left[\frac{2}{a} + \frac{a}{6}\right]\right\}$$

$$= \frac{1}{J}\left\{\tfrac{1}{3}(J+\tfrac{1}{2})^2 a - \frac{a}{12}\right\}$$

$$= \frac{a}{3J}(J^2 + J + \tfrac{1}{4} - \tfrac{1}{4}),$$

and finally,

$$B_J(a) = \frac{J+1}{3}a \qquad (\text{for } a \ll 1).$$

Small values of a are achieved when

$$\frac{\mathcal{H}}{T} \ll 6 \times 10^5 \text{ A/m/K}.$$

When this is the case,

$$M = Ng\mu_B J\frac{J+1}{3}a = \frac{Ng^2\mu_B^2\mu_0 J(J+1)\mathcal{H}}{3kT},$$

or

$$\boxed{M = \frac{Ng^2\mu_B^2\mu_0 J(J+1)}{3k}\frac{\mathcal{H}}{T}} \qquad \left(\text{for } \frac{\mathcal{H}}{T} \ll 6 \times 10^5 \text{ A/m/K}\right). \tag{18-18}$$

This is Curie's equation first mentioned in Art. 2-12, and the Curie constant C'_C is seen to be

$$\boxed{C'_C = \frac{N\mu_B^2\mu_0 g^2 J(J+1)}{3k}.} \tag{18-19}$$

A mass of crystal containing exactly N_A (Avogadro's number) magnetic ions is known as the molar mass $\mathcal{M}$. The molar masses of the four crystals listed in Table 18-1 are given in Table 18-2. The Curie constant per mole is given by

$$C'_C = \frac{6.02 \times 10^{23} \text{ 1/mol} \times (9.27)^2 \times 10^{-48} \text{ J}^2/\text{T}^2 \times 4\pi \times 10^{-7} \text{ H/m}}{3 \times 1.38 \times 10^{-23} \text{ J/K}} g^2 J(J+1).$$

$$C'_C = 1.57 \times 10^{-6}\frac{\text{m}^3\text{ K}}{\text{mol}} g^2 J(J+1) \tag{18-20}$$

and provides the calculated values in Table 18-2. The agreement with the measured values is very good.

Table 18-2 Curie constants and heat-capacity constants

Paramagnetic salt	$\mathcal{M}$ kg	$J(J+1)$	C'_C (measured) $10^{-5}\,\frac{m^3 \cdot K}{mol}$	C'_C (calculated) $10^{-5}\,\frac{m^3 \cdot K}{mol}$	A/R K^2
$Cr_2(SO_4)_3 \cdot K_2SO_4 \cdot 24H_2O$	0.499	3.75	2.31	2.36	0.018
$Fe_2(SO_4)_3 \cdot (NH_4)_2SO_4 \cdot 24H_2O$	0.482	8.75	5.52	5.50	0.013
$Gd_2(SO_4)_3 \cdot 8H_2O$	0.373	15.75	9.80	9.89	0.35
$2Ce(NO_3)_3 \cdot 3Mg(NO_3)_2 \cdot 24H_2O$	0.765		~ 0 (‖) 0.398 (⊥)		6.1×10^{-6}

18-4 THERMAL PROPERTIES OF A MAGNETIC-ION SUBSYSTEM

The internal energy of a magnetic-ion subsystem is given by Eq. (18-8), namely,

$$U = NkT^2 \left(\frac{\partial \ln Z}{\partial T} \right)_{\mathcal{H}} + \mu_0 \mathcal{H} M.$$

We have

$$\begin{aligned} \left(\frac{\partial \ln Z}{\partial T} \right)_{\mathcal{H}} &= \frac{d \ln Z_{\mathcal{H}}}{da} \left(\frac{\partial a}{\partial T} \right)_{\mathcal{H}} + \frac{d \ln Z_{\text{int}}}{dT} \\ &= JB_J(a) \left(-\frac{g\mu_B \mu_0 \mathcal{H}}{kT^2} \right) + \frac{d}{dT} [\ln (g_0 + g_1 e^{-\delta_1/kT})] \\ &= -\frac{g\mu_B \mu_0 \mathcal{H}}{kT^2} JB_J(a) + \frac{(g_1 \delta_1 / kT^2) e^{-\delta_1/kT}}{g_0 + g_1 e^{-\delta_1/kT}}. \end{aligned}$$

Therefore,
$$U = -Ng\mu_B \mu_0 \mathcal{H} JB_J(a) + \frac{N\delta_1}{1 + (g_0/g_1) e^{\delta_1/kT}} + \mu_0 \mathcal{H} M. \quad (18\text{-}21)$$

According to Eq. (18-16), the first term on the right is $-\mu_0 \mathcal{H} M$, which cancels the third term, so that

$$\boxed{U = \frac{N\delta_1}{1 + (g_0/g_1) e^{\delta_1/kT}}.} \quad (18\text{-}22)$$

Since
$$T\,dS = dU - \mu_0 \mathcal{H}\,dM,$$

$$T \frac{dS}{dT} = \frac{dU}{dT} - \mu_0 \mathcal{H} \frac{dM}{dT}. \quad (18\text{-}23)$$

At constant M, the left-hand member becomes C_M, and since U is a function of T only, dU/dT is also a function of T only. Therefore,

$$C_M = \frac{dU}{dT}, \tag{18-24}$$

and from the expression for U in Eq. (18-22),

$$C_M = \frac{-N\delta_1(g_0/g_1)(-\delta_1/kT^2)e^{\delta_1/kT}}{[1 + (g_0/g_1)e^{\delta_1/kT}]^2}.$$

If $N = N_A$, then, since $N_A k = R$, we have the heat capacity per mole c_M given by

$$\frac{c_M}{R} = \frac{\delta_1^2}{k^2T^2}\frac{(g_0/g_1)e^{\delta_1/kT}}{[1 + (g_0/g_1)e^{\delta_1/kT}]^2}. \tag{18-25}$$

This is known as *Schottky's equation* and is plotted in Fig. 18-4 for three different values of the ratio g_1/g_0.

To appreciate the numerical aspects of Schottky's equation, consider a magnetic-ion subsystem in which $\delta_1/k = 0.1$ K and $g_0/g_1 = 1$. From the middle curve of Fig. 18-4, we see that, at $T = \delta_1/k = 0.1$ K, $c_M/R = 0.2$. Since $R = 8.3$ J/mol K, $c_M \approx 1700$ mJ/mol K. If the rest of the crystal had a Debye Θ of about 250 K and, say, 100 times as many particles, the heat capacity would be, from Eq. (12-21),

$$c \approx 100\left(\frac{125}{\Theta}\right)^3 T^3 \approx 100 \times \tfrac{1}{8} \times (0.1)^3 \approx 0.013 \text{ mJ/mol} \cdot \text{K}.$$

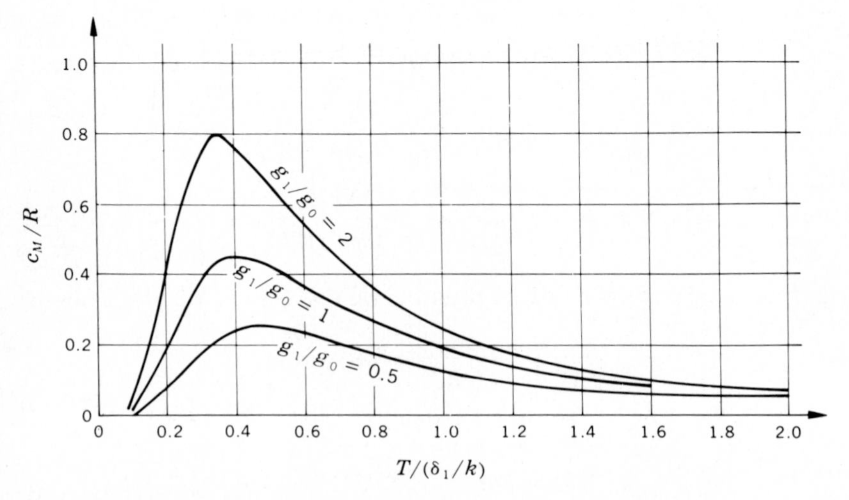

Figure 18-4 Schottky heat capacity of a magnetic-ion subsystem whose lowest energy level, in the absence of an external magnetic field, is split into two states with degeneracies g_0 and g_1 for several values of g_1/g_0.

The heat capacity of the magnetic-ion subsystem is therefore over 100,000 times as large as the rest of the crystal!

Many subsystems have values of δ_1/k less than 0.1 K but are of interest and of use at temperatures above this value. When $T \gg \delta_1/k$, Eq. (18-25) reduces to

$$\frac{c_M}{R} = \frac{g_0/g_1}{(1 + g_0/g_1)^2} \frac{(\delta_1/k)^2}{T^2} = \frac{\text{const.}}{T^2}.$$

The "tail" of the Schottky curve is therefore an inverse square curve. It is an interesting and important fact that this inverse-square behavior is true at *higher temperatures* not only when Stark-effect splitting gives rise to two degenerate levels but also when there are *any number* of closely spaced levels. Also, other causes of splitting when $\mathscr{H} = 0$, such as magnetic interactions among the ions, are found to give rise to a $1/T^2$ curve. It is therefore the custom to represent the heat capacity per gram-ion of a paramagnetic salt at temperatures above the Schottky maximum by the equation

$$\frac{c_M}{R} = \frac{A/R}{T^2}. \tag{18-26}$$

Values of A/R are listed in the last column of Table 18-2 for four paramagnetic salts, and the $1/T^2$ law is verified in Fig. 18-5 for copper potassium sulfate.

The last salt, cerium magnesium nitrate (abbreviated CMN), plays an important role in low temperature physics. It is a complicated crystal with a

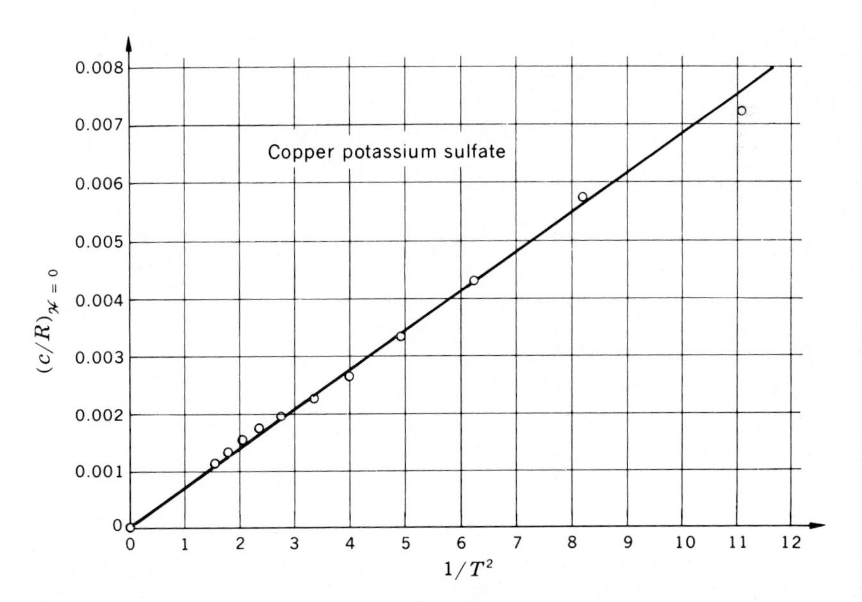

Figure 18-5 Test of the relation $c/R \propto 1/T^2$. (*Ashmead.*)

structure that is highly anisotropic. Parallel to the axis of symmetry (the so-called trigonal axis), the Landé splitting factor g is so small that the Curie constant is almost zero. Perpendicular to this axis, however, the Curie constant has an appreciable value. If a single crystal is placed with its trigonal axis parallel to an external field, its magnetization is very small. A simple rotation through 90 deg is all that is needed to magnetize the crystal. Moreover, the heat capacity constant A/R is remarkably small, indicating a very small value for the various factors that split the lowest energy level of the cerium ion in the absence of an external magnetic field. It is believed that this splitting is due to ionic interaction rather than the Stark effect.

The most important thermal property of a magnetic-ion subsystem is the entropy. According to Eq. (18-7),

$$S = NkT\left(\frac{\partial \ln Z}{\partial T}\right)_{\mathscr{H}} + Nk \ln Z;$$

and from Eq. (18-14),

$$\ln Z = \ln \sinh (J + \tfrac{1}{2})a - \ln \sinh \frac{a}{2} + \ln (g_0 + g_1 e^{-\delta_1/kT}).$$

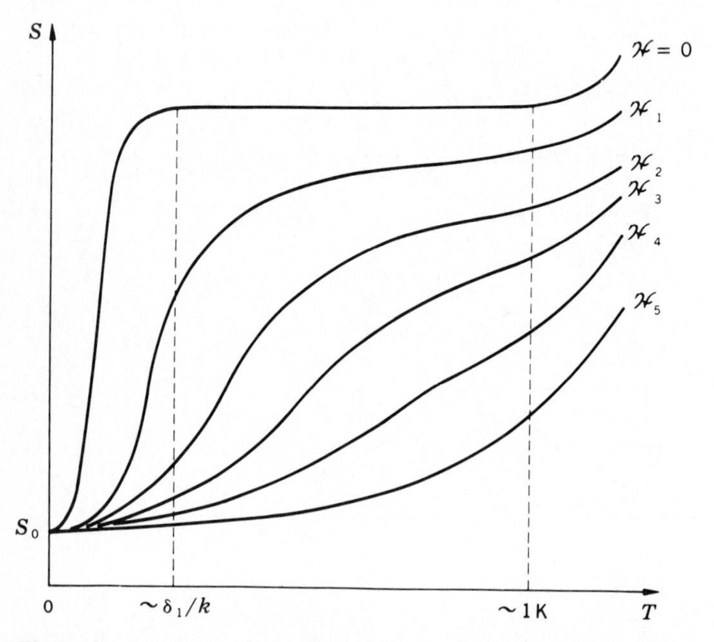

Figure 18-6 Entropy of a magnetic-ion subsystem as a function of temperature for various values of the magnetic field. In the region between the two dashed lines, $S_{\mathscr{H}=0}$ is constant, U is constant, and $C_M = 0$.

When ln Z is introduced into the expression for S, the resulting equation looks rather formidable but has simple properties when the temperature range is divided into two parts:

$$(1) \quad T \gg \delta_1/k \quad \text{and} \quad S = S_1(\mathscr{H}/T). \tag{18-27}$$

$$(2) \quad T \leq \delta_1/k \quad \text{and} \quad S = S_2(T, \mathscr{H}).$$

If S is plotted as a function of T, at various values of $\mathscr{H}$, the resulting curves resemble those of Fig. 18-6. The ST curve for $\mathscr{H} = 0$ is particularly significant. Within the range indicated by vertical dashed lines, $S_{\mathscr{H}=0}$ is constant because the temperature is too low to provide an appreciable heat capacity of the nonmagnetic particles, but too high to allow crystalline field splitting or ionic magnetic interaction to show an appreciable effect. Since the only term involving the dependence of *internal* energy on temperature [Eq. (18-22)] is provided by the interaction of the ions with either the crystalline field or with one another, the internal energy is also constant in this temperature region. Finally, since $C_M = T\, dS/dT = dU/dT$, it follows that $C_M = 0$ in this range.

18-5 PRODUCTION OF MILLIDEGREE TEMPERATURES BY ADIABATIC REDUCTION OF THE MAGNETIC FIELD

Almost every property of matter shows interesting changes or variations in the temperature range below about 20 K. Temperatures down to about 1 K are easily obtained by causing ^{4}He to evaporate rapidly. With ^{3}He it is possible to reach about 0.3 K. In 1926 it was suggested independently by Giauque and by Debye that much lower temperatures could be achieved with the aid of paramagnetic salts. The principle of the method is presented graphically in Fig. 18-7,

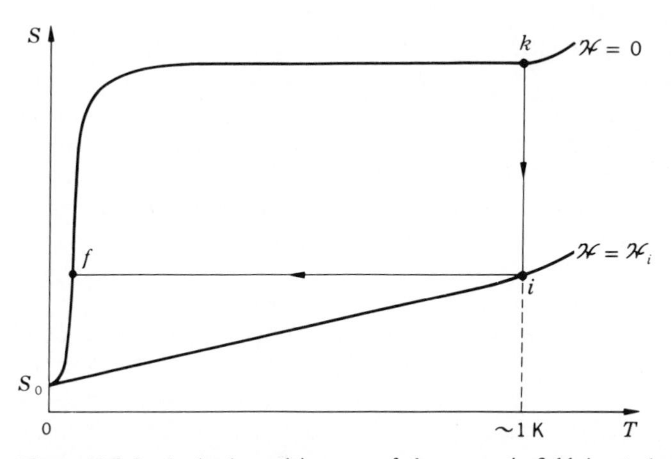

Figure 18-7 In the isothermal increase of the magnetic field $k \to i$, the entropy decreases. In the adiabatic decrease of the magnetic field $i \to f$ (to zero field), the temperature decreases.

where the entropy of a magnetic-ion subsystem is plotted against the temperature for two values of the external magnetic field, 0 and $\mathcal{H}_i$. At a temperature in the neighborhood of 1 K, where the lattice heat capacity of the nonmagnetic particles is negligibly small, the salt is magnetized isothermally in the process $k \to i$. For this process the second $T\,dS$ equation gives

$$T\,dS = 0 + \mu_0 T\left(\frac{\partial M}{\partial T}\right)_{\mathcal{H}} d\mathcal{H}.$$

The derivative $(\partial M/\partial T)_{\mathcal{H}}$ is a measure of the change of alignment of a system of magnets accompanying a rise of temperature (disordering effect) when the external field (ordering effect) is kept constant. The derivative is therefore *negative* for all substances of the type dealt with in this chapter. Since $(\partial M/\partial T)_{\mathcal{H}}$ is negative, heat goes *out* during an isothermal magnetization.

The second step, $i \to f$, is a reversible adiabatic reduction of $\mathcal{H}$ in which

$$0 = C_{\mathcal{H}}\,dT + \mu_0 T\left(\frac{\partial M}{\partial T}\right)_{\mathcal{H}} d\mathcal{H}.$$

Since $(\partial M/\partial T)_{\mathcal{H}}$ and $d\mathcal{H}$ are both negative, it follows that dT is negative. This change in temperature is called the *magnetocaloric effect*. Experiments of this sort were first performed by Giauque in America and were then taken up by Kurti and Simon in England and by de Haas and Wiersma in Holland. In these experiments a paramagnetic salt is cooled to as low a temperature as possible with the aid of liquid helium. A strong magnetic field is then applied, producing a rise of temperature in the substance and a consequent flow of heat to the surrounding helium, some of which is thereby evaporated. After a while, the substance is both strongly magnetized and as cold as possible. At this moment, the space surrounding the substance is evacuated. The magnetic field is now reduced to zero, and the temperature of the paramagnetic salt drops to a low value.

The paramagnetic salt is either a single crystal, a pressed powder, or a mixture of small crystals in the form of either a sphere, a cylinder, or a spheroid. It is placed in a space that may be connected at one time to a pump or, at another time, to a gas supply. This space is surrounded by liquid helium whose pressure (and therefore temperature) may be controlled. Surrounding the liquid helium is liquid nitrogen, and the intervening space is evacuated. Helium gas is admitted into the space containing the paramagnetic salt before the magnet is switched on. The rise of temperature produced by switching on the magnet causes a flow of heat through this helium gas into the liquid helium. In other words, the helium gas is used as a conductor of heat to enable the paramagnetic salt to come to temperature equilibrium rapidly. It is therefore called the *exchange gas*. As soon as temperature equilibrium is attained, the exchange gas is pumped out, leaving the paramagnetic salt thermally insulated. In many past experiments, the adiabatic reduction of field was accomplished by swinging the entire cryostat out of the magnetic field produced by a huge electromagnet such as that depicted in Fig. 18-8. In recent years, solenoids have been made of many thousands of turns of fine superconducting wire, through which from 10 to 100 A may be sent. There is no power dissipation in the coil so long as it is maintained at a temperature below which the wire is superconducting. The conventional experimental apparatus is compared roughly with the more

Figure 18-8 The famous electromagnet of the Laboratoire Aimé Cotton at Bellevue, near Paris, used by Simon and Kurti in some of their pioneer experiments. (*Courtesy of N. Kurti.*)

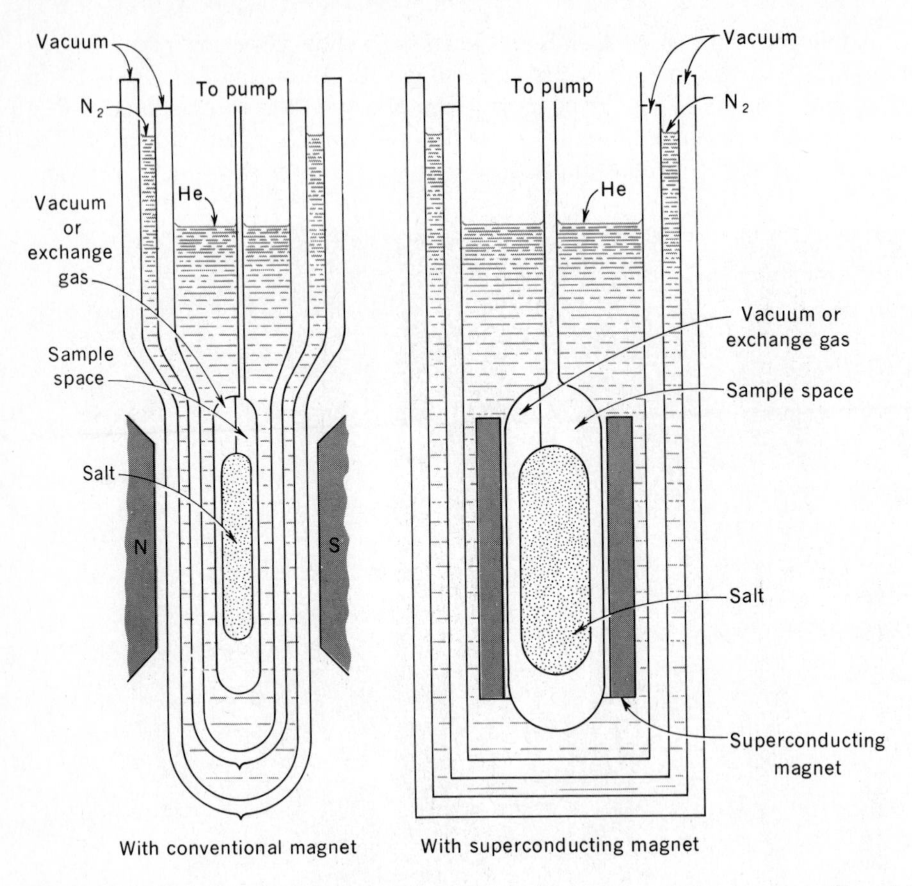

Figure 18-9 Adiabatic reduction of magnetic field experiments are carried out more readily with a superconducting magnet.

modern one in Fig. 18-9. With a superconducting magnet, (1) the sample space may be made larger; (2) the magnetic field may be made stronger; (3) the cost of the magnet may be one-quarter that of a conventional magnet; and (4) the power requirement is negligible.

The next step is to estimate the temperature. For this purpose, separate coils of wire surrounding the paramagnetic salt are used. The paramagnetic susceptibility $M/\mathcal{H}$, which is a function of the temperature, is measured by means of a special bridge circuit. A new temperature scale is now defined with the aid of Curie's equation. The new temperature T^*, called the *magnetic temperature*, is defined as

$$T^* = \frac{\text{Curie constant}}{\text{susceptibility}} = \frac{C'_C \mathcal{H}}{M}.$$

It is seen that, in the region where Curie's law holds, T^* is the real Kelvin temperature, whereas in the region around absolute zero T^* is expected to differ somewhat from the Kelvin temperature.

Table 18-3 Pioneer results in the magnetic production of low temperatures

Experimenters	Date	Paramagnetic salt	Initial field, 10^6 A/m	Initial temp., K	Final magnetic temp., T^*
Giauque and MacDougall	1933	Gadolinium sulfate	0.64	1.5	0.25
De Haas, Wiersma, and Kramers	1933	Cerium fluoride	2.20	1.35	0.13
		Dysprosium ethyl sulfate	1.55	1.35	0.12
		Cerium ethyl sulfate	2.20	1.35	0.085
De Haas and Wiersma	1934	Chromium potassium alum	1.96	1.16	0.031
	1935	Iron ammonium alum	1.92	1.20	0.018
		Alum mixture	1.92	1.29	0.0044
		Cesium titanium alum	1.92	1.31	0.0055
Kurti and Simon	1935	Gadolinium sulfate	0.43	1.15	0.35
		Manganese ammonium sulfate	0.64	1.23	0.09
		Iron ammonium alum	1.12	1.23	0.038
		Iron ammonium alum	0.66	1.23	0.072
		Iron ammonium alum	0.39	1.23	0.114

The early results of the great pioneers in this field are listed in Table 18-3 (compiled by Burton, Grayson-Smith, and Wilhelm).

There is a set of conditions, easily achieved in the laboratory, under which it is possible to calculate the final temperature achieved after an adiabatic demagnetization:

1. The initial temperature is low enough to make the contributions of the non-magnetic particles negligible.
2. The temperature never gets lower than the tail of the Schottky curve where $C_M = A/T^2$.
3. Values of $\mathscr{H}/T$ are always small enough that Curie's law is obeyed, $M/\mathscr{H} = C'_C/T$.
4. The external magnetic field is reduced to a low value at which the magnetization is M_f. Both $\mathscr{H}_f$ and M_f may be made equal to zero.

Under these conditions we may use the first $T\,dS$ equation as follows:

$$T\,dS = C_M\,dT - T\mu_0\left(\frac{\partial \mathscr{H}}{\partial T}\right)_M dM.$$

From Curie's equation, we get

$$\left(\frac{\partial \mathscr{H}}{\partial T}\right)_M = \frac{M}{C'_C},$$

and we also have

$$C_M = \frac{A}{T^2}.$$

Substituting these values into the first $T\,dS$ equation and setting $dS = 0$ for an adiabatic demagnetization, we get

$$0 = \frac{A}{T^2}\,dT - T\mu_0 \frac{M}{C'_C}\,dM,$$

or

$$A \int_{T_i}^{T_f} \frac{dT}{T^3} = \frac{\mu_0}{C_C} \int_{M_i}^{M_f} M\,dM,$$

where the substance is demagnetized to magnetization M_f. Upon performing the integrations,

$$\frac{A}{2}\left(\frac{1}{T_f^2} - \frac{1}{T_i^2}\right) = \frac{\mu_0}{2C'_C}(M_i^2 - M_f^2),$$

and using Curie's equation,

$$\frac{1}{T_f^2} - \frac{1}{T_i^2} = \frac{\mu_0 C'_c}{A}\left(\frac{\mathcal{H}_i^2}{T_i^2} - \frac{\mathcal{H}_f^2}{T_f^2}\right),$$

or

$$\frac{T_f}{T_i} = \sqrt{\frac{\mu_0(C'_C/A)\mathcal{H}_f^2 + 1}{\mu_0(C'_C/A)\mathcal{H}_i^2 + 1}}. \tag{18-28}$$

When $\mathcal{H}_f = 0$,

$$\left(\frac{T_i}{T_f}\right)^2 = 1 + \frac{\mu_0 C'_C}{A}\mathcal{H}_i^2, \tag{18-29}$$

which shows that the lowest temperatures are reached by salts in which A is small and C'_C is large. From the values of A/R and C'_C in Table 18-2, it may be seen that cerium magnesium nitrate is the most favorable salt by a factor of over 100. The graph in Fig. 18-10 shows that the conditions under which Eq. (18-29) was derived hold well for CMN down to 0.01 K. The remainder of the graph (down to 0.002 K) is the result of experiments by Daniels and Robinson; Hudson, Kaeser, and Radford; and de Klerk, combined with experiments and theoretical calculations of Frankel, Shirley, and Stone. The entire graph enables CMN to be used as a thermometer down to 0.002 K.

Most experiments on adiabatic field reduction are performed in order to achieve the lowest possible temperature, in which case the field reduction proceeds until $\mathcal{H}_f = 0$. If, instead of reducing to zero field, we stay within the region of field and temperature designated by the space between the vertical dashed lines of Fig. 18-6 (e.g., $\mathcal{H}_5 \to \mathcal{H}_2$, or $\mathcal{H}_3 \to \mathcal{H}_1$), then curious results are obtained. First of all, the entropy is a function of $\mathcal{H}/T$ *only*, therefore, during an isentropic reduction of $\mathcal{H}$ (*not to zero*), since S is constant, *the ratio $\mathcal{H}/T$ must remain constant*. That this is true under the conditions specified is demonstrated

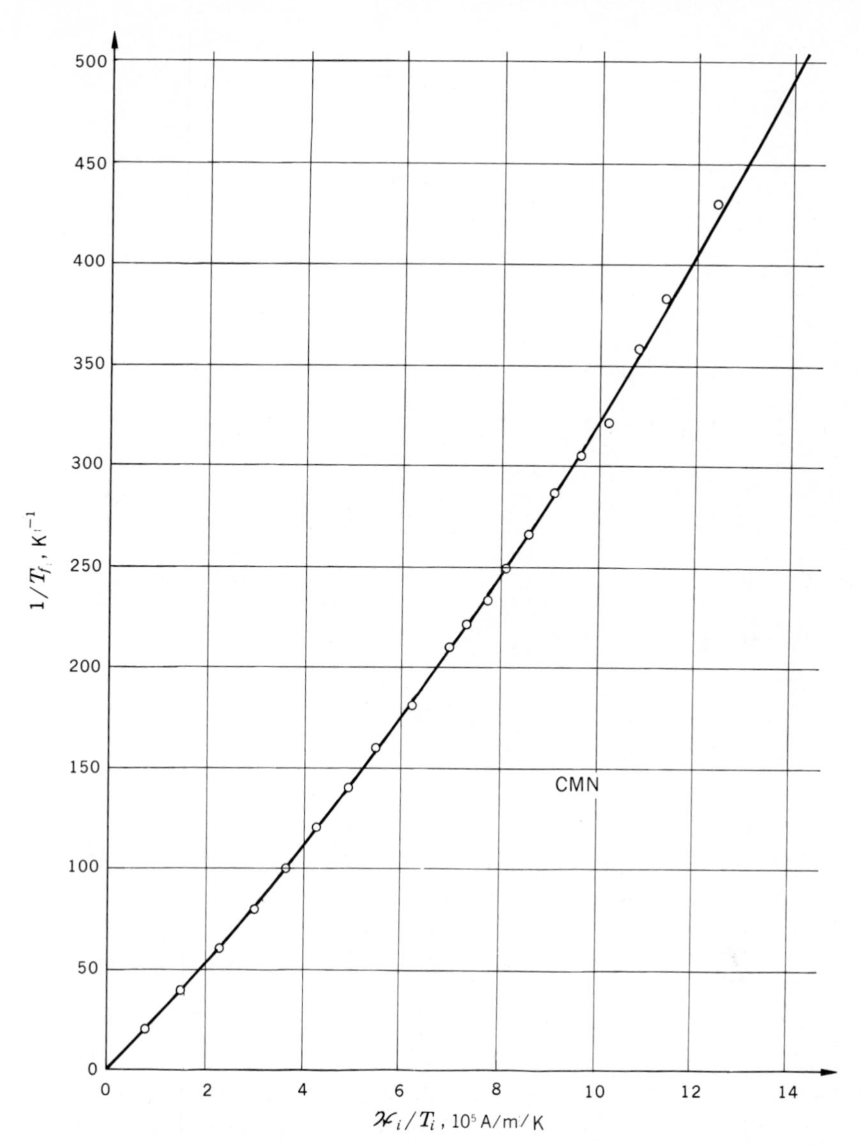

Figure 18-10 Adiabatic reduction of magnetic field to zero for cerium magnesium nitrate. (*R. B. Frankel, D. A. Shirley, and N. J. Stone, 1965.*)

by the experimental results of Hill and Milner, shown in Fig. 18-11. Second, if $\mathscr{H}/T$ is constant, since M is a function of $\mathscr{H}/T$ only, *M is constant!* If, therefore, the magnetization does not change, the term adiabatic "demagnetization" is a misnomer. As a result, the word "degaussing" has been suggested. Because the gauss is a cgs unit, we shall refer to the misnamed process of adiabatic "demagnetization" as the *adiabatic reduction of the magnetic field* in this book. If M does

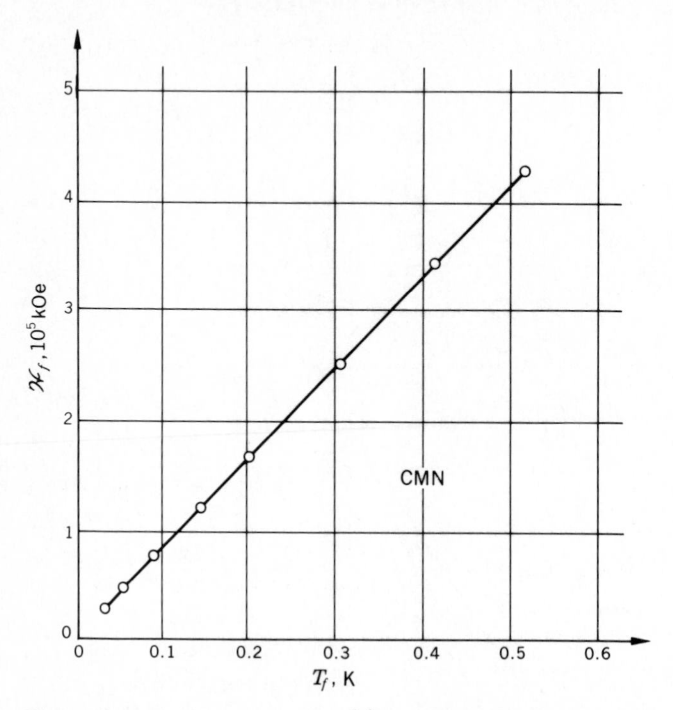

Figure 18-11 Constancy of the ratio $\mathcal{H}_f/T_f$ during adiabatic reduction of the magnetic field of CMN from $\mathcal{H}_i$, T_i (always the same) to $\mathcal{H}_f$, T_f. *(J. S. Hill and J. H. Milner, 1957.)*

not change (and this has been experimentally verified by de Haas, Wiersma, and Casimir, 1936 and 1940), then *the work* $\mu_0 \int \mathcal{H}\, dM$ *is zero.* But in an adiabatic process, the work done is equal to the change of internal energy, so that ΔU is zero. Since the heat capacity at constant magnetization C_M is equal to $(\partial U/\partial T)_M$, then C_M is zero. If the first $T\,dS$ equation is applied to this process,

$$T\,dS = C_M\,dT - T\mu_0\left(\frac{\partial \mathcal{H}}{\partial T}\right)_M dM,$$

we have dS, C_M, and dM all equal to zero. Of course, it must be understood that these results apply only when temperatures are not too low and the field is not reduced to zero. These conditions will be found to be important in certain experiments on nuclear magnetic subsystems.

Let ϵ_i be the energy of an ion with total quantum number m_i in a field $\mathcal{H}$. When the field changes a small amount $d\mathcal{H}$, the energy change $d\epsilon_i$ is

$$d\epsilon_i = -g\mu_B\,\mu_0\,m_i\,d\mathcal{H};$$

and if N_i is the population of the m_ith state,

$$\sum N_i\,d\epsilon_i = -\left(\sum N_i g\mu_B m_i\right)\mu_0\,d\mathcal{H},$$

or

$$\sum N_i\,d\epsilon_i = -M\mu_0\,d\mathcal{H}. \tag{18-30}$$

Now $\sum N_i \epsilon_i$ is the *total* energy of the ions, that is, the sum of the internal energy U and the magnetic potential energy $-\mu_0 \mathscr{H} M$. Thus,

$$\sum N_i \epsilon_i = U - \mu_0 \mathscr{H} M,$$

and

$$\sum \epsilon_i \, dN_i + \sum N_i \, d\epsilon_i = dU - \mu_0 \mathscr{H} \, dM - M\mu_0 \, d\mathscr{H}.$$

Since $\sum N_i \, d\epsilon_i = -M\mu_0 \, d\mathscr{H}$, this reduces to

$$\sum \epsilon_i \, dN_i = dU - \mu_0 \mathscr{H} \, dM,$$

or

$$\boxed{\sum \epsilon_i \, dN_i = T \, dS.} \tag{18-31}$$

It follows that an isothermal decrease of entropy, which takes place in the first step of any experiment for producing low temperatures, must involve changes in N_i, that is, population changes. During the second step, however, in which an adiabatic magnetic field reduction takes place, dS is zero, and therefore the dN_i's are zero—that is, the populations remain unchanged.

In Fig. 18-12k, the magnetic-ion subsystem is in zero field, and therefore the magnetic energy states are extremely close together, with some of the degeneracy being removed by internal electric and magnetic effects. Since the states are so close together, they are equally populated. From $k \to i$ the populations change so that the low-energy states become highly populated, as shown in Fig. 18-12i. In the last step, $i \to f$, the populations remain constant in order to keep the entropy constant.

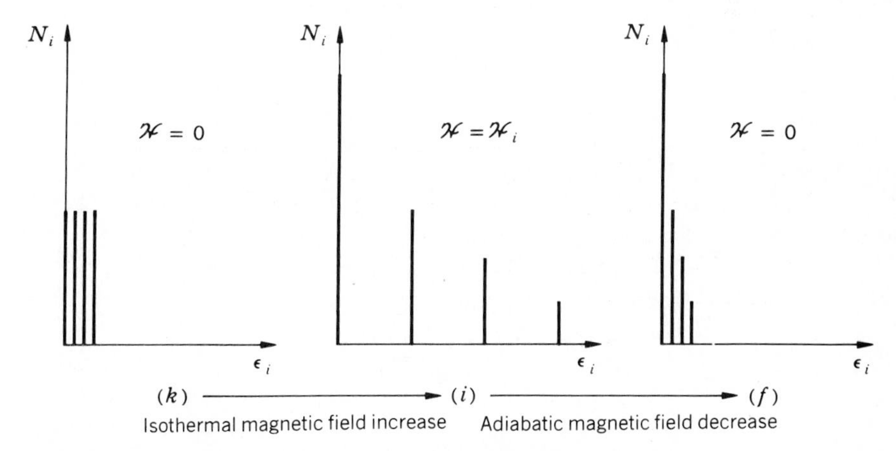

Figure 18-12 Changes in energy states and ionic populations during isothermal magnetic field production $k \to i$ and adiabatic magnetic field reduction $i \to f$.

18-6 LOW-TEMPERATURE THERMOMETRY

When a gas is liquefied, the normal boiling point (or, in the case of carbon dioxide, the normal sublimation point) and the triple-point temperatures and pressures must be determined with great accuracy. Temperatures are usually measured with a helium-gas thermometer according to the principles developed in Chap. 1. The bulb of the gas thermometer is often incorporated within the liquefaction apparatus. Pressures are usually measured with a mercury manometer, sighted through the telescope of a cathetometer. Once these temperatures and pressures are measured, they may be used as fixed points for the calibration of gas thermometers of simpler design. Some of the most important fixed points of low-temperature physics are listed in Table 18-4.

The lowest temperature that can be reached easily with liquid ^{4}He is about 1 K. This is achieved by pumping the vapor away as fast as possible through as wide a tube as possible. With special high-speed pumps, temperatures as low as 0.7 K have been reached, but this is rare. Temperatures lower than 0.7 K cannot be reached by pumping liquid ^{4}He because a film of liquid helium II creeps up the walls of the pumping tube, vaporizes, and then recondenses.

One of the principal methods of reaching temperatures significantly below 1 K is by using a ^{3}He evaporation cryostat. The lower limit of pumping on a liquid bath of ^{4}He is a temperature of about 0.7 K, but the lower limit with ^{3}He is slightly below 0.3 K. There are two reasons for the better performance with ^{3}He. First, the vapor pressure of ^{3}He is at any temperature higher than the vapor pressure of ^{4}He at the same temperature. Second, at these temperatures there is no superfluid ^{3}He film to creep up the walls as in the case of the liquid helium II phase of ^{4}He. Thus, a ^{3}He bath may be pumped using a wide-diameter tube without recourse to narrow constructions to suppress film creep. The criti-

Table 18-4 Useful fixed points in low-temperature physics

Gas	Critical point	Normal boiling point	Triple point	Normal sublimation point
Carbon dioxide	304 K 7.4 MPa		216.6 K 0.518 MPa	194.7 K 101 kPa
Nitrogen	126.26 K 3.398 MPa	77.35 K 101 kPa	63.14 K 12.5 kPa	
Hydrogen—normal	33 K 1.297 MPa	20.39 K 101 kPa	13.96 K 7.21 kPa	
Helium-4	5.20 K 229 kPa	4.224 K 101 kPa	2.172 K 5.05 kPa	
Helium-3	3.32 K 116 kPa	3.195 K 101 kPa	0.026 K 3.4 MPa	

cal temperature of ^{3}He is 3.32 K and its critical pressure is 116 kPa, so pumping on liquid ^{3}He provides a range of operating temperatures for the ^{3}He evaporation cryostat from 3.3 to 0.3 K.

In studying the properties of matter at low temperatures, the experimental apparatus is usually surrounded by a bath of liquid ^{4}He, and a measurement of the pressure of the vapor in equilibrium either with the liquid bath itself or of some liquid helium in a separate bulb, in conjunction with a vapor-pressure-temperature table, serves to determine the temperature of the apparatus and of any secondary thermometer mounted on the apparatus.

The gas thermometer and the vapor-pressure thermometer are elaborate, exacting, and sluggish devices. To measure heat capacities, thermal conductivities, and several other physical quantities of interest at low temperatures, many measurements of small temperature changes must be made quickly and accurately. For these purposes, secondary thermometers must be used.

One of the first to be employed was a resistance thermometer made of carbon. Pieces of paper with carbon deposited on them or strips of carbon prepared by painting with colloidal suspensions have two advantages. They have extremely small heat capacities and can therefore follow temperature changes quickly, and their electric resistance, which increases rapidly as the temperature is reduced, is insensitive to the presence of a magnetic field. The main disadvantage of such thermometers is their lack of reproducibility. They must be calibrated anew each time they are used.

In 1951, Clement and Quinell discovered that carbon composition radio resistors, made by Allen-Bradley and rated from $\frac{1}{2}$ to 1 W, had all the properties most desired in a low-temperature secondary thermometer—namely, high sensitivity, reproducibility, and insensitivity to magnetic fields. The reasons for these desirable properties are not understood, but to such thermometers are attributed the accuracy of much of the work done in low-temperature physics since 1951. In using small radio resistors as thermometers, the plastic coating is first removed and replaced by a thin coat of lacquer. The thermometer is then attached to the experimental apparatus or placed in a hole drilled for that purpose. The resistance is measured at a number of temperatures that are known from measurements of the helium vapor pressure. The greatest number of such measurements has been made by J. R. Clement, who tried several empiric equations to represent the relation between the resistance R' and T. One of the most satisfactory equations is

$$\sqrt{\log R'/T} = a + b \log R', \tag{18-32}$$

which fits the experimental results within a few millidegrees. If a carbon resistance thermometer is calibrated in the liquid helium range and *then kept below* 20 K, it will retain its calibration indefinitely. After warming to room temperature, however, it will have to be recalibrated when brought back to helium temperatures.

Higher reproducibility is attainable with the aid of germanium resistance thermometers developed by Kunzler, Geballe, and Hull at the Bell Telephone

Laboratories. The germanium is "doped" with excess arsenic atoms and then sealed in helium-filled capsules. In an analysis of several of these semiconductor thermometers along with radio resistor thermometers, Lindenfeld gives as an upper limit to the variability of the temperature corresponding to a given resistance the value ± 0.4 mK. This is better than that possible with carbon resistors. The main advantage of germanium thermometers is their reproducibility. They retain their calibration after any number of cycles between helium and room temperatures.

Some paramagnetic salts obey Curie's law very closely down to 1 K and even lower. If the Curie constant is determined by measurements at known temperatures, then the magnetic salt may be used as a thermometer. In the liquid helium range, the change of magnetization of all paramagnetic salts is rather small. To measure temperature to 1 mK, the magnetization must be measured with tremendous accuracy with the aid of a SQUID magnetometer. The superconducting *qu*antum *i*nterference *d*evice (SQUID), based on the concept of fluxoid quantization in a superconductor and on Josephson tunneling through a "weak link" between two superconductors, can be used as an extraordinarily sensitive low-temperature magnetometer. The magnetic thermometer is of most importance in the temperature range below 1 K, where its magnetization changes by a larger amount for a small temperature change.

At temperatures below the region in which Curie's law holds and which have been represented by the so-called magnetic temperature $T^* = C'_C \mathcal{H}/M$, the corresponding Kelvin temperatures must be determined. To understand how this is done, consider the ST curves of Fig. 18-13 showing a number of isothermal increases of the magnetic field ($1 \to 2$, $1 \to 4$, or $1 \to 6$, etc.) and a number of adiabatic decreases of the magnetic field ($2 \to 3$, $4 \to 5$, or $6 \to 7$, etc.). By measuring the magnetic susceptibility at each of the points numbered 3, 5, 7, etc., and by using the known value of the Curie constant, the T^* values may be obtained at these points. Taking the process $1 \to 2$ as a typical isothermal increase of $\mathcal{H}$, we may calculate the entropy change $S_2 - S_1$ by applying the second $T\,dS$ equation,

$$T\,dS = C_{\mathcal{H}}\,dT + \mu_0 T\left(\frac{\partial M}{\partial T}\right)_{\mathcal{H}} d\mathcal{H}.$$

Integrating from 1 to 2, we get

$$S_1 - S_2 = -\mu_0 \int_0^{\mathcal{H}_{\max}} \left(\frac{\partial M}{\partial T}\right)_{\mathcal{H}} d\mathcal{H}. \tag{18-33}$$

Since the paramagnetic salt in question obeys Brillouin's equation accurately, the right-hand integral may be evaluated, and therefore after a series of such calculations we may obtain $S_4 - S_2$, $S_6 - S_2$, $S_8 - S_2$, etc. But these entropy changes are equal to $S_5 - S_3$, $S_7 - S_3$, $S_9 - S_3$, etc. Since the T^* values are known at the points 3, 5, 7, etc., a graph of $S - S_3$ can be plotted against T^*, as shown in Fig. 18-14*a*.

Now let us imagine that we have performed the adiabatic decrease of the

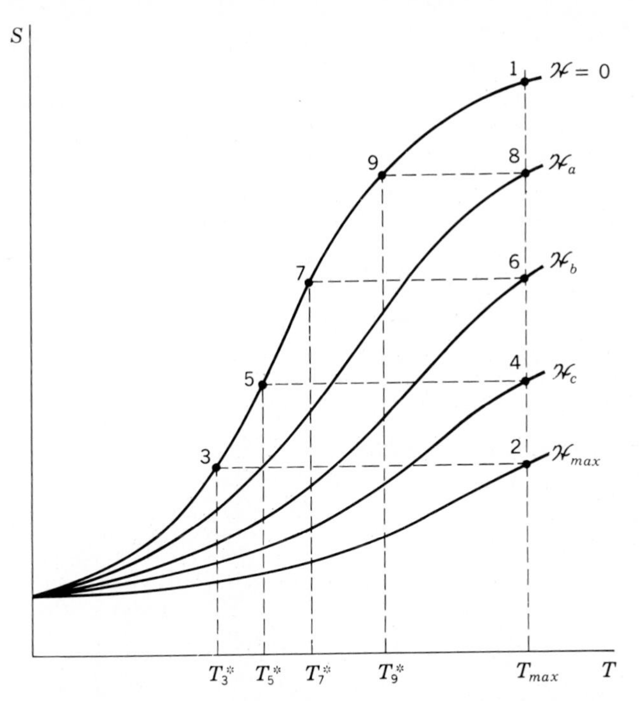

Figure 18-13 Isothermal increases in magnetic field and adiabatic decreases in magnetic field on an entropy-temperature diagram. Above $T_{\max}$ are Kelvin temperatures; below $T_{\max}$ are magnetic temperatures.

magnetic field $2 \to 3$ and are at the minimum temperature, characterized by the magnetic temperature T_3^*. With the aid of some suitable heating device (such as a gold heating coil used by Giauque, the absorption of gamma rays used by Kurti and Simon, or magnetic hysteresis used by de Klerk), let us measure the heat absorbed at zero magnetic field in going from 3 to 5, from 3 to 7, from 3 to 9, and from 3 to 1. Since these values of heat are measured in zero magnetic field, no work is done, and hence the heat is equal to the internal-energy change $U - U_3$. These internal-energy changes are plotted against T^* in Fig. 18-14*b*.

Combining graphs (*a*) and (*b*), we get the graph of Fig. 18-14*c*, in which $U - U_3$ is plotted against $S - S_3$. Since the point 3 is always the same, U_3 and S_3 are constants, and therefore the *slope* of the curve at any point is $(\partial U/\partial S)_{\mathcal{H}=0}$. Since $T\,dS = dU - \mu_0 \mathcal{H}\,dM$ and $\mathcal{H} = 0$,

$$(\partial U/\partial S)_{\mathcal{H}=0} = T.$$

Once this procedure has been followed and a table of values of T^* and T, such as Table 18-5, has been obtained, other thermometers such as radio resisters, germanium crystals, or other paramagnetic salts, can be calibrated. The most widely used thermometric salt is a single crystal of CMN cut into the shape

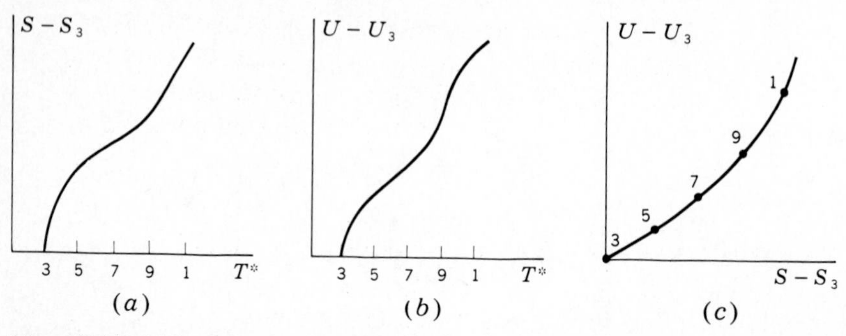

Figure 18-14 Kelvin temperatures at the points 3, 5, 7, etc., are obtained by measuring the slope of the curve in (c) at the designated points.

Table 18-5 Calibration data for a sphere of CMN

$\mathscr{H}_i/T_i$, 10^5 A/m · K	S/R (calculated)	$1/T^*$	$1/T$, K^{-1}	$\mathscr{H}_i/T_i$, 10^5 A/m · K	S/R (calculated)	$1/T^*$	$1/T$, K^{-1}
0.80	0.691	20	20	6.21	0.590	180	181
1.51	0.686	40	40	6.96	0.567	200	210
2.31	0.678	60	60	8.11	0.529	230	249
3.02	0.667	80	80	9.63	0.477	260	305
3.70	0.654	100	100	11.30	0.420	290	383
4.30	0.640	120	120	12.33	0.384	300	430
4.93	0.625	140	140	14.32	0.321	310	500
5.49	0.610	160	160				

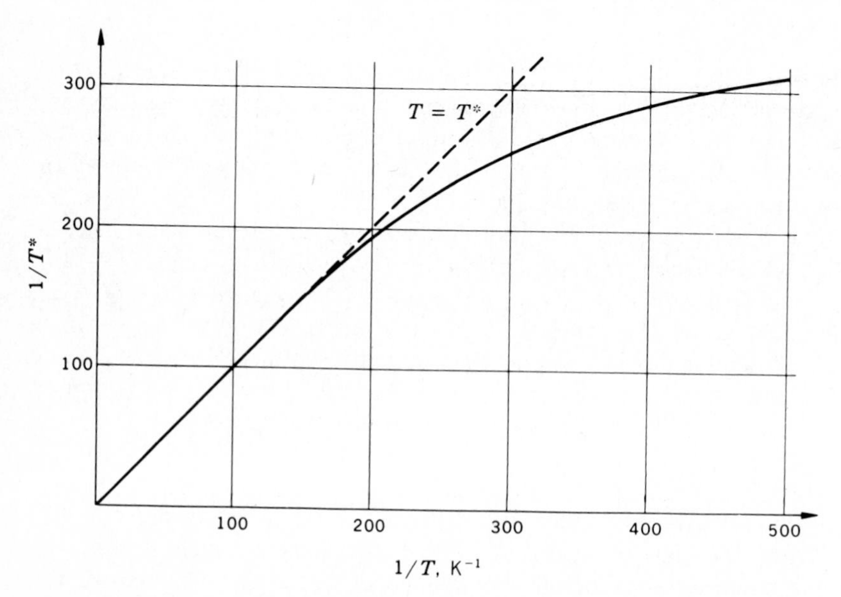

Figure 18-15 Relation between T^* and T for a sphere of CMN. (*Frankel, Shirley, and Stone, 1965.*)

of a sphere. The calibration curve between T^* and T, shown in Fig. 18-15, is the combined effort of many physicists, but the final corrections were made by Frankel, Shirley, and Stone, who used radioactive, oriented cerium ions whose temperature was measured by noting the asymmetry of γ-ray emission.

18-7 ^{3}He/^{4}He DILUTION REFRIGERATOR

The principle of the dilution refrigerator was first proposed by London in 1951 and built by Das, De Bruyn Ouboter, and Taconis in 1965. By 1978, the dilution refrigerator was able to maintain a temperature of 2 mK in continuous operation, thereby replacing refrigerators based on the adiabatic reduction of the magnetic field in ionic paramagnetic salts for many areas of research.

The operation of the dilution refrigerator is based on the properties of liquid mixtures of ^{3}He and ^{4}He below the lambda point of ^{4}He. Figure 18-16 shows the phase diagram at saturated vapor pressure, where $x = n_3/(n_3 + n_4)$ is the concentration of ^{3}He and n_3 and n_4 refer to the number of atoms of ^{3}He and ^{4}He, respectively.

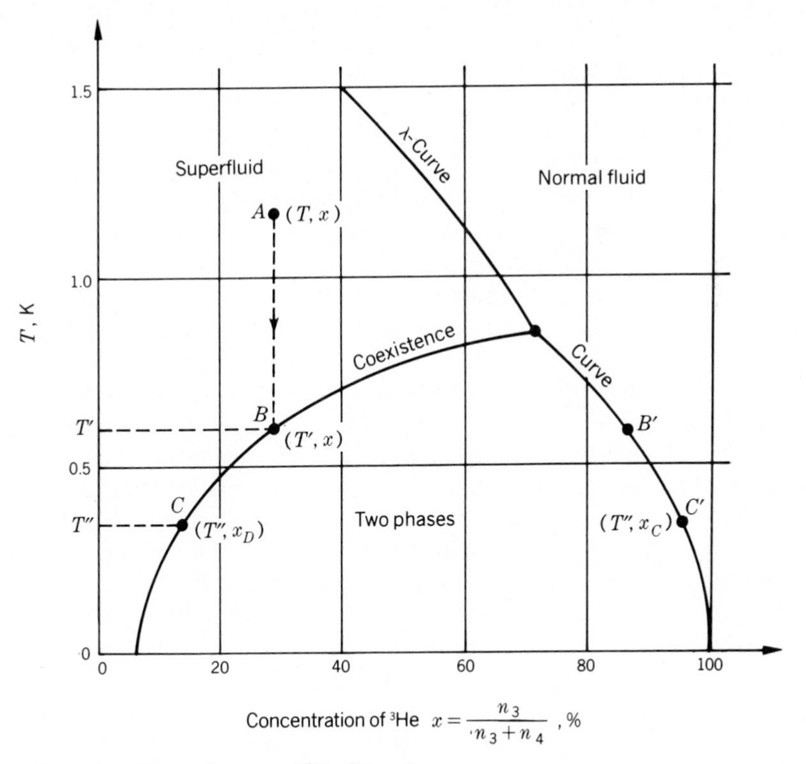

Figure 18-16 Phase diagram of ^{3}He/^{4}He mixtures.

Consider the point A (T, x) in the single-phase superfluid region where ^{3}He and ^{4}He are mixed, then allow the temperature to be lowered while the concentration x remains constant. At point B (T', x) on the coexistence curve the mixture spontaneously separates into two phases, one being concentrated in ^{3}He (point B') and the other dilute in ^{3}He (point B). Because of its lower density, the concentrated ^{3}He phase floats on top of the dilute ^{3}He (concentrated ^{4}He) phase. As the temperature is lowered still further, from T' to T'', the upper phase becomes more concentrated in ^{3}He while the lower phase becomes more dilute in ^{3}He. A point representing the concentrated phase moves down the right branch of the coexistence curve to (T'', x_C) while a point representing the dilute phase moves down the left branch of the coexistence curve to (T'', x_D). As the temperature drops below 100 mK, the concentration of the upper or concentrated phase x_C approaches 100 percent while the concentration of the lower or dilute phase x_D approaches 6.4 percent. In other words, the upper, floating phase is almost pure ^{3}He while the lower, sunken phase is 6.4 percent ^{3}He and 93.6 percent ^{4}He.

The thermal properties of liquid ^{3}He and liquid ^{4}He differ at low temperatures because of the different quantum statistics of the two liquids. An atom of ^{4}He has zero nuclear spin and therefore obeys Bose-Einstein statistics. Consequently, at temperatures below 500 mK, liquid ^{4}He is effectively in its quantum mechanical ground state (so-called "Bose condensation"). Very few phonons are excited, so liquid ^{4}He is thermodynamically inert. The lighter isotope ^{3}He has a nuclear spin of $\frac{1}{2}$ and thus obeys Fermi-Dirac statistics. It does not exhibit Bose condensation and behaves very differently. The specific heat and entropy of ^{3}He are relatively large in the temperature range of dilution refrigerators. So it is possible to transport a large amount of ^{3}He from the upper, concentrated phase to the lower, dilute phase. As a result, the ^{4}He in the dilute phase has been compared to a "mechanical vacuum" for the "dilute gas" of active ^{3}He atoms.

The operation of the dilution refrigerator can be understood by comparing it to an ordinary evaporation refrigerator or cryostat. The upper, concentrated phase of ^{3}He atoms, which are densely packed, is analogous to the liquid phase in an evaporation refrigerator. The lower, dilute phase of ^{3}He atoms, which are dispersed among the ^{4}He atoms, is similar to the vapor phase in an evaporation refrigerator. Cooling is produced when ^{3}He from the concentrated phase crosses the phase boundary into the dilute phase, analogous to the cooling in an evaporation refrigerator when molecules evaporate from the liquid phase into the vapor phase. By means of a vacuum pump ^{3}He atoms are continuously removed from the dilute phase, thereby cooling the system in the dilution refrigerator. Unlike an evaporation refrigerator, in which the vapor pressure rapidly approaches zero as the temperature is reduced, the transport of ^{3}He across the phase boundary will continue because of the 6.4 percent solubility of ^{3}He in ^{4}He even in the limit of absolute zero.

Figure 18-17 is a simplified schematic drawing of a typical dilution refrigerator. The incoming ^{3}He gas is first precooled to 4.2 K and then condensed by contact with a pumped ^{4}He evaporator at approximately 1.3 K. The pressure of

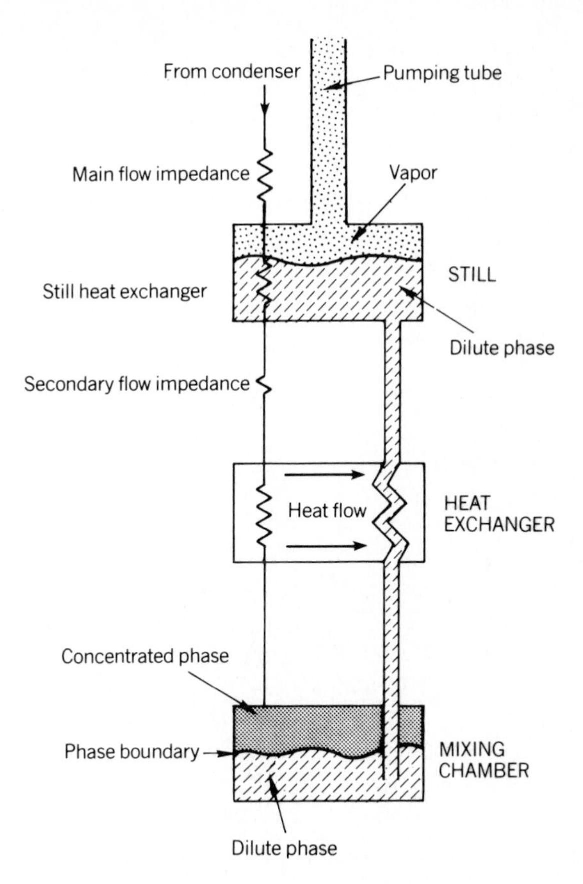

Figure 18-17 Schematic diagram of the $^3He/^4He$ dilution refrigerator that is inserted into the low-temperature cryostat.

the ^3He is kept sufficiently high for condensation by means of a flow-limiting constriction known as the main flow impedance. The liquid then enters the still heat exchanger, where it is cooled to the temperature of the still. Next, it passes through the secondary flow impedance and into another heat exchanger, to cool further the incoming ^3He before it enters the mixing chamber.

The phase separation and dilution refrigeration occur in the mixing chamber. Continuous operation, the result of continuously dissolving ^3He from the concentrated phase at the top of the mixing chamber into the dilute phase at the bottom, is obtained by circulating ^3He in the system. After crossing the phase boundary inside the mixing chamber, the ^3He atoms are driven by an osmotic pressure gradient along a tube leading back to the still. A heater within the still raises the temperature to approximately 0.5 K in order to evaporate ^3He into a tube connected to the pumping station. There the ^3He is compressed and recycled back into the dilution refrigerator, thereby providing continuous, closed-circuit operation. Commercial dilution refrigerators are manufactured by the S.H.E. Corporation in San Diego, California.

PROBLEMS

18-1 In a magnetic field $\mathscr{H}$, where the total energy of an ion is

$$\epsilon_i = -g\mu_B \mu_0 \mathscr{H} m_i + \delta_i,$$

the number of ions N_i with magnetic quantum number m_i is given by the Boltzmann equation. The total magnetization M is

$$M = \sum N_i g\mu_B \mu_0 m_i.$$

Show that

$$M = \frac{NkT}{\mu_0}\left(\frac{\partial \ln Z}{\partial \mathscr{H}}\right)_T.$$

18-2 (*a*) Derive Eq. (18-22) directly from the relation $U = \sum N_i \delta_i$ and Boltzmann's equation.
(*b*) Prove that, when $J = \frac{1}{2}$,

$$B_J(a) = \tanh \frac{a}{2}.$$

18-3 Prove that, when splitting of the lowest energy level by the crystalline field is neglected,

(*a*)
$$\frac{S}{R} = \ln \frac{e^{(J+1/2)a} - e^{-(J+1/2)a}}{e^{(1/2)a} - e^{-(1/2)a}} - JaB_J(a).$$

(*b*) When $a \ll 1$,

$$\frac{S}{R} = \ln(2J+1) - \frac{J(J+1)a^2}{3}$$

$$= \ln(2J+1) - \frac{C'_C}{R}\frac{\mathscr{H}^2}{T^2},$$

where C'_C is the Curie constant.

18-4 Convert the second energy equation (Chap. 9) into a form appropriate to the coordinates $\mathscr{H}$, M, and T. Prove that, since $M = f(\mathscr{H}/T)$, U of a magnetic-ion subsystem must be a function of T only.

18-5 For a paramagnetic salt obeying Brillouin's equation, prove that:

(*a*)
$$C_{\mathscr{H}} - C_M = -\mu_0 \mathscr{H}\left(\frac{\partial M}{\partial T}\right)_{\mathscr{H}}.$$

(*b*) As $T \to 0$ at constant $\mathscr{H}$, $C_{\mathscr{H}} \to C_M$.

(*c*)
$$\left(\frac{\partial H^*}{\partial \mathscr{H}}\right)_T = T\left(\frac{\partial M}{\partial T}\right)_{\mathscr{H}} - M.$$

18-6 Between absolute zero and the Curie point T_C, an *antiferromagnetic* oxide has a magnetic susceptibility that is roughly proportional to the temperature; thus,

$$\frac{M}{\mathscr{H}} = C''_C T.$$

Show that (*a*) the magnetic enthalpy is a function of temperature only and that (*b*) the heat capacity at constant field is a function of temperature only.

18-7 (*a*) Prove that, for a paramagnetic gas,

$$T\,dS = C_{V,M}\,dT + T\left(\frac{\partial P}{\partial T}\right)_{V,M} + T\mu_0\left(\frac{\partial \mathscr{H}}{\partial T}\right)_{V}, M,\, dM.$$

(*b*) If the gas is ideal and obeys Curie's law, show that

$$T\,dS = C_{V,M}\,dT + nRT\frac{dV}{V} - \frac{T}{C_C}M\,dM.$$

(*c*) Sketch a reversible adiabatic surface on a TVM diagram, assuming $C_{V,M}$ to be constant.

18-8 A paramagnetic salt obeys Curie's law and also $C_M = A/T^2$. Show that:

(*a*)
$$S = -\frac{1}{2}\frac{A}{T^2} - \frac{M^2}{2C_C} + \text{const.}$$

(*b*)
$$C_{\mathcal{H}} - C_M = C_C\frac{\mathcal{H}^2}{T^2}.$$

(*c*) The equation of an adiabat is $M = \text{const.}\ \dfrac{\mathcal{H}}{\mathcal{H}\sqrt{1 + (C_C/A)\mathcal{H}^2}}$

18-9 Make a graph of the heat capacity of gadolinium sulfate as a function of T from 1.5 to 5.0 K and for values of   equal to 0, 1.6×10^5, 4×10^5, and 8×10^5 A/m.

18-10 Take a paramagnetic solid obeying Curie's law through a Carnot cycle, and verify that

$$\frac{Q_H}{Q_C} = \frac{T_H}{T_C}.$$

18-11 From the microscopic point of view, the disorder of a paramagnetic solid may be increased in two ways: by increasing the temperature and by decreasing the field. From this point of view, explain why a reversible reduction of the magnetic field should be accompanied by a decrease in temperature.

18-12 Using Schottky's equation, Eq. (18-25), and setting $\delta_1/kT = x$,

(*a*) Show that the heat capacity is a maximum when

$$\frac{\dfrac{x}{2} - 1}{\dfrac{x}{2} + 1} = \frac{g_1}{g_0}e^{-x}$$

and find the maximizing value of x when $g_1/g_0 = \frac{3}{2}$.

(*b*) What is the maximum value of c_M/R when

$$\frac{g_1}{g_0} = \frac{3}{2}?$$

18-13 Two moles of gadolinium sulfate obeying Curie's law are in a magnetic field of 15×10^5 A/m and at a temperature of 15 K. The field is reduced reversibly and isothermally to zero.

(*a*) How much heat is transferred?

(*b*) How much work is done?

(*c*) What is the internal energy change?

(*d*) What would be the final temperature if the process were performed reversibly and adiabatically?

18-14 At what initial temperature should iron ammonium alum be in order that an adiabatic decrease of magnetic field from an initial field of 8×10^5 A/m to zero final field yield a final temperature of 0.2 K?

18-15 CMN originally at 1.5 K is magnetized isothermally from zero field to 4×10^5 A/m.

(*a*) What is the heat of magnetization per mole?

(*b*) After an adiabatic reduction of the magnetic field to a field of 8×10^4 A/m, what is the final temperature?

18-16 What initial field is required in order that an adiabatic reduction to zero field should lower the temperature from 1 to 10^{-2} K, using the following:

(*a*) Chromium potassium alum?

(*b*) CMN?

18-17 A solid at temperature T is placed in an external magnetic field of 2.4×10^6 A/m.

(*a*) If the solid contains weakly interacting magnetic ions with $J = \frac{1}{2}$ and $g = 1$, to what temperature must one cool the solid so that 75 percent of the atoms are polarized with their magnetic moments parallel to the external magnetic field?

(*b*) If, on the other hand, the solid were paraffin with many protons (but no magnetic ions), with $I = \frac{1}{2}$, $g_N = 2$, and a magnetic moment equal to 1.77×10^{-32} J · m/A, to what temperature must the paraffin be lowered to achieve a nuclear polarization of 75 percent?

18-18 A sample of mineral oil is placed in an external magnetic field $\mathscr{H}$. Each proton has $I = \frac{1}{2}$ and a magnetic moment μ. An applied radio-frequency field can induce transitions between the two energy states if its frequency ν is equal to $2\mu_0\mathscr{H}$. The power absorbed from this radiation field is proportional to the *difference* in the number of nuclei in the two energy levels. Assuming that $kT \gg 2\mu_0\mathscr{H}$, how does the absorbed power depend on T?

18-19 (*a*) One mole of chromium potassium alum is at 1 K and at 8×10^5 A/m. After undergoing an adiabatic decrease of magnetic field to zero field, it is immersed in 36 mol of liquid ^{4}He at 4.2 K whose specific heat capacity equals 2.5 kJ/kg · K. What is the final temperature of the combination?

(*b*) If, at the conclusion of the decrease of magnetic field in part (*a*), the alum is immersed in 36 mol of liquid ^{4}He at 2 K, whose specific heat capacity equals $0.104T^{6.2}$ kJ/kg · K, what is the final temperature of the combination?

18-20 In a single-cycle dilution refrigerator the rate of cooling below 40 mK is given by $\dot{Q} = 84\dot{n}_3 T_M^2$ J/mol · K^2, where $\dot{n}_3$ is the molar flow rate of ^{3}He and T_M is the temperature of the mixing chamber. What is the cooling rate for a circulation of 30 μmol/sec of ^{3}He and a mixing chamber temperature of 12 mK?

CHAPTER

NINETEEN

NUCLEAR MAGNETISM, NEGATIVE TEMPERATURES, AND THE THIRD LAW OF THERMODYNAMICS

19-1 POLARIZATION OF MAGNETIC NUCLEI

The magnetic moments of the chromium, iron, and gadolinium ions are due to uncompensated spins of the electrons that surround the respective nuclei. When a paramagnetic salt containing, for example, the chromium ion with $J = \frac{3}{2}$ is placed in a magnetic field of the order of 10^6 A/m and the temperature is lowered to about 1 K (so that the ratio $\mathcal{H}/T$ is of the order of 10^6 A/m/K), the ionic magnets become partially oriented in the direction of the field. A convenient measure of this partial orientation is provided by the ratio of the magnetic moment M to the maximum value or saturation value M_{sat}. When this ratio is expressed in percent, it is often called the *magnetic polarization*. From Eq. (18-16),

$$\frac{M}{M_{\text{sat}}} = B_J\left(\frac{g\mu_B\mu_0\mathcal{H}}{kT}\right),$$

and under the conditions specified,

$$\frac{g\mu_B\mu_0\mathcal{H}}{kT} \approx \frac{2 \times 9.3 \times 10^{-24} \times 4\pi \times 10^{-7} \times 10^6}{1.4 \times 10^{-23} \times 1} \approx 1.7.$$

In Fig. 18-2, $B_{3/2}(1.7) = 0.86$; that is, the ionic magnets are 86 percent polarized.

The particles inside the nucleus of an atom also have spins that give rise to *nuclear magnetism*. The resultant nuclear spin is characterized by a quantum number I that plays the same role as the atomic quantum number J. A nuclear magnetic moment is much smaller than that of an atom, by a factor of almost 2000. In the expression for a Bohr magneton $\mu_B = eh/4\pi m$, if we substitute the mass of a proton, we get a *nuclear magneton* where

$$\mu_N = \frac{\mu_B}{m_p/m_e} = \frac{9 \times 10^{-24}}{1840}$$
$$\approx 5 \times 10^{-27} \text{ J/T}.$$

When nuclear magnets are subjected to an external magnetic field, the lowest energy level of the nucleus splits into a number of separate states, each of which is characterized by a nuclear magnetic quantum number that may take on discrete values corresponding to discrete orientations of the nuclear magnetic moment with respect to the external magnetic field. These values range from $-I$ to $+I$ in integral steps—an exact replica of the atomic situation, where m took on values from $-J$ to $+J$ in integral steps.

The calculation of the nuclear magnetic moment M_N parallels perfectly that for the atomic moment, so that

$$M_N = N\mu_N g_N I B_I\left(\frac{g_N \mu_N \mu_0 \mathcal{H}}{T}\right), \tag{19-1}$$

where g_N is the nuclear splitting factor, which we shall take as equal to 2. Suppose that we have a nucleus with $I = \frac{3}{2}$ and want to produce a nuclear polarization of 86 percent. Then,

$$B_I(g_N \mu_N \mu_0 \mathcal{H}/kT) = 0.86,$$

and

$$\frac{g_N \mu_N \mu_0 \mathcal{H}}{kT} = 1.7,$$

or

$$\frac{\mathcal{H}}{T} = \frac{1.7 \times 1.4 \times 10^{-23}}{2 \times 5 \times 10^{-27} \times 4\pi \times 10^{-7}} \approx 2 \times 10^9 \text{ A/m/K}.$$

If, therefore, we used a field of 2×10^6 A/m, the nuclear-spin system would have to be at a temperature of 10^{-3} K in order to provide a nuclear magnetic polarization of 86 percent. If we could settle for a smaller polarization, we could start at, say, 10^{-2} K. Lining up nuclei in the same direction is a very valuable procedure for the physicist. If the nucleus is radioactive and emits alpha particles or beta particles or gamma rays, it is important to know whether these emanations are emitted as abundantly in one direction relative to the magnetic moment as in another. If they are emitted equally in all directions, one refers to their *symmetry* or *isotropy*. A preferred direction indicating *asymmetry* or *anisotropy* enables the nuclear physicist to accomplish the following:

1. Test various *conservation laws* that are supposed to hold during nuclear disintegrations.

2. Obtain information about the *shape* of the nucleus.
3. Obtain numerical values of certain constants needed in the theory of nuclear processes.

Nuclear polarization has been achieved by using a very low temperature and a very large field; but this procedure, known as the *brute-force method*, has not been available in many laboratories.

The method of nuclear polarization that has been most fruitful was suggested first by Gorter and Rose and involves the behavior of a nucleus in the very strong *local* magnetic field (over 8×10^6 A/m) produced by *its own* electronic structure. To understand this, let us consider the implications of the theoretically derived and experimentally verified result that the magnetization of an isolated spin system (whether atomic or nuclear, provided that neither $\mathscr{H}$ nor T is close to zero) is a function of $\mathscr{H}/T$ *only*. If, for example, 50 percent magnetic saturation is achieved with a particular paramagnetic salt at a temperature of 1 K with a field of 8×10^5 A/m, then at a temperature of 0.01 K a field of only 8×10^3 A/m would be needed to produce the same magnetic polarization. In other words, if the proper material were used, starting at 1 K and decreasing the field from 8×10^5 A/m to a field of 8×10^3 A/m, the temperature would drop ($\mathscr{H}/T = \text{const.}$) to about 0.01 K, but *the magnetic polarization would still be the same*. One could also demagnetize to zero field, thereby reaching a somewhat lower temperature, and then raise the field to 8×10^3 A/m and still have available about 50 percent polarization. Gorter and Rose realized that these polarized ionic magnets would give rise to a unidirectional local field at the nucleus of each ion which would be much larger than any field then achievable in the laboratory (before the advent of superconducting magnets)—of the order of 8×10^6 A/m or more.

This experiment was carried out several times by Roberts and his coworkers at Oak Ridge Laboratories. They polarized Mn and Sm nuclei and detected their polarization by measuring their ability to scatter polarized slow neutrons from the Oak Ridge reactor.

When radioactive cobalt nuclei ^{60}Co are polarized, γ rays are emitted with different intensities in different directions. By comparing the intensity in one direction with that at right angles to this direction, a quantity called the *anisotropy* is defined. Experiment shows that the anisotropy is a sensitive function of the temperature, so that, if this function has been determined from previous experiments, a measurement of γ-ray anisotropy enables one to obtain the temperature.

One of the most spectacular experiments in nuclear cryogenics was performed by Ambler, Hudson, and Wu at the National Bureau of Standards in Washington, D.C., in 1957. Lee and Yang, who were later awarded the Nobel prize, suggested that the nuclei of ^{60}Co, in undergoing radioactive decay, might emit beta particles (electrons) more abundantly toward one magnetic pole than toward the other. To test this hypothesis, polarization of the cobalt nuclei was necessary. Beta-ray counters were then used to show whether there was a differ-

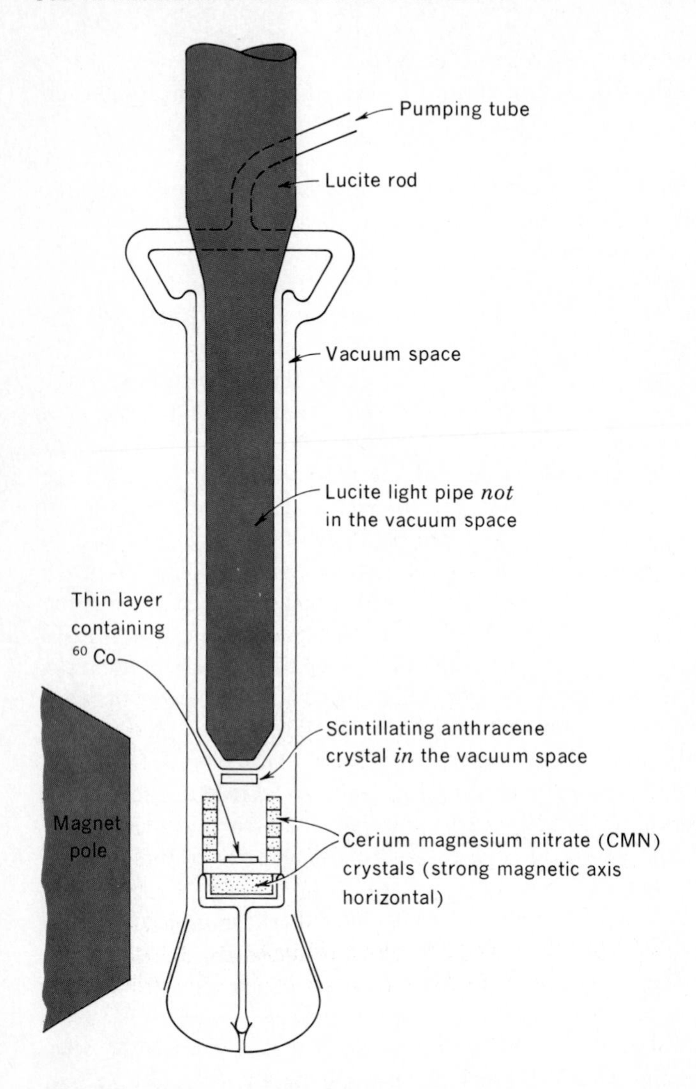

Figure 19-1 Apparatus of Ambler, Hudson, and Wu to measure β-particle emission from polarized ^{60}Co nuclei. (Liquid He and N_2 dewars and β-particle counters are not shown.)

ent reading in the direction of the north poles than in the reverse direction. The apparatus is shown in Fig. 19-1. The ^{60}Co, obtained by neutron bombardment of nonradioactive ^{59}Co, was introduced into the crystal lattice of cerium magnesium nitrate in the form of a thin layer lying in the bottom of a cup-shaped housing of this material. The beta particles emitted by ^{60}Co (in decaying to ^{60}Ni) produced scintillations in a small crystal of anthracene. The light flashes

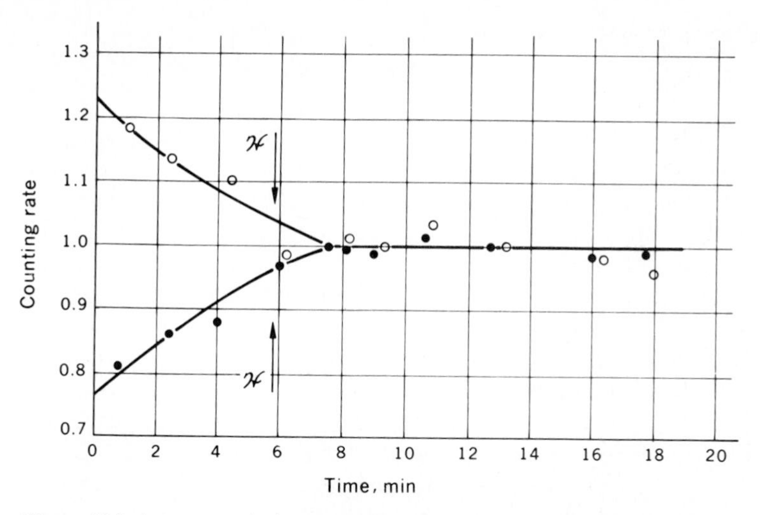

Figure 19-2 Asymmetry in beta-particle emission from polarized nuclei. Increasing time denotes increasing temperature. Maximum polarization is at the lowest temperature.

traveled up a *light pipe*, consisting of a 4-ft Lucite rod whose upper end communicated with a photomultiplier tube and counter.

The strong magnetic axis of the CMN crystals was horizontal, so that isothermal magnetization and adiabatic reduction to zero field could be accomplished with a horizontal magnetic field. By this means, the CMN housing and ^{60}Co layer were cooled below 0.01 K. A solenoid was then slipped over the outer dewar, and a vertical field of about 8×10^3 A/m was used to polarize the ^{60}Co ions without appreciable warming of the CMN. The strong local fields at the nuclei of the ions polarized the ^{60}Co nuclei, the direction of polarization depending upon the direction of the current in the solenoid. The north poles could therefore be made to face either toward the anthracene crystal or away from it.

The beta-particle ejection from ^{60}Co is accompanied by gamma-ray emission. Although they are not shown in Fig. 19-1, there were two gamma-ray counters to detect and to measure any gamma-ray anisotropy. Since gamma-ray anisotropy had already been measured as a function of temperature, it served as a convenient thermometer. The curve in Fig. 19-2 shows that *there are many more beta particles emitted in a direction opposite to the solenoid field*, that is, the south poles of the ^{60}Co nuclei emit beta particles more abundantly than the north poles. This result is in direct contradiction to a nuclear principle, known as the *principle of the conservation of parity*, which had assumed that certain nuclear processes should behave the same way for one configuration of the nucleus as for its mirror image. The experimental proof that parity is not conserved in beta decay has had a profound effect on theoretical and experimental physics.

19-2 PRODUCTION OF NANODEGREE TEMPERATURES BY NUCLEAR MAGNETIC FIELD REDUCTION

Since nuclear magnets are only about one one-thousandth as strong as ionic magnets, their polarization requires temperatures around 0.01 K and fields from 4×10^6 to 8×10^6 A/m. We have seen how local fields of this magnitude may be provided by the uncompensated spins of the electrons circulating outside each nucleus itself. If these polarized nuclei could then be made to undergo a reversible adiabatic field reduction, they would cool off to a temperature in the neighborhood of 10^{-5} K. Since these nuclei occupy a thin layer on a large crystal at 0.01 K, however, any loss of polarization would be much more isothermal than adiabatic. You cannot expect a few nuclei to cool off a big crystal.

One method that has been used so far to achieve temperatures below 10^{-3} K involves a double process, consisting of an ionic field reduction followed by a nuclear field reduction. Two separate magnetic fields supplied by two separate magnets are used, as shown schematically in Fig. 19-3, a diagram prepared by Kurti of Oxford, in whose laboratory such experiments have been carried out. In each of the four parts of this figure, the electronic stage represents a mass of chromium potassium alum in which are embedded 1500 enameled copper wires, each with a diameter of 0.0003 in. The copper wires continue for a distance of about 8 in. and are then bent over and bound together to form the nuclear stage itself. The first part of the cooling is done with the aid of chromium

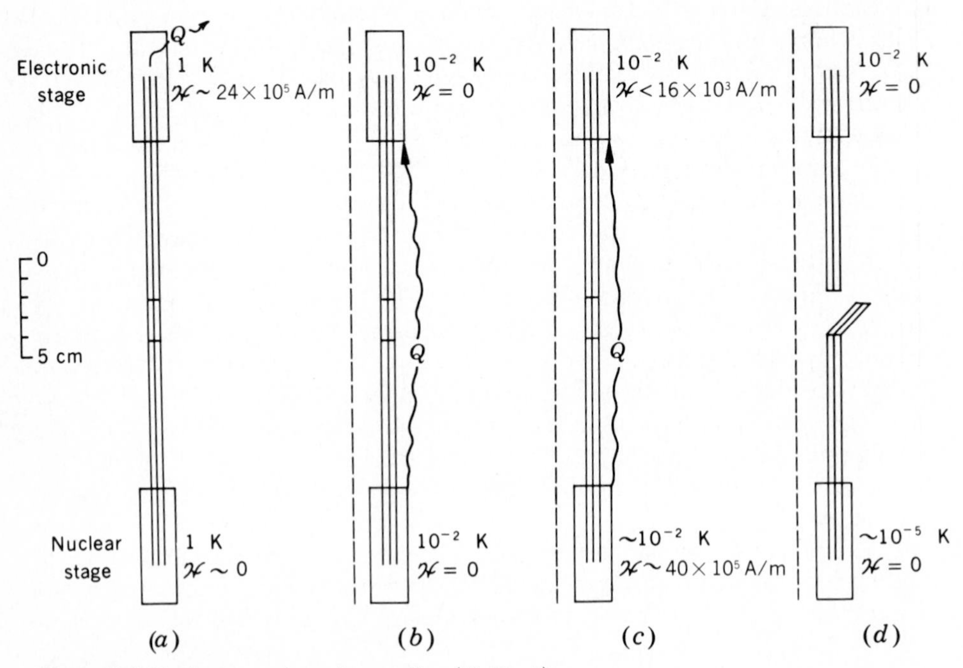

Figure 19-3 The four steps in nuclear cooling. (*N. Kurti.*)

ions, and the second part by *copper nuclei*. The fine, insulated copper wires serve three purposes:

1. Provide a heat-conducting medium between the nuclear and the electronic stages.
2. Minimize eddy currents induced by reduction of the magnetic field.
3. Produce a low temperature by nuclear reduction of the magnetic field.

The four steps in Fig. 19-3 are as follows:

(*a*) Isothermal magnetization of the electronic stage.
(*b*) Adiabatic field reduction of the electronic stage and cooling of nuclear stage to 10^{-2} K.
(*c*) Isothermal magnetization of the nuclear stage.
(*d*) Adiabatic field reduction of the nuclear stage, with an accompanying temperature drop to about 10^{-5} K.

The experiment is not as simple as it sounds. To quote Kurti, "The stringency of the conditions to be satisfied can be illustrated by remarking that even

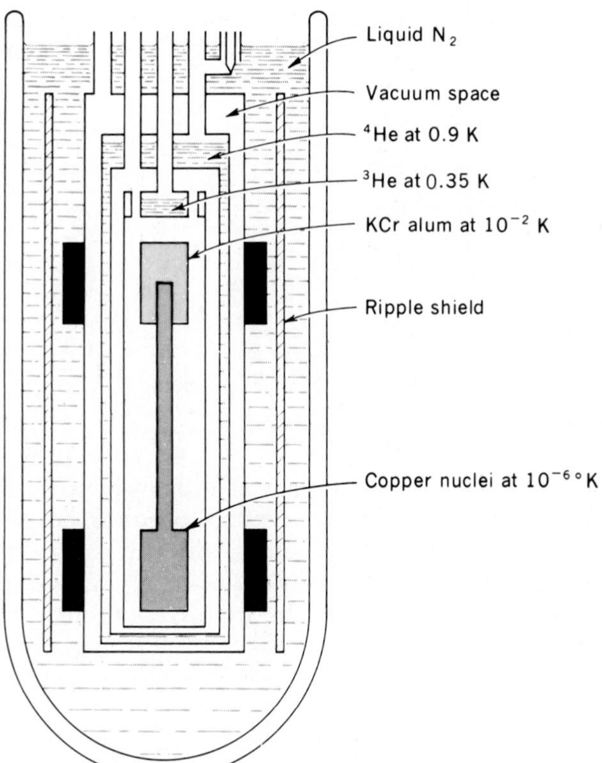

Figure 19-4 Cryostat for nuclear cooling (symbolic).

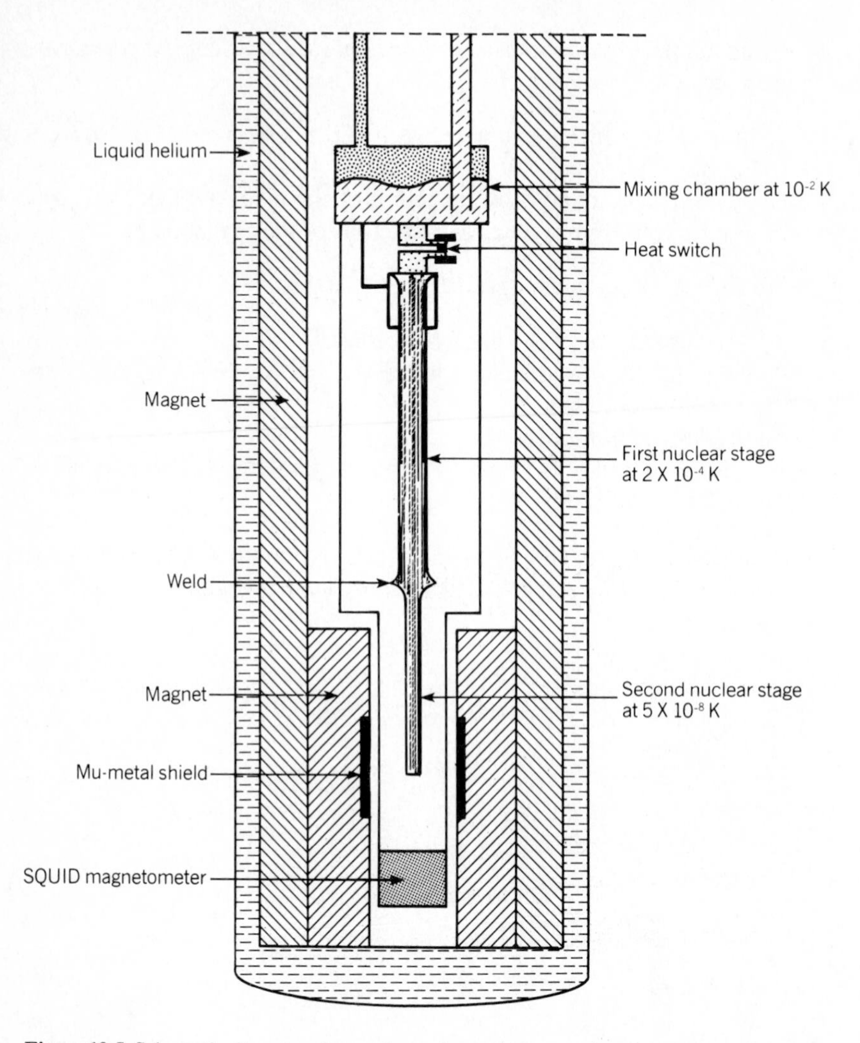

Figure 19-5 Schematic diagram of cascade nuclear refrigerator. (*O. V. Lounasmaa.*)

a minute amount of heating such as results from a small pin dropping through a height of one-eighth of an inch would warm a bulky specimen of several ounces from one-millionth of a degree to the starting temperature of one one-hundredth of a degree and thereby spoil the experiment." Even the eddy currents induced in the copper wires by virtue of slight variations of current (ripples) in the magnet coils must be prevented with the aid of the metal ripple shield shown in Fig. 19-4. The magnetic fields were supplied by solenoids in which currents of thousands of amperes were maintained.

One of the biggest experimental difficulties to overcome is the heat transfer between the nuclear and electronic stages. During the isothermal increase of $\mathcal{H}$

of the nuclear stage, this transfer must be good. During the following reduction of field, it must be poor. In the experiments of Kurti's group, the fine copper wires represent a compromise that served both purposes only moderately well. Another difficulty is to separate the electronic from the nuclear stage by a distance great enough to confine each magnetic field to its own paramagnetic particles. Both of these problems can be partially solved by a clever method conceived by Blaisse. Suppose the nuclear stage constitutes a *core* completely surrounded by a crystal (or a group of crystals identically oriented) of cerium magnesium nitrate with its strong magnetic axis pointing toward, let us say, the x axis. Suppose that we perform the following operations:

1. Magnetize isothermally at 1 K *in the x direction.*
2. Insulate thermally.
3. *Rotate the field to the y direction.* Since CMN is practically nonmagnetic in this direction, it therefore undergoes an adiabatic decrease of the magnetic field and its temperature drops, *even though the field is still there.*
4. Wait until the cold CMN has cooled the nuclear core, and then reduce the magnetic field to zero, thereby cooling the core by its own adiabatic decrease of the magnetic field.

To achieve even lower temperatures, two or more nuclear cooling stages must be operated in series. Ehnholm, Ekström, Jacquinot, Loponen, Lounasmaa, and Soini at the Helsinki University of Technology built a cryostat with two nuclear cooling stages attached to the mixing chamber of a $^3He/^4He$ dilution refrigerator as shown in Fig. 19-5. In their study of the magnetic properties of copper nuclei, they achieved a temperature of 50 nanokelvin.

The lowest temperature ever achieved anywhere (1979) *is* 5×10^{-8} K!

19-3 NEGATIVE KELVIN TEMPERATURES

Let us recall the original definition of the Kelvin scale of temperature: Two Kelvin temperatures are to each other as the heats transferred during isothermal processes at these temperatures, provided that these isothermal processes terminate on the *same* adiabatic surfaces. If Q and Q_s are the absolute values of the heats transferred at temperatures T and T_s, respectively, the original Kelvin definition provides the relation

$$T = T_s \frac{Q}{Q_s}.$$

If T_s refers to an arbitrary standard, the choice of a number for T_s is also arbitrary. If it is chosen to be negative, then all temperatures would be expressed by negative numbers. Whether T_s is chosen positive or negative, as Q is made smaller and smaller in any unordered way, the limiting value of Q is zero (i.e., the least amount of heat that can be transferred is no heat at all), and therefore *the lowest value of T is zero.* In other words, the lowest temperature is absolute zero, and if negative temperatures have any meaning at all, *they cannot mean temperatures colder than absolute zero!* But what is meant when the Kelvin scale is defined in the usual way with $T_s = +273.16$ K?

A clue as to the meaning of negative Kelvin temperatures is provided by the expression for temperature used in statistical thermodynamics,

$$T = \left(\frac{\partial U}{\partial S}\right)_V.$$

The most familiar thermodynamic systems, such as a mole of ideal gas or a mole of crystal, *have an infinite number of energy levels.* As the temperature is raised, more and more atoms are raised to higher levels. This requires more and more energy, and results in greater and greater disorder as the atoms are distributed over more and more states. As the energy goes up (positive dU), the entropy also goes up (positive dS); hence the ratio dU/dS is positive. For T to be negative, *an increase of energy would have to be accompanied by a decrease of entropy!* This obviously cannot take place when a system has an infinite number of energy levels.

Another way of looking at the matter is with the aid of the Boltzmann equation,

$$\frac{N_2}{N_1} = e^{-(\epsilon_2 - \epsilon_1)/kT}.$$

If the system has an infinite number of energy levels, an increase of temperature produces increased populations of higher and higher energy levels, but no energy level ever gets populated more than the one below it, so that the ratio N_2/N_1 is always less than 1 and T is positive. At $T = \infty$, N_2 would be equal to N_1, but this would require an infinite amount of energy because of the infinite number of energy levels! Evidently, for T to be negative, N_2 would have to be *larger* than N_1; that is, the upper energy levels would have to be populated *more* than the lower ones. This would require even more than infinite energy—which is even more than nonsensical. We conclude, therefore, that *in the case of an ordinary system which has an infinite number of energy levels, negative temperatures are an absurdity.*

But what about a system which has only a finite number of energy levels? Suppose, for the sake of argument, that a system were capable of existing in only two energy levels. Let the system consist of N particles and the levels have energies 0 and ϵ, where ϵ is an atomic constant, independent of any external field. The curve showing the relation between entropy S and internal energy U is shown in Fig. 19-6. At zero energy, all N atoms are in the lower energy level, which is a state of minimum disorder, or zero entropy. When the two energy levels are equally populated, the internal energy of the system is $N\epsilon/2$ and there

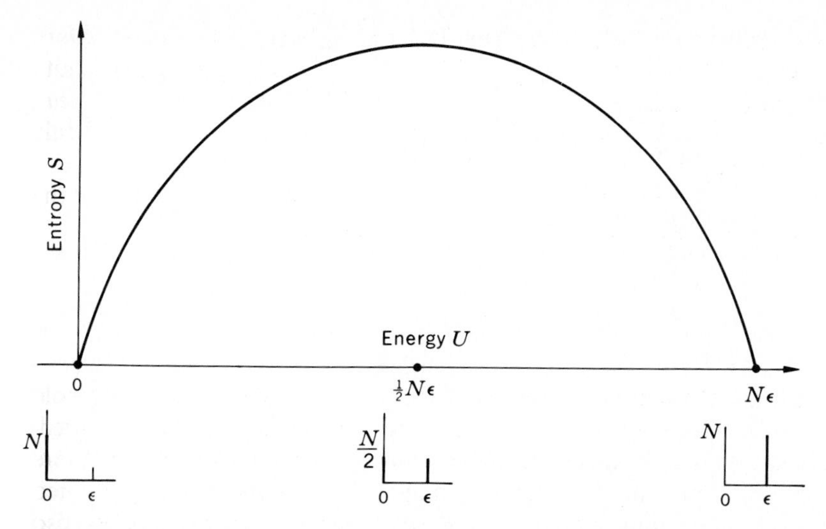

Figure 19-6 Relation between entropy and energy of a system of particles that can exist in only two energy levels.

is maximum disorder and, hence, maximum entropy. If and when all N atoms are in the upper energy level, $U = N\epsilon$, and again we have minimum disorder, or zero entropy. The left half of the curve has a positive slope, and therefore $(\partial U/\partial S)$ is positive. *The right half, with negative slope, is the region of negative temperatures.*

To achieve negative Kelvin temperatures, we must find a system with a finite number of energy levels and, somehow or other, succeed in producing a *population inversion*—that is, an equilibrium (or near-equilibrium) state in which there are more particles in upper states than in lower ones.

To reduce the temperature of a substance far below 1 K, the magnetic and thermal properties of a magnetic subsystem (ionic or nuclear) were used. The purpose of such experiments was to cool the *entire substance*, not just the subsystem. To accomplish this, it was necessary to satisfy the following conditions:

1. The magnetic ions must interact among themselves with sufficient strength and speed that (like the molecules of a gas) statistical equilibrium can be assumed and a definite temperature can be attributed to the ionic subsystem.
2. The nonmagnetic particles (called, for simplicity, the *lattice*) must have practically no heat capacity in the low-temperature region under consideration.
3. Equilibrium between the magnetic-ion subsystem and the lattice must be attained fairly rapidly.

To achieve negative temperatures, we must make use of the magnetic and thermal properties of a *nuclear* magnetic subsystem under the following conditions (one of which is the same as, while two are entirely different from, those just listed):

1. The nuclear magnetic subsystem comes to equilibrium with itself very rapidly.
2. The lattice is at room temperature, with a large heat capacity.
3. Equilibrium between the nuclear magnetic subsystem and the lattice is attained slowly enough (from, say, 2 min to several hours) that experiments can be performed on the nuclear subsystem in this time interval, as though the subsystem were isolated.

The system found by Pound, Purcell, and Ramsey in 1951 to satisfy the conditions for the production of negative temperatures is the subsystem consisting of the nuclei of the lithium ions in a LiF crystal. These were found to come to equilibrium among themselves in 10^{-5} sec, to require about 2 min or more to come to equilibrium with the lattice, and each to have a lowest energy level that is split into only four nuclear magnetic states ($I = \frac{3}{2}$) by an external magnetic field. The weak interactions among these magnetic nuclei involved emission and absorption of photons produced by transitions between some of the four states. That is, one nucleus would go from an upper to a lower state in emitting a photon, and the nucleus absorbing this photon would go from a lower to an upper state. These interactions play the same role in the achievement and maintenance of equilibrium as the collisions between gas molecules.

Experiments on the nuclear magnetic subsystem of LiF take place in a region of temperature and field similar to that enclosed within the vertical dashed lines of Fig. 18-6. In this region both S and M are functions of $\mathcal{H}/T$ only; therefore, in an adiabatic reduction of field, since S is constant, both $\mathcal{H}/T$ and M are constant. The internal energy is a function of T only and has an appreciable value and variation with T only at temperatures very much lower than those at which the experiments were made. During an adiabatic change of field, dU and dS are both zero, and therefore the expression for the temperature dU/dS becomes indeterminate. The previous analysis of negative temperatures presupposed that the energy-level spacing was an atomic constant. With a nuclear (or ionic) magnetic subsystem, however, the level spacing $\Delta\epsilon$ depends upon $\mathcal{H}$, and the magnetic energy $g\mu\mu_0\mathcal{H}$ is not internal but external potential energy. To obtain a useful and appropriate expression for T, we use the *magnetic enthalpy* H^*, where

$$H^* = U - \mu_0 \mathcal{H} M. \tag{19-2}$$

Since

$$dH^* = dU - \mu_0 \mathcal{H}\, dM - M\mu_0\, d\mathcal{H}$$

$$= T\, dS - M\mu_0\, d\mathcal{H}$$

we get

$$T = \left(\frac{\partial H^*}{\partial S}\right)_{\mathcal{H}}. \tag{19-3}$$

Choosing U to have the value zero, $H^* = -\mu_0 \mathcal{H} M$. Since both M and S are functions of $\mathcal{H}/T$, then M is a function of S, and $-\mu_0 \mathcal{H} M$ is a function of both S

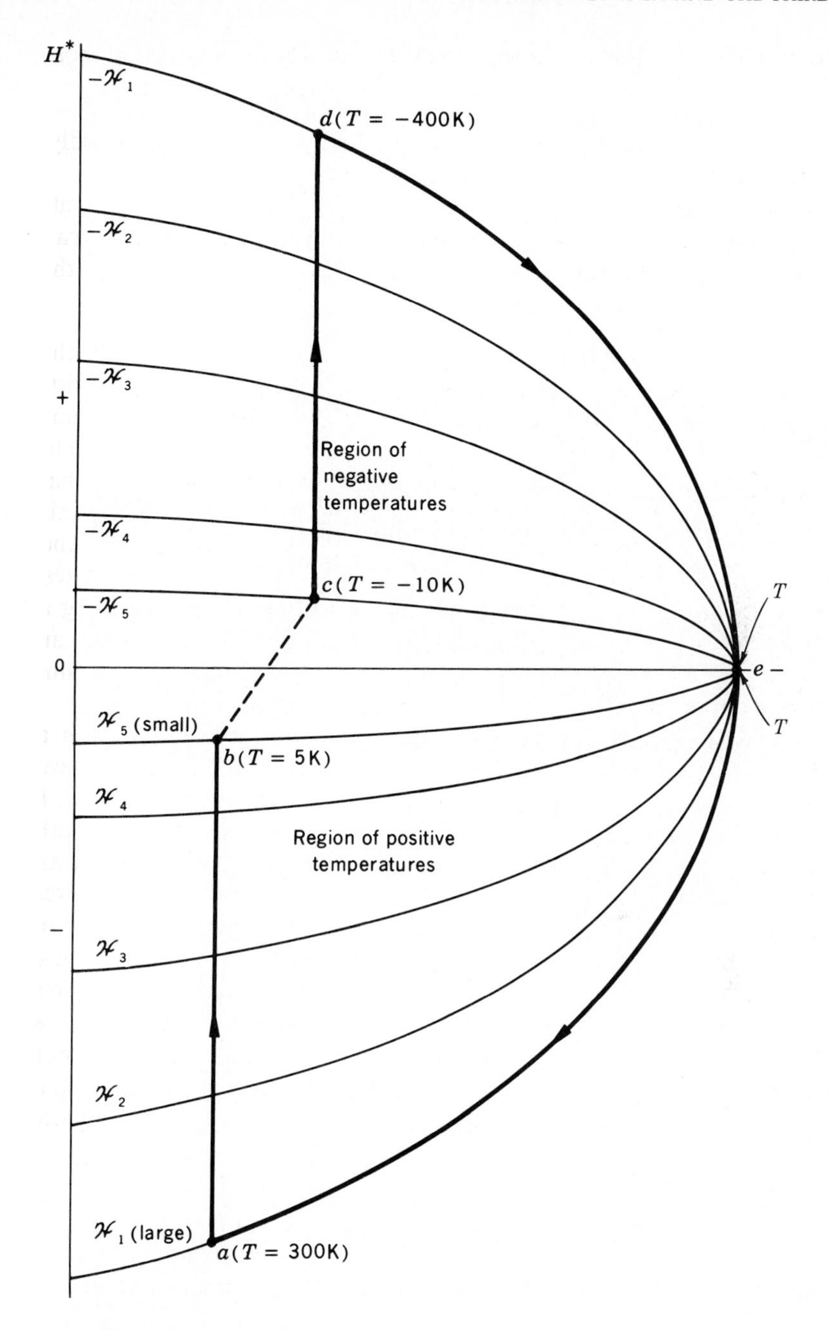

Figure 19-7 Magnetic enthalpy H^* vs. entropy S of nuclear magnetic subsystem at five different magnetic fields and for the same fields reversed. $a \to b$, adiabatic decrease of field; $b \to c$, rapid magnetic field reversal; $c \to d$, adiabatic increase of field; $d \to e \to a$, cooling through infinity at constant large magnetic field.

and $\mathscr{H}$. In Fig. 19-7, the lower half of the graph is a plot of $-\mu_0 \mathscr{H} M$ vs. S at five different values of $\mathscr{H}$, with the lowest curve (at $\mathscr{H}_1$) referring to the largest field. Notice that $(\partial H^*/\partial S)_{\mathscr{H}}$, the slope of any constant-field curve, is the temperature T and that the vertical line $a \rightarrow b$, at constant S, represents an adiabatic decrease from a large field $\mathscr{H}_1$ to a small one $\mathscr{H}_5$ during which M and $\mathscr{H}/T$ remain constant. In the upper half of Fig. 19-7, $+\mu_0 \mathscr{H} M$ is plotted against S for reversed fields, that is, for negative values of $\mathscr{H}$. The slope of every upper curve at every point is negative, and the process $c \rightarrow d$ represents an adiabatic increase of $\mathscr{H}$ during which the nuclear subsystem *cools* from -10 to -400 K.

19-4 THE EXPERIMENT OF POUND, PURCELL, AND RAMSEY

In the experiment of Pound, Purcell, and Ramsey, the crystal was placed in a magnetic field of about 5×10^5 A/m and allowed to come to thermal equilibrium at room temperature, 300 K. Under these circumstances,

$$a = \frac{g_N \mu_N \mu_0 \mathscr{H}}{kT} = \frac{2 \times 5 \times 10^{-27} \times 4\pi \times 10^{-7} \times 5 \times 10^5}{1.4 \times 10^{-23} \times 300}$$

$$= 1.5 \times 10^{-6}.$$

At such values of a, the Brillouin function reduces to

$$B_I(a) = \frac{I+1}{3} a,$$

and the fractional polarization is

$$\frac{M}{M_{\text{sat}}} = \frac{\frac{3}{2}+1}{3} \times 1.5 \times 10^{-6}$$

$$= 1.3 \times 10^{-6}.$$

Although this is a very small value, the methods of nuclear magnetic resonance (nmr) are still effective in showing the difference between the number of nuclei whose magnetic moments are in the direction of the field and those whose moments are opposite. The crystal is placed inside a small coil that is connected in series with a variable capacitor. The coil and capacitor form the resonant circuit in a radio-frequency oscillator, the frequency of which may be varied by adjusting the variable capacitor. The output of the oscillator is observed with an ordinary AM receiver. If the frequency of the oscillator is adjusted to a value $\nu_{\mathscr{H}}$, where

$$h\nu_{\mathscr{H}} = \Delta\epsilon = g_N \mu_N \mu_0 \mathscr{H}$$

$$\text{and} \quad \nu(\mathscr{H} = 5 \times 10^5 \text{ A/m}) \approx \frac{2 \times 5 \times 10^{-27} \times 4\pi \times 10^{-7} \times 5 \times 10^5}{6.6 \times 10^{-34}} \approx 10^7 \text{ Hz},$$

then some of the Li nuclei with their spins parallel to the field will be flipped so that their spins become antiparallel, with an *absorption* of energy, and some of the Li nuclei with their spins antiparallel to the field will be flipped to the parallel position, with an *emission* of energy. But since these two processes occur with equal probability, and since there are slightly more nuclei with their spins parallel than antiparallel, there is a *net absorption* of energy, which is observed as a drop in the amplitude of the oscillator output, and hence as a drop in the

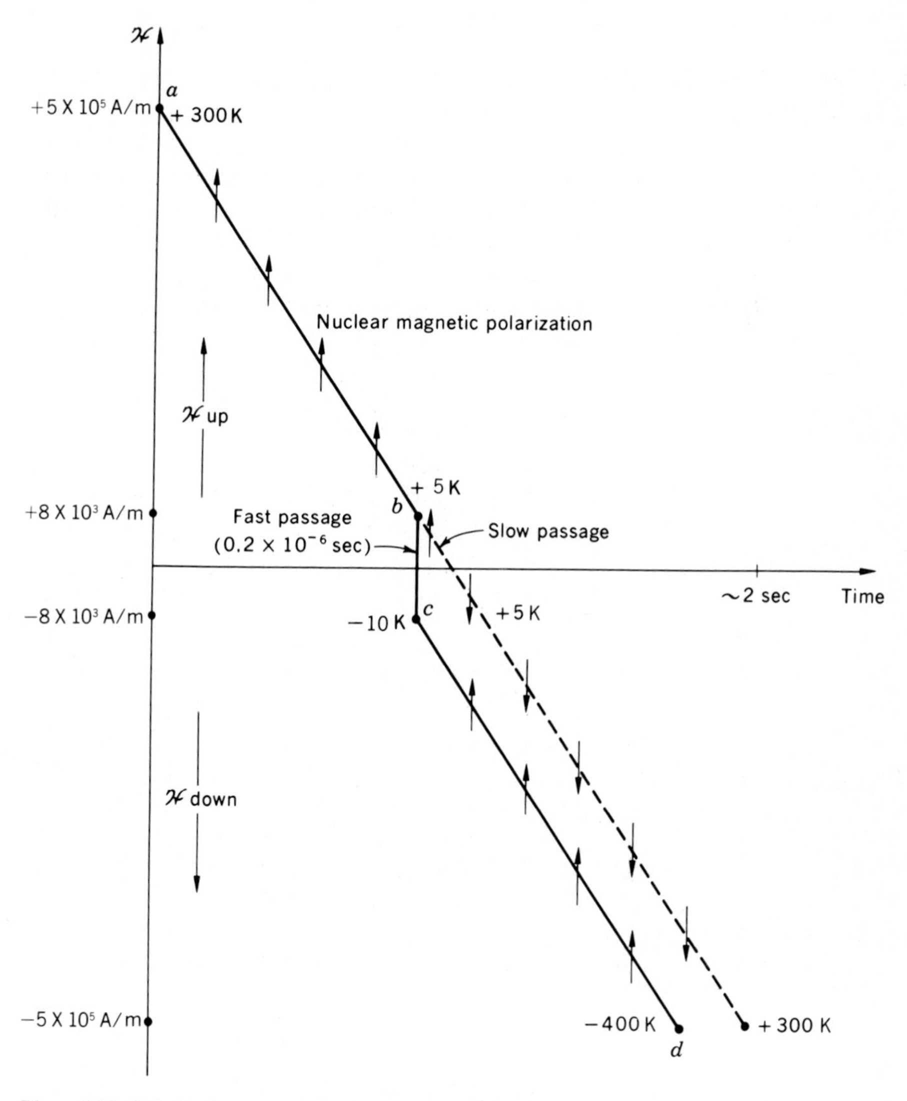

Figure 19-8 Relation between nuclear magnetic polarization and the magnetic field for slow passage compared with that for fast passage.

output of the AM receiver. This drop in output corresponded to a positive temperature for the nuclear subsystem at 300 K.

The next step, $a \to b$ in Fig. 19-7, was to remove the crystal from the magnetic field of 5×10^5 A/m to a coil in a field of about 8×10^3 A/m, reversibly (slowly) and adiabatically, during which the polarization (parallel to the field) remained constant and the temperature presumably fell to about 5 K, although no attempt was made to measure this nuclear-spin temperature. In a field of 8×10^3 A/m, a lithium nuclear "magnetic top" undergoes precession with a period of about 1 μs. By discharging a capacitor through the coil containing the LiF crystal, *the magnetic field was reversed to a value of about* -8×10^3 A/m *in a time of* 0.2 μs, *during which time the nuclear magnets could not follow the field.* (Such a field reversal is *highly irreversible.*) In this process, $b \to c$ in Fig. 19-7, the slight polarization parallel to the field at b (due to more nuclear magnets in lower states than in upper ones) became a polarization *opposite* the field (more nuclei in upper states than in lower ones), with a temperature about -10 K. The next process, $c \to d$, represents the adiabatic increase of the magnetic field accomplished by putting the crystal back into the reversed field of 5×10^5 A/m, during which the temperature cooled from -10 to -400 K. The last step, $d \to e \to a$, was the inevitable cooling due to interaction with the lattice, in which the temperature decreased from -400 K to $-\infty$ (which is the same as $+\infty$) and then went back to $+300$ K.

The success of the experiment depended on performing the field reversal in a time less than the Larmor precession period and bringing the crystal back to its place in a time less than the relaxation time for equilibrium between nuclear subsystem and lattice. A symbolic field-vs.-time graph is shown in Fig. 19-8, with

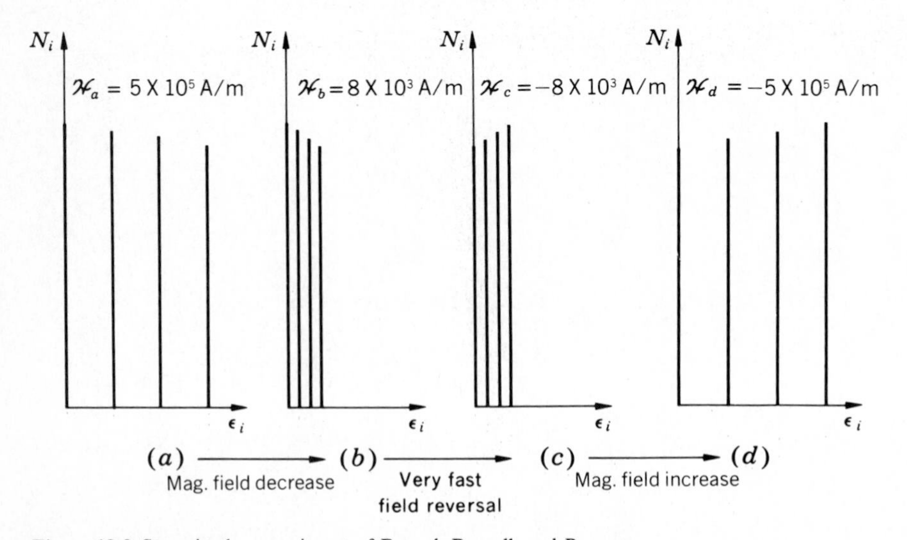

Figure 19-9 Steps in the experiment of Pound, Purcell, and Ramsey.

a comparison between the *fast-passage* (negative temperature) result and a *slow-passage* one. The small arrows indicate the magnetic polarization of the Li nuclear subsystem.

In the 2-min period in which the nuclear subsystem was at negative temperatures, the AM receiver of the nmr apparatus showed an increase in signal at $\nu\mathcal{H}$, indicating a net emission of energy from the Li nuclei and thereby proving the existence of negative temperatures.

Figure 19-9 shows the changes in the spacing and the populations of the energy levels during the experiment of Pound, Purcell, and Ramsey, which in the opinion of the authors ranks among the most significant experiments of modern times.

19-5 THERMODYNAMICS AT NEGATIVE TEMPERATURES

Classical thermodynamics takes some peculiar twists at negative temperatures, but much remains the same as at positive temperatures. Take, for example, the entropy principle, which states that the sum of all the entropy changes accompanying a natural irreversible process is positive. Suppose that Q units of heat leave a hot reservoir at a temperature, say, of -50 K and enter a colder reservoir at, say, -100 K, as shown in Fig. 19-10*a*. (Recall that the hottest negative temperature is -0 and the coldest is $-\infty$.) Since heat leaves the hotter reservoir,

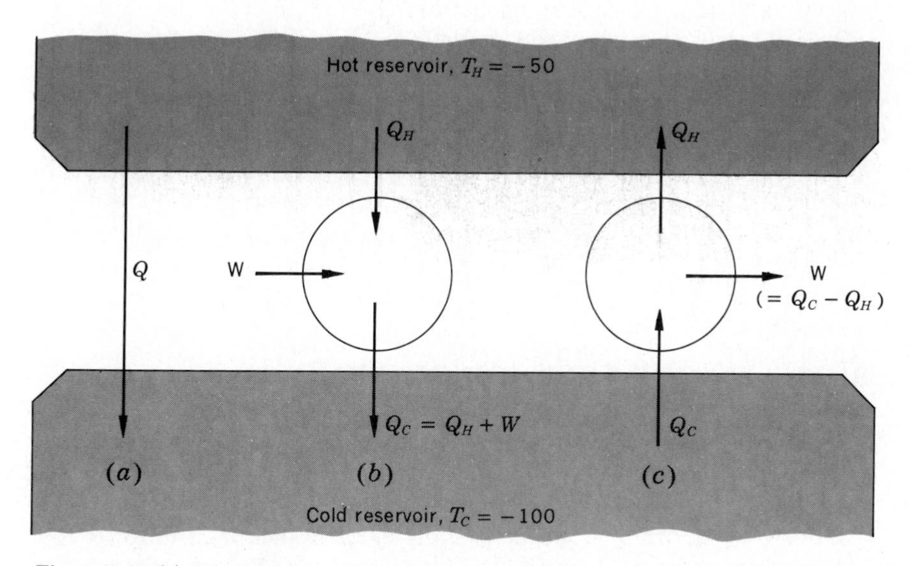

Figure 19-10 (*a*) Spontaneous flow of heat in the direction of decreasing temperature. (*b*) A costly device for doing something that requires no device. (*c*) A heat engine that could be used to convert heat $Q_C - Q_H$ from the cold reservoir completely into work.

the entropy change is $-Q/-50$; whereas the entropy change of the colder reservoir is $+Q/-100$. The total entropy change is

$$\frac{-Q}{-50} + \frac{Q}{-100} = \frac{Q}{100},$$

which is positive, just as it is with positive temperatures.

In Fig. 19-10*b* is shown an attempt to imitate the way a heat engine behaves, but at negative temperatures. Since, by definition of the Kelvin scale,

$$\frac{Q_H}{Q_C} = \frac{T_H}{T_C} = \frac{-50}{-100} = \frac{1}{2},$$

when Q_H units of heat leave the hotter reservoir, *twice as much* heat must enter the colder one. Therefore, instead of work W being done *by* the engine, work would have to be done *on* the engine in order not to violate the principle of the conservation of energy. But the device depicted in Fig. 19-10*b* is an expensive gadget for doing a job that requires no device at all. If all you want is to push heat into a cold reservoir, it is sufficient merely to allow Q_H to flow *naturally* from the hot to the cold reservoir.

To get W units of work *out* of a heat engine operating between reservoirs at negative temperatures, you would have to make use of the device shown in Fig. 19-10*c*, where Q_C units of heat are taken *from* the cold reservoir (as though it were a refrigerator). Then a *smaller* quantity Q_H would go into the hotter reservoir, and the rest would be available for work. But the hot reservoir could be dispensed with, for the Q_H units of heat would naturally flow back to the colder reservoir. The net result would be that $Q_C - Q_H$ units of heat were extracted from the *colder* reservoir and converted completely into work, in violation of the Kelvin-Planck statement of the second law. This is the only principle of classical physics that is violated by systems at negative temperatures—but it is an important and interesting one.

Up to the present time, the only real use for systems at negative temperatures has been in the rapidly expanding field of masers and lasers. Perhaps, in the future, experiments on heat engines and refrigerators will be performed at negative temperatures. Then it will truly be fun to be an engineer.

19-6 THIRD LAW OF THERMODYNAMICS

We have seen how the Joule-Kelvin effect is employed to produce liquid helium at a temperature below 5 K. The rapid adiabatic vaporization of liquid helium then results in a further lowering of the temperature to about 1 K with ^{4}He and to about 0.3 K with ^{3}He. The magnetocaloric effect is then used to lower the temperature of a paramagnetic compound (magnetic-ion subsystem *plus* lattice) to about 0.001 K. In principle, it is possible to achieve still lower temperatures of matter by repeated applications of the magnetocaloric effect. Thus, after the original isothermal increase in magnetic field, the first adiabatic decrease in magnetic field might be used to provide a large amount of material at temperature T_{f1} to serve as a heat reservoir for the next isothermal increase in magnetic field of a smaller amount of material. A second adiabatic magnetic field reduction might then give rise to a lower temperature T_{f2}, and so on. The question which naturally arises at this point is whether the magnetocaloric effect may be used to cool a substance to absolute zero.

Experiment shows that the fundamental feature of all cooling processes is that the lower the temperature achieved, the more difficult it is to go lower. For example, the colder a liquid is, the lower the vapor pressure, and the harder it is to produce further cooling by pumping away the vapor. The same is true for the magnetocaloric effect: if one decrease in magnetic field produces a temperature T_{f1}, say, one-tenth the original T_i, then a second decrease from the same original field will produce a temperature T_{f2} which is also approximately one-tenth of T_{f1}. Under these circumstances, an infinite number of adiabatic field reductions would be required to attain absolute zero. Generalizing from experience, we may state the following:

By no finite series of processes is the absolute zero attainable.

This is known as either *the principle of the unattainability of absolute zero* or *the unattainability statement of the third law of thermodynamics.* Just as in the case of the second law of thermodynamics, the third law has a number of alternative or equivalent statements. Another statement of the third law is the result of experiments leading to calculations of the way the entropy change of a condensed system during a *reversible, isothermal* process ΔS_T behaves as T approaches zero. For example, the entropy change of a solid during a reversible isothermal compression may either be measured at different T's or be calculated from the second $T\,dS$ equation, leading to

$$\Delta S_T = S(T, P_1) - S(T, 0) = -\int_0^{P_1} \left(\frac{\partial V}{\partial T}\right)_P dP.$$

Since β decreases as T decreases, this example of ΔS_T decreases as T decreases. The entropy change of a paramagnetic salt during a reversible isothermal increase of magnetic field also decreases as T decreases, because

$$\Delta S_T = S(T, \mathscr{H}_1) = S(T, 0) = \mu_0 \int_0^{\mathscr{H}_1} \left(\frac{\partial M}{\partial T}\right)_{\mathscr{H}} d\mathscr{H},$$

and $\mu_0(\partial M/\partial T)_{\mathscr{H}}$ decreases with T. Experimental evidence is very strong to support the view that, as T decreases, ΔS_T decreases provided that the system is a solid or a liquid—that is, a condensed system. The following principle is therefore accepted:

The entropy change associated with any isothermal reversible process of a condensed system approaches zero as the temperature approaches zero.

Let us call this theorem the *Nernst-Simon statement of the third law of thermodynamics.* Both this statement and the unattainability statement have had a long and checkered career since the original paper by Nernst in 1907. It took 30 years of experimental and theoretical research, during which time there were periods of great confusion, before all differences of opinion were resolved and the statement was agreed upon. Nernst originally stated, as the third law, that the temperature derivative of the change of Helmholtz function during an isothermal process approaches zero as the temperature approaches zero. He did not think in terms of entropy and, moreover, was of the opinion that this statement and also the unattainability statement could be derived from the second law with the additional assumption that the heat capacities of all materials approached zero as the temperature approached zero. Nernst also maintained that both statements were true for all kinds of processes, both reversible and irreversible. It was mainly the experiments and arguments of Simon in the period from 1927 to 1937 that made precise the region of validity of the third law.

In order to show that the Nernst-Simon statement and the unattainability statement are equivalent, it is necessary to derive an equation for the limiting value of the entropy change accompanying an isothermal reversible process. Let us return to a paramagnetic salt and consider any isentropic increase of magnetic field, $i \rightarrow f$ of Fig. 19-11. The entropy change between the point $(T = 0,$

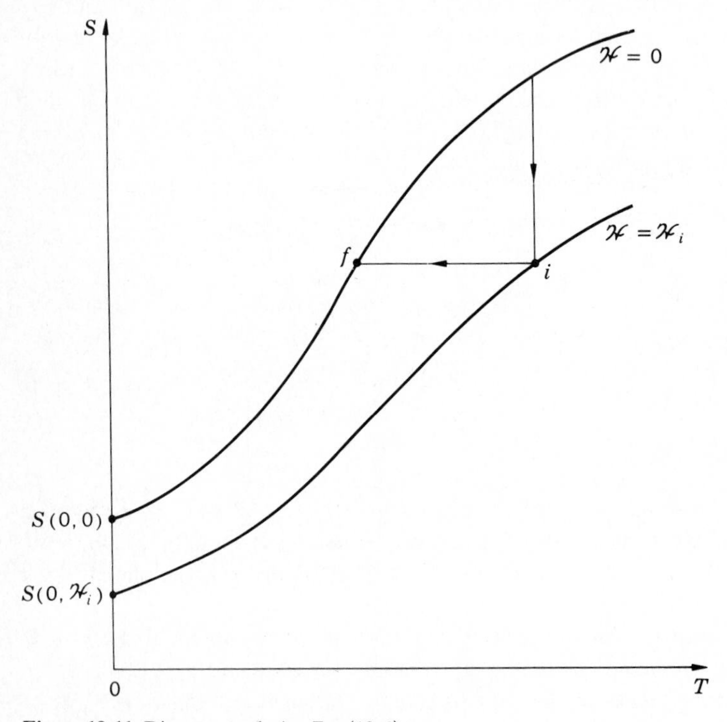

Figure 19-11 Diagram to derive Eq. (19-4).

$\mathscr{H} = \mathscr{H}_i$) and the state i is

$$S_i - S(0, \mathscr{H}_i) = \int_0^{T_i} \frac{C_{\mathscr{H}=\mathscr{H}_i}}{T}\,dT,$$

where $C_{\mathscr{H}}$ is the heat capacity at constant field, a positive quantity for all values of $\mathscr{H}$. The change in entropy between the point ($T = 0$, $\mathscr{H} = 0$) and f is

$$S_f - S(0, 0) = \int_0^{T_f} \frac{C_{\mathscr{H}=0}}{T}\,dT.$$

Since $S_i = S_f$ and $S(0, \mathscr{H}_i) - S(0, 0) = \lim\,[S(T, \mathscr{H}_i) - S(T, 0)]$, we have

$$\lim_{T\to 0}\,[S(T, \mathscr{H}_i) - S(T, 0)] = \int_0^{T_f} \frac{C_{\mathscr{H}=0}}{T}\,dT - \int_0^{T_i} \frac{C_{\mathscr{H}=\mathscr{H}_i}}{T}\,dT. \tag{19-4}$$

To prove the equivalence of the unattainability and Nernst-Simon statements of the third law, we proceed in the same manner as in the case of the Kelvin-Planck and Clausius statements of the second law.

Let U = truth of the unattainability statement;
$-U$ = falsity of the unattainability statement;
N = truth of the Nernst-Simon statement;
$-N$ = falsity of the Nernst-Simon statement.

As before, $U \equiv N$

when $-U \supset -N$ and $-N \supset -U$.

1. To prove that $-U \supset -N$, suppose that it is possible to find a value of T_i which makes $T_f = 0$, thereby violating the unattainability statement. Then, from Eq. (19-4), the left-hand member would be negative, thereby violating the Nernst statement.
2. To prove that $-N \supset -U$, suppose that the left-hand member of Eq. (19-4) had any negative value, thereby violating the Nernst-Simon statement. Then it would be possible to find a value of T_i in Eq. (19-4) that would make the second integral equal to this negative number. As a result, the first integral would vanish and T_f would be zero, thereby violating the unattainability statement.

The fact that $-N \supset -U$ may also be readily seen from Fig. 19-12*a*. If the point $(0, \mathscr{H}_i)$ lies below the point $(0, 0)$, then the adiabatic reduction of magnetic field $6 \to 7$ could be used to lower the system to absolute zero.

To complete the proof of the equivalence of the unattainability and the Nernst-Simon statements of the third law, one ought to consider a type of system that undergoes a decrease of entropy during an isothermal decrease of magnetic field and a decrease of temperature during an adiabatic increase of magnetic field, such as a *superconductor in the intermediate state.* Since the

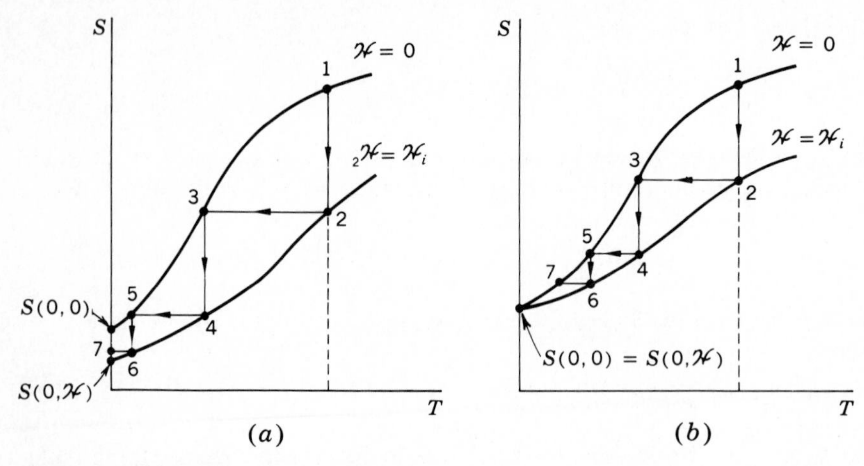

Figure 19-12 (*a*) If the Nernst-Simon statement of the third law were not true, the processes $5 \rightarrow 6$ and $6 \rightarrow 7$ could be used to attain absolute zero. (*b*) Diagram to show the equivalence of the three statements of the third law.

proof proceeds in exactly the same way as before, however, nothing would be gained by repeating the details.

A paramagnetic substance was invoked in the proof of the equivalence of *U* and *N* only for convenience and concreteness. By means of a slight change in symbols, the same proof may be applied to any system whatever, since all systems are capable of undergoing an isothermal reversible decrease of entropy followed by a reversible adiabatic decrease of temperature. Furthermore, the Nernst-Simon statement also applies to materials in frozen metastable equilibrium, provided that the isothermal process in question does not disturb this frozen equilibrium.

Referring to Fig. 19-12*b*, we see that, in the isothermal process $1 \rightarrow 2$, there is a decrease of entropy and that in $3 \rightarrow 4$ there is another decrease, and so on. If the entropy of the system at absolute zero is called the *zero-point entropy*, we see that a third equivalent statement of the third law is as follows:

By no finite series of processes can the entropy of a system be reduced to its zero-point value.

The equivalence of all three statements of the third law is clearly displayed in Fig. 19-12*b*.

There are many physical and chemical facts which substantiate the third law. For example, using Clapeyron's equation,

$$\frac{dP}{dT} = \frac{s^{(f)} - s^{(i)}}{v^{(f)} - v^{(i)}}$$

in conjunction with a phase change that takes place at low temperature, the statement that

$$\lim_{T \to 0} \left(s^{(f)} - s^{(i)}\right) = 0$$

implies that

$$\lim_{T \to 0} \frac{dP}{dT} = 0,$$

since $v^{(f)} - v^{(i)}$ is not zero for a first-order phase transition. This is substantiated by the melting curve of solid helium shown in Figs. 13-24 and 13-26. As a matter

Table 19-1 The journey toward absolute zero

Date	Investigator	Country	Development	Temp., K
1860	Kirk	Scotland	*First step toward deep refrigeration:* reached temperatures below freezing point of Hg.	234.0
1877	Cailletet	France	*First liquefied oxygen:* used throttling process from pressure vessel, obtaining fine mist only.	90.2
1884	Wroblewski and Olzewski	Poland	*First property measurements at low temperatures:* used small quantities of liquid N_2 and O_2.	77.3
1898	Dewar	England	*First liquefied hydrogen:* used Joule-Kelvin effect and counterflow heat exchanger.	20.4
1908	Kamerlingh-Onnes	The Netherlands	*First liquefied helium:* used same method as Dewar; shortly thereafter, lowered pressure over liquid to get 1 K.	4.2
1927	Simon	Germany and England	*Developed helium liquefier:* used adiabatic expansion from pressure vessel with liquid H_2 precooling.	4.2
1933	Giauque and MacDougall	United States	*First adiabatic field reduction:* Principle first proposed by Giauque and Debye in 1926.	0.25
1934	Kapitza	England and USSR	*Developed helium liquefier using expansion engine:* Made possible liquefaction of helium without liquid H_2 precooling.	4.2
1946	Collins	United States	*Developed commercial helium liquefier:* used expansion engines and counterflow heat exchangers.	2.0
1956	Simon and Kurti	England	*First nuclear experiments:* used adiabatic field reduction of nuclear stage of a paramagnetic salt.	10^{-5}
1979	Ehnholm and others	Finland	*Reached lowest temperature so far:* cascade nuclear refrigerator.	5×10^{-8}

of fact, dP/dT of solid ^{4}He approaches zero very rapidly, as shown by the experimental result of Simon and Swenson, that

$$\frac{dP}{dT} = 0.425T^7.$$

There are many other applications of the third law in the fields of physical chemistry and statistical mechanics. For further study, the writings of Simon and of Guggenheim are recommended.

The fact that absolute zero cannot be attained is no cause for misgiving. A temperature of 5×10^{-8} K represents a fraction of room temperature (300 K) equal to

$$\frac{5 \times 10^{-8}}{3 \times 10^2} \approx 10^{-10}.$$

Cryogenics has therefore enabled us to get to about a ten-billionth of room temperature. The surface temperature of the sun, 6000 K, is only 20 times room temperature, and the temperature in the interior of the hottest star, about 3×10^9 K, is 10 million times room temperature. Cryogenics is still ahead by a factor of a thousand.

A chronological account of the progress toward lower temperatures is given in Table 19-1.

PROBLEMS

19-1 Sketch a Carnot engine cycle on a $\mathcal{H}M$ diagram:

(*a*) At positive temperatures.

(*b*) At negative temperatures.

(*c*) Why is it impossible to operate a Carnot engine between a reservoir at a negative temperature and one at a positive temperature?

19-2 Suppose that the magnetization of a paramagnetic solid is given by

$$M = M_{\text{sat}} f(a),$$

where $a = g\mu\mu_0 \mathcal{H}/kT$ and $f(a) = 0$ when $a = 0$, and $f(a) \to 1$ when $a \to \infty$. The change of entropy during an isothermal reversible increase of field from 0 to $\mathcal{H}_i$ is given by

$$\Delta S_T = \mu_0 \int_0^{\mathcal{H}_i} \left(\frac{\partial M}{\partial T}\right)_{\mathcal{H}} d\mathcal{H}$$

$$= \mu_0 \int_0^{\mathcal{H}_i} M_{\text{sat}} f'(a) \left(\frac{-g\mu\mu_0 \mathcal{H}}{kT^2}\right) \frac{kT}{g\mu\mu_0}\, da$$

$$= -\frac{M_{\text{sat}} k}{g\mu} \int_0^{a_i} a f'(a)\, da.$$

Determine whether the Nernst-Simon statement of the third law is true in the following cases:

(*a*) Curie's equation: $f(a) = a$.

(*b*) Langevin's equation: $f(a) = \coth a - \dfrac{1}{a}$.

(*c*) Brillouin's equation, with $J = \frac{1}{2}$: $f(a) = \tanh \dfrac{a}{2}$.

(*d*) Brillouin's equation, with any J: $f(a) = B_J(a)$.

19-3 Using the Nernst-Simon statement of the third law, prove that:

(*a*) For a reversible cell or a thermocouple,

$$\lim_{T \to 0} \left(\frac{d\mathcal{E}}{dT} \right) = 0.$$

(*b*) For a surface film of liquid ^4He or ^3He,

$$\lim_{T \to 0} \left(\frac{d\mathcal{S}}{dT} \right) = 0.$$

19-4 (*a*) At $T = 0$, $(\partial S/\partial V)_{T=0} = 0$, and also

$$\frac{\partial}{\partial V} \left[\left(\frac{\partial S}{\partial V} \right)_T \right]_{T=0} = 0.$$

From this fact, prove that

$$\lim_{T \to 0} \left[\frac{\partial(1/\kappa)}{\partial T} \right]_V = 0,$$

where κ is the isothermal compressibility.

(*b*) In the case of a solid whose equation of state is

$$Pv + G(v) = \Gamma u,$$

where $G(v)$ is a function of volume only and Γ is a constant, prove that c_V approaches zero as T approaches zero.

APPENDIX

A

PHYSICAL CONSTANTS

Constant	Symbol	Rounded value
Electronic charge	e	1.602×10^{-19} C
Electronic rest mass	m	9.109×10^{-31} kg
Speed of light in vacuum	c	2.998×10^{8} m/s
Permeability of vacuum	μ_0	1.257×10^{-6} H/m
Permittivity of vacuum	ϵ_0	8.854×10^{-12} F/m
Planck's constant	h	6.626×10^{-34} J · s
Boltzmann's constant	k	1.381×10^{-23} J/K
Avogadro's number	N_A	6.022×10^{23} mol^{-1}
Faraday's constant	N_F	9.648×10^{4} C/mol
Bohr magneton	μ_B	9.274×10^{-24} J/T
Nuclear magneton	μ_N	5.051×10^{-27} J/T
Proton rest mass	m_P	1.673×10^{-27} kg
Universal gas constant	$R = N_A k$	8.314 J/mol K
Stefan-Boltzmann constant	σ	5.670×10^{-8} W/m^2 K^4
One electron volt	1 eV	1.602×10^{-19} J
One atomic mass unit	1u	1.660×10^{-27} kg
One atmosphere	1 atm	1.013×10^{5} Pa
One millimeter of mercury	1 mm Hg	133.3 Pa

Permeability of free space μ_0 $4\pi \times 10^{-7} \frac{Wb}{A\,m}$

APPENDIX

B

LAGRANGE'S METHOD OF MULTIPLIERS

Let us consider for the sake of simplicity a function f of only four variables y_1, y_2, y_3, and y_4 that is to be rendered an extremum, subject to the equations of constraint

$$\psi_1(y_1, y_2, y_3, y_4) = 0,$$

$$\psi_2(y_1, y_2, y_3, y_4) = 0.$$

Since there are two equations of constraint, only two of the four y's are independent. Taking the differential of the function f and equating it to zero, we get

$$\frac{\partial f}{\partial y_1} dy_1 + \frac{\partial f}{\partial y_2} dy_2 + \frac{\partial f}{\partial y_3} dy_3 + \frac{\partial f}{\partial y_4} dy_4 = 0.$$

Taking the differential of the equations of constraint, we get

$$\frac{\partial \psi_1}{\partial y_1} dy_1 + \frac{\partial \psi_1}{\partial y_2} dy_2 + \frac{\partial \psi_1}{\partial y_3} dy_3 + \frac{\partial \psi_1}{\partial y_4} dy_4 = 0,$$

$$\frac{\partial \psi_2}{\partial y_1} dy_1 + \frac{\partial \psi_2}{\partial y_2} dy_2 + \frac{\partial \psi_2}{\partial y_3} dy_3 + \frac{\partial \psi_2}{\partial y_4} dy_4 = 0.$$

Multiplying the first of the above two equations by λ_1 and the second by λ_2, we have the three equations

$$\frac{\partial f}{\partial y_1}\,dy_1 + \frac{\partial f}{\partial y_2}\,dy_2 + \frac{\partial f}{\partial y_3}\,dy_3 + \frac{\partial f}{\partial y_4}\,dy_4 = 0,$$

$$\lambda_1 \frac{\partial \psi_1}{\partial y_1}\,dy_1 + \lambda_1 \frac{\partial \psi_1}{\partial y_2}\,dy_2 + \lambda_1 \frac{\partial \psi_1}{\partial y_3}\,dy_3 + \lambda_1 \frac{\partial \psi_1}{\partial y_4}\,dy_4 = 0,$$

$$\lambda_2 \frac{\partial \psi_2}{\partial y_1}\,dy_1 + \lambda_2 \frac{\partial \psi_2}{\partial y_2}\,dy_2 + \lambda_2 \frac{\partial \psi_2}{\partial y_3}\,dy_3 + \lambda_2 \frac{\partial \psi_2}{\partial y_4}\,dy_4 = 0,$$

where λ_1 and λ_2 are unknown arbitrary functions of y_1, y_2, y_3, and y_4, known as *Lagrangian multipliers.* Adding the three equations, we get

$$\left(\frac{\partial f}{\partial y_1} + \lambda_1 \frac{\partial \psi_1}{\partial y_1} + \lambda_2 \frac{\partial \psi_2}{\partial y_1}\right) dy_1 + \left(\frac{\partial f}{\partial y_2} + \lambda_1 \frac{\partial \psi_1}{\partial y_2} + \lambda_2 \frac{\partial \psi_2}{\partial y_2}\right) dy_2$$
$$+ \left(\frac{\partial f}{\partial y_3} + \lambda_1 \frac{\partial \psi_1}{\partial y_3} + \lambda_2 \frac{\partial \psi_2}{\partial y_3}\right) dy_3 + \left(\frac{\partial f}{\partial y_4} + \lambda_1 \frac{\partial \psi_1}{\partial y_4} + \lambda_2 \frac{\partial \psi_2}{\partial y_4}\right) dy_4 = 0.$$

Now the values to be ascribed to the multipliers λ_1 and λ_2 may be chosen at will. Let us choose λ_1 and λ_2 such that the first two parentheses vanish. This provides two equations,

$$\frac{\partial f}{\partial y_1} + \lambda_1 \frac{\partial \psi_1}{\partial y_1} + \lambda_2 \frac{\partial \psi_2}{\partial y_1} = 0,$$

$$\frac{\partial f}{\partial y_2} + \lambda_1 \frac{\partial \psi_1}{\partial y_2} + \lambda_2 \frac{\partial \psi_2}{\partial y_2} = 0,$$

which serve to determine the values of λ_1 and λ_2. We are then left with the equation

$$\left(\frac{\partial f}{\partial y_3} + \lambda_1 \frac{\partial \psi_1}{\partial y_3} + \lambda_2 \frac{\partial \psi_2}{\partial y_3}\right) dy_3 + \left(\frac{\partial f}{\partial y_4} + \lambda_1 \frac{\partial \psi_1}{\partial y_4} + \lambda_2 \frac{\partial \psi_2}{\partial y_4}\right) dy_4 = 0.$$

Since two of the four y's are independent, let us regard y_3 and y_4 as the independent variables. It follows then that

$$\frac{\partial f}{\partial y_3} + \lambda_1 \frac{\partial \psi_1}{\partial y_3} + \lambda_2 \frac{\partial \psi_2}{\partial y_3} = 0,$$

$$\frac{\partial f}{\partial y_4} + \lambda_1 \frac{\partial \psi_1}{\partial y_4} + \lambda_2 \frac{\partial \psi_2}{\partial y_4} = 0.$$

These two equations plus the two equations of constraint constitute four equations that determine the extremal values of y_1, y_2, y_3, and y_4.

It is obvious that this method may be applied to a function of any number of coordinates subject to any number of equations of constraint. The method of Lagrangian multipliers can be summarized as follows:

1. Write down the differential of the function, and equate it to zero.
2. Take the differential of each equation of constraint, and multiply by as many different Lagrangian multipliers as there are equations of constraint.
3. Add all the equations, factoring the sum so that each differential appears only once.
4. Equate the coefficient of each differential to zero.

APPENDIX

C

EVALUATION OF THE INTEGRAL $\int_0^\infty e^{-ax^2}\,dx$

$$\int_0^\infty e^{-ax^2}\,dx = \frac{1}{2}\sqrt{\frac{\pi}{a}} \tag{1}$$

To obtain Eq. (1) it must first be recognized that the indefinite integral $\int e^{-ax^2}\,dx$ cannot be evaluated in terms of elementary functions. An appropriate change of variables will allow the definite integral to be evaluated. Equation (1) could equally well be written

$$I = \int_0^\infty e^{-ay^2}\,dy \tag{2}$$

so that the product of Eqs. (1) and (2) is

$$I^2 = \int_0^\infty e^{-ax^2}\,dx \int_0^\infty e^{-ay^2}\,dy,$$

$$I^2 = \int_0^\infty \int_0^\infty e^{-a(x^2+y^2)}\,dx\,dy. \tag{3}$$

Now change the integration variables to polar coordinates r and θ. Then $x^2 + y^2 = r^2$ and $dx\,dy = r\,dr\,d\theta$. The area being integrated is the first quadrant, so Eq. (3) becomes

$$\begin{aligned} I^2 &= \int_0^\infty \int_0^{\pi/2} e^{-ar^2} r\, dr\, d\theta \\ &= \frac{\pi}{2}\int_0^\infty e^{ar^2} r\, dr \\ &= \frac{\pi}{2}\int_0^\infty \left(-\frac{1}{2a}\right) d(e^{-ar^2}) \\ &= -\frac{\pi}{4a} e^{-ar^2}\Big|_0^\infty \\ &= \frac{\pi}{4a}. \end{aligned} \tag{4}$$

Equation (4) is the square of the definite integral, which is thus

$$I = \int_0^\infty e^{-ax^2}\, dx = \frac{1}{2}\sqrt{\frac{\pi}{\alpha}}.$$

APPENDIX

D

RIEMANN ZETA FUNCTIONS

$$\sum_{n=1}^{\infty} \frac{(-1)^{n-1}}{n^2} = \frac{\pi^2}{12}, \tag{1}$$

$$\zeta(2) = \sum_{n=1}^{\infty} \frac{1}{n^2} = \frac{\pi^2}{6}, \tag{2}$$

$$\zeta(4) = \sum_{n=1}^{\infty} \frac{1}{n^4} = \frac{\pi^4}{90}. \tag{3}$$

To obtain Eqs. (1) and (2), expand the function

$$f(x) = x^2 \qquad (-\pi \leq x \leq \pi)$$

in a Fourier series:

$$f(x) = \sum_{n=0}^{\infty} a_n \cos nx,$$

$$a_0 = \frac{1}{\pi} \int_0^{\pi} x^2 \, dx = \frac{\pi^2}{3},$$

$$a_n = \frac{2}{\pi} \int_0^{\pi} x^2 \cos nx \, dx = (-1)^n \frac{4}{n^2}.$$

Hence,
$$x^2 = \frac{\pi^2}{3} + 4 \sum_{n=1}^{\infty} \frac{(-1)^n}{n^2} \cos nx.$$

Setting $x = 0$ gives Eq. (1), and setting $x = \pi$ gives Eq. (2).

To obtain Eq. (3), expand

$$f(x) = x^4 \qquad (-\pi \le x \le \pi)$$

in a Fourier series:

$$a_0 = \frac{1}{\pi}\int_0^{\pi} x^4\,dx = \frac{\pi^4}{5},$$

$$a_n = \frac{2}{\pi}\int_0^{\pi} x^4 \cos nx\,dx = -(-1)^n\frac{8\pi^2}{n^2} - (-1)^n\frac{48}{n^4}.$$

Hence,
$$x^4 = \frac{\pi^4}{5} + 8\pi^2\sum_{n=1}^{\infty}\frac{(-1)^n}{n^2}\cos nx - 48\sum_{n=1}^{\infty}\frac{(-1)^n}{n^4}\cos nx.$$

Setting $x = \pi$ gives

$$\pi^4 = \frac{\pi^4}{5} + 8\pi^2\sum_{n=1}^{\infty}\frac{1}{n^2} - 48\sum_{n=1}^{\infty}\frac{1}{n^4}.$$

Substituting Eq. (2) and solving for $\sum_{n=1}^{\infty}\frac{1}{n^4}$ gives Eq. (3).

BIBLIOGRAPHY

TEMPERATURE

Plumb, H. H.: "Temperature—Its Measurement and Control in Sciences and Industry," Instrument Society of America, Washington, D.C., 1973.

THERMAL PHYSICS (Thermodynamics and Statistical Mechanics)

Girifalco, L. A.: "Statistical Physics of Materials," John Wiley & Sons, Inc., New York, 1973.
Kelly, D. C.: "Thermodynamics and Statistical Physics," Academic Press, Inc., New York, 1973.
Kittel, C., and H. Kroemer: "Thermal Physics," W. H. Freeman and Company, San Francisco, 1980.
Landsberg, P. T.: "Thermodynamics and Statistical Mechanics," Oxford University Press, New York, 1978.
Mandl, F.: "Statistical Physics," John Wiley & Sons, Inc., New York, 1971.
Sears, F. W., and G. L. Salinger: "Thermodynamics, Kinetic Theory and Statistical Mechanics," Addison-Wesley Publishing Co., Inc., Reading, Mass., 1975.

THERMAL PHYSICS (with Applications to Engineering and Chemistry)

Guggenheim, E. A.: "Thermodynamics," North-Holland Publishing Co., Amsterdam, 1967.
Haywood, R. W.: "Equilibrium Thermodynamics for Engineers and Scientists," John Wiley & Sons, Inc., New York, 1980.
Kestin, J.: "A Course in Thermodynamics," 2 vols., Blaisdell Publishing Company, Lexington, Mass., 1968.
Reynolds, W. C., and H. C. Perkins: "Engineering Thermodynamics," McGraw-Hill Book Company, New York, 1977.
Rock, P. A.: "Chemical Thermodynamics," Macmillan Publishing Co., Inc., New York, 1969.
Zemansky, M. W., M. M. Abbott, and H. C. Van Ness: "Basic Engineering Thermodynamics," McGraw-Hill Book Company, New York, 1975.

STATISTICAL PHYSICS

Baierlein, R.: "Atoms and Information Theory," W. H. Freeman and Company, San Francisco, 1971.
Reed, R. D., and R. R. Roy: "Statistical Physics for Students of Science and Engineering," Intext Education Publishers, Scranton, Pa., 1971.
Rapp, D.: "Statistical Mechanics," Holt, Rinehart and Winston, New York, 1972.

LOW-TEMPERATURE PHYSICS

Lounasmaa, O. V.: "Principles and Methods below 1 K," Academic Press, Inc., New York, 1974.
White, G. K.: "Experimental Techniques in Low-Temperature Physics," Oxford University Press, Inc., New York, 1979.
Wilks, J.: "Introduction to Liquid Helium," Oxford University Press, Inc., New York, 1970.

ANSWERS TO SELECTED PROBLEMS

Chapter 1

1-1 (*a*) $P(V - nb)$; $\dfrac{P'V'}{1 - nB'/V'}$; $P''V''$.

(*b*) $P(V - nb) = \dfrac{P'V'}{1 - nB'/V'}$.

1-2 (*a*) $\dfrac{C\mathscr{H}}{M}$; $\Theta + \dfrac{C'\mathscr{H}'}{M'}$; $\dfrac{PV}{nR}$.

(*b*) $\dfrac{M}{\mathscr{H}} = \dfrac{C}{\theta}$, Curie's equation.

$\dfrac{M'}{\mathscr{H}'} = \dfrac{C'}{\theta - \Theta}$, Weiss' equation.

$PV = nR\theta$, ideal-gas equation.

1-3 419.57 K.

1-4 (*a*) 4.00 K.

1-5 (*a*) 4.00 K.

Chapter 2

2-4 (*a*) 5.076×10^7 Pa; (*b*) 48.8°C.

2-5 4.235×10^7 Pa.

2-11 50.6 N.

2-12 21.3 Hz; 33.8 Hz.

2-13 33.6 N.

Chapter 3

3-3 -929 J.
3-6 0.34 J.
3-7 (*b*) 0.198 J.
3-12 (*a*) 125.7 J.
(*b*) 0.8 J at 300 K; 236.9 J at 1 K.
(*c*) 126.5 J at 300 K; 362.6 J at 1 K.

Chapter 4

4-7 1.23 V.
4-13 $3.75 \times 10^{-8} a\Theta + 1.50 \times 10^{-4} b\Theta^2$.
4-23 230 mW/m · K.
4-26 0.0046 K.
4-27 7.16 g.
4-28 1.38×10^{-4} m^2.
4-29 1360 K.
4-30 5750 K.
4-31 (*c*) 2.32.
4-32 13 min.

Chapter 5

5-1 (*b*) 49.8 K.
5-2 0.76 m.
5-3 1.25 m.
5-4 1.7 g.
5-5 Three times the initial value.
5-11 (*b*) 0.221 ft^3; 1020 ft · lb.
5-12 (*b*) 500 ft.
5-13 (*b*) 571 K.
5-14 (*a*) 10.2 kJ; (*b*) 614 K; (*c*) 3530 K; (*d*) 107.5 kJ.
5-17 (*c*) -9.5 K/km.
5-18 (*a*) 1.18 s; (*b*) 0.69 m.
5-19 (*a*) 0.966 s; (*b*) 0.471 m.
5-20 1.27.
5-23 320 m/s.
5-24 1.27.
5-25 80 percent He; 20 percent Ne.
5-26 Diatomic
5-27 13.8057 K.

Chapter 6

6-5 (*a*) 148.5 J; 148.5 J.
(*b*) -148.5 J; 0.
(*c*) 6.9 J; -6.9J; 95.4 percent.

6-11 (*a*) $\theta = \dfrac{a}{nR}V^2 + \dfrac{b}{nR}V.$

(*b*) V (at $\theta = \theta_{max}$) = 32.5 m^3.
(*c*) $\theta_0 = 308$ K; $\theta_{max} = 720$ K; $\theta_1 = 77$ K.

(*d*) $Q(V_0 \text{ to } V) = 4a\left(\dfrac{V^2}{2} - 32\right) + \frac{5}{2}b(V - 8).$

(*e*) Q = max: $P = 13.7$ Pa; $V = 41.2$ m^3.
(*f*) Q from V_0 to $V(Q = Q_{max}) = 1270$ J (in).
(*g*) Q from $V(Q = Q_{max})$ to $V_1 = -581$ J (out).

6-12 (*a*) 288 J; (*b*) 636 J; (*c*) 689 J; (*d*) 0.50; (*e*) 0.90.

Chapter 7

7-11 $\dfrac{R}{p}(c_p - c'_p) = 0.$

Chapter 8

8-6 (*a*) 0; (*b*) 8.33 J/K; (*c*) 58 J/K; (*d*) 58 J/K.
8-7 (*a*) 1310 J/K; −1120 J/K; 190 J/K.
(*b*) 1310 J/K; −1210 J/K; 90 J/K.
8-9 0.0300 R.
8-10 (*a*) 6.3 J/K; (*b*) 1.39 J/K; (*c*) 3.80 J/K.
8-11 (*a*) 0; (*b*) 1.83×10^{-5} J/K.
8-12 277 J/K.
8-13 −16.0 J/K.
8-14 1.5×10^5 Pa; 300 K; 0.0570 J/K.
8-15 (*a*) $4V_0/9$; (*b*) $3T_0/2$; (*c*) $21T_0/4$; (*d*) $\frac{19}{4}nC_V T_0$;
(*e*) $\frac{1}{2}nC_V T_0$; (*f*) 0; (*g*) $nC_p \ln \frac{21}{4} - nR \ln \frac{27}{8}$;
(*h*) $nC_p \ln \frac{63}{8} - 2nR \ln \frac{27}{8}$.
8-23 (*a*) 3740 J/mol; (*b*) 3740 J/mol;
(*d*) −2296 J/mol; (*e*) −1444 J/mol; (*f*) 3.82 J/mol · K.
8-26 About 400 K.

Chapter 9

9-6 (*a*) −87.5 J; (*b*) 50.6 J; (*c*) −36.9 J; (*d*) 0.684 K.
9-7 (*a*) 1.11 kJ; (*b*) 3.96 kJ; (*c*) 5.07 kJ.
9-8 −0.436 K; +0.099 K; +3.63 K.

Chapter 10

10-3 600.8 K.
10-4 (*c*) 6.7 percent.

10-5 (*b*) 4.34×10^7 Pa/K; (*c*) -2.29°C, 3.1×10^7 Pa.
10-6 (*d*) 0.77 percent.
10-7 (*c*) 8.26 μ/s.
10-8 7.2.
10-13 (*a*) 63.6 kJ/mol; (*b*) 69.5 kJ/mol; (*c*) 65.4 kJ/mol.
10-14 (*a*) 132 kJ/mol; (*b*) 1.21.
10-15 (*a*) 195 K; (*b*) $l_{SU} = 31.2$ kJ/mol; $l_{FU} = 5.80$ kJ/mol; $l_{VA} = 25.4$ kJ/mol.
10-16 (*a*) $l_{SU}/R = 137$ K; $l_{FU}/R = 19.2$ K; $l_{VA}/R = 118$ K; (*b*) 4.69×10^3 Pa/K.

Chapter 11

11-1 2×10^{11}.
11-11 A krypton-86 lamp at 77 K.
11-12 (*a*) 7730 K; (*b*) 7.73×10^6 K; (*c*) 7.73×10^9 K.
11-13 2.41×10^{15} atoms/s.
11-14 $k' = A\langle w \rangle / 4V$.
11-15 3.25 s.
11-16 2.47×10^{-2} Pa.

Chapter 13

13-3 (*a*) 37 K; (*b*) $P = 34.9$ atm, $T = 20.6$ K.

Chapter 14

14-2 1250 J.

Chapter 15

15-5 (*a*) 0.315 atm; (*b*) 0.613; (*c*) 61.2 kJ/mol.
15-6 94.6 kJ/mol.
15-7 39 percent; 81 percent.
15-8 100 percent.
15-10 709 J/K.
15-11 0.00129 K^{-1}.

Chapter 17

17-1 (*a*) 3.6 J; (*b*) 3.18 J; (*c*) 6.78 J; (*d*) -1.21 K.
17-4 (*b*) -0.900 K/m, -0.336 K/m, $+1.40$ K/m, $+1.37$ K/m.
17-7 -214 kJ.
17-8 -188 kJ.
17-12 0.25 mK, 1.00 mK, 25.0 mK, 100 mK.

17-14 (*a*) -6.67 mV; (*b*) -33.8 mV, -101 J, out; (*c*) $+0.152$ mV/K.
17-23 (*a*) 6000 K; (*b*) 0.984 J/m^3; 0.328 N/m^2.
17-24 (*a*) 1.6×10^{-5} J; (*b*) 1.2×10^{-5} J; (*c*) 4.0×10^{-6} J; (*d*) 431 K.

Chapter 18

18-13 (*a*) 18.5 J; (*b*) -18.5 J; (*c*) 0; (*d*) 1.53 K.
18-14 4.06 K.
18-15 (*a*) -0.267 J; (*b*) 0.3 K.
18-16 (*a*) 7.17×10^6 A/m; (*b*) 3.18×10^5 A/m.
18-17 (*a*) 1.03 K; (*b*) 0.00315 K.
18-19 (*a*) 2.00 K; (*b*) 0.14 K.
18-20 0.363 μW.

INDEX